Solutions Manual for

Introduction to General, Organic, and Biochemistry

Sixth Edition

Morris Hein
Mount San Antonio College

Scott Pattison
Ball State University

Susan Arena
University of Illinois, Urbana-Champaign

Brooks/Cole Publishing Company

I(T)P® An International Thomson Publishing Company

Pacific Grove • Albany • Belmont • Bonn • Boston • Cincinnati • Detroit
Johannesburg • London • Madrid • Melbourne • Mexico City
New York • Paris • Singapore • Tokyo • Toronto • Washington

Assistant Editor: *Beth Wilbur*
Cover Design: *Vernon T. Boes*
Cover Photo: *Photonica International*
Editorial Associate: *Nancy Conti*
Marketing Team: *Kathleen Sharp, Christine Davis, Carrie Beckwith*
Production Editor: *Mary Vezilich*
Printing and Binding: *Courier Westford*

For more information, contact:

BROOKS/COLE PUBLISHING COMPANY
511 Forest Lodge Road
Pacific Grove, CA 93950
USA

International Thomson Publishing Europe
Berkshire House 168-173
High Holborn
London WC1V 7AA
England

Thomas Nelson Australia
102 Dodds Street
South Melbourne, 3205
Victoria, Australia

Nelson Canada
1120 Birchmount Road
Scarborough, Ontario
Canada M1K 5G4

International Thomson Editores
Seneca 53
Col. Polanco
11560 México, D. F., México

International Thomson Publishing Japan
Hirakawacho Kyowa Building, 3F
2-2-1 Hirakawacho
Chiyoda-ku, Tokyo 102
Japan

International Thomson Publishing Asia
221 Henderson Road
#05-10 Henderson Building
Singapore 0315

International Thomson Publishing GmbH
Königswinterer Strasse 418
53227 Bonn
Germany

Printed in the United States of America

5 4 3 2 1

ISBN 0-534-34180-2

TO THE STUDENT

This Solutions Manual contains answers to all the exercises and solutions to all problems at the end of each chapter in the text.

This book will be valuable to you if you use it properly. It is not intended to be a substitute for answering the exercises by yourself. It is important that you go through the process of writing down the answers to the exercises and the solutions to problems before you see them in the solutions manual. Once you have seen the answer, much of the value of whether or not you have learned the material is lost. This does not mean that you cannot learn from the answers. If you absolutely cannot answer a question or problem, study the answer carefully to see where you are having difficulty. You should spend sufficient time answering each assigned exercise. Only after you have made a serious attempt to answer a question or problem should you resort to the solutions manual.

Chemistry is one of the most challenging courses of study you will encounter. The core of the study of chemistry is problem solving. Skill at solving problems is achieved through effective and consistent practice. Watching and listening to others solve problems may be of some use but will not result in facility with chemistry problems. The methods used for solving problems in this manual are essentially the same as those used in the text. The basic steps in problem solving are fairly universal:

1. Read the problem carefully and determine the type of problem.
2. Develop a plan for solving the problem.
3. Write down the given information, including units, in an organized fashion.
4. Write a complete set-up and solution for the problem.
5. Use the solutions manual to check the answer and solution.

Your solution might be correct and yet will vary from the manual. Usually, these differences are the result of completing operations in a different order, or using separate steps instead of a single line approach. If you make an error, take the time to analyze what went wrong. An error caused by an improper calculator entry can be frustrating, but it is not as serious as the inability to properly set up the problem. Do not become discouraged. Once you understand the steps in a problem, go back and rework it without looking at the answer.

The area of organic and biochemistry presents a different type of learning. Learning organic chemistry takes a great amount of practice. You need to put in a lot of time writing formulas, names, and reaction equations if you are to be successful. Initially there is a certain amount of memorization of material. For example, you must learn the names and formulas of the first ten aliphatic hydrocarbons: methane, ethane, propane, butane, pentane, hexane, heptane, octane, nonane, and decane. Say them to yourself several times daily; write the names together with their formulas over and over again, and soon you will know them very well. The names of many other classes of organic compounds are derived from the names of the aliphatic hydrocarbons. Remember, **practice** is the key word.

We have made every attempt to produce a manual with as few errors as possible. Please let us know of errors that you encounter, so that they can be corrected. Allow this manual to help you have success as you begin the great adventure of studying chemistry.

MORRIS HEIN
SUSAN ARENA

CONTENTS

Chapter 2 Standards for Measurement 1

Chapter 3 Classification of Matter 13

Chapter 4 Properties of Matter 20

Chapter 5 Early Atomic Theory and Structure 25

Chapter 6 Nomenclature of Inorganic Compounds 31

Chapter 7 Quantitative Composition of Compounds 40

Chapter 8 Chemical Equations 61

Chapter 9 Calculations from Chemical Equations 68

Chapter 10 Modern Atomic Theory 81

Chapter 11 Chemical Bonds: The Formation of Compounds from Atoms 89

Chapter 12 The Gaseous State of Matter 100

Chapter 13 Water and the Properties of Liquids 120

Chapter 14 Solutions 133

Chapter 15 Acids, Bases, and Salts 157

Chapter 16 Chemical Equilibrium 176

Chapter 17 Oxidation-Reduction 202

Chapter 18 Nuclear Chemistry 226

Chapter 19 Chemistry of Selected Elements 236

Chapter	20	Organic Chemistry: Saturated Hydrocarbons	245
Chapter	21	Unsaturated Hydrocarbons	257
Chapter	22	Alcohols, Ethers, Phenols, and Thiols	275
Chapter	23	Aldehydes and Ketones	292
Chapter	24	Carboxylic Acids and Esters	306
Chapter	25	Amides and Amines: Organic Nitrogen Compounds	324
Chapter	26	Polymers: Macromolecules	335
Chapter	27	Stereoisomerism	344
Chapter	28	Carbohydrates	358
Chapter	29	Lipids	376
Chapter	30	Amino Acids, Polypeptides, and Proteins	386
Chapter	31	Enzymes	397
Chapter	32	Nucleic Acids and Heredity	401
Chapter	33	Nutrition	410
Chapter	34	Bioenergetics	415
Chapter	35	Carbohydrate Metabolism	420
Chapter	36	Metabolism of Lipids and Proteins	426

CHAPTER 2

STANDARDS FOR MEASUREMENT

1. 100 cm = 1 m
   1000 m = 1 km
   100,000 cm = 1 km

2. 7.6 cm

3. The volumetric flask is a more precise measuring instrument than the graduated cylinder. This is because the narrow opening at the point of measurement (calibration mark in the neck of the flask means that a small error made in not filling exactly to the mark will make a smaller percentage error in total volume than will a similar error with the cylinder.

4. The three materials would sort out according to their densities with the most dense (mercury) at the bottom and the least dense (glycerin) at the top. In the cylinder, the solid magnesium would sink in the glycerin and float on the liquid mercury.

5. Order of increasing density: ethyl alcohol, vegetable oil, salt, lead.

6. The density of ice must be less than 0.91 g/mL and greater than 0.789 g/mL.

7. Heat is a form of energy, while temperature is a measure of the intensity of heat (how hot the system is).

8. Density is the ratio of the mass of a substance to the volume occupied by that mass. Density has the units of mass over volume. Specific gravity is the ratio (no units) of the density of a substance to the density of a reference substance (usually water at a specific temperature for solids and liquids). Specific gravity has no units.

9. Mass is not a measure of the amount of body fat, but rather the total composition of the body. Health risk is associated with an excess of fat in proportion to the total body mass. Leanness or obesity is indicated by the amount of fat compared to the rest of the body material.

10. Body density is determined by using the difference in mass in air and underwater. A lean person weighs more in water than an obese person and has a higher density as a result.

11. The principal advantages include:
    (a) conversion factors are multiples of 10, eliminating the necessity of fractions as measurements.
    (b) prefixes are the same throughout the metric system regardless of units. Therefore, there are fewer units to remember.
    (c) The metric system is used internationally.

(d) The metric system uses mass, which remains constant, regardless of position (including international travel), while the American system uses weight, which changes relative to gravitational attraction.

12. Rule 1. When the first digit after those you want to retain is 4 or less, that digit and all others to its right are dropped. The last digit retained is not changed.

Rule 2. When the first digit after those you want to retain is 5 or greater, that digit and all others to the right of it are dropped and the last digit retained is increased by one.

13. The number of degrees between the freezing and boiling point of water are
Fahrenheit 180°F
Celsius 100°C
Kelvin 100 K

14. The correct statements are: a, c, d, e, g, h, i, j, l, n, p, q.
(b) The length 10 cm is equal to 100 mm.
(f) The sum of 32.276 + 2.134 should contain 5 significant figures.
(k) One centimeter is shorter than one inch.
(m) The number 0.0002983 in exponential notation is 2.983×10^{-4}
(o) Temperature is a measure of energy of motion in a sample.

15. (a) gram = g
(b) microgram = μg
(c) centimeter = cm
(d) micrometer = μm
(e) milliliter = mL
(f) deciliter = dL

16. (a) milligram = mg
(b) kilogram = kg
(c) meter = m
(d) nanometer = nm
(e) angstrom = Å
(f) microliter = μL

17. (a) 503 zero is significant
(b) 0.007 zeros are not significant
(c) 4200 zeros are not significant
(d) 3.0030 zeros are significant
(e) 100.00 zeros are significant
(f) 8.00×10^2 zeros are significant

18. (a) 63,000 zeros are not significant
(b) 6.004 zeros are significant
(c) 0.00543 zeros are not significant
(d) 8.3090 zeros are significant
(e) 60. zero is significant
(f) 5.0×10^{-4} zero is significant

19. (a) 0.025 (2)
(b) 22.4 (3)
(c) 0.0404 (3)
(d) 5.50×10^3 (3)

20. (a) 40.0 (3)
(b) 0.081 (2)
(c) 129,042 (6)
(d) 4.090×10^{-3} (4)

21. (a) 93.2
(b) 0.0286
(c) 4.64
(d) 34.3

22. (a) 8.87
(b) 21.3
(c) 130. (1.30×10^2)
(d) 2.00×10^6

23. (a) 2.9×10^6
(b) 5.87×10^{-1}
(c) 8.40×10^{-3}
(d) 5.5×10^{-6}

24. (a) 4.56×10^{-2}
(b) 4.0822×10^3
(c) 4.030×10^1
(d) 1.2×10^7

25. (a)
$$\begin{array}{r} 12.62 \\ 1.5 \\ 0.25 \\ \hline 14.37 \end{array} = 14.4$$

(b) $(2.25 \times 10^3)(4.80 \times 10^4) = 10.8 \times 10^7 = 1.08 \times 10^8$

(c) $\dfrac{(462)(6.2)}{14.3} = 195.97 = 2.0 \times 10^2$

(d) $(0.0394)(12.8) = 0.504$

(e) $\dfrac{0.4278}{59.6} = 0.00718 = 7.18 \times 10^{-3}$

(f) $10.4 + 3.75(1.5 \times 10^4) = 5.6 \times 10^4$

26. (a)
$$\begin{array}{r} 15.2 \\ -2.75 \\ 15.67 \\ \hline 28.1 \end{array}$$

(b) $(4.68)(12.5) = 58.5$

(c) $\dfrac{182.6}{4.6} = 4.0 \times 10^1$

(d)
$$\begin{array}{r} 1986 \\ 23.84 \\ 0.012 \\ \hline 2009.852 \end{array} = 2010. = 2.010 \times 10^3$$

(e) $\dfrac{29.3}{(284)(415)} = 2.49 \times 10^{-4}$

(f) $(2.92 \times 10^{-3})(6.14 \times 10^5) = 1.79 \times 10^3$

27. Fractions to decimals (3 significant figures)

(a) $\frac{5}{6} = 0.833$

(b) $\frac{3}{7} = 0.429$

(c) $\frac{12}{16} = 0.750$

(d) $\frac{9}{18} = 0.500$

28. Decimals to fractions

(a) $0.25 = \frac{1}{4}$

(b) $0.625 = \frac{5}{8}$

(c) $1.67 = 1\frac{2}{3}$ or $\frac{5}{3}$

(d) $0.8888 = \frac{8}{9}$

29. (a) $3.42x = 6.5$

$$\frac{3.42x}{3.42} = \frac{6.5}{3.42}$$

$$x = \frac{6.6}{3.42} = 1.9$$

(b) $\frac{x}{12.3} = 7.05$

$$x = (7.05)(12.3) = 86.7$$

(c) $\frac{0.525}{x} = 0.25$

$$0.525 = 0.25x$$

$$x = \frac{0.525}{0.25} = 2.1$$

30. (a) $x = \frac{212 - 32}{1.8}$

$$x = 1.00 x 10^2$$

(b) $8.9\frac{\text{g}}{\text{mL}} = \frac{40.90\text{ g}}{x}$

$$\left(8.9\frac{\text{g}}{\text{mL}}\right)x = 40.90\text{ g}$$

$$x = \frac{40.90\text{ g}}{8.9\ \frac{\text{g}}{\text{mL}}} = 4.6\text{ mL}$$

(c) $72 = 1.8x + 32$

$$72 - 32 = 1.8x$$

$$40 = 1.8x$$

$$\frac{40}{1.8} = x$$

$$22 = x$$

31. (a) $(28.0\text{ cm})\left(\frac{1\text{ m}}{100\text{ cm}}\right) = 0.280\text{ m}$

(b) $(1000.\text{ m})\left(\frac{1\text{ km}}{1000\text{ m}}\right) = 1.000\text{ km}$

(c) $(9.28 \text{ cm})\left(\frac{10 \text{ mm}}{1 \text{ cm}}\right) = 92.8 \text{ mm}$

(d) $(10.68 \text{ g})\left(\frac{1000 \text{ mg}}{1 \text{ g}}\right) = 1.068 \times 10^4 \text{ mg}$

(e) $(6.8 \times 10^4 \text{ mg})\left(\frac{1 \text{ g}}{1000 \text{ mg}}\right)\left(\frac{1 \text{ kg}}{1000 \text{ g}}\right) = 6.8 \times 10^{-2} \text{ kg}$

(f) $(8.54 \text{ g})\left(\frac{1 \text{ kg}}{1000 \text{ g}}\right) = 0.00854 \text{ kg}$

(g) $(25.0 \text{ mL})\left(\frac{1 \text{ L}}{1000 \text{ mL}}\right) = 2.50 \times 10^{-2} \text{ L}$

(h) $(22.4 \text{ L})\left(\frac{10^6 \ \mu\text{L}}{1 \text{ L}}\right) = 2.24 \times 10^7 \ \mu\text{L}$

32. (a) $(4.5 \text{ cm})\left(\frac{1 \text{ m}}{100 \text{ cm}}\right)\left(\frac{1 \text{ Å}}{10^{-10} \text{ m}}\right) = 4.5 \times 10^8 \text{ Å}$

(b) $(12 \text{ nm})\left(\frac{10^{-9} \text{ m}}{1 \text{ nm}}\right)\left(\frac{100 \text{ cm}}{1 \text{ m}}\right) = 1.2 \times 10^{-6} \text{ cm}$

wrong

(c) $(8.0 \text{ km})\left(\frac{1000 \text{ m}}{1 \text{ km}}\right)\left(\frac{1000 \text{ nm}}{1 \text{ m}}\right) = 8.0 \times 10^6 \text{ mm}$

(d) $(164 \text{ mg})\left(\frac{1 \text{ g}}{1000 \text{ mg}}\right) = 0.164 \text{ g}$

(e) $(0.65)\left(\frac{1000 \text{ g}}{1 \text{ kg}}\right)\left(\frac{1000 \text{ mg}}{1 \text{ g}}\right) = 6.5 \times 10^5 \text{ mg}$

(f) $(5.5 \text{ kg})\left(\frac{1000 \text{ g}}{1 \text{ kg}}\right) = 5.5 \times 10^3 \text{ g}$

(g) $(0.468 \text{ L})\left(\frac{1000 \text{ mL}}{1 \text{ L}}\right) = 468 \text{ mL}$

(h) $(9.0 \ \mu\text{L})\left(\frac{1 \text{ L}}{10^6 \ \mu\text{L}}\right)\left(\frac{1000 \text{ mL}}{1 \text{ L}}\right) = 9.0 \times 10^{-3} \text{ mL}$

33. (a) $(42.2 \text{ in.})\left(\frac{2.54 \text{ cm}}{1 \text{ in.}}\right) = 107 \text{ cm}$

(b) $(0.64 \text{ mi})\left(\frac{5280 \text{ ft}}{1 \text{ mi}}\right)\left(\frac{12 \text{ in.}}{1 \text{ ft}}\right) = 4.1 \times 10^4 \text{ in.}$

(c) $(2.00\ \text{in.}^2)\left(\frac{2.54\ \text{cm}}{1\ \text{in.}}\right)^2 = 12.9\ \text{cm}^2$

(d) $(42.8\ \text{kg})\left(\frac{2.205\ \text{lb}}{\text{kg}}\right) = 94.2\ \text{lb}$

(e) $(3.5\ \text{qt})\left(\frac{946\ \text{mL}}{1\ \text{qt}}\right) = 3.3 \times 10^3\ \text{mL}$

(f) $(20.0\ \text{gal})\left(\frac{4\ \text{qt}}{1\ \text{gal}}\right)\left(\frac{0.946\ \text{L}}{1\ \text{qt}}\right) = 75.7\ \text{L}$

34. (a) $(35.6\ \text{m})\left(\frac{100\ \text{cm}}{1\ \text{m}}\right)\left(\frac{1\ \text{in.}}{2.54}\right)\left(\frac{1\ \text{ft}}{12\ \text{in.}}\right) = 117\ \text{ft}$

(b) $(16.5\ \text{km})\left(\frac{1\ \text{mi.}}{1.609\ \text{km}}\right) = 10.3\ \text{mi}$

(c) $(4.5\ \text{in.}^3)\left(\frac{2.54\ \text{cm}}{1\ \text{in.}}\right)^3\left(\frac{10\ \text{mm}}{1\ \text{cm}}\right)^3 = 7.4 \times 10^4\ \text{mm}^3$

(d) $(95\ \text{lb})\left(\frac{453.6\ \text{g}}{1\ \text{lb}}\right) = 4.3 \times 10^4\ \text{g}$

(e) $(20.0\ \text{gal})\left(\frac{4\ \text{qt}}{1\ \text{gal}}\right)\left(\frac{0.946\ \text{L}}{1\ \text{qt}}\right) = 75.7\ \text{L}$

(f) $(4.5 \times 10^4\ \text{ft}^3)\left(\frac{12\ \text{in.}}{1\ \text{ft}}\right)^3\left(\frac{2.54\ \text{cm}}{1\ \text{in.}}\right)^3\left(\frac{1\ \text{m}}{1000\ \text{m}}\right)^3 = 1.3 \times 10^3\ \text{m}^3$

35. $\left(55\ \frac{\text{mi}}{\text{hr}}\right)\left(1.609\ \frac{\text{km}}{\text{mi}}\right) = 88\ \frac{\text{km}}{\text{hr}}$

36. $\left(55\ \frac{\text{km}}{\text{hr}}\right)\left(\frac{0.62137\ \text{mi}}{1\ \text{km}}\right)\left(\frac{5280\ \text{ft}}{1\ \text{mi}}\right)\left(\frac{1\ \text{hr}}{3600\ \text{s}}\right) = 50.\ \frac{\text{ft}}{\text{s}}$

37. $\left(\frac{100.\ \text{m}}{9.92\ \text{s}}\right)\left(\frac{100\ \text{cm}}{1\ \text{m}}\right)\left(\frac{1\ \text{in.}}{2.54\ \text{cm}}\right)\left(\frac{1\ \text{ft}}{12\ \text{in.}}\right) = 33.1\ \frac{\text{ft}}{\text{s}}$

38. $\left(\frac{229\ \text{mi}}{1\ \text{hr}}\right)\left(\frac{1.609\ \text{km}}{\text{mi}}\right)\left(\frac{1\ \text{hr}}{3600\ \text{s}}\right) = 0.102\ \frac{\text{km}}{\text{s}}$

39. $\left(\frac{27{,}000\ \text{mi}}{\text{hr}}\right)\left(\frac{1.609\ \text{km}}{\text{mi}}\right)\left(\frac{1\ \text{hr}}{3600\ \text{s}}\right) = 12\ \frac{\text{km}}{\text{s}}$

40. $(9.3 \times 10^7\ \text{mi})\left(\frac{1.609\ \text{km}}{\text{mi}}\right)\left(\frac{1000\ \text{m}}{1\ \text{km}}\right)\left(\frac{1\ \text{s}}{3.00 \times 10^8\ \text{m}}\right) = 5.0 \times 10^2\ \text{s}$

41. $(176\ \text{lb})\left(\frac{453.6\ \text{g}}{1\ \text{lb}}\right)\left(\frac{1\ \text{kg}}{1000\ \text{g}}\right) = 79.8\ \text{kg}$

42. $(1\ \text{oz})\left(\frac{1\ \text{lb}}{16\ \text{oz}}\right)\left(\frac{453.6\ \text{g}}{1\ \text{lb}}\right)\left(\frac{1000\ \text{mg}}{1\ \text{g}}\right) = 3 \times 10^4\ \text{mg}$

43. $(5.0\ \text{grains})\left(\frac{1\ \text{lb}}{7000\ \text{grains}}\right)\left(\frac{453.6\ \text{g}}{1\ \text{lb}}\right) = 0.32\ \text{g}$

44. $(21\ \text{lb})\left(\frac{453.6\ \text{g}}{1\ \text{lb}}\right) = 9.5 \times 10^3\ \text{g} = \text{mass condor}$

$\frac{9.5 \times 10^3\ \text{g (condor)}}{3.2\ \text{g (hummingbird)}} = 3.0 \times 10^3$ hummingbirds to equal the mass of one (1) condor

45. $\left(\frac{\$1.49}{283.5\ \text{g}}\right)\left(\frac{453.6\ \text{g}}{1\ \text{lb}}\right) = \frac{\$2.38}{\text{lb}}$

46. $\left(\frac{\$350}{1\ \text{oz}}\right)\left(\frac{14.58\ \text{oz}}{1\ \text{lb}}\right)\left(\frac{1\ \text{lb}}{453.6\ \text{g}}\right)(250\ \text{g}) = \$2800$

47. $\left(\frac{\$0.35}{1\ \text{L}}\right)\left(\frac{0.946\ \text{L}}{1\ \text{qt}}\right)\left(\frac{4\ \text{qt}}{1\ \text{gal}}\right)(15.8\ \text{gal}) = \$21$

48. $(525\ \text{mi})\left(\frac{1\ \text{gal}}{35\ \text{mi}}\right)\left(\frac{4\ \text{qt}}{1\ \text{gal}}\right)\left(\frac{0.946\ \text{L}}{1\ \text{qt}}\right) = 57\ \text{L}$

49. $\left(\frac{20.\ \text{drops}}{\text{mL}}\right)\left(\frac{946\ \text{mL}}{\text{qt}}\right)\left(\frac{4\ \text{qt}}{\text{gal}}\right)(1\ \text{gal}) = 7.6 \times 10^4\ \text{drops}$

50. $(42\ \text{gal})\left(\frac{4\ \text{qt}}{\text{gal}}\right)\left(\frac{0.946\ \text{L}}{\text{qt}}\right) = 160\ \text{L}$

51. $(1.00\ \text{ft}^3)\left(\frac{12\ \text{in.}}{\text{ft}}\right)^3\left(\frac{2.54\ \text{cm}}{\text{in.}}\right)^3\left(\frac{1\ \text{mL}}{1\ \text{cm}^3}\right) = 2.83 \times 10^4\ \text{mL}$

52. $V = A \times h \quad A = \text{area} \quad h = \text{height}$

$$A = \frac{V}{h} = \left(\frac{200\ \text{cm}^3}{0.5\ \text{nm}}\right)\left(\frac{1\ \text{nm}}{10^{-9}\ \text{m}}\right)\left(\frac{1\ \text{m}}{100\ \text{cm}}\right)^3 = 4 \times 10^5\ \text{m}^2$$

53. (a) $(27\ \text{cm})(21\ \text{cm})(4.4\ \text{cm}) = 2.5 \times 10^3\ \text{cm}^3$

(b) $(2.5 \times 10^3\ \text{cm}^3)\left(\frac{1\ \text{m}}{100\ \text{cm}}\right)^3\left(\frac{1\ \text{L}}{10^{-3}\ \text{m}^3}\right) = 2.5\ \text{L}$

(c) $(2.5 \times 10^3\ \text{cm}^3)\left(\frac{1\ \text{in.}}{2.54\ \text{cm}}\right)^3 = 1.5 \times 10^2\ \text{in.}^3$

54. $(16 \text{ in.})(8 \text{ in.})(10 \text{ in.})\left(\frac{2.54 \text{ cm}}{1 \text{ in.}}\right)^3\left(\frac{1 \text{ m}}{100 \text{ cm}}\right)^3\left(\frac{1 \text{ L}}{10^{-3} \text{ m}^3}\right) = 2 \times 10^1 \text{ L}$

$(2 \times 10^1 \text{ L})\left(\frac{1 \text{ qt}}{0.946\text{L}}\right)\left(\frac{1 \text{ gal}}{4 \text{ qt}}\right) = 5 \text{ gal}$

55. $\frac{98.6 - 32}{1.8} = 37.0°\text{C}$

56. $(1.8)(45) + 32 = 113°\text{F}$ Summer!

57. (a) $\frac{162 - 32}{1.8} = 72.2°\text{C}$

(b) $\frac{0.0 - 32}{1.8} + 273 = 255.2 \text{ K}$

Remember to express the answer to the same precision as the original measurement.

(c) $1.8(-18) + 32 = -0.40°\text{F}$

(d) $212 - 273 = -61°\text{C}$

58. (a) $1.8(32) + 32 = 90.°\text{F}$

(b) $\frac{-8.6 - 32}{1.8} = -22.6°\text{C}$

(c) $273 + 273 = 546 \text{ K}$

(d) $(100 - 273)(1.8) + 32 = -300 \text{ K}$ (1 significant figure in 100)

59.

$$\begin{aligned} °\text{F} &= °\text{C} \\ °\text{F} &= 1.8(°\text{C}) + 32 \\ °\text{F} &= 1.8(°\text{F}) + 32 \\ -32 &= 0.8(°\text{F}) \\ \frac{-32}{0.8} &= °\text{F} \\ -40 &= °\text{F} \\ -40°\text{F} &= -40°\text{C} \end{aligned}$$

60.

$$°\text{F} = -°\text{C}$$

$$\begin{aligned} °\text{F} &= 1.8(°\text{C}) + 32 \\ -°\text{C} &= 1.8(°\text{C}) + 32 \\ 2.8(°\text{C}) &= -32 \\ °\text{C} &= \frac{-32}{2.8} \\ °\text{C} &= -11.4 \\ -11.4°\text{C} &= 11.4°\text{F} \end{aligned}$$

61. $D = \frac{m}{V} = \frac{78.26 \text{ g}}{50.00 \text{ mL}} = 1.565 \frac{\text{g}}{\text{mL}}$

62. $D = \frac{m}{V} = \frac{39.9 \text{ g}}{12.8 \text{ mL}} = 3.12 \frac{\text{g}}{\text{mL}}$

63. 29.6 mL – 25.0 mL = 4.6 mL (V of chromium)

$D = \frac{m}{V} = \frac{32.7 \text{ g}}{4.6 \text{ mL}} = 7.1 \frac{\text{g}}{\text{mL}}$

64. 106.773 g – 42.817 g = 63.956 g (mass of the liquid)

$D = \frac{m}{V} = \frac{63.956 \text{ g}}{50.0 \text{ mL}} = 1.28 \frac{\text{g}}{\text{mL}}$

65. $D = \frac{m}{V}$

$m = DV = \left(1.19 \frac{\text{g}}{\text{mL}}\right)(250.0 \text{ mL}) = 298 \text{ g}$

66. $D = \frac{m}{V}$

$m = DV = \left(13.6 \frac{\text{g}}{\text{mL}}\right)(25.0 \text{ mL}) = 3.40 \times 10^2 \text{ g}$

67. $\left(\frac{1032 \text{ g}}{1 \text{ L}}\right)\left(\frac{1 \text{ L}}{1000 \text{ mL}}\right) = 1.032 \frac{\text{g}}{\text{mL}}$

$\left(\frac{1032 \text{ g}}{1 \text{ L}}\right)\left(\frac{1 \text{ kg}}{1000 \text{ g}}\right) = 1.032 \frac{\text{kg}}{\text{L}}$

68. $(3.1 \text{ L})\left(\frac{1000 \text{ cm}^3}{1 \text{ L}}\right)\left(1.03 \frac{\text{g}}{\text{cm}^3}\right)\left(\frac{1 \text{ lb}}{453.6 \text{ g}}\right) = 7.0 \text{ lbs}$

69. area per line $= (2.5 \text{ ft})(4.0 \text{ in.})\left(\frac{1 \text{ ft}}{12 \text{ in.}}\right) = 0.83 \text{ ft}^2$

$\left(\frac{1 \text{ line}}{0.83 \text{ ft}^2}\right)\left(\frac{43 \text{ ft}^2}{1 \text{ qt}}\right)\left(\frac{4 \text{ qt}}{1 \text{ gal}}\right)(15 \text{ gal}) = 3100 \text{ lines}$

70. $V = s^3 = (0.50 \text{ m})^3 = 0.125 \text{ m}^3$

$(0.125 \text{ m}^3)\left(\frac{1000 \text{ L}}{\text{m}^3}\right) = 125 \text{ L}$

Yes, the cube will hold the solution. 125 L – 8.5 L = 116.5 L additional solution is necessary to fill the container.

71. $\left(\frac{180\ \mu\text{g}}{1 \text{ m}^3}\right)\left(\frac{1 \text{ m}^3}{1000 \text{ L}}\right)\left(2 \times 10^4 \frac{\text{L}}{\text{day}}\right) = 4000\ \mu\text{g/day ingested}$

Yes, the technician is at risk. This is well over the toxic limit.

72. $\dfrac{(4.5-32)}{1.8} = -15.3^\circ\text{C} \rightarrow 4.5^\circ\text{F}$

$-15^\circ\text{C} > -15.3^\circ\text{C}$
$-15^\circ\text{C} > 4.5^\circ\text{F}$
-15°C is the higher temperature

73. $m_{\text{pan 1}} = m_{\text{pan 2}}$ (when balanced)

$m_{\text{pan 1}} = m_{\text{flask}} + m_{\text{alcohol}} = m_{\text{flask}} + (100.\ \text{mL})(0.789\ \text{g/mL})$

$= m_{\text{flask}} + 78.9\ \text{g}$

$m_{\text{pan 2}} = (m_{\text{flask}} + 11.0\ \text{g}) + m_{\text{turpentine}} = (m_{\text{flask}} + 11.0\ \text{g}) + x\left(0.87\ \dfrac{\text{g}}{\text{mL}}\right)$

Since $m_{\text{pan 1}} = m_{\text{pan 2}}$

$$m_{\text{flask}} + 78.9\ \text{g} = (m_{\text{flask}} + 11.0\ \text{g}) + x\left(0.87\ \frac{\text{g}}{\text{mL}}\right)$$

$$78.9\ \text{g} = 11.0\ \text{g} + x\left(0.87\ \frac{\text{g}}{\text{mL}}\right)$$

$$67.9\ \text{g} = x\left(0.87\ \frac{\text{g}}{\text{mL}}\right)$$

$$78\ \text{mL} = x$$

$80\ \text{mL} = x$ (1 significant figure in 100 mL)

74. $D = \frac{m}{V}$

$V = \frac{m}{D}$

$$V_A = \frac{25\ \text{g}}{10.0\ \frac{\text{g}}{\text{mL}}} = 2.5\ \text{mL}$$

$$V_B = \frac{65\ \text{g}}{4.0\ \frac{\text{g}}{\text{mL}}} = 16.25\ \text{mL} = 16\ \text{mL}$$

$16\ \text{mL} - 2.5\ \text{mL} = 13.5\ \text{mL}$

B is 14 mL larger than A

75.

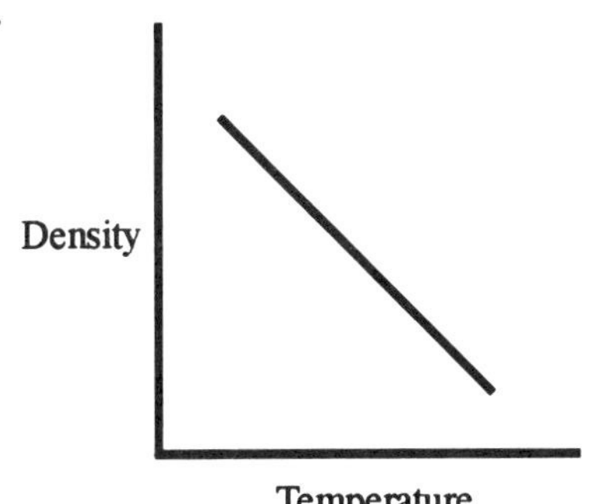

Since $D = \frac{m}{V}$, as the volume decreases, density decreases.
As solids are heated the density decreases.

76. m = DV = (0.789 g/mL)(35.0 mL) = 27.6 g ethyl alcohol
27.6 g + 49.28 g = 76.9 g (mass of cylinder and alcohol)

77. $D = \frac{m}{V}$ The cube with the largest V has the smallest D. Use Table 2.5.

Cube A – lowest density	1.74 g/mL – magnesium
Cube B	2.70 g/mL – aluminum
Cube C – highest density	10.5 g/mL – silver

78. The volume of the aluminum cube is:

$$V = \frac{m}{D} = \frac{500.\ \text{g}}{2.70\ \frac{\text{g}}{\text{mL}}} = 185\ \text{mL}$$

This is the same volume as the gold cube thus:

$$m = DV = (185\ \text{mL})(19.3\ \text{g/mL}) = 3.57 \times 10^3\ \text{g}$$

79. $D = \frac{m}{V} = \frac{24.12\ \text{g}}{25.0\ \text{mL}} = \frac{0.965\ \text{g}}{\text{mL}}$

80. 150.50 g – 88.25 g = 62.25 g (mass of liquid)

$$D = \frac{m}{V} \text{ thus } V = \frac{m}{D} = \frac{62.25\ \text{g}}{1.25\ \frac{\text{g}}{\text{mL}}} = 49.8\ \text{mL}$$

The container must hold at least 50 mL.

81. H_2O $\quad \frac{50\ \text{g}}{1.0\ \frac{\text{g}}{\text{mL}}} = 50\ \text{mL}$

alcohol $\quad \frac{50\ \text{g}}{0.789\ \frac{\text{g}}{\text{mL}}} = 60\ \text{mL}$

alcohol has the greater volume

82. $V = (2.00\ \text{cm})(15.0\ \text{cm})(6.00\ \text{cm})\left(\frac{1\ \text{mL}}{1\ \text{cm}^3}\right) = 180.\ \text{mL}$

$d = \frac{m}{V} = 3300\ \text{g}/180.\ \text{mL} = 18.3\ \text{g/mL}$

The density of gold is 19.3 g/mL from Table 2.5, therefore, the gold bar is not pure gold, since its density is only 18.3 g/mL, or it is hollow inside.

83. $\left(\frac{9.33 \times 10^4\ \text{g}}{19.3\ \text{g/mL}}\right) = 4.83 \times 10^3\ \text{mL} = 4.83 \times 10^3\ \text{cm}^3$

$(9.33 \times 10^4\ \text{g})\left(\frac{1\ \text{lb}}{453.6\ \text{g}}\right)\left(\frac{14.58\ \text{oz}}{1\ \text{lb}}\right) = 3.00 \times 10^3\ \text{oz}$

$(3.00 \times 10^3\ \text{oz})(\$345/\text{oz}) = \$1{,}035{,}000$

84. Volume of slug $30.7\ \text{mL} - 25.0\ \text{mL} = 5.7\ \text{mL}$

Density of slug $D = \frac{m}{V} = \frac{15.434\ \text{g}}{5.7\ \text{mL}} = 2.7\ \text{g/mL}$

Mass of liquid, cylinder, and slug	125.934 g
Mass of slug (subtract)	−15.434 g
Mass of cylinder (subtract)	−89.450 g
Mass of the liquid	21.050 g

Density of liquid $D = \frac{m}{V} = \frac{21.050\ \text{g}}{25.0\ \text{mL}} = 0.842\ \text{g/mL}$

CHAPTER 3

CLASSIFICATION OF MATTER

1. Answers will vary

2. (a) The attractive forces among the ultimate particles of a solid (atoms, ions, or molecules) are strong enough to hold these particles in fixed position within the solid and thus maintain the solid in a definite shape. The attractive forces among the ultimate particles of a liquid (usually molecules) are sufficiently strong enough to hold them together (preventing the liquid from rapidly becoming a gas) but are not strong enough to hold the particles in fixed positions (as in a solid).

   (b) The ultimate particles in a liquid are quite closely packed (essentially in contact with each other) and thus the volume of the liquid is fixed at a given temperature. But, the ultimate particles in a gas are relatively far apart and essentially independent of each other. Consequently, the gas does not have a definite volume.

   (c) In a gas the particles are relatively far apart and are easily compressed, but in a solid the particles are closely packed together and are virtually incompressible.

3. The water in the beaker does not fill the test tube. Since the tube is filled with air (a gas) and two objects cannot occupy the same space at the same time, the gas is shown to occupy space.

4. Mercury and water are the only liquids in the table which are not mixtures.

5. Air is the only gas mixture found in the table.

6. Three phases are present within the bottle; solid and liquid are observed visually, while gas is detected by the immediate odor.

7. The system is heterogeneous as multiple phases are present.

8. A system containing only one substance is not necessarily homogeneous. Two phases may be present. Example: ice in water.

9. A system containing two or more substances is not necessarily heterogeneous. In a solution only one phase is present. Examples: sugar dissolved in water, dilute sulfuric acid.

10. Silicon 25.67% Hydrogen 0.87%

    In 100 g $\frac{25.67\text{ g Si}}{0.87\text{ g H}} = 30\text{ g Si/1 g H}$

    Si is 28 times heavier than H, thus since 30 > 28, there are more Si atoms than H atoms.

11. The symbol of an element represents the element itself. It may stand for a single atom or a given quantity of the element.

12.

phosphorus	P	sodium	Na
aluminum	Al	nitrogen	N
hydrogen	H	nickel	Ni
potassium	K	silver	Ag
magnesium	Mg	plutonium	Pu

13. (a) Si – 1 atom silicon — SI – System International or 1 atom sulfur 1 atom iodine

(b) Pb – 1 atom lead — PB – 1 atom phosphorus 1 atom boron

(c) 4P – 4 atoms phosphorus — P_4 – 1 molecule phosphorus (made of 4 phosphorus atoms)

14.

Na	sodium	Ag	silver
K	potassium	W	tungsten
Fe	iron	Au	gold
Sb	antimony	Hg	mercury
Sn	tin	Pb	lead

15.

H	hydrogen	S	sulfur
B	boron	K	potassium
C	carbon	V	vanadium
N	nitrogen	Y	yttrium
O	oxygen	I	iodine
F	fluorine	W	tungsten
P	phosphorus	U	uranium

16. In an element all atoms are alike, while a compound contains two or more elements (different atoms) which are chemically combined. Compounds may be decomposed into simpler substances while elements cannot.

17. 4 metals — 7 metalloids — 18 nonmetals

18. 7 metals — 1 metalloid — 2 nonmetals

19. 1 metal — 0 metalloids — 5 nonmetals

20. The symbol for gold is based upon the Latin word for gold, aurum.

21. (a) iodine
(b) bromine

22. A compound is composed of two or more elements which are chemically combined in a definite proportion by mass. Its properties differ from those of its components. A mixture is the physical combining of two or more substances (not necessarily elements). The composition may vary, the substances retain their properties, and they may be separated by physical means.

23. Molecular compounds exist as molecules formed from two or more elements bonded together. Ionic compounds exist as cations and anions held together by electrical attractions.

24. Compounds are distinguished from one another by their characteristic physical and chemical properties.

25. (a) H_2 – 2 atoms

(b) H_2O – 3 atoms

(c) H_2SO_4 – 7 atoms

26. Cations are positively charged, while anions are charged negatively.

27. H_2 – hydrogen

N_2 – nitrogen

O_2 – oxygen

F_2 – fluorine

Cl_2 – chlorine

Br_2 – bromine

I_2 – iodine

28. Homogeneous mixtures contain only one phase, while heterogeneous mixtures contain two or more phases.

29.

Metals	Nonmetals
solid at room T (except Hg)	solids, liquids or gas at room T
luster	dull (no luster)
conduct heat & electricity	insulator (does not conduct electricity)
malleable	react with each other forming compounds
react with nonmetals to form compounds	

30. diatomic molecules (a) H_2, (c) HCl, (e) NO

31. The major difficulty in using hydrogen as a fuel is the storage problem. Since the density of hydrogen is very low, a bulk container is needed to store it as a gas. Its boiling point is so low refrigeration is needed to store hydrogen as a liquid.

32. Sponge iron reacts easily. During this process hydrogen is generated and colleced. The sponge iron can be regenerated from the rust formed and the process repeated indefinitely.

33. graphite, diamond, buckminsterfullerene

34. Charcoal – destructive distillation of wood.
Bone black – destructive distillation of bones or wastes.
Carbon black – residue from burning natural gas.

35. Uses for graphite include: lubricants, "lead" in pencils, electrodes in dry cells, paints, water purification systems, gas masks, decolorizing carbon, carbon paper, printer's ink, and tires.

36. Correct statements are: c, f, g, j, m, o, q, t, u, w, x, y.

(a) Gases are the least compact state of matter.

(b) Liquids have a definite volume but no definite shape.

(d) Wood is heterogeneous.

(e) Wood is a mixture.

(h) Many systems made of only one substance are homogeneous.

(i) Some systems containing two or more substances are heterogeneous.

(k) The smallest unit of an element that can exist and enter into a chemical reaction is called an atom.

(l) The basic building blocks of all substances, that cannot be decomposed into simpler substances by ordinary chemical change, are elements.

(n) The most abundant element in the human body, by mass, is oxygen.

(p) The symbol for copper is Cu.

(r) The symbol for potassium is K.

(s) The symbol for lead is Pb.

(v) The smallest uncharged individual unit of a compound formed by the union of two or more atoms is called a molecule.

(z) A general property of metals is that they are good conductors of heat and electricity.

37. (a) Potassium, iodine
(b) Sodium, carbon, oxygen
(c) Aluminum, oxygen
(d) Calcium, bromine
(e) Hydrogen, carbon, oxygen

38. (a) Magnesium, bromine
(b) Carbon, chlorine
(c) Hydrogen, nitrogen, oxygen
(d) Barium, sulfur, oxygen
(e) Aluminum, phosphorus, oxygen

39. (a) ZnO
(b) $KClO_3$
(c) $NaOH$
(d) C_2H_6O

40. (a) $AlBr_3$
(b) CaF_2
(c) $PbCrO_4$
(d) C_6H_6

41. (a) 2 atoms H, 1 atom O
(b) 2 atoms Na, 1 atom S, 4 atoms O
(c) 4 atoms H, 2 atoms C, 2 atoms O

42. (a) 1 atom Al, 3 atoms Br
(b) 1 atom Ni, 2 atoms N, 6 atoms O
(c) 12 atoms C, 22 atoms H, 11 atoms O

43. (a) 2 atoms
(b) 5 atoms
(c) 11 atoms
(d) 8 atoms
(e) 16 atoms

44. (a) 2 atoms
(b) 2 atoms
(c) 9 atoms
(d) 5 atoms
(e) 17 atoms

45. (a) 1 atom O
(b) 4 atoms O
(c) 2 atoms O
(d) 3 atoms O
(e) 9 atoms O

46. (a) 2 atoms H
(b) 6 atoms H
(c) 12 atoms H
(d) 4 atoms H
(e) 8 atoms H

47. (a) mixture
(b) element
(c) compound
(d) mixture

48. (a) element
(b) compound
(c) element
(d) mixture

49. (a) mixture
(b) compound
(c) element
(d) mixture

50. (a) mixture
(b) element
(c) mixture
(d) compound

51. (a) CH_2O
(b) C_4H_9
(c) $C_{25}H_{52}$

52. (a) HO
(b) C_2H_6O
(c) $Na_2Cr_2O_7$

53. Yes. The gaseous elements are all found on the extreme right of the periodic table. They are the entire last column and in the upper right corner of the table. Hydrogen is the exception and located at the upper left of the table.

54. No. The only common liquid elements (at room temperature) are mercury and bromine.

55. $\frac{18 \text{ metals}}{36 \text{ elements}} \times 100 = 50\% \text{ metals}$

56. $\frac{27 \text{ solids}}{36 \text{ elements}} \times 100 = 75\% \text{ solids}$

57. (a) 181 atoms/molecule

63 C
88 H
1 Co
14 N
14 O
1 P
———
181 atoms

(b) $\frac{63\ \text{C}}{181\ \text{atoms}} \times 100 = 35\%$ C atoms

(c) $\frac{1\ \text{Co}}{181\ \text{atoms}} = \frac{1}{181}$ metals

58. HNO_3 has 5 atoms/molecule

7 dozen = 84

(84 molecule)(5 atoms/molecule) = 420 atoms

or $(7\ \text{dz})\left(\frac{12\ \text{molecules}}{\text{dz}}\right)\left(\frac{5\ \text{atoms}}{\text{molecule}}\right) = 420$ atoms

59. Each represents eight units of sulfur. In 8 S the atoms are separate and distinct. In S_8 the atoms are joined as a unit (molecule).

60. $Ca(H_2PO_4)_2$

$(10\ \text{molecules})\left(\frac{4\ \text{atoms H}}{\text{molecule}}\right) = 40$ atoms H

61. $C_{145}H_{293}O_{168}$

145 C
293 H
168 O
———
606 atoms/molecule

62. (a) magnesium, manganese, molybdenum, mendeleevium, mercury
(b) carbon, phosphorus, sulfur, selenium, iodine, astatine, boron
(c) sodium, potassium, iron, silver, tin, antimony

63. $\left(4 \times 10^{-4}\ \frac{\text{mg}}{\text{L}}\right)\left(\frac{1\ \text{L}}{1000\ \text{cm}^3}\right)\left(1 \times 10^{15}\ \text{cm}^3\right)\left(\frac{\$19.40}{\text{g}}\right) = \$8 \times 10^9$

64.

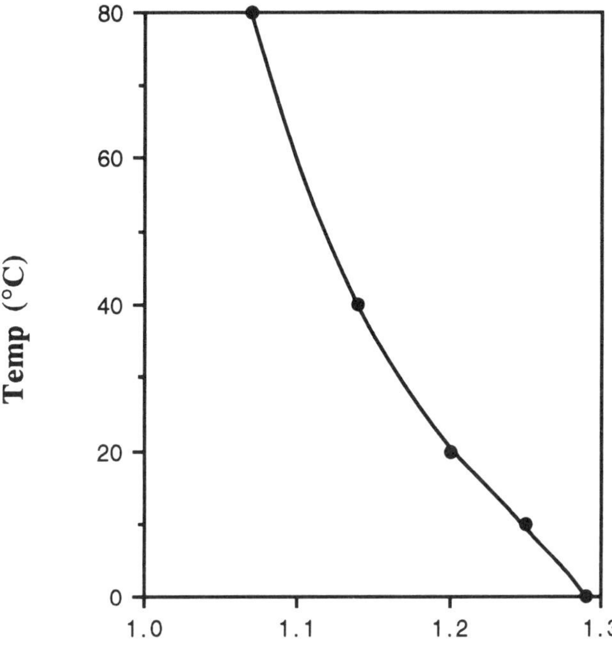

(a) As temperature decreases, density increases.

(b) approximately 1.28 g/L 5°C
approximately 1.17 g/L 25°C
approximately 1.08 g/L 70°C

CHAPTER 4

PROPERTIES OF MATTER

1. Solid

2. Solid 102 K = −171°C melting point is −101.6°C

3. Small bubbles appear at each electrode. Gas collects above the electrodes. The system now contains water and gas.

4. Water disappears. Gas appears above each electrode and as bubbles in solution.

5. Physical properties are characteristics which may be determined without altering the composition of the substance. Chemical properties describe the ability of a substance to form new substances by reaction or decomposition.

6. A new substance is always formed during a chemical change, but never formed during physical changes.

7. Foods are assigned an energy value by burning them in a calorimeter and measuring the heat released. Since the products of combustion and metabolism are the same, the caloric content of food is assigned by this method.

8. The hot pack contains a solution of sodium acetate or sodium thiosulfate. A small crystal is added by squeezing a corner of the bag or by bending a small metal activator. The solution crystallizes and heat is released to the surroundings. To reuse it, the pack is heated in boiling water until the crystals dissolve. Then, it is slowly cooled and stored till needed.

9. In a chemical change, any change of mass that occurs is so small as to be undetectable in an experiment, though it can be calculated from the energy change. Since it is undetectable, we say that mass is nether lost nor gained.

10. Potential energy is the energy of position. By the position of an object, it has the potential of movement to a lower energy state. Kinetic energy is the energy matter possesses due to its motion.

11. (a) $118.0°\text{C} + 273 = 391.0\ \text{K}$
    (b) $(118.0°\text{C})1.8 + 32 = 244.4°\text{F}$

12. The correct statements are a, f, h, i.

    (b) When heated in the air, a platinum wire has a constant mass.
    (c) When heated in the air, a copper wire gains mass.
    (d) 4.184 Joules is the equivalent of 1.0 cal of energy.

(e) Boiling water represents a physical changes because a change of state occurs without a change in composition.
(g) All of the following represent physical changes: breaking a stick, melting wax, folding a napkin.
(j) A stretched rubber band possesses potential energy.

13. (a) physical
(b) physical
(c) physical
(d) chemical
(e) chemical
(f) chemical

14. (a) chemical
(b) physical
(c) physical
(d) chemical
(e) chemical
(f) physical

15. Although the appearance of the platinum wire changed during the heating, the original appearance was restored when the wire cooled. No change in the composition of the platinum could be detected.

16. The copper wire, like the platinum wire, changed to a glowing red color when heated (physical change). Upon cooling, the original appearance of the copper wire was not restored, but a new substance, black copper(II) oxide, had appeared in its stead.

17. Reactants: copper, oxygen
Product: copper II oxide

18. Reactant: water
Product: hydrogen, oxygen

19. The kinetic energy is converted to thermal energy (heat), chiefly in the brake system, and eventually dissipated into the atmosphere.

20. The transformation of kinetic energy to thermal energy (heat) is responsible for the fiery reentry of a space vehicle.

21. (a) +
(b) –
(c) +
(d) +
(e) –

22. (a) +
(b) –
(c) –
(d) +
(e) –

23. $E = (m)(\text{specific heat})(\Delta t)$
$= (75\text{ g})(4.184\text{ J/g°C})(70.0\text{°C} - 20.0\text{°C})$
$= 1.6 \times 10^4\text{ J}$

24. $E = (m)(\text{specific heat})(\Delta t)$
$= (65\text{ g})(0.473\text{ J/g°C})(95°\text{C} - 25°\text{C})$
$= 2.2 \times 10^3\text{ J}$

25. $E = (m)(\text{specific heat})(\Delta t)$

$$\text{specific heat} = \frac{E}{m(\Delta t)} = \frac{5.866 \times 10^3\text{ J}}{(250.0\text{ g})(100.0°\text{C} - 22°\text{C})} = 0.30\text{ J/g°C}$$

26. $E = (m)(\text{specific heat})(\Delta t)$

$$\text{specific heat} = \frac{E}{m(\Delta t)} = \frac{3.07 \times 10^4\text{ J}}{(1.00 \times 10^3\text{ g})(630.0°\text{C} - 20.0°\text{C})} = 5.03 \times 10^{-2}\text{ J/g°C}$$

27. heat lost by gold = heat gained by water x = final temperature

$$\begin{aligned}(m)(\text{specific heat})(\Delta t) &= (m)(\text{specific heat})(\Delta t)\\ (325\text{ g})(0.131\text{ J/g°C})(427°\text{C} - x) &= (200.0\text{ g})(4.184\text{ J/g°C})(x - 22.0°\text{C})\\ 18179\text{ J} - 42.575x\text{ J/°C} &= 836.8x\text{ J/°C} - 18409.6\text{ J}\\ 18179\text{ J} + 18409.6\text{ J} &= 836.8x + 42.575x\\ 36588\text{ J} &= 879.375x\text{ J/°C}\\ 41.6°\text{C} &= x\end{aligned}$$

28. heat lost by iron = heat gained by water x = final temperature

$$m = VD = (2.0\text{ L})\left(\frac{1000\text{ mL}}{1\text{ L}}\right)\left(1.0\ \frac{\text{g}}{\text{mL}}\right) = 2.0 \times 10^3\text{ g } H_2O$$

$$\begin{aligned}(m)(\text{specific heat})(\Delta t) &= (m)(\text{specific heat})(\Delta t)\\ (500.0\text{ g})(0.473\text{ J/g°C})(212°\text{C} - x) &= (2.0 \times 10^3\text{ g})(4.184\text{ J/g°C})(x - 24.0°\text{C})\\ 50138\text{ J} - 236.5x\text{ J/°C} &= 8368x\text{ J/°C} - 200832\text{ J}\\ 250970\text{ J} &= 8604.5x\ \frac{\text{J}}{°\text{C}}\\ x &= 29°\text{C}\\ \Delta t &= 29°\text{C} - 24°\text{C} = 5°\text{C}\end{aligned}$$

29. $E = (m)(\text{specific heat})(\Delta t)$
$= (250.\text{ g})(0.096\text{ cal/g°C})(150.0°\text{C} - 24°\text{C})$
$= 3.0 \times 10^3\text{ cal}$

30. $E = (m)(\text{specific heat})(\Delta t)$ x = final temperature
$4.00 \times 10^4\text{ J} = (500.0\text{ g})(4.184\text{ J/g°C})(x - 10°\text{C})$
$4.00 \times 10^4\text{ J} = 2092x\text{ J/°C} - 20920\text{ J}$
$60{,}920\text{ J} = 2092x\text{ J/°C}$
$29.1°\text{C} = x$

31. heat lost by coal = heat gained by water $\quad x$ = mass of coal in g
(5500 cal/g)x = (500.0 g)(1.00 cal/g°C)(90.0°C – 20.0°C) = 35,000 cal
x = 6.36 g coal

32. (7000. cal)(4.184 J/cal) = 29290 J = 29.29 kJ
heat lost by coal = heat gained by water $\quad x$ = mass of coal in g

4.0 L H_2O = 4.0 × 10^3 g H_2O

$$\left(2.929 \times 10^4 \frac{J}{g}\right)x = \left(4.0 \times 10^3 \text{ g}\right)\left(4.184 \frac{J}{g°C}\right)(100.0°C - 20.0°C)$$

$$\left(2.929 \times 10^4 \frac{J}{g}\right)x = 1.34 \times 10^6 \text{ J}$$

x = 45.7 g coal

33. (a) (110.0 g)(0.0921 cal/g°C)(100.0°C – 10.0°C) = 829 cal
(b) 829 cal = (100.0 g)(0.215 cal/g°C)(x – 10.0°C)
x = 48.6°C (final temperature for aluminum)
Therefore the copper gets hotter since it ended up at 100.0°C.
Note: You can figure this out without calculation if you consider the specific heats of the metals. Since the specific heat of copper is much less than aluminum the copper heats more easily.

34. heat lost by iron = heat gained by water $\quad x$ = initial temperature of iron

$$(m)(\text{specific heat})(\Delta t) = (m)(\text{specific heat})(\Delta t)$$

$$(500.0 \text{ g})(0.113 \text{ cal/g°C})(x - 90.0°C) = (400. \text{ g})(1.00 \text{ cal/g°C})(90.0°C - 10.0°C)$$

$$\left(56.5 \frac{cal}{°C}\right)x - 5085 \text{ cal} = 32{,}000 \text{ cal}$$

$$\left(56.5 \frac{cal}{°C}\right)x = 37085 \text{ cal}$$

$$x = 656°C$$

35. heat lost by metal = heat gained by water $\quad x$ = specific heat of metal

$$(m)(\text{specific heat})(\Delta t) = (m)(\text{specific heat})(\Delta t)$$
$$(20.0 \text{ g})(x)(203°C - 29.0°C) = (100.0 \text{ g})(4.184 \text{ J/g°C})(29.0°C - 25.0°C)$$
$$4060x \text{ g°C} - 580x \text{ g°C} = 12133.6 \text{ J} - 10460 \text{ J}$$
$$(3480 \text{ g°C})x = 1674 \text{ J}$$
$$x = 0.481 \text{ J/g°C}$$

36. heat lost = heat gained $\qquad\qquad x$ = final temperature

(m)(specific heat)(Δt) = (m)(specific heat)(Δt) (specific heats are the same)

$$(10.0\ \text{g})\left(\cancel{4.184\,\frac{\text{J}}{\text{g°C}}}\right)(50.0\text{°C} - x) = (50.0\ \text{g})\left(\cancel{4.184\,\frac{\text{J}}{\text{g°C}}}\right)(x - 10.0\text{°C})$$
$$500.0 - 10x = 50.0x - 500.0$$
$$1000.0 = 60.0x$$
$$16.7\text{°C} = x$$

37. Specific heats for the metals are Fe: 0.473J/g°C; Cu: 0.385 J/g°C; Al: 0.900 J/g°C. The metal with the lowest specific heat will warm most quickly, therefore the copper pan heats fastest, and fries the egg fastest.

38. In order for the water to boil both the pan and water must reach 100.0°C.

$$(300.0\ \text{g})\left(0.0921\ \frac{\text{cal}}{\text{g°C}}\right)(100.\ \text{°C} - 25\text{°C}) + (800.0\ \text{g})\left(1.00\ \frac{\text{cal}}{\text{g°C}}\right)(100.\ \text{°C} - 25\text{°C})$$

= 2072 cal + 60000 cal
= 62000 cal needed to heat the pan and water

$$(62{,}000)\left(\frac{1\ \text{s}}{150\ \text{cal}}\right) = 410\ \text{s} = 6.90\ \text{min}$$

At 6:06 and 54 sec. the water boils.

39. Heat is transferred from the molecules of coffee on the surface to the air above them. As you blow you move the warmed air molecules away from the surface replacing them with cooler ones which are warmed by the coffee and cool it in a repeating cycle. Inserting a spoon into hot coffee cools the coffee by heat transfer as well. Heat is transferred from the coffee to the spoon lowering the temperature of the coffee and raising the temperature of the spoon.

40. The potatoes will cook at the same rate whether the water boils vigorously or slowly. Once the boiling point is reached the water temperature remains constant. The energy available is the same so the cooking times should be equal.

41. (250 mL)(0.04) = 10 mL fat
(10 mL)(0.8 g/mL) = 8 g fat in a glass of milk

42. mercury + sulfur → compound
The mercury and sulfur react to form a compound since the properties of the product are different from the properties of either reactant.

$$(100.0\ \text{mL})\left(13.6\ \frac{\text{g}}{\text{mL}}\right) = 1.36 \times 10^3\ \text{g mercury}$$

1360 g	+	100.0 g		1460 g
mercury	+	sulfur	→	compound

This supports the Law of Conservation of Matter since the mass of the product is equal to the mass of the reactants.

CHAPTER 5

EARLY ATOMIC THEORY AND STRUCTURE

1.

	Element	Atomic number
(a)	copper	29
(b)	nitrogen	7
(c)	phosphorus	15
(d)	radium	88
(e)	zinc	30

2. The neutron is about 1840 times heavier than an electron.

3.

Particle	charge	mass
proton	+1	1 amu
neutron	0	1 amu
electron	−1	0

4. An atom is electrically neutral, containing equal numbers of protons and electrons.
   An ion has a charge resulting from an imbalance between the numbers of protons and electrons.

5. Isotopic notation $^{A}_{Z}X$

   Z represents the atomic number
   A represents the mass number

6. A wintergreen lifesaver is an example of a substance which exihibits triboluminescence. The wintergreen molecules absorb some of the energy and re-emit it as blue-green light.

7. Synthetic and natural vanilla contain exactly the same molecules. The difference between them is in the source of the molecules of vanillin. Natural vanilla comes from the vanilla bean, while the synthetic variety is made from lignin. Both extracts are used as flavorings.

8. The SIRA technique is stable isotope ratio analysis. The carbon atoms in a molecule are analyzed to determine the ratio of C-12 and C-13. The ratio is different for synthetic and natural vanillin and so can be used to differentiate between these types.

9. Isotopes contain the same number of protons and the same number of electrons.
   Isotopes have different numbers of neutrons and thus different atomic masses.

10. The correct statements are a, b, c, f, g
    (d) Chadwick stated that atoms are composed of protons, neutrons, and electrons.

(e) Portions of Dalton's theory are still considered valid today.
(h) The nucleus of an atom contains protons and neutrons.

11. The correct statesments are b, c, f, g, h, k.

(a) An element with an atomic number of 29 has 29 protons and 29 electrons.

(d) An atom of $^{31}_{15}P$ contain 15 protons, 16 neutrons, and 15 electrons.

(e) In the isotope of $^{6}_{3}Li$ Z = 3 and A = 6

(i) $^{24}_{11}Na$ has one more neutron than $^{23}_{11}Na$

(j) $^{24}_{11}Na$ has one more neutron than $^{23}_{11}Na$

(l) Most elements exist in nature as mixtures of isotopes.

12. The correct statements are b and d.

(a) One atomic mass unit has a mass $\frac{1}{12}$ that of one carbon- atom.

(c) $^{23}_{11}Na$ and $^{24}_{11}Na$ have different atomic masses.

13. Gold nuclei are very massive (compared to an alpha particle) and have a large positive charge. As the alpha particles approach the atom, some are deflected by this positive charge. Those approaching a gold nucleus directly are deflected backwards by the massive positive nucleus.

14. (a) The nucleus of the atom contains most of the mass since only a collision with a very dense, massive object would cause an alpha particle to be deflected back towards the source.

(b) The deflection of the alpha particles from their initial flight indicates the nucleus of the atom is also positively charged.

(c) Most alpha particles pass through the gold foil undeflected leading to the conclusion that the atom is mostly empty space.

15. In the atom, protons and neutrons are found within the nucleus. Electrons occupy the remaining space within the atom outside the nucleus.

16. The nucleus of an atom contains nearly all of its mass.

17. (a) Dalton contributed the concept that each element is composed of atoms which are unique, and can combine in ratios of small whole numbers.

(b) Thomson discovered the electron, determined its properties, and found the mass of a proton is 1840 times the mass of the electron. He developed the Thomson model of the atom.

(c) Rutherford devised the model of a nuclear atom with a positive charge and mass concentrated in the nucleus. Most of the atom is empty space.

18. Electrons: Dalton – electons are not part of his model
Thomson – electrons are scattered throughout the positive mass of matter in the atom
Rutherford – electrons are located out in space away from the central positive mass

Positive matter: Dalton – no positive matter in his model
Thomson – positive matter is distributed throughout the atom
Rutherford – positive matter is concentrated in a small central nucleus

19. Atomic mass are not whole numbers because:

(a) the neutron and proton do not have identical masses and neither is exactly 1 amu.

(b) most elements exist in nature as a mixture of isotopes. The atomic mass is the average of all these isotopes.

20. The isotope of C with a mass of 12 is an exact number by definition. The mass of other isotopes such as ${}^{63}_{29}Cu$ will not be an exact number for reasons given in Exercise 19.

21. The isotopes of hydrogen are protium, deuterium, and tritium.

22. All three isotopes of hydrogen have the same number of protons (1) and electrons (1). They differ in the number of neutrons (0, 1, and 2).

23. ${}^{52}_{24}Cr$ chromium-52

24. (a) $201 - 121 = 80$ protons; electrical charge of the nucleus is +80
(b) Hg, mercury

25. All six isotopes have 20 protons and 20 electrons. The number of neutrons are

Isotope	Neutrons
40	20
42	22
43	23
44	24
46	26
48	28

26. The most abundant isotope is 40. This is certain as 40 is the average of all isotopes and is lowest on the list. An arithmetic average would be between 40 and 48. Since the atomic mass is 40.48, there must be only small amounts of the other isotopes.

27. (a) ${}^{55}_{26}Fe$ (c) ${}^{6}_{3}Li$

(b) ${}^{26}_{12}Mg$ (d) ${}^{188}_{79}Au$

28. (a) $^{59}_{27}Co$ Nucleus contains 27 protons and 32 neutrons

(b) $^{31}_{15}P$ Nucleus contains 15 protons and 16 neutrons

(c) $^{184}_{74}W$ Nucleus contains 74 protons and 110 neutrons

(d) $^{235}_{92}U$ Nucleus contains 92 protons and 143 neutrons

29. (0.2360)(205.9745 amu) + (0.2260)(206.9759 amu) +
(0.5230)(207.9766 amu) + (0.01480)(203.973 amu)
= 48.61 amu + 46.78 amu + 108.8 amu + 3.019 amu
= 207.2 amu = average atomic mass Pb

30. (0.7899)(23.985 amu) + (0.1000)(24.986 amu) + (0.1101)(25.983 amu)
= 18.95 amu + 2.500 amu + 2.861 amu
= 24.31 amu = average atomic mass Mg

31. (0.604)(68.9257 amu) + (1.00 – 0.604)(70.9249 amu)
= 41.6 amu + 28.1 amu
= 69.7 amu = average atomic mass
The element is gallium.

32. (0.3000)(6.015 amu) + (0.7000)(7.016 amu)
= 1.805 amu + 4.911 amu = 6.716 amu = average atomic mass Li

33. $\dfrac{3.0 \times 10^{-8}\ \text{cm}}{2.0 \times 10^{-13}\ \text{cm}} = 1.5 \times 10^{5} : 1$ (ratio of diameter to Al nucleus)

34. $V_{sphere} = \frac{4}{3}\pi r^3$ r_A = radius of atom, r_N = radius of nucleus

$$\frac{V_{atom}}{V_{nucleus}} = \frac{\frac{4}{3}\pi r_A^3}{\frac{4}{3}\pi r_N^3} = \frac{r_A^3}{r_N^3} = \frac{(1.0 \times 10^{-8})^3}{(1.0 \times 10^{-13})^3} = 1.0 \times 10^{15} : 1$$

(ratio of atomic volume to nuclear volume)

35. (a) In Rutherford's experiment the majority of alpha particles passed through the gold foil without deflection. This shows that the atom is mostly empty space and the nucleus is very small.

(b) In Thomson's experiments with the cathode ray tube rays were observed coming from both the anode and the cathode.

(c) In Rutherford's experiment an alpha particle was occasionally dramatically deflected by the nucleus of a gold atom. The direction of deflection showed the nucleus to be positive.

36. (a) These atoms are isotopes

(b) These atoms are adjacent to each other on the periodic table. The atoms have the same mass.

37. The nucleus of the atoms will have 2 less protons (lowering the nuclear charge by 2). The nuclear mass will be reduced by 4 amu.

38. (a) The amount of deflection is inversely related to the particle's mass, therefore the (+) particles coming from the anode are composed of much lighter mass particles than the (–) particles.

(b) This nucleus must be negatively charged since it attracts and captures (+) alpha particles.

39. $\dfrac{1.5 \text{ cm}}{0.77 \times 10^{-8} \text{ cm}} = 1.9 \times 10^{8} : 1,\ 1.9 \times 10^{8}$ enlargement

40. The properties of an element are related to the number of protons and electrons. If the number of neutrons differs, isotopes result. Isotopes of the same element are still the same element even though the nuclear composition of the atoms are different.

41. ^{210}Bi has $210 - 83 = 127$ neutrons $\rightarrow$ largest number/atom
^{210}Po has $210 - 84 = 126$ neutrons
^{210}At has $210 - 85 = 125$ neutrons
^{211}At has $211 - 85 = 126$ neutrons

42. percent of sample $^{60}\text{Q} = x$
percent of sample $^{63}\text{Q} = 1 - x$

$$\begin{aligned} (x)(60.\text{ amu}) + (1 - x)(63\text{ amu}) &= 61.5\text{ amu} \\ 60.\ x + 63\text{ amu} - 63x &= 61.5\text{ amu} \\ 63\text{ amu} - 61.5\text{ amu} &= 63x - 60x \\ 1.5 &= 3x \\ 0.50 &= x \\ ^{60}\text{Q} &= 50\% \\ ^{63}\text{Q} &= 50\% \end{aligned}$$

43. $\left(\dfrac{2.18 \times 10^{-22} \text{ g}}{1.9927 \times 10^{-23} \text{ g C}}\right)(12.0 \text{ g C}) = 131 \text{ g}$

44. $(40.0 \text{ g})\left(\dfrac{1 \text{ atom}}{6.63 \times 10^{-24} \text{ g}}\right) = 6.03 \times 10^{24}$ atoms

45.

	protons	neutrons	electrons
He	2	2	2
C	6	6	6
N	7	7	7
O	8	8	8
Ne	10	10	10
Mg	12	12	12
Si	14	14	14
S	16	16	16
Ca	20	20	20

46.

	Atomic number	Mass number	Symbol	Protons	Neutrons
(a)	8	16	O	8	8
(b)	28	58	Ni	28	30
(c)	80	199	Hg	80	119

47.

	Element	Symbol	Atomic #	Protons	Neutrons	Electrons
(a)	platinum	^{195}Pt	78	78	117	78
(b)	phosphorus	^{30}P	15	15	15	15
(c)	iodone	^{127}I	53	53	74	53
(d)	krypton	^{84}Kr	36	36	48	36
(e)	selenium	^{79}Se	34	34	45	34
(f)	calcium	^{40}Ca	20	20	20	20

CHAPTER 6

NOMENCLATURE OF INORGANIC COMPOUNDS

1. (a) $NaClO_3$
   (b) H_2SO_4
   (c) $Sn(C_2H_3O_2)_2$
   (d) Cu_2O
   (e) $Zn(HCO_3)_2$
   (f) $Fe_2(CO_3)_3$

2. No, if elements combine in a one-to-one ratio the charges on their ions must be equal and opposite in sign. They could be +1, −1, or +2, −2 or +3, −3 etc.

3. Hard water contains calcium and magnesium ions which form precipitates in the presence of soap. These precipitates stick to the clothing fiber, dulling the appearance of white laundry.

4. Seaborg is still alive. Elements can only be named for a person after he is dead according to IUPAC rules.

5. Plutonium, neptunium, americium, fermium, mendelevium, nobelium, berkelium, californium, curium, and einsteinium.

6. The correct statements are: a, b, c, e, f, h, i, k, n, p, q.
   (d) The formula for the compound between Fe^{3+} and O^{2-} is Fe_2O_3.
   (g) The name for CuO is copper(II) oxide.
   (j) The name for Na_2O is sodium oxide.
   (l) If the name of an anion ends with ite, the corresponding acid name will end in ous.
   (m) If the name of an acid ends in ous, the corresponding salt will end with ite.
   (o) In Cu_2SO_4, the copper ion is copper(I) because two copper ions are conbined with a −2 sulfate ion.
   (r) $Sn(CrO_4)_2$ is called tin(IV) chromate.

7. Formulas of compounds.

(a) Na and I	NaI		(d) K and S	K_2S
(b) Ba and F	BaF_2		(e) Cs and Cl	CsCl
(c) Al and O	Al_2O_3		(f) Sr and Br	$SrBr_2$

8. Formulas of compounds.

(a) Ba and O	BaO		(d) Be and Br	$BeBr_2$
(b) H and S	H_2S		(e) Li and Si	Li_4Si
(c) Al and Cl	$AlCl_3$		(f) Mg and P	Mg_3P_2

9.

sodium	Na^+	cobalt(II)	Co^{2+}
magnesium	Mg^{2+}	barium	Ba^{2+}
aluminum	Al^{3+}	hydrogen	H^+
copper(II)	Cu^{2+}	mercury(II)	Hg^{2+}
iron(II)	Fe^{2+}	tin(II)	Sn^{2+}
iron(III)	Fe^{3+}	chromium(III)	Cr^{3+}
lead(II)	Pb^{2+}	tin(IV)	Sn^{4+}
silver	Ag^+	manganese(II)	Mn^{2+}
		bismuth(III)	Bi^{3+}

10.

chloride	Cl^-	hydrogen sulfate	HSO_4^-
bromide	Br^-	hydrogen sulfite	HSO_3^-
fluoride	F^-	chromate	CrO_4^{2-}
iodide	I^-	carbonate	CO_3^{2-}
cyanide	CN^-	hydrogen carbonate	HCO_3^-
oxide	O^{2-}	acetate	$C_2H_3O_2^-$
hydroxide	OH^-	chlorate	ClO_3^-
sulfide	S^{2-}	permanganate	MnO_4^-
sulfate	SO_4^{2-}	oxalate	$C_2O_4^{2-}$

11.

Ion	Br^-	O^{2-}	NO_3^-	PO_4^{3-}	CO_3^{2-}
K^+	KBr	K_2O	KNO_3	K_3PO_4	K_2CO_3
Mg^{2+}	$MgBr_2$	MgO	$Mg(NO_3)_2$	$Mg_3(PO_4)_2$	$MgCO_3$
Al^{3+}	$AlBr_3$	Al_2O_3	$Al(NO_3)_3$	$AlPO_4$	$Al_2(CO_3)_3$
Zn^{2+}	$ZnBr_2$	ZnO	$Zn(NO_3)_2$	$Zn_3(PO_4)_2$	$ZnCO_3$
H^+	HBr	H_2O	HNO_3	H_3PO_4	H_2CO_3

12.

Ion	SO_4^{2-}	Cl^-	AsO_4^{3-}	$C_2H_3O_2^-$	CrO_4^{2-}
NH_4^+	$(NH_4)_2SO_4$	NH_4Cl	$(NH_4)_3AsO_4$	$NH_4C_2H_3O_2$	$(NH_4)_2CrO_4$
Ca^{2+}	$CaSO_4$	$CaCl_2$	$Ca_3(AsO_4)_2$	$Ca(C_2H_3O_2)_2$	$CaCrO_4$
Fe^{3+}	$Fe_2(SO_4)_3$	$FeCl_3$	$FeAsO_4$	$Fe(C_2H_3O_2)_3$	$Fe_2(CrO_4)_3$
Ag^+	Ag_2SO_4	$AgCl$	Ag_3AsO_4	$AgC_2H_3O_2$	Ag_2CrO_4
Cu^{2+}	$CuSO_4$	$CuCl_2$	$Cu_3(AsO_4)_2$	$Cu(C_2H_3O_2)_2$	$CuCrO_4$

13. Nonmetal binary compound formulas

(a) carbon monoxide, CO
(b) sulfur trioxide, SO_3
(c) carbon tetrabromide, CBr_4
(d) phosphorus trichloride, PCl_3
(e) nitrogen dioxide, NO_2
(f) dinitrogen pentoxide, N_2O_5
(g) iodine monobromide, IBr
(h) silicon tetrachloride, $SiCl_4$
(i) phosphorus pentiodide, PI_5
(j) diboron trioxide, B_2O_3

14. Naming binary nonmetal compounds:

(a) CO_2 carbon dioxide
(b) N_2O dinitrogen oxide
(c) PCl_5 phosphorus pentachloride
(d) CCl_4 carbon tetrachloride
(e) SO_2 sulfur dioxide
(f) N_2O_4 dinitrogen tetroxide
(g) P_2O_5 diphosphorus pentoxide
(h) OF_2 oxygen difluoride
(i) NF_3 nitrogen trifluoride
(j) CS_2 carbon disulfide

15. (a) sodium nitrate, $NaNO_3$
(b) magnesium fluoride, MgF_2
(c) barium hydroxide, $Ba(OH)_2$
(d) ammonium sulfate, $(NH_4)_2SO_4$
(e) silver carbonate, Ag_2CO_3
(f) calcium phosphate, $Ca_3(PO_4)_2$
(g) potassium nitrite, KNO_2
(h) strontium oxide, SrO

16. (a) K_2O, potassium oxide
(b) NH_4Br, ammonium bromide
(c) CaI_2, calcium iodide
(d) $BaCO_3$, barium carbonate
(e) Na_3PO_4, sodium phosphate
(f) Al_2O_3, aluminum oxide
(g) $Zn(NO_3)_2$, zinc nitrate
(h) Ag_2SO_4, silver sulfate

17. (a) $CuCl_2$ copper(II) chloride
(b) $CuBr$ copper(I) bromide
(c) $Fe(NO_3)_2$ iron(II) nitrate
(d) $FeCl_3$ iron(III) chloride
(e) SnF_2 tin(II) fluoride
(f) $HgCO_3$ mercury(II) carbonate

18. Formulas:

(a) tin(IV) bromide $SnBr_4$
(b) copper(I) sulfate Cu_2SO_4
(c) iron(III) carbonate $Fe_2(CO_3)_3$
(d) mercury(II) nitrite $Hg(NO_2)_2$
(e) titanium(IV) sulfide TiS_2
(f) iron(II) acetate $Fe(C_2H_3O_2)_2$

19. Acid formulas:

(a) hydrochloric acid, HCl
(b) chloric acid, $HClO_3$
(c) nitric acid, HNO_3
(d) carbonic acid, H_2CO_3
(e) sulfurous acid, H_2SO_3
(f) phosphoric acid, H_3PO_4

20. Formulas of acids:

(a) acetic acid, $HC_2H_3O_2$
(b) hydrofluoric acid, HF
(c) hypochlorous acid, $HClO$
(d) boric acid, H_3BO_3
(e) nitrous acid, HNO_2
(f) hydrosulfuric acid, H_2S

21. Naming acids:

(a) HNO_2, nitrous acid
(b) H_2SO_4, sulfuric acid
(c) $H_2C_2O_4$, oxalic acid
(d) HBr, hydrobromic acid
(e) H_3PO_3, phosphorous acid
(f) $HC_2H_3O_2$, acetic acid
(g) HF, hydrofluoric acid
(h) $HBrO_3$, bromic acid

22. Naming acids:

(a) H_3PO_4, phosphoric acid
(b) H_2CO_3, carbonic acid
(c) HIO_3, iodic acid
(d) HCl, hydrochloric acid
(e) HClO, hypochlorous acid
(f) HNO_3, nitric acid
(g) HI, hydroiodic acid
(h) $HClO_4$, perchloric acid

23. Formulas for:

(a) silver sulfite — Ag_2SO_3
(b) cobalt(II) bromide — $CoBr_2$
(c) tin(II) hydroxide — $Sn(OH)_2$
(d) aluminum sulfate — $Al_2(SO_4)_3$
(e) manganese(II) fluoride — MnF_2
(f) ammonium carbonate — $(NH_4)_2CO_3$
(g) chromium(III) oxide — Cr_2O_3
(h) cupric chloride — $CuCl_2$
(i) potassium permanganate — $KMnO_4$
(j) barium nitrite — $Ba(NO_2)_2$
(k) sodium peroxide — Na_2O_2
(l) iron(II) sulfate — $FeSO_4$
(m) potassium dichromate — $K_2Cr_2O_7$
(n) bismuth(III) chromate — $Bi_2(CrO_4)_3$

24. Formulas for:

(a) sodium chromate — Na_2CrO_4
(b) magnesium hydride — MgH_2

(c) nickel(II) acetate	$Ni(C_2H_3O_2)_2$
(d) calcium chlorate	$Ca(ClO_3)_2$
(e) lead(II) nitrate	$Pb(NO_3)_2$
(f) potassium dihydrogen phosphate	KH_2PO_4
(g) manganese(II) hydroxide	$Mn(OH)_2$
(h) cobalt(II) hydrogen carbonate	$Co(HCO_3)_2$
(i) sodium hypochlorite	$NaClO$
(j) arsenic(V) carbonate	$As_2(CO_3)_5$
(k) chromium(III) sulfite	$Cr_2(SO_3)_3$
(l) antimony(III) sulfate	$Sb_2(SO_4)_3$
(m) sodium oxalate	$Na_2C_2O_4$
(n) potassium thiocyanate	KSCN

25.

Formula	Name
(a) $ZnSO_4$	zinc sulfate
(b) $HgCl_2$	mercury(II) chloride
(c) $CuCO_3$	copper(II) carbonate
(d) $Cd(NO_3)_2$	cadmium nitrate
(e) $Al(C_2H_3O_2)_3$	aluminum acetate
(f) CoF_2	cobalt(II) fluoride
(g) $Cr(ClO_3)_3$	chromium(III) chlorate
(h) Ag_3PO_4	silver phosphate
(i) NiS	nickel(II) sulfide
(j) $BaCrO_4$	barium chromate

26.

Formula	Name
(a) $Ca(HSO_4)_2$	calcium hydrogen sulfate
(b) $As_2(SO_3)_3$	arsenic(III) sulfite
(c) $Sn(NO_2)_2$	tin(II) nitrite
(d) $FeBr_3$	iron(III) bromide
(e) $KHCO_3$	potassium hydrogen carbonate

(f) $BiAsO_4$ bismuth(III) arsenate

(g) $Fe(BrO_3)_2$ iron(II) bromate

(h) $(NH_4)_2HPO_4$ ammonium monohydrogen phosphate

(i) NaClO sodium hypochlorite

(j) $KMnO_4$ potassium permanganate

27. Formulas for:

(a) baking soda $NaHCO_3$

(b) lime CaO

(c) Epsom salts $MgSO_4 \cdot 7\ H_2O$

(d) muriatic acid HCl

(e) vinegar $HC_2H_3O_2$

(f) potash K_2CO_3

(g) lye NaOH

28. Formulas for:

(a) fool's gold FeS_2

(b) saltpeter $NaNO_3$

(c) limestone $CaCO_3$

(d) cane sugar $C_{12}H_{22}O_{11}$

(e) milk of magnesia $Mg(OH)_2$

(f) washing soda $Na_2CO_3 \cdot 10\ H_2O$

(g) grain alcohol C_2H_5OH

29. Naming compounds

(a) $Ba(NO_3)_2$, barium nitrate

(b) $NaC_2H_3O_2$, sodium acetate

(c) PbI_2, lead(II) iodide

(d) $MgSO_4$, magnesium sulfate

(e) $CdCrO_4$, cadmium chromate

(f) $BiCl_3$, bismuth(III) chloride

(g) NiS, nickel(II) sulfide

(h) $Sn(NO_3)_4$, tin(IV) nitrate

(i) $Ca(OH)_2$, calcium hydroxide

30. ide: suffix is used to indicate a binary compound except for hydroxides, cyanides, and ammonium compounds.

ous: used as a suffix to name an acid that has a lower oxygen content than the -ic acid (e.g. HNO_2, nitrous acid and HNO_3, nitric acid); also used as a suffix to name the lower ionic charge of a multivalent metal (e.g. Fe^{2+}, ferrous and Fe^{3+}, ferric).

hypo: used as a prefix in naming an acid that has a lower oxygen content that the -ous acid when there are more than two oxyacids with the same elements (e.g. HClO, hypochlorous acid and $HClO_2$, chlorous acid).

per: used as a prefix in naming an acid that has a higher oxygen content than the -ic acid when there are more than two oxyacids with the same elements (e.g. $HClO_4$, perchloric acid and $HClO_3$, chloric acid).

ite: the suffix of a salt derived from an -ous acid.

ate: the suffix of a salt derived from an -ic acid.

Roman numerals: In the Stock System Roman numerals are used in naming compounds that contain metals that may exist in more than one type of cation. The charge of a metal is indicated by a Roman numeral written in parentheses immediately after the name of the metal.

31. (a) $AgNO_3 + NaCl \rightarrow AgCl + NaNO_3$
(b) $Fe_2(SO_4)_3 + Ca(OH)_2 \rightarrow Fe(OH)_3 + CaSO_4$
(c) $KOH + H_2SO_4 \rightarrow K_2SO_4 + H_2O$

32. (a) $50e^-$, 50 p (b) $48e^-$, 50 p (c) $46e^-$, 50 p

33. The formula for a compound must be electrically neutral. Therefore $X = +3$ and $Y = -2$ since in X_2Y_3 this would give $2(+3) + 3(-2) = 0$.

34. $Li_3Fe(CN)_6$
$AlFe(CN)_6$
$Zn_3[Fe(CN)_6]_2$

35. (a) N^{3-} nitride
NO_2^- nitrite
One has oxygen the other does not, charges on the ions differ.

(b) NO_2^- nitrite
NO_3^- nitrate
The number of oxygens differ, but the charge is the same.

(c) HNO_2 nitrous acid
HNO_3 nitric acid
The number of oxygens in the componds differ but they both have only one hydrogen.

36.

$(NH_4)_2O$	ammonium oxide
$(NH_4)_2CO_3$	ammonium carbonate
NH_4Cl	ammonium chloride
$NH_4C_2H_3O_2$	ammonium acetate
ZnO	zinc oxide
$ZnCO_3$	zinc carbonate
$ZnCl_2$	zinc chloride
$Zn(C_2H_3O_2)_2$	zinc acetate
H_2CO_3	carbonic acid
$HC_2H_3O_2$	acetic acid
HCl	hydrochloric acid

CHAPTER 7

QUANTITATIVE COMPOSITION OF COMPOUNDS

1. A mole is an amount of substance containing the same number of particles as there are atoms in exactly 12 g of carbon–12.

It is Avogadro's number (6.022×10^{23}) of anything (atoms, molecules, ping-pong balls, etc).

2. A mole of gold (197.0 g) has a higher mass than a mole of potassium (39.10 g).

3. Both samples (Au and K) contain the same number of atoms. (6.022×10^{23}).

4. A mole of gold atoms contains more electrons than a mole of potassium atoms, as each Au atom has 79 e^-, while each K atom has only 19 e^-.

5. No. Avogadro's number is a constant. The mole is defined as Avogadro's number of C–12 atoms. Changing the atomic mass to 50 amu would change only the size of the atomic mass unit, not Avogadro's number.

6. 6.022×10^{23}

7. There are Avogadro's number of particles in one mole of substance.

8. (a) A mole of oxygen atoms (O) contains **6.022×10^{23}** atoms.

(b) A mole of oxygen molecules (O_2) contains **6.022×10^{23}** molecules.

(c) A mole of oxygen molecules (O_2) contains **1.204×10^{24}** atoms.

(d) A mole of oxygen atoms (O) has a mass of **16.00** g.

(e) A mole of oxygen molecules (O_2) has a mass of **32.00** g.

9. 6.022×10^{23} molecules in one molar mass of H_2SO_4.
4.215×10^{24} atoms in one molar mass of H_2SO_4.

10. Choosing 100 g of a compound allows us to simply drop the % sign and use grams for each percent.

11. Food additives are generally preservatives, colorings, flavorings, antioxidants or sweeteners. They are used to enhance the flavor or appearance of food or to help preserve it for later use.

12. The maximum daily tolerable intake of an additive is determined by calculation based on body mass in mg additive/kg mass/day.

13. It is difficult to determine safe levels of food additives because testing is done on animals which are not always the same as humans. Variation also exists between species regarding sensitivity for an additive. Children with different metabolic rates also pose a problem in determining safe levels.

14. An empirical formula gives the smallest whole number ratio of the atoms present in a compound. The molecular formula represents the actual number of atoms of each element in a molecule of the compound. It may be the same as the empirical formula of may be a multiple of the empirical formula.

15. The correct answers are a, b, c, d, h.

(e) A mole of chlorine molecules (Cl_2) contains 1.204×10^{24} atoms of chlorine.

(f) A mole of aluminum atoms has a different mass than a mole of tin atoms.

(g) A mole of H_2O contains 1.807×10^{24} atoms.

16. The correct answers are a, d, f, g, j, k, m.

(b) One mole of nitrogen gas (N_2) has a mass of 28.02 g.

(c) The percent of oxygen is higher in Na_2CrO_4 than it is in K_2CrO_4.

(e) K_2CrO_4 and Na_2CrO_4 have different percentages of Cr by mass.

(h) The empirical formula for sucrose is $C_{12}H_{22}O_{11}$.

(i) A hydrocarbon that has a molar mass of 280 and an empirical formula of CH_2 has a molecular formula of $C_{20}H_{40}$.

(l) If the molecular formula and the empirical formula of a compound are not the same, the molecular formula will be an integral multiple of the empirical formula.

(n) A compound having an empirical formula of CH_2O, and a molar mass of 60, has a molecular formula of $C_2H_4O_2$.

17. Molar masses

(a) KBr	1	K	39.10 g
	1	Br	79.90 g
			119.0 g
(b) Na_2SO_4	2	Na	45.98 g
	1	S	32.06 g
	4	O	64.00 g
			142.0 g
(c) $Pb(NO_3)_2$	1	Pb	207.2 g
	2	N	28.02 g
	6	O	96.00 g
			331.2 g

(d) C_2H_5OH

2	C	24.02 g
6	H	6.048 g
1	O	16.00 g
		46.07 g

(e) $HC_2H_3O_2$

4	H	4.032 g
2	C	24.02 g
2	O	32.00 g
		60.05 g

(f) Fe_3O_4

3	Fe	167.6 g
4	O	64.00 g
		231.6 g

(g) $C_{12}H_{22}O_{11}$

12	C	144.1 g
22	H	22.18 g
11	O	176.0 g
		342.3 g

(h) $Al_2(SO_4)_3$

2	Al	53.96 g
3	S	96.18 g
12	O	192.0 g
		342.1 g

(i) $(NH_4)_2HPO_4$

9	H	9.072 g
2	N	28.02 g
1	P	30.97 g
4	O	64.00 g
		132.1 g

18. Molar masses

(a) NaOH

1	Na	22.99 g
1	O	16.00 g
1	H	1.008 g
		40.00 g

(b) Ag_2CO_3

2	Ag	215.8 g
1	C	12.01 g
3	O	48.00 g
		275.8 g

(c) Cr_2O_3

2	Cr	104.0 g
3	O	48.00 g
		152.0 g

(d) $(NH_4)_2CO_3$	2	N	28.02 g
	8	H	8.064 g
	1	C	12.01 g
	3	O	48.00 g
			96.09 g
(e) $Mg(HCO_3)_2$	1	Mg	24.31 g
	2	H	2.016 g
	2	C	24.02 g
	6	O	96.00 g
			146.3 g
(f) C_6H_5COOH	7	C	84.07 g
	6	H	6.048 g
	2	O	32.00 g
			122.1 g
(g) $C_6H_{12}O_6$	6	C	72.06 g
	12	H	12.10 g
	6	O	96.00 g
			180.2 g
(h) $K_4Fe(CN)_6$	4	K	156.4 g
	1	Fe	55.85 g
	6	C	72.06 g
	6	N	84.06 g
			368.4 g
(i) $BaCl_2 \cdot 2\ H_2O$	1	Ba	137.3 g
	2	Cl	70.90 g
	4	H	4.032 g
	2	O	32.00 g
			244.2 g

19. Moles of atoms.

(a) $(22.5\text{ g Zn})\left(\frac{1\text{ mol}}{65.38\text{ g}}\right) = 0.344\text{ mol Zn}$

(b) $(0.688\text{ g Mg})\left(\frac{1\text{ mol}}{24.31\text{ g}}\right) = 2.83 \times 10^{-2}\text{ mol Mg}$

(c) $(4.5 \times 10^{22}\text{ atoms Cu})\left(\frac{1\text{ mol}}{6.022 \times 10^{23}\text{atoms}}\right) = 7.5 \times 10^{-2}\text{ mol Cu}$

(d) $(382\text{ g Co})\left(\frac{1\text{ mol}}{58.93\text{ g}}\right) = 6.48\text{ mol Co}$

(e) $(0.055\text{ g Sn})\left(\frac{1\text{ mol}}{118.7\text{ g}}\right) = 4.6 \times 10^{-4}\text{ mol Sn}$

(f) $(8.5 \times 10^{24}\text{ molecules } N_2)\left(\frac{2\text{ atoms N}}{1\text{ molecule } N_2}\right)\left(\frac{1\text{ mol N atoms}}{6.022\ 10^{23}\text{ atoms N}}\right) = 28\text{ mol N atoms}$

20. Number of moles.

(a) $(25.0 \text{ g NaOH})\left(\frac{1 \text{ mol}}{40.00 \text{ g}}\right) = 0.625 \text{ mol NaOH}$

(b) $(44.0 \text{ g Br}_2)\left(\frac{1 \text{ mol}}{159.8 \text{ g}}\right) = 0.275 \text{ mol Br}_2$

(c) $(0.684 \text{ g MgCl}_2)\left(\frac{1 \text{ mol}}{95.21 \text{ g}}\right) = 7.18 \times 10^{-3} \text{ mol MgCl}_2$

(d) $(14.8 \text{ g CH}_3\text{OH})\left(\frac{1 \text{ mol}}{32.04 \text{ g}}\right) = 0.462 \text{ mol CH}_3\text{OH}$

(e) $(2.88 \text{ g Na}_2\text{SO}_4)\left(\frac{1 \text{ mol}}{142.0 \text{ g}}\right) = 2.03 \times 10^{-2} \text{ mol Na}_2\text{SO}_4$

(f) $(4.20 \text{ lb ZnI}_2)\left(\frac{453.6 \text{ g}}{1 \text{ lb}}\right)\left(\frac{1\text{mol}}{319.2 \text{ g}}\right) = 5.97 \text{ mol ZnI}_2$

21. Number of grams.

(a) $(0.550 \text{ mol Au})\left(\frac{197.0 \text{ g}}{1 \text{ mol}}\right) = 108 \text{ g Au}$

(b) $(15.8 \text{ mol H}_2\text{O})\left(\frac{18.02 \text{ g}}{\text{mol}}\right) = 285 \text{ g H}_2\text{O}$

(c) $(12.5 \text{ mol Cl}_2)\left(\frac{70.90 \text{ g}}{\text{mol}}\right) = 886 \text{ g Cl}_2$

(d) $(3.15 \text{ mol NH}_4\text{NO}_3)\left(\frac{80.05 \text{ g}}{\text{mol}}\right) = 252 \text{ g NH}_4\text{NO}_3$

22. Number of grams.

(a) $(4.25 \times 10^{-4} \text{ mol H}_2\text{SO}_4)\left(\frac{98.08 \text{ g}}{\text{mol}}\right) = 0.0417 \text{ g H}_2\text{SO}_4$

(b) $(4.5 \times 10^{22} \text{ molecules CCl}_4)\left(\frac{1 \text{ mol}}{6.022 \times 10^{23} \text{molecules}}\right)\left(\frac{153.8 \text{ g}}{\text{mol}}\right) = 11 \text{ g CCl}_4$

(c) $(0.00255 \text{ mol Ti})\left(\frac{47.90 \text{ g}}{\text{mol}}\right) = 0.122 \text{ g Ti}$

(d) $(1.5 \times 10^{16} \text{atoms S})\left(\frac{32.06 \text{ g}}{6.022 \times 10^{23} \text{ atoms}}\right) = 8.0 \times 10^{-7} \text{ g S}$

23. Number of molecules.

(a) $(1.26 \text{ mol } O_2)\left(\frac{6.022 \times 10^{23} \text{ molecules}}{\text{mol}}\right) = 7.59 \times 10^{23} \text{molecules } O_2$

(b) $(0.56 \text{ mol } C_6H_6)\left(\frac{6.022 \times 10^{23} \text{ molecules}}{\text{mol}}\right) = 3.4 \times 10^{23} \text{ molecules } C_6H_6$

(c) $(16.0 \text{ g } CH_4)\left(\frac{6.022 \times 10^{23} \text{ molecules}}{16.04 \text{ g}}\right) = 6.01 \times 10^{23} \text{ molecules } CH_4$

(d) $(1000. \text{ g HCl})\left(\frac{6.022 \times 10^{23} \text{ molecules}}{36.46 \text{ g}}\right) = 1.652 \times 10^{25} \text{ molecules HCl}$

24. (a) $(1.75 \text{ mol } Cl_2)\left(\frac{6.022 \times 10^{23} \text{ molecules}}{\text{mol}}\right) = 1.05 \times 10^{24} \text{ molecules } Cl_2$

(b) $(0.27 \text{ mol } C_2H_6O)\left(\frac{6.022 \times 10^{23} \text{ molecules}}{\text{mol}}\right) = 1.6 \times 10^{23} \text{ molecules } C_2H_6O$

(c) $(12.0 \text{ g } CO_2)\left(\frac{6.022 \times 10^{23} \text{ molecules}}{44.01 \text{ g}}\right) = 1.64 \times 10^{23} \text{ molecules } CO_2$

(d) $(100. \text{ g } CH_4)\left(\frac{6.022 \times 10^{23} \text{ molecules}}{16.04 \text{ g}}\right) = 3.75 \times 10^{24} \text{ molecules } CH_4$

25. Number of grams.

(a) $(1 \text{ atom Pb})\left(\frac{207.2 \text{ g}}{6.022 \times 10^{23} \text{ atoms}}\right) = 3.441 \times 10^{-22} \text{ g Pb}$

(b) $(1 \text{ atom Ag})\left(\frac{107.9 \text{ g}}{6.022 \times 10^{23} \text{ atoms}}\right) = 1.792 \times 10^{-22} \text{ g Ag}$

(c) $(1 \text{ molecule } H_2O)\left(\frac{18.02 \text{ g}}{6.022 \times 10^{23} \text{ molecules}}\right) = 2.992 \times 10^{-23} \text{ g } H_2O$

(d) $(1 \text{ molecule } C_3H_5(NO_3)_3)\left(\frac{227.1 \text{ g}}{6.022 \times 10^{23} \text{ molecules}}\right) = 3.771 \times 10^{-22} \text{ g } C_3H_5(NO_3)_3$

26. (a) $(1 \text{ atom Au})\left(\frac{197.0 \text{ g}}{6.022 \times 10^{23} \text{ atoms}}\right) = 3.271 \times 10^{-22} \text{ g Au}$

(b) $(1 \text{ atom U})\left(\frac{238.0 \text{ g}}{6.022 \times 10^{23} \text{ atoms}}\right) = 3.952 \times 10^{-22} \text{ U}$

(c) $(1 \text{ molecule } NH_3)\left(\frac{17.03 \text{ g}}{6.022 \times 10^{23} \text{ molecules}}\right) = 2.828 \times 10^{-23} \text{ g } NH_3$

(d) $(1 \text{ molecule } C_6H_4(NH_2)_2)\left(\frac{108.1 \text{ g}}{6.022 \times 10^{23} \text{ molecules}}\right) = 1.795 \times 10^{-22} \text{ g } C_6H_4(NH_2)_2$

27. (a) $(8.66 \text{ mol Cu})\left(\frac{63.55 \text{ g}}{\text{mol}}\right) = 550. \text{ g Cu}$

(b) $(125 \text{ mol Au})\left(\frac{197.0 \text{ g}}{\text{mol}}\right)\left(\frac{1 \text{ kg}}{1000 \text{ g}}\right) = 24.6 \text{ kg Au}$

(c) $(10 \text{ atoms C})\left(\frac{1 \text{ mol}}{6.022 \times 10^{23} \text{ atoms}}\right) = 2 \times 10^{-23} \text{ mol C}$

(d) $(5000 \text{ molecules } CO_2)\left(\frac{1 \text{ mol}}{6.022 \times 10^{23} \text{ molecules}}\right) = 8 \times 10^{-21} \text{ mol } CO_2$

28. (a) $(28.4 \text{ g S})\left(\frac{1 \text{ mol}}{32.06 \text{ g}}\right) = 0.886 \text{ mol S}$

(b) $(2.50 \text{ kg NaCl})\left(\frac{1000 \text{ g}}{\text{kg}}\right)\left(\frac{1 \text{ mol}}{58.44 \text{ g}}\right) = 42.8 \text{ mol NaCl}$

(c) $(42.4 \text{ g Mg})\left(\frac{6.022 \times 10^{23} \text{ atoms}}{24.31 \text{ g}}\right) = 1.05 \times 10^{24} \text{ atoms Mg}$

(d) $(485 \text{ mL } Br_2)\left(\frac{3.12 \text{ g}}{\text{mL}}\right)\left(\frac{1 \text{ mol}}{159.8 \text{ g}}\right) = 9.47 \text{ mol } Br_2$

29. One mole of carbon disulfide (CS_2) contains:

(a) 6.022×10^{23} molecules of CS_2

(b) $(6.022 \times 10^{23} \text{ molecules } CS_2)\left(\frac{1 \text{ C atom}}{1 \text{ molecule } CS_2}\right) = 6.022 \times 10^{23} \text{ C atoms}$

(c) $(6.022 \times 10^{23} \text{ molecules } CS_2)\left(\frac{2 \text{ S atoms}}{1 \text{ molecule } CS_2}\right) = 1.204 \times 10^{24} \text{ S atoms}$

(d) $(6.022 \times 10^{23} \text{ atoms}) + (1.204 \times 10^{24}) \text{ atoms} = 1.806 \times 10^{24} \text{ atoms}$

30. One mole of ammonia (NH_3) contains

(a) 6.022×10^{23} molecules of NH_3

(b) $(6.022 \times 10^{23} \text{ molecules } NH_3)\left(\frac{1 \text{ N atom}}{\text{molecule } NH_3}\right) = 6.022 \times 10^{23} \text{ N atoms}$

(c) $(6.022 \times 10^{23} \text{ molecules } NH_3)\left(\frac{3 \text{ H atoms}}{\text{molecule } NH_3}\right) = 1.807 \times 10^{24} \text{ H atoms}$

(d) $(6.022 \times 10^{23} \text{ atoms}) + (1.807 \times 10^{24} \text{ atoms}) = 2.409 \times 10^{24} \text{ atoms}$

31. Atoms of oxygen in:

(a) $(16.0 \text{ g } O_2)\left(\frac{1 \text{ mol}}{32.00 \text{ g}}\right)\left(\frac{2 \text{ mol O}}{1 \text{ mol } O_2}\right)\left(\frac{6.022 \times 10^{23} \text{ atom}}{\text{mol}}\right) = 6.02 \times 10^{23} \text{ atoms O}$

(b) $(0.622 \text{ mol MgO})\left(\frac{1 \text{ mol O}}{\text{mol MgO}}\right)\left(\frac{6.022 \times 10^{23} \text{ atoms}}{\text{mol O}}\right) = 3.75 \times 10^{23} \text{ atoms O}$

(c) $(6.00 \times 10^{22} \text{ molecules } C_6H_{12}O_6)\left(\frac{6 \text{ atoms O}}{\text{molecule } C_6H_{12}O_6}\right) = 3.60 \times 10^{23} \text{ atoms O}$

32. Atoms of oxygen in:

(a) $(5.0 \text{ mol } MnO_2)\left(\frac{2 \text{ mol O}}{\text{mol } MnO_2}\right)\left(\frac{6.022 \times 10^{23} \text{ atoms}}{\text{mol}}\right) = 6.0 \times 10^{24} \text{ atoms O}$

(b) $(255 \text{ g } MgCO_3)\left(\frac{1 \text{ mol}}{84.32 \text{ g}}\right)\left(\frac{3 \text{ mol O}}{\text{mol } MgCO_3}\right)\left(\frac{6.022 \times 10^{23} \text{ atoms O}}{\text{mol}}\right) = 5.46 \times 10^{24} \text{ atoms O}$

(c) $(5.0 \times 10^{18} \text{ molecules } H_2O)\left(\frac{1 \text{ atom O}}{\text{molecule } H_2O}\right) = 5.0 \times 10^{18} \text{ atoms O}$

33. The number of grams of:

(a) silver in 25.0 g AgBr

$(25.0 \text{ g AgBr})\left(\frac{107.9 \text{ g Ag}}{187.8 \text{ g AgBr}}\right) = 14.4 \text{ g Ag}$

(b) nitrogen in 6.34 mol $(NH_4)_3PO_4$

$(6.34 \text{ mol } (NH_4)_3PO_4)\left(\frac{42.03 \text{ g N}}{\text{mol } (NH_4)_3PO_4}\right) = 266 \text{ g N}$

(c) oxygen in 8.45×10^{22} molecules SO_3

$(8.45 \times 10^{22} \text{ molecules } SO_3)\left(\frac{1 \text{ mol}}{6.022 \times 10^{23} \text{ molecules}}\right)\left(\frac{48.00 \text{ g O}}{\text{mol } SO_3}\right) = 6.74 \text{ g O}$

34. The number of grams of:

(a) chlorine in 5.00 g $PbCl_2$

$(5.00 \text{ g } PbCl_2)\left(\frac{70.90 \text{ g Cl}}{278.1 \text{ g } PbCl_2}\right) = 1.27 \text{ g Cl}$

(b) hydrogen in 4.50 g H_2SO_4

$(4.50 \text{ g } H_2SO_4)\left(\frac{2.016 \text{ g H}}{\text{mol } H_2SO_4}\right) = 9.07 \text{ g H}$

(c) iodine in 5.45×10^{22} molecules CaI_2

$$(5.45 \times 10^{22} \text{ molecules } CaI_2)\left(\frac{1 \text{ mol}}{6.022 \times 10^{23} \text{ molecules}}\right)\left(\frac{253.8 \text{ g I}}{\text{mol } CaI_2}\right) = 23.0 \text{ g I}$$

35. Percent composition

(a) NaBr

Na	22.99 g	$\left(\frac{22.99 \text{ g}}{102.9 \text{ g}}\right)(100) = 22.34\%$ Na
Br	79.90 g	$\left(\frac{79.90 \text{ g}}{102.9 \text{ g}}\right)(100) = 77.65\%$ Br
	102.9 g	

(b) $KHCO_3$

K	39.10 g	$\left(\frac{39.10 \text{ g}}{100.1 \text{ g}}\right)(100) = 39.06\%$ K
H	1.008 g	$\left(\frac{1.008 \text{ g}}{100.1 \text{ g}}\right)(100) = 1.007\%$ H
3 O	48.00 g	$\left(\frac{12.01 \text{ g}}{100.1 \text{ g}}\right)(100) = 12.00\%$ C
C	12.01 g	$\left(\frac{48.00 \text{ g}}{100.1 \text{ g}}\right)(100) = 47.95\%$ O
	100.1 g	

(c) $FeCl_3$

Fe	55.85 g	$\left(\frac{55.85 \text{ g}}{162.3 \text{ g}}\right)(100) = 34.41\%$ Fe
3 Cl	106.4 g	$\left(\frac{106.4 \text{ g}}{162.3 \text{ g}}\right)(100) = 65.56\%$ Cl
	162.3 g	

(d) $SiCl_4$

Si	28.09 g	$\left(\frac{28.09 \text{ g}}{169.9 \text{ g}}\right)(100) = 16.53\%$ Si
4 Cl	141.8 g	$\left(\frac{141.8 \text{ g}}{169.9 \text{ g}}\right)(100) = 83.46\%$ Cl
	169.9 g	

(e) $Al_2(SO_4)_3$

2 Al	53.96 g	$\left(\frac{53.96 \text{ g}}{342.1 \text{ g}}\right)(100) = 15.77\%$ Al
3 S	96.18 g	$\left(\frac{96.18 \text{ g}}{342.1 \text{ g}}\right)(100) = 28.11\%$ S
12 O	192.0 g	$\left(\frac{192.0 \text{ g}}{342.1 \text{ g}}\right)(100) = 56.12\%$ O
	342.1 g	

(f) $AgNO_3$

Ag	107.9 g	$\left(\frac{107.9 \text{ g}}{169.9 \text{ g}}\right)(100) = 63.51\%$ Ag
N	14.01 g	$\left(\frac{14.01 \text{ g}}{169.9 \text{ g}}\right)(100) = 8.246\%$ N
3 O	48.00 g	$\left(\frac{48.00 \text{ g}}{169.9 \text{ g}}\right)(100) = 28.25\%$ O
	169.9 g	

36. Percent composition

(a) $ZnCl_2$

Zn	65.38 g	$\left(\frac{65.38\ g}{136.3\ g}\right)(100)$	= 47.97% Zn
2 Cl	70.90 g	$\left(\frac{70.90\ g}{136.3\ g}\right)(100)$	= 52.02% Cl
	136.3 g		

(b) $NH_4C_2H_3O_2$

N	14.01 g	$\left(\frac{14.01\ g}{77.09\ g}\right)(100)$	= 18.17% N
7 H	7.056 g	$\left(\frac{7.056\ g}{77.09\ g}\right)(100)$	= 9.153% H
2 C	24.02 g	$\left(\frac{24.02\ g}{77.09\ g}\right)(100)$	= 31.16% C
2 O	32.00 g	$\left(\frac{32.00\ g}{77.09\ g}\right)(100)$	= 41.51% O
	77.09 g		

(c) MgP_2O_7

Mg	24.31 g	$\left(\frac{24.31\ g}{198.3\ g}\right)(100)$	= 12.26% Mg
2 P	61.94 g	$\left(\frac{61.94\ g}{198.3\ g}\right)(100)$	= 31.24% P
7 O	112.0 g	$\left(\frac{112.0\ g}{198.3\ g}\right)(100)$	= 56.48% O
	198.3 g		

(d) $(NH_4)_2SO_4$

2 N	28.02 g	$\left(\frac{28.02\ g}{132.1\ g}\right)(100)$	= 21.21% N
8 H	8.064 g	$\left(\frac{8.064\ g}{132.1\ g}\right)(100)$	= 6.104% H
S	32.06 g	$\left(\frac{32.06\ g}{132.1\ g}\right)(100)$	= 24.27% S
4 O	64.00 g	$\left(\frac{64.00\ g}{132.1\ g}\right)(100)$	= 48.45% O
	132.1 g		

(e) $Fe(NO_3)_3$

Fe	55.85 g	$\left(\frac{55.85\ g}{241.9\ g}\right)(100)$	= 23.09% Fe
3 N	42.03 g	$\left(\frac{42.03\ g}{241.9\ g}\right)(100)$	= 17.37% N
9 O	144.0 g	$\left(\frac{144.0\ g}{241.9\ g}\right)(100)$	= 59.53% O
	241.9 g		

(f) ICl_3

I	126.9 g	$\left(\frac{126.9\ g}{233.3\ g}\right)(100)$	= 54.39% I
3 Cl	106.4 g	$\left(\frac{106.4\ g}{233.3\ g}\right)(100)$	= 45.61% Cl
	233.3 g		

37. Percent of iron

(a) FeO

Fe	55.85 g	$\left(\frac{55.85\text{ g}}{71.85\text{ g}}\right)(100) = 77.73\%$ Fe
O	16.00 g	
	71.85 g	

(b) Fe_2O_3

2 Fe	111.7 g	$\left(\frac{111.7\text{ g}}{159.7\text{ g}}\right)(100) = 69.94\%$ Fe
3 O	48.00 g	
	159.7 g	

(c) Fe_3O_4

3 Fe	167.4 g	$\left(\frac{167.6\text{ g}}{231.6\text{ g}}\right)(100) = 72.37\%$ Fe
4 O	64.00 g	
	231.6 g	

(d) $K_4Fe(CN)_6$

Fe	55.85 g	$\left(\frac{55.85\text{ g}}{368.4\text{ g}}\right)(100) = 15.16\%$ Fe
4 K	156.4 g	
6 C	72.06 g	
6 N	84.06 g	
	368.4 g	

38. Percent chlorine

(a) KCl

K	39.10 g	$\left(\frac{35.45\text{ g}}{74.55\text{ g}}\right)(100) = 47.55\%$ Cl
Cl	35.45 g	
	74.55 g	

(b) $BaCl_2$

Ba	137.3 g	$\left(\frac{70.90\text{ g}}{208.2\text{ g}}\right)(100) = 34.05\%$ Cl
2 Cl	70.90 g	
	208.2 g	

(c) $SiCl_4$

Si	28.09 g	$\left(\frac{141.8\text{ g}}{169.9\text{ g}}\right)(100) = 83.46\%$ Cl
4 Cl	141.8 g	
	169.9 g	

(d) $LiCl$

Li	6.941 g	$\left(\frac{35.45\text{ g}}{42.39\text{ g}}\right)(100) = 83.63\%$ Cl
Cl	35.45 g	
	42.39 g	

Highest % Cl is in LiCl: lowest % Cl is in $BaCl_2$

39. Percent composition of oxide

14.20 g oxide	$\left(\frac{6.20\text{ g}}{14.20\text{ g}}\right)(100) = 43.7\%$ P
−6.20 g P	
8.00 g oxygen	$\left(\frac{8.00\text{ g}}{14.20\text{ g}}\right)(100) = 56.3\%$ O

40. Percent composition of ethylene chloride

$$\begin{array}{rll} 6.00 & g & C \\ 1.00 & g & H \\ 17.75 & g & Cl \\ \hline 24.75 & g & \text{total} \end{array}$$

$$\left(\frac{6.00\ g}{24.75\ g}\right)(100) = 24.2\%\ C$$

$$\left(\frac{1.00\ g}{24.75\ g}\right)(100) = 4.04\%\ H$$

$$\left(\frac{17.75\ g}{24.75\ g}\right)(100) = 71.72\%\ Cl$$

41. (a) H_2O

(b) N_2O_3

(c) equal

42. (a) $KClO_3$

(b) $KHSO_4$

(c) Na_2CrO_4

43. Empirical formulas from percent composition.

(a) Step 1. Express each element as grams/100 g material

63.6% N = 63.6 g N/100 g material

36.4% O = 36.4 g O/100 g material

Step 2. Calculate the relative moles of each element.

$$(63.6\ g\ N)\left(\frac{1\ mol}{14.01\ g}\right) = 4.54\ mol\ N$$

$$(36.4\ g\ O)\left(\frac{1\ mol}{16.00\ g}\right) = 2.28\ mol\ O$$

Step 3. Change these moles to whole numbers by dividing each by the smaller number.

$$\frac{4.54\ mol\ N}{2.28} = 1.99\ mol\ N$$

$$\frac{2.28\ mol\ O}{2.28} = 1.00\ mol\ O$$

The simplest ratio of N:O is 2:1. The empirical formula, therfore, is N_2O

(b) 46.7% N, 53.3% O

$$(46.7\ g\ N)\left(\frac{1\ mol}{14.01\ g}\right) = 3.33\ mol\ N \qquad \frac{3.33}{3.33} = 1.00\ mol\ N$$

$$(53.3\ g\ O)\left(\frac{1\ mol}{16.00\ g}\right) = 3.33\ mol\ O \qquad \frac{3.33}{3.33} = 1.00\ mol\ O$$

The empirical formula is NO.

(c) 25.9% N, 71.4% O

$(25.9 \text{ g N})\left(\frac{1 \text{ mol}}{14.01 \text{ g}}\right) = 1.85 \text{ mol N} \qquad \frac{1.85}{1.85} = 1.00 \text{ mol N}$

$(74.1 \text{ g O})\left(\frac{1 \text{ mol}}{16.00 \text{ g}}\right) = 4.63 \text{ mol O} \qquad \frac{4.63}{1.85} = 2.5 \text{ mol O}$

Since these values are not whole numbers, mutiply each by 2 to change them to whole numbers.

(1.00 mol N)(2) = 2.00 mol N; (2.5 mol O)(2) = 5.00 mol O

The empirical formula is N_2O_5.

(d) 43.4% Na, 11.3% C, 45.3% O

$(43.4 \text{ g Na})\left(\frac{1 \text{ mol}}{22.99 \text{ g}}\right) = 1.89 \text{ mol Na} \qquad \frac{1.89}{0.941} = 2.01 \text{ mol Na}$

$(11.3 \text{ g C})\left(\frac{1 \text{ mol}}{12.01 \text{ g}}\right) = 0.941 \text{ mol C} \qquad \frac{0.941}{0.941} = 1.00 \text{ mol C}$

$(45.3 \text{ g O})\left(\frac{1 \text{ mol}}{16.00 \text{ g}}\right) = 2.83 \text{ mol O} \qquad \frac{2.83}{0.941} = 3.00 \text{ mol O}$

The empirical formula is Na_2CO_3.

(e) 18.8 % Na, 29.0%Cl, 52.3% O

$(18.8 \text{ g Na})\left(\frac{1 \text{ mol}}{22.99 \text{ g}}\right) = 0.818 \text{ mol Na} \qquad \frac{0.818}{0.818} = 1.00 \text{ mol Na}$

$(29.0 \text{ g Cl})\left(\frac{1 \text{ mol}}{35.45 \text{ g}}\right) = 0.818 \text{ mol Cl} \qquad \frac{0.818}{0.818} = 1.00 \text{ mol Cl}$

$(52.3 \text{ g O})\left(\frac{1 \text{ mol}}{16.00 \text{ g}}\right) = 3.27 \text{ mol O} \qquad \frac{3.27}{0.818} = 4.00 \text{ mol O}$

The empirical formula is $NaClO_4$.

(f) 72.02% Mn, 27.98% O

$(72.02 \text{ g Mn})\left(\frac{1 \text{ mol}}{54.94 \text{ g}}\right) = 1.311 \text{ mol Mn} \qquad \frac{1.311}{1.311} = 1.000 \text{ mol Mn}$

$(27.98 \text{ g O})\left(\frac{1 \text{ mol}}{16.00 \text{ g}}\right) = 1.749 \text{ mol O} \qquad \frac{1.749}{1.311} = 1.334 \text{ mol O}$

Multiply both values by 3 to give whole numbers.

(1.000 mol Mn)(3) = 3.000 mol Mn; (1.334 mol O)(3) = 4.002 mol O

The empirical formula is Mn_3O_4.

44. Empirical formulas from percent composition

(a) 64.1% Cu, 35.9% Cl

$$(64.1\text{ g Cu})\left(\frac{1\text{ mol}}{63.55\text{ g}}\right) = 1.01\text{ mol Cu} \qquad \frac{1.01}{1.01} = 1.00\text{ mol Cu}$$

$$(35.9\text{ g Cl})\left(\frac{1\text{ mol}}{35.45\text{ g}}\right) = 1.01\text{ mol Cl} \qquad \frac{1.01}{1.01} = 1.00\text{ mol Cl}$$

The empirical formula is CuCl.

(b) 47.2% Cu, 52.8% Cl

$$(47.2\text{ g Cu})\left(\frac{1\text{ mol}}{63.55\text{ g}}\right) = 0.743\text{ mol Cu} \qquad \frac{0.743}{0.743} = 1.00\text{ mol Cu}$$

$$(52.8\text{ g Cl})\left(\frac{1\text{ mol}}{35.45\text{ g}}\right) = 1.49\text{ mol Cl} \qquad \frac{1.49}{0.743} = 2.01\text{ mol Cl}$$

The empirical formula is $CuCl_2$.

(c) 51.9% Cr, 48.1% S

$$(51.9\text{ g Cr})\left(\frac{1\text{ mol}}{52.00\text{ g}}\right) = 0.998\text{ mol Cr} \qquad \frac{0.998}{0.998} = 1.00\text{ mol Cr}$$

$$(48.1\text{ g S})\left(\frac{1\text{ mol}}{32.06\text{ g}}\right) = 1.50\text{ mol S} \qquad \frac{1.50}{0.998} = 1.50\text{ mol S}$$

Multiply both values by 2 to give whole numbers.

(1.00 mol Cr)(2) = 2.00 mol Cr; (1.50 mol S)(2) = 3.00 mol S

The empirical formula is Cr_2S_3.

(d) 55.3% K, 14.6% P, 30.1% O

$$(55.3\text{ g K})\left(\frac{1\text{ mol}}{39.10\text{ g}}\right) = 1.41\text{ mol K} \qquad \frac{1.41}{0.471} = 2.99\text{ mol K}$$

$$(14.6\text{ g P})\left(\frac{1\text{ mol}}{30.97\text{ g}}\right) = 0.471\text{ mol P} \qquad \frac{0.471}{0.471} = 1.00\text{ mol P}$$

$$(30.1\text{ g O})\left(\frac{1\text{ mol}}{16.00\text{ g}}\right) = 1.88\text{ mol O} \qquad \frac{1.88}{0.471} = 3.99\text{ mol O}$$

The empirical formula is K_3PO_4.

(e) 38.9% Ba, 29.4% Cr, 31.7% O

$$(38.9 \text{ g Ba})\left(\frac{1 \text{ mol}}{137.3 \text{ g}}\right) = 0.283 \text{ mol Ba} \qquad \frac{0.283}{0.283} = 1.00 \text{ mol Ba}$$

$$(29.4 \text{ g Cr})\left(\frac{1 \text{ mol}}{52.00 \text{ g}}\right) = 0.565 \text{ mol Cr} \qquad \frac{0.565}{0.283} = 2.00 \text{ mol Cr}$$

$$(31.7 \text{ g O})\left(\frac{1 \text{ mol}}{16.00 \text{ g}}\right) = 1.98 \text{ mol O} \qquad \frac{1.98}{0.283} = 7.00 \text{ mol O}$$

The empirical formula is $BaCr_2O_7$.

(f) 3.99% P, 82.3% Br, 13.7% Cl

$$(3.99 \text{ g P})\left(\frac{1 \text{ mol}}{30.97 \text{ g}}\right) = 0.129 \text{ mol P} \qquad \frac{0.129}{0.129} = 1.00 \text{ mol P}$$

$$(82.3 \text{ g Br})\left(\frac{1 \text{ mol}}{79.90 \text{ g}}\right) = 1.03 \text{ mol Br} \qquad \frac{1.03}{0.129} = 7.98 \text{ mol Br}$$

$$(13.7 \text{ g Cl})\left(\frac{1 \text{ mol}}{35.45 \text{ g}}\right) = 0.386 \text{ mol Cl} \qquad \frac{0.386}{0.129} = 2.99 \text{ mol Cl}$$

The empirical formula is PBr_8Cl_3.

45. Empirical formula

$$(3.996 \text{ g Sn})\left(\frac{1 \text{ mol}}{118.7 \text{ g}}\right) = 0.0337 \text{ mol Sn} \qquad \frac{0.0337}{0.0337} = 1.00 \text{ mol Sn}$$

$$(1.077 \text{ g O})\left(\frac{1 \text{ mol}}{16.00 \text{ g}}\right) = 0.0673 \text{ mol O} \qquad \frac{0.0673}{0.0337} = 2.00 \text{ mol O}$$

The empirical formula is SnO_2.

46. Empirical formula

5.454 g product – 3.054 g V = 2.400 g O

$$(3.054 \text{ g V})\left(\frac{1 \text{ mol}}{50.94 \text{ g}}\right) = 0.0600 \text{ mol V} \qquad \frac{0.0600}{0.0600} = 1.00 \text{ mol V}$$

$$(2.400 \text{ g O})\left(\frac{1 \text{ mol}}{16.00 \text{ g}}\right) = 0.1500 \text{ mol O} \qquad \frac{0.1500}{0.0600} = 2.50 \text{ mol O}$$

Multiplying both by 2 gives the empirical formula V_2O_5.

47. 65.45% C, 5.45% H, 29.09% O; molar mass = 110.1

$$(65.45 \text{ g C})\left(\frac{1 \text{ mol}}{12.01 \text{ g}}\right) = 5.450 \text{ mol C} \qquad \frac{5.450}{1.818} = 2.998 \text{ mol C}$$

$$(5.45 \text{ g H})\left(\frac{1 \text{ mol}}{1.008 \text{ g}}\right) = 5.41 \text{ mol H} \qquad \frac{5.41}{1.818} = 2.98 \text{ mol H}$$

$$(29.09 \text{ g O})\left(\frac{1 \text{ mol}}{16.00 \text{ g}}\right) = 1.818 \text{ mol O} \qquad \frac{1.818}{1.818} = 1.000 \text{ mol O}$$

The empirical formula is C_3H_3O making the empirical mass 55.05.

$$\frac{\text{molar mass}}{\text{empirical mass}} = \frac{110.1}{55.05} = 2$$

The molecular formula is twice that of the empirical formula.

Molecular formula = $(C_3H_3O)_2$ = $C_6H_6O_2$

48. 40.0% C, 6.7% H, 53.3% O; molar mass = 180.1

$$(40.0 \text{ g C})\left(\frac{1 \text{ mol}}{12.01 \text{ g}}\right) = 3.33 \text{ mol C} \qquad \frac{3.33}{3.33} = 1.00 \text{ mol C}$$

$$(6.7 \text{ g H})\left(\frac{1 \text{ mol}}{1.008 \text{ g}}\right) = 6.6 \text{ mol H} \qquad \frac{6.6}{3.33} = 2.0 \text{ mol H}$$

$$(53.3 \text{ g O})\left(\frac{1 \text{ mol}}{16.00 \text{ g}}\right) = 3.33 \text{ mol O} \qquad \frac{3.33}{3.33} = 1.00 \text{ mol O}$$

The empirical formula is CH_2O making the empirical mass 30.03.

$$\frac{\text{molar mass}}{\text{empirical mass}} = \frac{180.1}{30.03} = 5.994$$

The molecular formula is six times that of the empirical formula.

Molecular formula = $(CH_2O)_6$ = $C_6H_{12}O_6$

49. $(0.350 \text{ mol } P_4)\left(\frac{6.022 \times 10^{23} \text{ molecules}}{\text{mol}}\right)\left(\frac{4 \text{ atoms P}}{\text{molecule } P_4}\right) = 8.43 \times 10^{23}$ atoms P

50. $(10.0 \text{ g K})\left(\frac{1 \text{ mol}}{39.10 \text{ g}}\right)\left(\frac{1 \text{ mol Na}}{1 \text{ mol K}}\right)\left(\frac{22.99 \text{ g}}{\text{mol}}\right) = 5.88 \text{ g Na}$

51. $(1.79 \times 10^{-23}$ g/atom$)(6.022 \times 10^{23}$ atoms/atomic mass) = 10.8 g/atomic mass

52. $(6.022 \times 10^{23} \text{ sheets})\left(\frac{4.60 \text{ cm}}{500 \text{ sheets}}\right)\left(\frac{1 \text{ m}}{100 \text{ cm}}\right) = 5.54 \times 10^{19} \text{ m}$

53. $\left(\frac{6.022 \times 10^{23} \text{ dollars}}{5.0 \times 10^{9} \text{ people}}\right) = 1.2 \times 10^{14}$ dollars/person

54. (a) $(1 \text{ mi}^3)\left(\frac{5280 \text{ ft}}{\text{mile}}\right)^3\left(\frac{12.0 \text{ in.}}{\text{ft}}\right)^3\left(\frac{2.54 \text{ cm}}{\text{inch}}\right)^3\left(\frac{20 \text{ drops}}{1.0 \text{ cm}^3}\right) = 8 \times 10^{16}$ drops

(b) $(6.022 \times 10^{23} \text{ drops})\left(\frac{1 \text{ mi}^3}{8 \times 10^{16} \text{ drops}}\right) = 8 \times 10^{6} \text{ mi}^3$

55. 1 mol Ag = 107.9 g Ag

(a) $(107.9 \text{ g Ag})\left(\frac{1 \text{ cm}^3}{10.5 \text{ g}}\right) = 10.3 \text{ cm}^3$

(b) 10.3 cm^3 = volume of cube = $(\text{side})^3$

side = $\sqrt[3]{10.3 \text{ cm}^3} = 2.18 \text{ cm}$

56. $(1.00 \text{ L})\left(\frac{1000 \text{ mL}}{1 \text{ L}}\right)\left(\frac{1.55 \text{ g}}{1.00 \text{ mL}}\right)\left(\frac{0.650 \text{ g } H_2SO_4}{1.00 \text{ g}}\right)\left(\frac{1 \text{ mol}}{98.08 \text{ g}}\right) = 10.3 \text{ mol } H_2SO_4$

57. $(100. \text{ mL})\left(\frac{1.42 \text{ g}}{\text{mL}}\right)\left(\frac{0.720 \text{ g } HNO_3}{1.000 \text{ g}}\right)\left(\frac{1 \text{ mol}}{63.02 \text{ g } HNO_3}\right) = 1.62 \text{ mol } HNO_3$

58. (a) Determine the molar mass of each compound.

CO_2, 44.01 g: O_2, 32.00 g; H_2O, 18.02 g; CH_3OH, 32.04 g. The 1.00 gram sample with the lowest molar mass will contain the most moleclues. Thus, H_2O will contain the most molecules.

(b) $(1.00 \text{ g } H_2O)\left(\frac{1 \text{ mol}}{18.02 \text{ g}}\right)\left(\frac{(3)\ 6.022 \times 10^{23} \text{ atoms}}{\text{mol}}\right) = 1.00 \times 10^{23} \text{ atoms}$

$(1.00 \text{ g } CH_3OH)\left(\frac{1 \text{ mol}}{32.04 \text{ g}}\right)\left(\frac{(6)\ 6.022 \times 10^{23} \text{ atoms}}{\text{mole}}\right) = 1.13 \times 10^{23} \text{ atoms}$

$(1.00 \text{ g } CO_2)\left(\frac{1 \text{ mol}}{44.01 \text{ g}}\right)\left(\frac{(3)(6.022 \times 10^{23} \text{ atoms}}{\text{mol}}\right) = 4.24 \times 10^{22} \text{ atoms}$

$(1.00 \text{ g } O_2)\left(\frac{1 \text{ mol}}{32.00 \text{ g}}\right)\left(\frac{(2)(6.022 \times 10^{23} \text{ atoms}}{\text{mol}}\right) = 3.76 \times 10^{22} \text{ atoms}$

The 1.00 g sample of CH_3OH contains the most atoms

59. 1 mol Fe_2S_3 = 207.9 g Fe_2S_3 = 6.022×10^{23} formula units

$(6.022 \times 10^{23} \text{ atoms})\left(\frac{1 \text{ formula unit}}{5 \text{ atoms}}\right)\left(\frac{207.9 \text{ g } Fe_2S_3}{6.022 \times 10^{23} \text{ formula units}}\right) = 41.58 \text{ g } Fe_2S_3$

60. From the formula, 2 Li (13.88 g) combine with 1 S (32.06 g).

$\left(\frac{13.88 \text{ g Li}}{32.06 \text{ S}}\right)(20.0 \text{ g S}) = 8.66 \text{ g Li}$

61. (a) $HgCO_3$

Hg	200.6 g	$\left(\frac{200.6\ g}{260.6\ g}\right)(100) = 76.98\%$ Hg
C	12.01 g	
3 O	48.00 g	
	260.6 g	

(b) $Ca(ClO_3)_2$

6 O	96.00 g	$\left(\frac{96.00\ g}{207.0\ g}\right)(100) = 46.38\%$ O
2 Cl	70.90 g	
Ca	40.08 g	
	207.0 g	

(c) $C_{10}H_{14}N_2$

2 N	28.02 g	$\left(\frac{28.02\ g}{162.2\ g}\right)(100) = 17.28\%$ N
10 C	120.1 g	
14 H	14.11 g	
	162.2 g	

(d) $C_{55}H_{72}MgN_4O_5$

Mg	24.31 g	$\left(\frac{24.31\ g}{893.5\ g}\right)(100) = 2.721\%$ Mg
55 C	660.55 g	
72 H	72.58 g	
4 N	56.04 g	
5 O	80.00 g	
	893.5 g	

62. According to the formula, 1 mol (65.38 g) Zn combines with 1 mol (32.06 g) S.

$$(19.5\ g\ Zn)\left(\frac{32.06\ g\ S}{65.38\ g\ Zn}\right) = 9.56\ g\ S$$

19.5 g Zn require 9.56 g S for complete reaction. Therefore, there is not sufficient S present (9.40 g) to react with the Zn.

63. 60.0% C, 4.48% H, 35.5% O; molar mass of aspirin = 180.2

$$(60.0\ g\ C)\left(\frac{1\ mol}{12.01\ g}\right) = 5.00\ mol\ C \qquad \frac{5.00}{2.22} = 2.25\ mol\ C$$

$$(4.48\ g\ H)\left(\frac{1\ mol}{1.008\ g}\right) = 4.44\ mol\ H \qquad \frac{4.44}{2.22} = 2.00\ mol\ H$$

$$(35.5\ g\ O)\left(\frac{1\ mol}{16.00\ g}\right) = 2.22\ mol\ O \qquad \frac{2.22}{2.22} = 1.00\ mol\ O$$

Multiplying each by 4 gives the empirical formula $C_9H_8O_4$. The empirical mass is 180.2 Since the empirical mass equals the molar mass, the molecular formula is the same as the empirical formula, $C_9H_8O_4$.

64. Calculate the percent of oxygen in $Al_2(SO_4)_3$.

2	Al	53.96
3	S	96.18
12	O	192.0
		342.1

$$\left(\frac{192.0}{342.1}\right)(100) = 56.12\%\ O$$

Now take 56.12% of 8.50 g.
(8.50 g O)(0.5612) = 4.77 g O

65. Empirical formula of gallium arsenide; 48.2% Ga, 51.8% As

$$(48.2\ g\ Ga)\left(\frac{1\ mol}{69.72\ g}\right) = 0.691\ mol\ Ga \qquad \frac{0.691}{0.691} = 1.00\ mol\ Ga$$

$$(51.8\ g\ As)\left(\frac{1\ mol}{74.92\ g}\right) = 0.691\ mol\ As \qquad \frac{0.691}{0.691} = 1.00\ mol\ As$$

The empirical formula is GaAs.

66. (a) 7.79% C, 92.21% Cl

$$(7.79\ g\ C)\left(\frac{1\ mol}{12.01\ g}\right) = 0.649\ mol\ C \qquad \frac{0.649}{0.649} = 1.00\ mol\ C$$

$$(92.21\ g\ Cl)\left(\frac{1\ mol}{35.45\ g}\right) = 2.601\ mol\ Cl \qquad \frac{2.601}{0.649} = 4.01\ mol\ Cl$$

The empirical formula is CCl_4. The empirical mass is 153.8 which equals the molar mass, therefore the molecular formula is CCl_4.

(b) 10.13% C, 89.87% Cl

$$(10.13\ g\ C)\left(\frac{1\ mol}{12.01\ g}\right) = 0.8435\ mol\ C \qquad \frac{0.8435}{0.8435} = 1.000\ mol\ C$$

$$(89.87\ g\ Cl)\left(\frac{1\ mol}{35.45\ g}\right) = 2.535\ mol\ Cl \qquad \frac{2.535}{0.8435} = 3.005\ mol\ Cl$$

The empirical formula is CCl_3. The empirical mass is 118.4.

$$\frac{\text{molar mass}}{\text{empirical mass}} = \frac{236.7}{118.4} = 1.999$$

The molecular formula is twice that of the empirical formula.

Molecular formula = C_2Cl_6.

(c) 25.26% C, 74.74% Cl

$$(25.26 \text{ g C})\left(\frac{1 \text{ mol}}{12.01 \text{ g}}\right) = 2.103 \text{ mol C} \qquad \frac{2.103}{2.103} = 1.000 \text{ mol C}$$

$$(74.74 \text{ g Cl})\left(\frac{1 \text{ mol}}{35.45 \text{ g}}\right) = 2.108 \text{ mol Cl} \qquad \frac{2.108}{2.103} = 1.002 \text{ mol Cl}$$

The empirical formula is CCl. The empirical mass is 47.46.

$$\frac{\text{molar mass}}{\text{empirical mass}} = \frac{284.8}{47.46} = 6$$

The molecular formula is six times that of the empirical formula.

Molecular formula = C_6Cl_6.

(d) 11.25% C, 88.75% Cl

$$(11.25 \text{ g C})\left(\frac{1 \text{ mol}}{12.01 \text{ g}}\right) = 0.9367 \text{ mol C} \qquad \frac{0.9367}{0.9367} = 1.000 \text{ mol C}$$

$$(88.75 \text{ g Cl})\left(\frac{1 \text{ mol}}{35.45 \text{ g}}\right) = 2.504 \text{ mol Cl} \qquad \frac{2.504}{0.9367} = 2.673 \text{ mol Cl}$$

Multiplying each by 3 gives the empirical formula C_3Cl_8. The empirical mass is 319.6 Since the molar mass is also 319.6 the molecular formula is C_3Cl_8.

67. $(6.022 \times 10^{23} \text{ s})\left(\frac{1 \text{ min}}{60 \text{ s}}\right)\left(\frac{1 \text{ hr}}{60 \text{ min}}\right)\left(\frac{1 \text{ day}}{24 \text{ hr}}\right)\left(\frac{1 \text{ year}}{365 \text{ days}}\right) = 1.910 \times 10^{16} \text{ years}$

68. $(2.5 \text{ g Cu})\left(\frac{1 \text{ mol}}{63.55 \text{ g}}\right)\left(\frac{6.022 \times 10^{23} \text{ atoms}}{\text{mol}}\right) = 2.4 \times 10^{23} \text{ atoms Cu}$

69. $(1000. \times 10^{12} \text{ molecules } C_3H_8O_3)\left(\frac{1 \text{ mol}}{6.022 \times 10^{23} \text{ molecules}}\right)\left(\frac{92.09 \text{ g}}{\text{mol}}\right) = 1.529 \times 10^{-7} \text{ g } C_3H_8O_3$

70. $(5.0 \times 10^{9} \text{ people})\left(\frac{1 \text{ mol people}}{6.022 \times 10^{23} \text{ people}}\right) = 8.3 \times 10^{-15} \text{ mol people}$

71. 23.3% Co, 25.3% Mo, 51.4% Cl

$$(23.3 \text{ g Co})\left(\frac{1 \text{ mol}}{58.93 \text{ g}}\right) = 0.395 \text{ mol Co} \qquad \frac{0.395}{0.264} = 1.50$$

$$(23.3 \text{ g Mo})\left(\frac{1 \text{ mol}}{95.94 \text{ g}}\right) = 0.264 \text{ mol Mo} \qquad \frac{0.264}{0.264} = 1.00$$

$$(51.4 \text{ g Cl})\left(\frac{1 \text{ mol}}{35.45 \text{ g}}\right) = 1.45 \text{ mol Cl} \qquad \frac{1.45}{0.264} = 5.49$$

Multiplying by 2 gives the empirical formula $Co_3Mo_2Cl_{11}$.

72. $(18 \text{ g Al})\left(\frac{1 \text{ mol}}{26.98 \text{ g}}\right)\left(\frac{2 \text{ mol Mg}}{1 \text{ mol Al}}\right)\left(\frac{24.31 \text{ g}}{\text{mol}}\right) = 32 \text{ g Mg}$

73. $(10.0 \text{ g N})(0.177) = 1.77 \text{ g N}$

$(1.77 \text{ g N})\left(\frac{1 \text{ mol}}{14.01 \text{ g}}\right) = 0.126 \text{ mol N}$

$(3.8 \times 10^{23} \text{ atoms H})\left(\frac{1 \text{ mol}}{6.022 \times 10^{23} \text{ atoms}}\right) = 0.63 \text{ mol H}$

To determine mol C first find grams H

$(0.63 \text{ mol H})\left(\frac{1.008 \text{ g}}{\text{mol}}\right) = 0.64 \text{ g H}$

10.0 g	sample
−1.77 g	N
−0.63 g	H
7.6 g	C

$(7.6 \text{ g C})\left(\frac{1 \text{ mol}}{12.01 \text{ g}}\right) = 0.63 \text{ mol C}$

N $\frac{0.126}{0.126} = 1.00$

H $\frac{0.63}{0.126} = 5.0$

C $\frac{0.63}{0.126} = 5.0$

The formula is C_5H_5N

74. Let x = molar mass of A_2O

$$\begin{aligned} 0.400x &= 16.00 \text{ g O} \quad (\text{Since } A_2O \text{ has only one mol of O atoms}) \\ x &= 40.0 \text{ g/mol } A_2O \\ 40.0 &= 16.00 + 2y \qquad\qquad y = \text{molar mass A} \\ 40.0 - 16.00 &= 2y \\ 12.0 \frac{\text{g}}{\text{mol}} &= y \end{aligned}$$

carbon = y = mystery element

75. (a) CH_2O

(b) C_4H_9

(c) CH_2O

(d) $C_{25}H_{52}$

(e) $C_6H_2Cl_2O$

CHAPTER 8

CHEMICAL EQUATIONS

1. The purpose of balancing chemical equations is to conform to the Law of Conservation of Mass. Ratios of reactants and products can then be easily determined.

2. The coefficients in a balanced chemical equation represent the number of moles (or molecules or formula units) of each of the chemical species in the reaction.

3. (a) Yes. It is necessary to conserve atoms to follow the Law of Conservation of Mass.

   (b) No. Molecules can be taken apart and rearranged to form different molecules in reactions.

   (c) Moles of molecules are not conserved. (See (b)). Moles of atoms are conserved (See (a)).

4. The correct statements are a, d, e, f, h, i, j, l, n, o, p, q.

   (b) A balanced chemical equation is one that has the same number of atoms of each element on each side of the equation.

   (c) In a chemical equation, the symbol $\xrightarrow{\Delta}$ indicates that heat is required during the reaction.

   (g) The equation $H_2O \longrightarrow H_2 + O_2$ can be balanced by placing a 2 in front of the H_2O and a 2 in front of the H_2.

   (k) The products are the substances produced by the chemical reaction.

   (m) When a precipitate is formed in a chemical reaction, it can be indicated in the equation with the symbol *(s)* immediately following the formula of the substance precipitated.

5. (a) $2\ H_2 + O_2 \longrightarrow 2\ H_2O$

   (b) $3\ C + Fe_2O_3 \longrightarrow 2\ Fe + 3\ CO$

   (c) $H_2SO_4 + 2\ NaOH \longrightarrow 2\ H_2O + Na_2SO_4$

   (d) $Al_2(CO_3)_3 \xrightarrow{\Delta} Al_2O_3 + 3\ CO_2$

   (e) $2\ NH_4I + Cl_2 \longrightarrow 2\ NH_4Cl + I_2$

6. (a) $H_2 + Br_2 \longrightarrow 2\ HBr$

   (b) $4\ Al + 3\ C \xrightarrow{\Delta} Al_4C_3$

   (c) $Ba(ClO_3)_2 \xrightarrow{\Delta} BaCl_2 + 3\ O_2$

   (d) $CrCl_3 + 3\ AgNO_3 \longrightarrow Cr(NO_3)_3 + 3\ AgCl$

   (e) $2\ H_2O_2 \longrightarrow 2\ H_2O + O_2$

7. (a) combination
(b) single displacement
(c) double displacement
(d) decomposition
(e) single displacement

8. (a) combination
(b) combination
(c) decomposition
(d) double displacement
(e) decomposition

9. (a) $2\ SO_2 + O_2 \longrightarrow 2\ SO_3$
(b) $4\ Al + 3\ MnO_2 \xrightarrow{\Delta} 3\ Mn + 2\ Al_2O_3$
(c) $2\ Na + 2\ H_2O \longrightarrow 2\ NaOH + H_2$
(d) $2\ AgNO_3 + Ni \longrightarrow Ni(NO_3)_2 + 2\ Ag$
(e) $Bi_2S_3 + 6\ HCl \longrightarrow 2\ BiCl_3 + 3\ H_2S$
(f) $2\ PbO_2 \xrightarrow{\Delta} 2\ PbO + O_2$
(g) $2\ LiAlH_4 \xrightarrow{\Delta} 2\ LiH + 2\ Al + 3\ H_2$
(h) $2\ KI + Br_2 \longrightarrow 2\ KBr + I_2$
(i) $2\ K_3PO_4 + 3\ BaCl_2 \longrightarrow 6\ KCl + Ba_3(PO_4)_2$

10. (a) $2\ MnO_2 + CO \longrightarrow Mn_2O_3 + CO_2$
(b) $Mg_3N_2 + 6\ H_2O \longrightarrow 3\ Mg(OH)_2 + 2\ NH_3$
(c) $4\ C_3H_5(NO_3)_3 \longrightarrow 12\ CO_2 + 10\ H_2O + 6\ N_2 + O_2$
(d) $4\ FeS + 7\ O_2 \longrightarrow 2\ Fe_2O_3 + 4\ SO_2$
(e) $2\ Cu(NO_3)_2 \longrightarrow 2\ CuO + 4\ NO_2 + O_2$
(f) $3\ NO_2 + H_2O \longrightarrow 2\ HNO_3 + NO$
(g) $2\ Al + 3\ H_2SO_4 \longrightarrow Al_2(SO_4)_3 + 3\ H_2$
(h) $4\ HCN + 5\ O_2 \longrightarrow 2\ N_2 + 4\ CO_2 + 2\ H_2O$
(i) $2\ B_5H_9 + 12\ O_2 \longrightarrow 5\ B_2O_3 + 9\ H_2O$

11. (a) $2\ Cu + S \longrightarrow Cu_2S$
(b) $2\ H_3PO_4 + 3\ Ca(OH)_2 \xrightarrow{\Delta} Ca_3(PO_4)_2 + 6\ H_2O$
(c) $2\ Ag_2O \xrightarrow{\Delta} 4\ Ag + O_2$
(d) $FeCl_3 + 3\ NaOH \longrightarrow Fe(OH)_3 + 3\ NaCl$
(e) $Ni_3(PO_4)_2 + 3\ H_2SO_4 \longrightarrow 3\ NiSO_4 + 2\ H_3PO_4$

(f) $ZnCO_3 + 2\ HCl \longrightarrow ZnCl_2 + H_2O + CO_2$

(g) $3\ AgNO_3 + AlCl_3 \longrightarrow 3\ AgCl + Al(NO_3)_3$

12. (a) $2\ H_2O \longrightarrow 2\ H_2 + O_2$

(b) $HC_2H_3O_2 + KOH \longrightarrow KC_2H_3O_2 + H_2O$

(c) $2\ P + 3\ I_2 \longrightarrow 2\ PI_3$

(d) $2\ Al + 3\ CuSO_4 \longrightarrow 3\ Cu + Al_2(SO_4)_3$

(e) $(NH_4)_2SO_4 + BaCl_2 \longrightarrow 2\ NH_4Cl + BaSO_4$

(f) $SF_4 + 2\ H_2O \longrightarrow SO_2 + 4\ HF$

(g) $Cr_2(CO_3)_3 \xrightarrow{\Delta} Cr_2O_3 + 3\ CO_2$

13. (a) $Ag(s) + H_2SO_4(aq) \longrightarrow$ no reaction

(b) $Cl_2(g) + 2\ NaBr(aq) \longrightarrow Br_2(l) + 2\ NaCl(aq)$

(c) $Mg(s) + ZnCl_2(aq) \longrightarrow Zn(s) + MgCl_2(aq)$

(d) $Pb(s) + 2\ AgNO_3(aq) \longrightarrow 2\ Ag(s) + Pb(NO_3)_2(aq)$

14. (a) $Cu(s) + FeCl_3(aq) \longrightarrow$ no reaction

(b) $H_2(g) + Al_2O_3(s) \xrightarrow{\Delta}$ no reaction

(c) $2\ Al(s) + 6\ HBr(aq) \longrightarrow 3\ H_2(g) + 2\ AlBr_3(aq)$

(d) $I_2(s) + HCl(aq) \longrightarrow$ no reaction

15. (a) $H_2(g) + I_2(g)$ 700K $2\ HI(g)$

(b) $CaCO_3(s) \xrightarrow{\Delta} CaO(s) + CO_2(g)$

(c) $Mg(s) + H_2SO_4(aq) \longrightarrow H_2(g) + MgSO_4(s)$

(d) $FeCl_2(s) + 2\ NaOH(aq) \longrightarrow Fe(OH)_2(s) + 2\ NaCl(aq)$

16. (a) $SO_2 + H_2O \longrightarrow H_2SO_3$

(b) $SO_3 + H_2O \longrightarrow H_2SO_4$

(c) $Ca + 2\ H_2O \longrightarrow Ca(OH)_2 + H_2$

(d) $2\ Bi(NO_3)_3 + 3\ H_2S \longrightarrow Bi_2S_3 + 6\ HNO_3$

17. (a) $2\ Ba + O_2 \longrightarrow 2\ BaO$

(b) $2\ NaHCO_3 \xrightarrow{\Delta} Na_2CO_3 + H_2O + CO_2$

(c) $Ni + CuSO_4 \longrightarrow NiSO_4 + Cu$

(d) $MgO + 2\ HCl \longrightarrow MgCl_2 + H_2O$

(e) $H_3PO_4 + 3\ KOH \longrightarrow K_3PO_4 + 3\ H_2O$

18. (a) $C + O_2 \xrightarrow{\Delta} CO_2$

(b) $2\ Al(ClO_3)_3 \xrightarrow{\Delta} 9\ O_2 + 2\ AlCl_3$

(c) $CuBr_2 + Cl_2 \longrightarrow CuCl_2 + Br_2$

(d) $2\ SbCl_3 + 3\ (NH_4)_2S \longrightarrow Sb_2S_3 + 6\ NH_4Cl$

(e) $2\ NaNO_3 \xrightarrow{\Delta} 2\ NaNO_2 + O_2$

19. (a) One mole of $MgBr_2$ reacts with two moles of $AgNO_3$ to yield one mole of $Mg(NO_3)_2$ and two moles of AgBr.

(b) One mole of N_2 reacts with three moles of H_2 to produce two moles of NH_3.

(c) Two moles of C_3H_7OH react with nine moles O_2 to form six moles of CO_2 and eight moles of H_2O.

20. (a) Two moles of Na react with one mole of Cl_2 to produce two moles of NaCl and release 822 kJ of energy. The reaction is exothermic.

(b) One mole of PCl_5 absorbs 92.9 kJ of energy to produce one mole of PCl_3 and one mole of Cl_2. The reaction is endothermic.

21. (a) $CaO + H_2O \longrightarrow Ca(OH)_2 + 65.3\ kJ$

(b) $2\ Al_2O_3 + 3260\ kJ \longrightarrow 4\ Al + 3\ O_2$

22. (a) $2\ Al + 3\ I_2 \xrightarrow{H_2O} 2\ AlI_3 + \text{heat}$

(b) $4\ CuO + CH_4 + \text{heat} \longrightarrow 4\ Cu + CO_2 + 2\ H_2O$

(c) $Fe_2O_3 + 2\ Al \longrightarrow 2\ Fe + Al_2O_3 + \text{heat}$

23. (a) change in color and texture of the bread

(b) change in texture of the white and the yolk

(c) the flame, change in matchhead, odor

24. $P_4O_{10} + 12\ HClO_4 \longrightarrow 6\ Cl_2O_7 + 4\ H_3PO_4$

10 O + 12 (4 O)	6 (7 O) + 4 (4 O)
10 O + 48 O	42 O + 16 O
58 O	58 O

25. In 7 $Al_2(SO_4)_3$ there are:

(a) 14 atoms of Al

(b) 21 atoms of S

(c) 84 atoms of O

(d) 119 total atoms

26. A balanced equation tells us:
(1) the types of atoms/molecules involved in the reaction
(2) the relationship between quantities of the substances in the reaction

A balanced equation gives no information about
(1) the time required for the reaction
(2) odors or colors which may result

27.

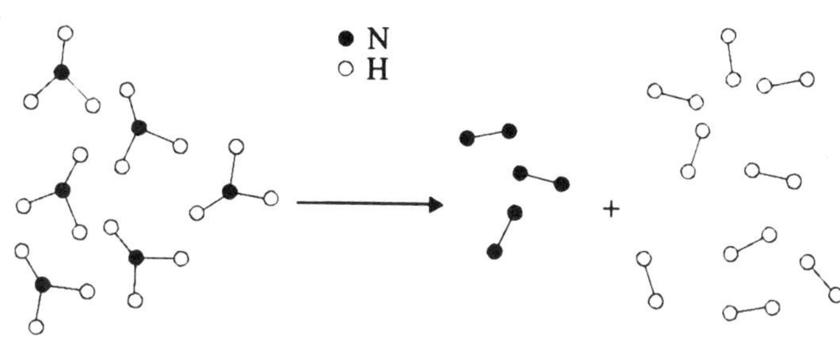

$6\ NH_3 \xrightarrow{\Delta} 3\ N_2 + 9\ H_2$

28. Zn metal falls below Mg on the activity series.

29. $Ti + Ni(NO_3)_2 \longrightarrow$ yes
$Ti + Pb(NO_3)_2 \longrightarrow$ yes
$Ti + Mg(NO_3)_2 \longrightarrow$ no

Ti is above Ni and Pb in the activity series since both react. Ti is below Mg in the series since it will not replace Mg. From the printed activity series in the chapter Ni lies above Pb so:

Mg
Ti
Ni
Pb

30. (a) $4\ K + O_2 \longrightarrow 2\ K_2O$
(b) $2\ Al + 3\ Cl_2 \longrightarrow 3\ AlCl_3$
(c) $CO_2 + H_2O \longrightarrow H_2CO_3$
(d) $CaO + H_2O \longrightarrow Ca(OH)_2$

31. (a) $2\ HgO \xrightarrow{\Delta} 2\ Hg + O_2$
(b) $2\ NaClO_3 \xrightarrow{\Delta} 2\ NaCl + 3\ O_2$
(c) $MgCO_3 \xrightarrow{\Delta} MgO + CO_2$
(d) $2\ PbO_2 \xrightarrow{\Delta} 2\ PbO + O_2$

32. (a) $Zn + H_2SO_4 \longrightarrow H_2 + ZnSO_4$
(b) $2\ AlI_3 + 3\ Cl_2 \longrightarrow 2\ AlCl_3 + 3\ I_2$

(c) $Mg + 2\ AgNO_3 \longrightarrow Mg(NO_3)_2 + 2\ Ag$

(d) $2\ Al + 3\ CoSO_4 \longrightarrow Al_2(SO_4)_3 + 3\ Co$

33. (a) $ZnCl_2 + 2\ KOH \longrightarrow Zn(OH)_2 + 2\ KCl$

(b) $CuSO_4 + H_2S \longrightarrow H_2SO_4 + CuS$

(c) $3\ Ca(OH)_2 + 2\ H_3PO_4 \longrightarrow 6\ H_2O + Ca_3(PO_4)_2$

(d) $2\ (NH_4)_3PO_4 + 3\ Ni(NO_3)_2 \longrightarrow 6\ NH_4NO_3 + Ni_3(PO_4)_2$

(e) $Ba(OH)_2 + 2\ HNO_3 \longrightarrow 2\ H_2O + Ba(NO_3)_2$

(f) $(NH_4)_2S + 2\ HCl \longrightarrow H_2S + 2\ NH_4Cl$

34. (a) $AgNO_3(aq) + KCl(aq) \longrightarrow AgCl(s) + KNO_3(aq)$

(b) $Ba(NO_3)_2(aq) + MgSO_4(aq) \longrightarrow Mg(NO_3)_2(aq) + BaSO_4(s)$

(c) $H_2SO_4(aq) + Mg(OH)_2(aq) \longrightarrow 2\ H_2O(l) + MgSO_4(aq)$

(d) $MgO(s) + H_2SO_4(aq) \longrightarrow 2\ H_2O(l) + MgSO_4(aq)$

(e) $Na_2CO_3(aq) + NH_4Cl(aq) \longrightarrow$ no reaction

35. (a) $2\ C_2H_6 + 7\ O_2 \longrightarrow 4\ CO_2 + 6\ H_2O$

(b) $2\ C_6H_6 + 15\ O_2 \longrightarrow 12\ CO_2 + 6\ H_2O$

(c) $C_7H_{16} + 11\ O_2 \longrightarrow 7\ CO_2 + 8\ H_2O$

36.

pigment	function
chlorophyll	photosynthesis (energy absorption)
carotenoids	absorb high energy oxygen and releases it appropriately
anthocyanins	diversity in leaf color

37. CO_2, water, sunlight and chlorophyll are requirements for successful photosynthesis.

38. leaves containing all pigments

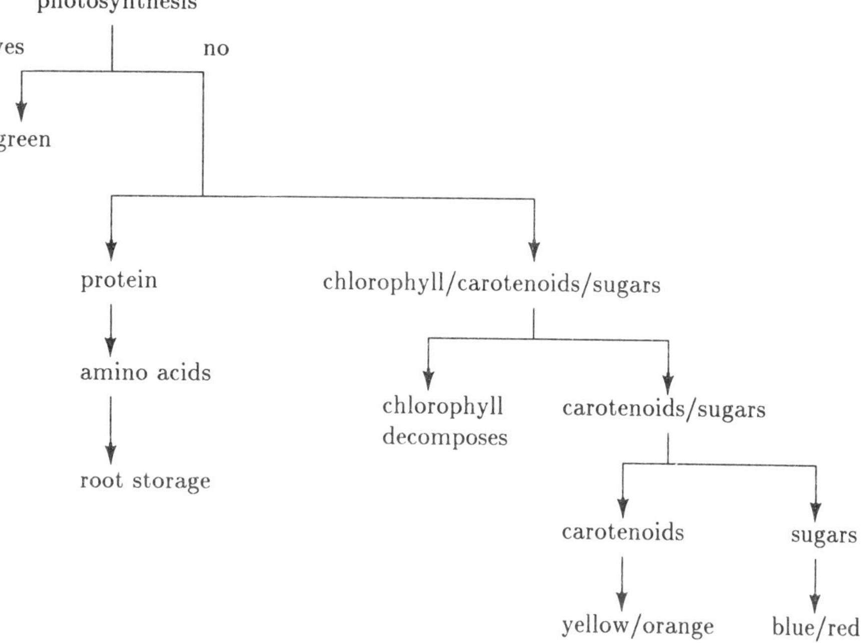

39.
1. combustion of fossil fuels
2. destruction of the rain forests by burning
3. increased population

40. Carbon dioxide, methane, and water are all considered to be greenhouse gases. They each act to trap the heat near the surface of the earth in the same manner in which a greenhouse is warmed.

41. The effects of global warming can be reduced by:

1. developing new energy sources (not dependent on fossil fuels)
2. conservation of energy resources
3. recycling
4. decreased destruction of the rain forests and other forests

42. About half the carbon dioxide released into the atmosphere remains in the air. The rest is absorbed by plants and used in photosynthesis or is dissolved in the oceans.

CHAPTER 9

CALCULATIONS FROM CHEMICAL EQUATIONS

1. The balanced equation is

$$Ca_3P_2 + 6\ H_2O \longrightarrow 3\ Ca(OH)_2 + 2\ PH_3$$

(a) Correct: $(1 \text{ mol } Ca_3P_2)\left(\frac{2 \text{ mol } PH_3}{1 \text{ mol } Ca_3P_2}\right) = 2 \text{ mol } PH_3$

(b) Incorrect: 1 g Ca_3P_2 would produce 0.4 g PH_3

$$(1 \text{ g } Ca_3P_2)\left(\frac{1 \text{ mol}}{182.2 \text{ g}}\right)\left(\frac{2 \text{ mol } PH_3}{1 \text{ mol } Ca_3P_2}\right)\left(\frac{33.99 \text{ g}}{\text{mol}}\right) = 0.4 \text{ g } PH_3$$

(c) Correct: see equation

(d) Correct: see equation

(e) Incorrect: 2 mol Ca_3P_2 requires 12 mol H_2O to produce 4.0 mol PH_3.

$$(2 \text{ mol } Ca_3P_2)\left(\frac{6 \text{ mol } H_2O}{1 \text{ mol } Ca_3P_2}\right) = 12 \text{ mol } H_2O$$

(f) Correct: 2 mol Ca_3P_2 will react with 12 mol H_2O (3 mol H_2O are present in excess) and 6 mol $Ca(OH)_2$ will be formed.

$$(2 \text{ mol } Ca_3P_2)\left(\frac{3 \text{ mol } Ca(OH)_2}{1 \text{ mol } Ca_3P_2}\right) = 6 \text{ mol } Ca(OH)_2$$

(g) Incorrect: $(200. \text{ g } Ca_3P_2)\left(\frac{1 \text{ mol}}{182.2 \text{ g}}\right)\left(\frac{6 \text{ mol } H_2O}{1 \text{ mol } Ca_3P_2}\right)\left(\frac{18.02 \text{ g}}{\text{mol}}\right) = 118 \text{ g } H_2O$

The amount of water present (100. g) is less than needed to react with 200. g Ca_3P_2. H_2O is the limiting reactant.

(h) Incorrect: H_2O is the limiting reactant.

$$(100. \text{ g } H_2O)\left(\frac{1 \text{ mol}}{18.02 \text{ g}}\right)\left(\frac{2 \text{ mol } PH_3}{6 \text{ mol } H_2O}\right)\left(\frac{33.99 \text{ g}}{\text{mol}}\right) = 63.0 \text{ g } PH_3$$

2. The balanced equation is

$$2\ CH_4 + 3\ O_2 + 2\ NH_3 \longrightarrow 2\ HCN + 6\ H_2O$$

(a) Correct

(b) Incorrect: $(16 \text{ mol } O_2)\left(\frac{2 \text{ mol HCN}}{3 \text{ mol } O_2}\right) = 10.7 \text{ mol HCN}$ (not 12 mol HCN)

(c) Correct

(d) Incorrect: $(12 \text{ mol HCN})\left(\frac{6 \text{ mol } H_2O}{2 \text{ mol HCN}}\right) = 36 \text{ mol } H_2O$ (not 4 mol H_2O)

(e) Correct

(f) Incorrect: O_2 is the limiting reactant.

$(3 \text{ mol } O_2)\left(\frac{2 \text{ mol HCN}}{3 \text{ mol } O_2}\right) = 2 \text{ mol HCN (not 3 mol HCN)}$

3. (a) $(25.0 \text{ g } KNO_3)\left(\frac{1 \text{ mol}}{101.1 \text{ g}}\right) = 0.247 \text{ mol } KNO_3$

(b) $(56 \text{ mmol NaOH})\left(\frac{1 \text{ mol}}{1000 \text{ mmol}}\right) = 0.056 \text{ mol NaOH}$

(c) $\left(5.4 \times 10^2 \text{ g } (NH_4)_2C_2O_4\right)\left(\frac{1 \text{ mol}}{124.1 \text{ g}}\right) = 4.4 \text{ mol } (NH_4)_2C_2O_4$

(d) $(16.8 \text{ mL solution})\left(\frac{1.727 \text{ g}}{\text{mL}}\right)\left(\frac{0.800 \text{ g } H_2SO_4}{\text{g solution}}\right)\left(\frac{1 \text{ mol}}{98.08 \text{ g}}\right) = 0.237 \text{ mol } H_2SO_4$

4. (a) $(2.10 \text{ kg } NaHCO_3)\left(\frac{1000 \text{ g}}{\text{kg}}\right)\left(\frac{1 \text{ mol}}{84.01 \text{ g}}\right) = 25.0 \text{ mol } NaHCO_3$

(b) $(525 \text{ mg } ZnCl_2)\left(\frac{1 \text{ g}}{1000 \text{ mg}}\right)\left(\frac{1 \text{ mol}}{136.3 \text{ g}}\right) = 3.85 \times 10^{-3} \text{ mol } ZnCl_2$

(c) $(9.8 \times 10^{24} \text{ molecules } CO_2)\left(\frac{1 \text{ mol}}{6.022 \times 10^{23} \text{ molecules}}\right) = 16 \text{ mol } CO_2$

(d) $(250 \text{ mL } C_2H_5OH)\left(\frac{0.789 \text{ g}}{\text{mL}}\right)\left(\frac{1 \text{ mol}}{46.06 \text{ g}}\right) = 4.3 \text{ mol } C_2H_5OH$

5. (a) $2.55 \text{ mol Fe(OH)}_3)\left(\frac{106.9 \text{ g}}{\text{mol}}\right) = 273 \text{ g Fe(OH)}_3$

(b) $(125 \text{ kg } CaCO_3)\left(\frac{1000 \text{ g}}{\text{kg}}\right) = 1.25 \times 10^5 \text{ g } CaCO_3$

(c) $(10.5 \text{ mol } NH_3)\left(\frac{17.03 \text{ g}}{\text{mol}}\right) = 179 \text{ g } NH_3$

(d) $(72 \text{ mmol HCl})\left(\frac{1 \text{ mol}}{1000 \text{ mmol}}\right)\left(\frac{36.46 \text{ g}}{\text{mol}}\right) = 2.6 \text{ g HCl}$

(e) $(500 \text{ mL } Br_2)\left(\frac{3.119 \text{ g}}{\text{mL}}\right) = 1559.5 \text{ g } Br_2 = 2 \times 10^3 \text{ g } Br_2$

6. (a) $(0.00844 \text{ mol } NiSO_4)\left(\frac{154.8 \text{ g}}{\text{mol}}\right) = 1.31 \text{ g } NiSO_4$

(b) $(0.0600 \text{ mol } HC_2H_3O_2)\left(\frac{60.05 \text{ g}}{\text{mol}}\right) = 3.60 \text{ g } HC_2H_3O_2$

(c) $(0.725 \text{ mol } Bi_2S_3)\left(\frac{514.2 \text{ g}}{\text{mol}}\right) = 373 \text{ g } Bi_2S_3$

(d) $(4.50 \times 10^{21} \text{ molecules } C_6H_{12}O_6)\left(\frac{1 \text{ mol}}{6.022 \times 10^{23} \text{ molecules}}\right)\left(\frac{180.2 \text{ g}}{\text{mol}}\right) = 1.35 \text{ g } C_6H_{12}O_6$

(e) $(75 \text{ mL solution})\left(\frac{1.175 \text{ g}}{\text{mL}}\right)\left(\frac{0.200 \text{ g } K_2CrO_4}{\text{g solution}}\right) = 18 \text{ g } K_2CrO_4$

7. 10.0 g H_2O or 10.0 g H_2O_2

Water has a lower molar mass than hydrogen peroxide. 10.0 grams of water contains more moles, and therefore more molecules than 10.0 g of H_2O_2.

8. 25.0 g HCl or 85.0 g $C_6H_{12}O_6$

$(25.0 \text{ g HCl})\left(\frac{1 \text{ mol}}{36.46 \text{ g}}\right)\left(\frac{6.022 \times 10^{23} \text{ molecules}}{\text{mol}}\right) = 4.13 \times 10^{23} \text{ molecules HCl}$

$(85.0 \text{ g } C_6H_{12}O_6)\left(\frac{1 \text{ mol}}{180.2 \text{ g}}\right)\left(\frac{6.022 \times 10^{23} \text{ molecules}}{\text{mol}}\right) = 2.84 \times 10^{23} \text{ molecules } C_6H_{12}O_6$

HCl contains more molecules

9. Mole ratios

$2\ C_3H_7OH + 9\ O_2 \longrightarrow 6\ CO_2 + 8\ H_2O$

(a) $\frac{6 \text{ mol } CO_2}{2 \text{ mol } C_3H_7OH}$

(b) $\frac{2 \text{ mol } C_3O_7OH}{9 \text{ mol } O_2}$

(c) $\frac{9 \text{ mol } O_2}{6 \text{ mol } CO_2}$

(d) $\frac{8 \text{ mol } H_2O}{2 \text{ mol } C_3H_7OH}$

(e) $\frac{6 \text{ mol } CO_2}{8 \text{ mol } H_2O}$

(f) $\frac{8 \text{ mol } H_2O}{9 \text{ mol } O_2}$

10. Mole ratios

$3\ CaCl_2 + 2\ H_3PO_4 \longrightarrow Ca_3(PO_4)_2 + 6\ HCl$

(a) $\frac{3 \text{ mol } CaCl_2}{1 \text{ mol } Ca_3(PO_4)_2}$

(b) $\frac{6 \text{ mol HCl}}{2 \text{ mol } H_3PO_4}$

(c) $\frac{3 \text{ mol } CaCl_2}{2 \text{ mol } H_3PO_4}$

(d) $\frac{1 \text{ mol } Ca_3(PO_4)_2}{2 \text{ mol } H_3PO_4}$

(e) $\frac{6 \text{ mol HCl}}{1 \text{ mol } Ca_3(PO_4)_2}$

(f) $\frac{2 \text{ mol } H_3PO_4}{6 \text{ mol HCl}}$

11. Moles of Cl_2

$$4\ HCl + O_2 \longrightarrow 2\ Cl_2 + 2\ H_2O$$

$$(5.60 \text{ mol HCl})\left(\frac{2 \text{ mol } Cl_2}{4 \text{ mol HCl}}\right) = 2.80 \text{ mol } Cl_2$$

12. $C_2H_5OH + 3\ O_2 \longrightarrow 2\ CO_2 + 3\ H_2O$

$$(7.75 \text{ mol } C_2H_5OH)\left(\frac{2 \text{ mol } CO_2}{1 \text{ mol } C_2H_5OH}\right) = 15.5 \text{ mol } CO_2$$

13. $Al_4C_3 + 12\ H_2O \longrightarrow 4\ Al(OH)_3 + 3\ CH_4$

(a) $(100.\text{ g } Al_4C_3)\left(\frac{1 \text{ mol}}{144.0 \text{ g}}\right)\left(\frac{12 \text{ mol } H_2O}{1 \text{ mol } Al_4C_3}\right) = 8.33 \text{ mol } H_2O$

(b) $(0.600 \text{ mol } CH_4)\left(\frac{4 \text{ mol } Al(OH)_3}{3 \text{ mol } CH_4}\right) = 0.800 \text{ mol } Al(OH)_3$

14. $MnO_2 + 4\ HCl \longrightarrow Cl_2 + MnCl_2 + 2\ H_2O$

$$(1.05 \text{ mol } MnO_2)\left(\frac{4 \text{ mol HCl}}{1 \text{ mol } MnO_2}\right) = 4.20 \text{ mol HCl}$$

15. Grams of NaOH

$$Ca(OH)_2 + Na_2CO_3 \longrightarrow 2\ NaOH + CaCO_3$$

$$(500 \text{ g } Ca(OH)_2)\left(\frac{1 \text{ mol}}{74.10 \text{ g}}\right)\left(\frac{2 \text{ mol NaOH}}{1 \text{ mol } Ca(OH)_2}\right)\left(\frac{40.00 \text{ g}}{\text{mol}}\right) = 5 \times 10^2 \text{ g NaOH}$$

16. Grams of $Zn_3(PO_4)_2$

$$3\ Zn + 2\ H_3PO_4 \longrightarrow Zn_3(PO_4)_2 + 3\ H_2$$

$$(10.0 \text{ g Zn})\left(\frac{1 \text{ mol}}{65.38 \text{ g}}\right)\left(\frac{1 \text{ mol } Zn_3(PO_4)_2}{3 \text{ mol Zn}}\right)\left(\frac{386.1 \text{ g}}{\text{mol}}\right) = 19.7 \text{ g } Zn_3(PO_4)_2$$

17. The balanced equation is $Fe_2O_3 + 3\ C \longrightarrow 2\ Fe + 3\ CO$

$$(125 \text{ kg } Fe_2O_3)\left(\frac{1 \text{ kmol}}{159.7 \text{ kg}}\right)\left(\frac{2 \text{ kmol Fe}}{1 \text{ kmol } Fe_2O_3}\right)\left(\frac{55.85 \text{ kg}}{\text{kmol}}\right) = 87.4 \text{ kg Fe}$$

18. The balanced equation is $3\ Fe + 4\ H_2O \longrightarrow Fe_3O_4 + 4\ H_2$

$$(375 \text{ g } Fe_3O_4)\left(\frac{1 \text{ mol}}{231.6 \text{ g}}\right)\left(\frac{4 \text{ mol } H_2O}{1 \text{ mol } Fe_3O_4}\right)\left(\frac{18.02 \text{ g}}{\text{mol}}\right) = 117 \text{ g } H_2O$$

$$(375 \text{ g } Fe_3O_4)\left(\frac{1 \text{ mol}}{231.6 \text{ g}}\right)\left(\frac{3 \text{ mol Fe}}{1 \text{ mol } Fe_3O_4}\right)\left(\frac{55.85 \text{ g}}{\text{mol}}\right) = 271 \text{ g Fe}$$

19. The balanced equation is $2\ C_2H_6 + 7\ O_2 \longrightarrow 4\ CO_2 + 6\ H_2O$

(a) $(15.0 \text{ mol } C_2H_6)\left(\frac{7 \text{ mol } O_2}{2 \text{ mol } C_2H_6}\right) = 52.5 \text{ mol } O_2$

(b) $(8.00 \text{ g } H_2O)\left(\frac{1 \text{ mol}}{18.02 \text{ g}}\right)\left(\frac{4 \text{ mol } CO_2}{6 \text{ mol } H_2O}\right)\left(\frac{44.01 \text{ g}}{\text{mol}}\right) = 13.0 \text{ g } CO_2$

(c) $(75.0 \text{ g } C_2H_6)\left(\frac{1 \text{ mol}}{30.07 \text{ g}}\right)\left(\frac{4 \text{ mol } CO_2}{2 \text{ mol } C_2H_6}\right)\left(\frac{44.01 \text{ g}}{\text{mol}}\right) = 2.20 \times 10^2 \text{ g } CO_2$

20. $4\ FeS_2 + 11\ O_2 \longrightarrow 2\ Fe_2O_3 + 8\ SO_2$

(a) $(1.00 \text{ mol } FeS_2)\left(\frac{2 \text{ mol } Fe_2O_3}{4 \text{ mol } FeS_2}\right) = 0.500 \text{ mol } Fe_2O_3$

(b) $(4.50 \text{ mol } FeS_2)\left(\frac{11 \text{ mol } O_2}{4 \text{ mol } FeS_2}\right) = 12.4 \text{ mol } O_2$

(c) $(1.55 \text{ mol } Fe_2O_3)\left(\frac{8 \text{ mol } SO_2}{2 \text{ mol } Fe_2O_3}\right) = 6.20 \text{ mol } SO_2$

(d) $(0.512 \text{ mol } FeS_2)\left(\frac{8 \text{ mol } SO_2}{4 \text{ mol } FeS_2}\right)\left(\frac{64.06 \text{ g}}{\text{mol}}\right) = 65.6 \text{ g } SO_2$

(e) $(40.6 \text{ g } SO_2)\left(\frac{1 \text{ mol}}{64.06 \text{ g}}\right)\left(\frac{11 \text{ mol } O_2}{8 \text{ mol } SO_2}\right) = 0.871 \text{ mol } O_2$

(f) $(221 \text{ g } Fe_2O_3)\left(\frac{1 \text{ mol}}{159.7 \text{ g}}\right)\left(\frac{4 \text{ mol } FeS_2}{2 \text{ mol } Fe_2O_3}\right)\left(\frac{120.0 \text{ g}}{\text{mol}}\right) = 332 \text{ g } FeS_2$

21. (a) $\underset{16.0\text{ g}}{KOH} + \underset{12.0\text{ g}}{HNO_3} \longrightarrow KNO_3 + H_2O$

Choose one of the products, calculate its mass using each given reactant. Using KNO_3 as the product:

$(16.0 \text{ g KOH})\left(\frac{1 \text{ mol}}{56.10 \text{ g}}\right)\left(\frac{1 \text{ mol } KNO_3}{1 \text{ mol KOH}}\right)\left(\frac{101.1 \text{ g}}{\text{mol}}\right) = 28.8 \text{ g } KNO_3$

$(12.0 \text{ g } HNO_3)\left(\frac{1 \text{ mol}}{63.02 \text{ g}}\right)\left(\frac{1 \text{ mol } KNO_3}{1 \text{ mol KOH}}\right)\left(\frac{101.1 \text{ g}}{\text{mol}}\right) = 19.3 \text{ g } KNO_3$

Since HNO_3 produces less KNO_3, it is the limiting reactant and KOH is in excess.

(b) $\underset{10.0\text{ g}}{2\ NaOH} + \underset{10.0\text{ g}}{H_2SO_4} \longrightarrow Na_2SO_4 + 2\ H_2O$

Choose one of the products, calculate its mass using each given reactant. Using H_2O as the product:

$$(10.0\text{ g NaOH})\left(\frac{1\text{ mol}}{40.00\text{ g}}\right)\left(\frac{2\text{ mol }H_2O}{2\text{ mol NaOH}}\right)\left(\frac{18.02\text{ g}}{\text{mol}}\right) = 4.51\text{ g }H_2O$$

$$(10.0\text{ g }H_2SO_4)\left(\frac{1\text{ mol}}{98.08\text{ g}}\right)\left(\frac{2\text{ mol }H_2O}{1\text{ mol }H_2SO_4}\right)\left(\frac{18.02\text{ g}}{\text{mol}}\right) = 3.67\text{ g }H_2O$$

Since H_2SO_4 produces less H_2O, it is the limiting reactant and NaOH is in excess.

22. (a) $\underset{50.0\text{ g}}{2\ Bi(NO_3)_3} + \underset{6.00\text{ g}}{3\ H_2S} \longrightarrow Bi_2S_3 + 6\ HNO_3$

Choose one of the products, calculate its mass using each given reactant. Using Bi_2S_3 as the product:

$$(50.0\text{ g }Bi(NO_3)_3)\left(\frac{1\text{ mol}}{395.0\text{ g}}\right)\left(\frac{1\text{ mol }Bi_2S_3}{2\text{ mol }Bi(NO_3)_3}\right)\left(\frac{514.2\text{ g}}{\text{mol}}\right) = 32.5\text{ g }Bi_2S_3$$

$$(6.00\text{ g }H_2S)\left(\frac{1\text{ mol}}{34.08\text{ g}}\right)\left(\frac{1\text{ mol }Bi_2S_3}{3\text{ mol }H_2S}\right)\left(\frac{514.2\text{ g}}{\text{mol}}\right) = 30.2\text{ g }Bi_2S_3$$

Since H_2S produces less Bi_2S_3, it is the limiting reactant and $Bi(NO_3)_3$ is in excess.

(b) $\underset{40.0\text{ g}}{3\ Fe} + \underset{16.0\text{ g}}{4\ H_2O} \longrightarrow Fe_3O_4 + 4\ H_2$

Choose one of the products, calculate its mass using each given reactant. Using H_2 as the product:

$$(40.0\text{ g Fe})\left(\frac{1\text{ mol}}{55.85\text{ g}}\right)\left(\frac{4\text{ mol }H_2}{3\text{ mol Fe}}\right)\left(\frac{2.016\text{ g}}{\text{mol}}\right) = 1.93\text{ g }H_2$$

$$(16.0\text{ g }H_2O)\left(\frac{1\text{ mol}}{18.02\text{ g}}\right)\left(\frac{4\text{ mol }H_2}{4\text{ mol }H_2O}\right)\left(\frac{2.016\text{ g}}{\text{mol}}\right) = 1.79\text{ g }H_2$$

Since H_2O produces less H_2, it is the limiting reactant and Fe is in excess.

23. Limiting reactant calculations

$$C_3H_8 + 5\ O_2 \longrightarrow 3\ CO_2 + 4\ H_2O$$

(a) Reaction between 5.0 mol C_3H_8 and 5.0 mol O_2

$$(5.0\text{ mol }C_3H_8)\left(\frac{3\text{ mol }CO_2}{1\text{ mol }C_3H_8}\right) = 15\text{ mol }CO_2$$

$$(5.0\text{ mol }O_2)\left(\frac{3\text{ mol }CO_2}{5\text{ mol }O_2}\right) = 3.0\text{ mol }CO_2$$

The O_2 is the limiting reactant; 3.0 mol CO_2 produced.

(b) Reaction between 3.0 mol C_3H_8 and 20.0 mol O_2

$$(3.0 \text{ mol } C_3H_8)\left(\frac{3 \text{ mol } CO_2}{1 \text{ mol } C_3H_8}\right) = 9.0 \text{ mol } CO_2$$

$$(20.0 \text{ mol } O_2)\left(\frac{3 \text{ mol } CO_2}{5 \text{ mol } O_2}\right) = 12.0 \text{ mol } CO_2$$

The C_3H_8 is the limiting reactant; 9.0 mol CO_2 produced.

(c) Reaction between 20.0 mol C_3H_8 and 3.0 mol O_2

According to the equation, 1 mol C_3H_8 to 5 mol O_2, O_2 is clearly the limiting reactant.

$$(3.0 \text{ mol } O_2)\left(\frac{3 \text{ mol } CO_2}{5 \text{ mol } O_2}\right) = 1.8 \text{ mol } CO_2 \text{ produced}$$

24. Limiting reactant calculations

$C_3H_8 + 5\ O_2 \longrightarrow 3\ CO_2 + 4\ H_2O$

(a) Reaction between 20.0 g C_3H_8 and 20.0 g O_2
Convert each amount to grams of CO_2

$$(20.0 \text{ g } C_3H_8)\left(\frac{1 \text{ mol}}{44.09 \text{ g}}\right)\left(\frac{3 \text{ mol } CO_2}{1 \text{ mol } C_3H_8}\right)\left(\frac{44.01 \text{ g}}{\text{mol}}\right) = 59.9 \text{ g } CO_2$$

$$(20.0 \text{ g } O_2)\left(\frac{1 \text{ mol}}{32.00 \text{ g}}\right)\left(\frac{3 \text{ mol } CO_2}{5 \text{ mol } O_2}\right)\left(\frac{44.01 \text{ g}}{\text{mol}}\right) = 16.5 \text{ g } CO_2$$

O_2 is the limiting reactant. The yield is 16.5 g CO_2.

(b) Reaction between 20.0 g C_3H_8 and 80.0 g O_2

Convert each amount to grams of CO_2

$$(20.0 \text{ g } C_3H_8)\left(\frac{1 \text{ mol}}{44.09 \text{ g}}\right)\left(\frac{3 \text{ mol } CO_2}{1 \text{ mol } C_3H_8}\right)\left(\frac{44.01 \text{ g}}{\text{mol}}\right) = 59.9 \text{ g } CO_2$$

$$(80.0 \text{ g } O_2)\left(\frac{1 \text{ mol}}{32.00 \text{ g}}\right)\left(\frac{3 \text{ mol } CO_2}{5 \text{ mol } O_2}\right)\left(\frac{44.01 \text{ g}}{\text{mol}}\right) = 66.0 \text{ g } CO_2$$

C_3H_8 is the limiting reactant. The yield is 59.9 g CO_2.

(c) Reaction between 2.0 mol C_3H_8 and 14.0 mol O_2
According to the equation, 2 mol C_3H_8 will react with10 mol O_2. Therefore, C_3H_8 is the limiting reactant and 4.0 mol O_2 will remain unreacted.

$$(2.0 \text{ mol } C_3H_8)\left(\frac{3 \text{ mol } CO_2}{1 \text{ mol } C_3H_8}\right) = 6.0 \text{ mol } CO_2 \text{ produced}$$

$$(2.0 \text{ mol } C_3H_8)\left(\frac{4 \text{ mol } H_2O}{1 \text{ mol } C_3H_8}\right) = 8.0 \text{ mol } H_2O \text{ produced}$$

When the reaction is completed, 6.0 mol CO_2, 8.0 mol H_2O, and 4.0 mol O_2 will be in the container.

25. $X + 2\ HCl \longrightarrow XCl_2 + H_2$

$$(2.42 \text{ g } H_2)\left(\frac{1 \text{ mol}}{2.016 \text{ g}}\right)\left(\frac{1 \text{ mol } X}{1 \text{ mol } H_2}\right) = 1.20 \text{ mol } X$$

$$\frac{78.5 \text{ g}}{1.20 \text{ mol}} = 65.4\ \frac{\text{g}}{\text{mol}}$$

The metal is zinc.

26. $X_8 + 12\ O_2 \longrightarrow 8\ XO_3$

$$(120.0 \text{ g } O_2)\left(\frac{1 \text{ mol}}{32.00 \text{ g}}\right)\left(\frac{1 \text{ mol } X_8}{12 \text{ mol } O_2}\right) = 0.3125 \text{ mol } X_8$$

$$\frac{80.0 \text{ g}}{0.3125 \text{ mol}} = 256 \text{ g/mol } X_8$$

$$\text{molar mass } X = \frac{256\ \frac{\text{g}}{\text{mol}}}{8} = 32.0\ \frac{\text{g}}{\text{mol}}$$

The element is sulfur.

27. Limiting reactant calculation

$2\ Al + 3\ Br_2 \longrightarrow 2\ AlBr_3$

Reaction between 25.0 g Al and 100. g Br_2

$$(25.0 \text{ g Al})\left(\frac{1 \text{ mol}}{26.98 \text{ g}}\right)\left(\frac{2 \text{ mol } AlBr_3}{2 \text{ mol Al}}\right)\left(\frac{266.7 \text{ g}}{\text{mol}}\right) = 247 \text{ g } AlBr_3$$

$$(100. \text{ g } Br_2)\left(\frac{1 \text{ mol}}{159.8 \text{ g}}\right)\left(\frac{2 \text{ mol } AlBr_3}{3 \text{ mol } Br_2}\right)\left(\frac{266.7 \text{ g}}{\text{mol}}\right) = 111 \text{ g } AlBr_3$$

Br_2 is limiting; 111 g $AlBr_3$ is the theoretical yield of product.

$$\text{Percent yield} = \left(\frac{\text{actual yield}}{\text{theoretical yield}}\right)(100) = \left(\frac{64.2 \text{ g}}{111 \text{ g}}\right)(100) = 57.8\%$$

28. Percent yield calculation

$$Fe_{(s)} + CuSO_{4(aq)} \longrightarrow Cu_{(s)} + FeSO_{4(aq)}$$

$$(400. \text{ g } CuSO_4)\left(\frac{1 \text{ mol}}{159.6 \text{ g}}\right)\left(\frac{1 \text{ mol Cu}}{1 \text{ mol } CuSO_4}\right)\left(\frac{63.55 \text{ g}}{\text{mol}}\right) = 159 \text{ g Cu (theoretical yield)}$$

$$\%\text{ yield} = \left(\frac{\text{actual yield}}{\text{theoretical yield}}\right)(100) = \left(\frac{151 \text{ g}}{159 \text{ g}}\right)(100) = 95.0\% \text{ yield of Cu}$$

29. The balanced equation is $3\ C + 2\ SO_2 \longrightarrow CS_2 + 2\ CO_2$

$$(950\ g\ CS_2)\left(\frac{1}{0.860}\right)\left(\frac{1\ mol}{76.13\ g}\right)\left(\frac{3\ mol\ C}{1\ mol\ CS_2}\right)\left(\frac{12.01\ g}{mol}\right) = 5.2 \times 10^2\ g\ C$$

30. The balanced equation is $CaC_2 + 2\ H_2O \longrightarrow C_2H_2 + Ca(OH)_2$

$$(0.540\ mol\ C_2H_2)\left(\frac{1\ mol\ CaC_2}{1\ mol\ C_2H_2}\right)\left(\frac{64.10\ g}{mol}\right) = 34.6\ g\ CaC_2 \text{ in the impure sample}$$

$$\%\ yield = \left(\frac{\text{actual yield}}{\text{theoretical yield}}\right)(100) = \left(\frac{34.6\ g}{44.5\ g}\right)(100)$$

$$= 77.8\%\ CaC_2 \text{ in the impure material}$$

31. No. There are not enough screwdrivers, wrenches or pliers. 2400 screwdrivers, 3600 wrenches and 1200 pliers are needed for 600 tool sets.

32. A subscript is used to indicate the number of atoms in a formula. It cannot be changed without changing the identity of the substance. Coefficients are used only to balance atoms in chemical equations. They may be changed as needed to achieve a balanced equation.

33. $4\ KO_2 + 2\ H_2O + 4\ CO_2 \longrightarrow 4\ KHCO_3 + 3\ O_2$

(a) $$\left(\frac{0.85\ g\ CO_2}{min}\right)\left(\frac{1\ mol}{44.01\ g}\right)\left(\frac{4\ mol\ KO_2}{4\ mol\ CO_2}\right) = \frac{0.019\ mol\ KO_2}{min}$$

$$\left(\frac{0.019\ mol\ KO_2}{min}\right)(10.0\ min) = 0.19\ mol\ KO_2$$

(b) $$\left(\frac{0.85\ g\ CO_2}{min}\right)\left(\frac{1\ mol}{44.01\ g}\right)\left(\frac{3\ mol\ O_2}{4\ mol\ CO_2}\right)\left(\frac{32.00\ g}{mol}\right)\left(\frac{60.0\ min}{1.0\ hr}\right) = \frac{28\ g\ O_2}{hr}$$

34. (a) $$(750\ g\ C_6H_{12}O_6)\left(\frac{1\ mol}{180.2\ g}\right)\left(\frac{2\ mol\ C_2H_5OH}{1\ mol\ C_6H_{12}O_6}\right)\left(\frac{46.07\ g}{mol}\right) = 380\ g\ C_2H_5OH$$

$$(750\ g\ C_6H_{12}O_6)\left(\frac{1\ mol}{180.2\ g}\right)\left(\frac{2\ mol\ CO_2}{1\ mol\ C_6H_{12}O_6}\right)\left(\frac{44.01\ g}{mol}\right) = 370\ g\ CO_2$$

(b) $$(380\ g\ C_2H_5OH)\left(\frac{1\ mL}{0.79\ g}\right) = 480\ mL\ C_2H_5OH$$

35. $4\ P + 5\ O_2 \longrightarrow P_4O_{10}$

$P_4O_{10} + 6\ H_2O \longrightarrow 4\ H_3PO_4$

In the first reaction:

$$(20.0\ g\ P)\left(\frac{1\ mol}{30.97\ g}\right) = 0.646\ mol\ P$$

$$(30.0\ g\ O_2)\left(\frac{1\ mol}{32.00\ g}\right) = 0.938\ mol\ O_2$$

Therefore P is the limiting reactant and the P_4O_{10} produces is:

$$(0.646 \text{ mol P})\left(\frac{1 \text{ mol } P_4O_{10}}{4 \text{ mol P}}\right) = 0.162 \text{ mol } P_4O_{10}$$

In the second reaction:

$$(15.0 \text{ g } H_2O)\left(\frac{1 \text{ mol}}{18.02 \text{ g}}\right) = 0.832 \text{ mol } H_2O$$

and we have 0.162 mol P_4O_{10}

Therefore H_2O is the limiting reactant and the H_3PO_4 produced is:

$$(0.832 \text{ mol } H_2O)\left(\frac{4 \text{ mol } H_3PO_4}{6 \text{ mol } H_2O}\right)\left(\frac{97.99 \text{ g}}{\text{mol}}\right) = 54.4 \text{ g } H_3PO_4$$

36. $2\ CH_3OH + 3\ O_2 \longrightarrow 2\ CO_2 + 4\ H_2O$

$$(60.0 \text{ mL } CH_3OH)\left(\frac{0.72 \text{ g}}{\text{mL}}\right)\left(\frac{1 \text{ mol}}{32.04 \text{ g}}\right)\left(\frac{3 \text{ mol } O_2}{2 \text{ mol } CH_3OH}\right)\left(\frac{32.00 \text{ g}}{\text{mol}}\right) = 65 \text{ g } O_2$$

37. $7\ H_2O_2 + N_2H_4 \longrightarrow 2\ HNO_3 + 8\ H_2O$

(a) $(0.33 \text{ mol } N_2H_4)\left(\frac{2 \text{ mol } HNO_3}{1 \text{ mol } N_2H_4}\right) = 0.66 \text{ mol } HNO_3$

(b) $(2.75 \text{ mol } H_2O)\left(\frac{7 \text{ mol } H_2O_2}{8 \text{ mol } H_2O}\right) = 2.41 \text{ mol } H_2O_2$

(c) $(8.72 \text{ mol } HNO_3)\left(\frac{8 \text{ mol } H_2O}{2 \text{ mol } HNO_3}\right) = 34.9 \text{ mol } H_2O$

(d) $(120 \text{ g } N_2H_4)\left(\frac{1 \text{ mol}}{32.05 \text{ g}}\right)\left(\frac{7 \text{ mol } H_2O_2}{1 \text{ mol } N_2H_4}\right)\left(\frac{32.04 \text{ g}}{\text{mol}}\right) = 890 \text{ g } H_2O_2$

38. $4\ Ag + 2\ H_2S + O_2 \longrightarrow 2\ Ag_2S + 2\ H_2O$

$$(1.1 \text{ g Ag})\left(\frac{1 \text{ mol}}{107.9 \text{ g}}\right)\left(\frac{2 \text{ mol } Ag_2S}{4 \text{ mol Ag}}\right)\left(\frac{247.9 \text{ g}}{\text{mol}}\right) = 1.3 \text{ g } Ag_2S$$

$$(0.14 \text{ g } H_2S)\left(\frac{1 \text{ mol}}{34.08 \text{ g}}\right)\left(\frac{2 \text{ mol } Ag_2S}{2 \text{ mol } H_2S}\right)\left(\frac{247.9 \text{ g}}{\text{mol}}\right) = 1.0 \text{ g } Ag_2S$$

$$(0.080 \text{ g } O_2)\left(\frac{1 \text{ mol}}{32.00 \text{ g}}\right)\left(\frac{2 \text{ mol } Ag_2S}{1 \text{ mol } O_2}\right)\left(\frac{247.9 \text{ g}}{\text{mol}}\right) = 1.2 \text{ g } Ag_2S$$

The H_2S is the limiting reactant so 1.0 g Ag_2S is formed.

39. The balanced equation is $Zn + 2\ HCl \longrightarrow ZnCl_2 + H_2$

180.0 g Zn – 35 g Zn = 145 g Zn reacted with HCl

(a) $(145 \text{ g Zn})\left(\frac{1 \text{ mol}}{65.38 \text{ g}}\right)\left(\frac{1 \text{ mol } H_2}{1 \text{ mol Zn}}\right) = 2.22 \text{ mol } H_2$

(b) $(145 \text{ g Zn})\left(\frac{1 \text{ mol}}{65.38 \text{ g}}\right)\left(\frac{2 \text{ mol HCl}}{1 \text{ mol Zn}}\right)\left(\frac{36.46 \text{ g}}{\text{mol}}\right) = 162 \text{ g HCl}$

40. $Fe + CuSO_4 \longrightarrow Cu + FeSO_4$
2.0 mol 3.0 mol

(a) 2.0 mol Fe react with 2.0 mol $CuSO_4$ to yield 2.0 mol Cu and 2.0 mol $FeSO_4$. 1.0 mol $CuSO_4$ is unreacted. At the completion of the reaction, there will be 2.0 mol Cu, 2.0 mol $FeSO_4$, and 1.0 mol $CuSO_4$.

(b) $(20.0 \text{ g Fe})\left(\frac{1 \text{ mol}}{55.85 \text{ g}}\right)\left(\frac{1 \text{ mol Cu}}{1 \text{ mol Fe}}\right)\left(\frac{63.55 \text{ g}}{\text{mol}}\right) = 22.8 \text{ g Cu}$

$(40.0 \text{ g } CuSO_4)\left(\frac{1 \text{ mol}}{159.6 \text{ g}}\right)\left(\frac{1 \text{ mol Cu}}{1 \text{ mol } CuSO_4}\right)\left(\frac{63.55 \text{ g}}{\text{mol}}\right) = 15.9 \text{ g Cu}$

Since $CuSO_4$ produces less Cu, it is the limiting reactant. Determine the mass of $FeSO_4$ produced from $CuSO_4$.

$(40.0 \text{ g } CuSO_4)\left(\frac{1 \text{ mol}}{159.6 \text{ g}}\right)\left(\frac{1 \text{ mol } FeSO_4}{1 \text{ mol } CuSO_4}\right)\left(\frac{151.9 \text{ g}}{\text{mol}}\right) = 38.1 \text{ g } FeSO_4$ produced

Determine the mass of Fe, which is unreacted.

$(40.0 \text{ g } CuSO_4)\left(\frac{1 \text{ mol}}{159.6 \text{ g}}\right)\left(\frac{1 \text{ mol Fe}}{1 \text{ mol } CuSO_4}\right)\left(\frac{55.85 \text{ g}}{\text{mol}}\right) = 14.0 \text{ g Fe}$ will react

Unreacted Fe = 20.0 g – 14.0 g = 6.0 g. Therefore, at the completion of the reaction, 15.9 g Cu, 38.1 g $FeSO_4$, 6.0 g Fe, and no $CuSO_4$ remain.

41. Limiting reactant calculation

$CO(g) + 2\ H_2(g) \longrightarrow CH_3OH(l)$

Reaction between 40.0 g CO and 10.0 g H_2

$(40.0 \text{ g CO})\left(\frac{1 \text{ mol}}{28.01 \text{ g}}\right)\left(\frac{1 \text{ mol } CH_3OH}{1 \text{ mol CO}}\right)\left(\frac{32.04 \text{ g}}{\text{mol}}\right) = 45.8 \text{ g } CH_3OH$

$(10.0 \text{ g } H_2)\left(\frac{1 \text{ mol}}{2.016 \text{ g}}\right)\left(\frac{1 \text{ mol } CH_3OH}{2 \text{ mol } H_2}\right)\left(\frac{32.04 \text{ g}}{\text{mol}}\right) = 79.5 \text{ g } CH_3OH$

CO is limiting; H_2 is in excess; 45.8 g CH_3OH will be produced.

$(40.0 \text{ g CO})\left(\frac{1 \text{ mol}}{28.01 \text{ g}}\right)\left(\frac{2 \text{ mol } H_2}{1 \text{ mol CO}}\right)\left(\frac{2.016 \text{ g}}{\text{mol}}\right) = 5.76 \text{ g } H_2$ react

$10.0 \text{ g } H_2 - 5.76 \text{ g } H_2 = 4.2 \text{ g } H_2$ remain unreacted

42. The balanced equation is $C_6H_{12}O_6 \longrightarrow 2\ C_2H_5OH + 2\ CO_2$

(a) $(750 \text{ g } C_6H_{12}O_6)\left(\frac{1 \text{ mol}}{180.2 \text{ g}}\right)\left(\frac{2 \text{ mol } C_2H_5OH}{1 \text{ mol } C_6H_{12}O_6}\right)\left(\frac{46.07 \text{ g}}{\text{mol}}\right)(0.846) = 3.2 \times 10^2 \text{ g } C_2H_5OH$

(b) $(475 \text{ g } C_2H_5OH)\left(\frac{1}{0.846}\right)\left(\frac{1 \text{ mol}}{46.07 \text{ g}}\right)\left(\frac{1 \text{ mol } C_6H_{12}O_6}{2 \text{ mol } C_2H_5OH}\right)\left(\frac{180.2 \text{ g}}{\text{mol}}\right) = 1.10 \times 10^3 \text{ g } C_6H_{12}O_6$

43. The balanced equations are:

$CaCl_2 + 2\ AgNO_3 \longrightarrow Ca(NO_3)_2 + 2\ AgCl$

$MgCl_2 + 2\ AgNO_3 \longrightarrow Mg(NO_3)_2 + 2\ AgCl$

1 mol of each salt will produce the same amount (2 mol) of AgCl. $MgCl_2$ has a higher percentage of Cl than $CaCl_2$ because Mg has a lower atomic mass than Ca. Therefore, on an equal mass basis, $MgCl_2$ will produce more AgCl than will $CaCl_2$.

Calculations show that 1.00 g $MgCl_2$ produces 3.01 g AgCl, and 1.00 g $CaCl_2$ produces 2.56 g AgCl.

44. The balanced equation is $Li_2O + H_2O \longrightarrow 2\ LiOH$

$$\left(\frac{2500\ g\ H_2O}{astronaut\ day}\right)\left(\frac{1\ mol}{18.02\ g}\right)\left(\frac{1\ mol\ Li_2O}{1\ mol\ H_2O}\right)\left(\frac{29.88\ g}{mol}\right)\left(\frac{1\ kg}{1000\ g}\right) = \frac{4.1\ kg\ Li_2O}{astronaut\ day}$$

$$\left(\frac{4.1\ kg\ Li_2O}{astronaut\ day}\right)(30\ days)(3\ astronauts) = 3.7 \times 10^2\ kg\ Li_2O$$

45. The balanced equation is

$H_2SO_4 + 2\ NaCl \longrightarrow Na_2SO_4 + 2\ HCl$

$$(20.0\ L\ HCl\ solution)\left(\frac{1000\ mL}{1\ L}\right)\left(\frac{1.20\ g}{1.00\ mL}\right)(0.420) = 1.0 \times 10^4\ g\ HCl$$

$$(9600\ g\ HCl)\left(\frac{1\ mol}{36.46\ g}\right)\left(\frac{1\ mol\ H_2SO_4}{2\ mol\ HCl}\right)\left(\frac{98.08\ g}{1\ mol}\right) = 1.4 \times 10^4\ g\ H_2SO_4$$

$$(1.4 \times 10^4\ g\ H_2SO_4)\left(\frac{1.00\ g\ H_2SO_4\ solution}{0.96\ g\ H_2SO_4}\right)\left(\frac{1kg}{1000\ g}\right) = 14\ kg\ concentrated\ H_2SO_4$$

46. The balanced equation is

$Al(OH)_{3(s)} + 3\ HCl_{(aq)} \longrightarrow AlCl_{3(aq)} + 3\ H_2O_{(l)}$

$$(2.5\ L)\left(\frac{3.0\ g\ HCl}{L}\right)\left(\frac{1\ mol}{36.46\ g}\right)\left(\frac{1\ mol\ Al(OH)_3}{3\ mol\ HCl}\right)\left(\frac{78.00\ g}{mol}\right) = 5.3\ g\ Al(OH)_3$$

$$(5.3\ g\ Al(OH)_3)\left(\frac{1000\ mg}{g}\right)\left(\frac{1\ tablet}{400.\ mg}\right) = 13\ tablets$$

47. The balanced equation is $2\ KClO_3 \longrightarrow 2\ KCl + 3\ O_2$

$12.82 \text{ g mixture} - 9.45 \text{ g residue} = 3.37 \text{ g } O_2$ lost by heating

Because the O_2 lost came only from $KClO_3$, we can use it to calculate the amount of $KClO_3$ in the mixture.

$$(3.37 \text{ g } O_2)\left(\frac{1 \text{ mol}}{32.00 \text{ g}}\right)\left(\frac{2 \text{ mol } KClO_3}{3 \text{ mol } O_2}\right)\left(\frac{122.6 \text{ g}}{\text{mol}}\right) = 8.61 \text{ g } KClO_3$$

$$\left(\frac{8.61 \text{ g } KClO_3}{12.82 \text{ g sample}}\right)(100) = 67.2\% \ KClO_3$$

48. The mask protects the parts of the machine which are not to be etched away.

49. The silicon dioxide is the material which interacts with the chemicals used to etch. The silicon dioxide is removed or sacrificed during this process.

50. Micromachines could be used as smart pills, drug reservoirs, or mini computers. To be effective in these applications, sensing imaging systems and means of locomotion must be devised.

CHAPTER 10

MODERN ATOMIC THEORY

1. An electron orbital is a region in space around the nucleus of an atom where an electron is most probably found.

2. A second electron may enter an orbital already occupied by an electron if its spin is opposite that of the electron already in the orbital and all other orbitals of the same sublevel contain an electron.

3. All the electrons in the atom are located in the orbitals closest to the nucleus.

4. Both 1s and 2s orbitals are spherical in shape and located symmetrically around the nucleus. The sizes of the spheres are different - the radius of 2s is larger than the 1s. The electrons in 2s orbitals are located further from the nucleus.

5. The energy sublevels are s, p, d, and f.

6. 1s, 2s, 2p, 3s, 3p, 4s, 3d, 4p

7. s - 2 per shell
   p - 6 per shell after the first level
   d - 10 per shell after the second level

8. The main difference is that the Bohr orbit has an electron traveling a specific path while an orbital is a region in space where the electron is most probably found.

9. Bohr's model was inadequate since it could not account for atoms more complex than hydrogen. It was modified by Schrodinger into the modern concept of the atom in which electrons exhibit wave and particle properties. The motion of electrons is determined only by probability functions as a region in space, or a cloud surrounding the nucleus.

10. s orbital

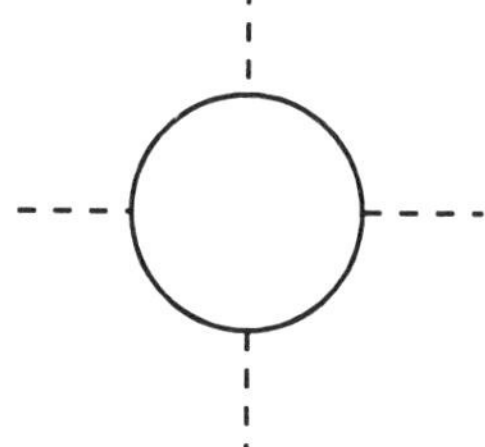

p orbitals

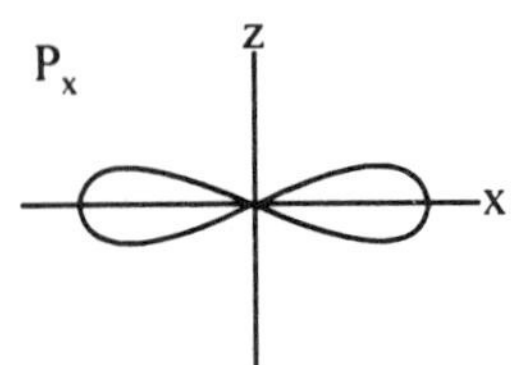

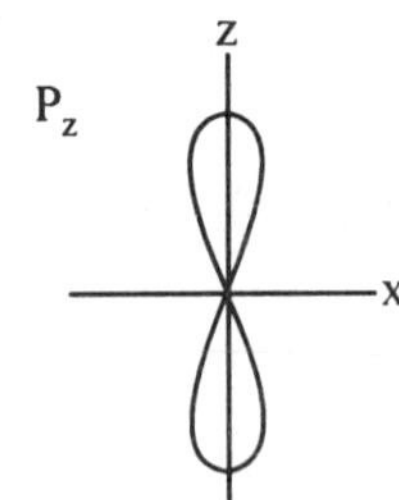

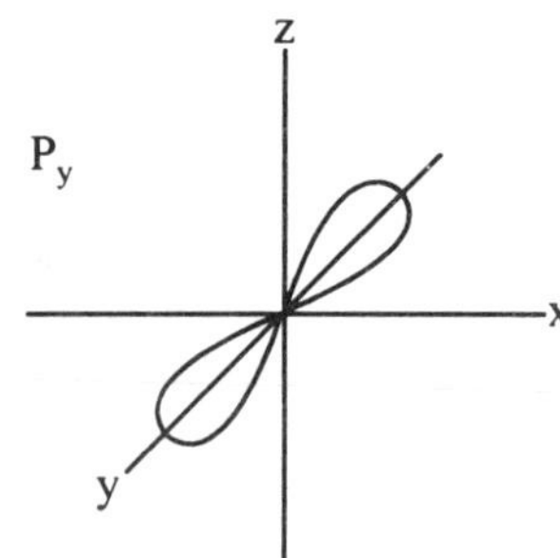

11. 3 is the third energy level
    d indicates an energy sublevel
    7 indicates the number of electrons in the d sublevel

12. Transition elements are found in the center of the periodic table. The last electrons for these elements are found in d orbitals.
    Representative elements are located on either side of the periodic table. The valence electrons for these elements are found in s and/or p orbitals.

13. Elements in the s-block all have one or two electrons in the outermost shell. These valence electrons are located in an s-orbital.

14.

Atomic #	Symbol
8	O
16	S
34	Se
52	Te
84	Po

All these elements have an outermost electron structure of s^2p^4.

15. F, Cl, Br, I, At

16. The greatest number of elements in any period is 32. The 6th period has this number.

17. The elements in Group A always have their last electrons in the outermost energy level, while the last electrons in Group B lie in an inner level.

18. Pairs of elements which are out of sequence with respect to atomic masses are: Ar and K; Co and Ni; Te and I; Th and Pa; U and Np; Es and Fm; Md and No.

19. The correct statements are a, c, d, e, f, h, i, j, k, m, o.

    (b) The maximum number of p electrons in the first energy level is 0. Electrons don't occupy p orbitals until level 2.

    (g) The electron structure of a calcium atom is $1s^22s^22p^63s^23p^64s^2$.

(l) An s orbital is spherically symmetrical around the nucleus.

(n) When an orbital contains two electrons, they have opposite spins.

(p) Rutherford concluded from his experiment that the positive charge and almost all the mass were concentrated in a very small nucleus.

20. The correct statements are a, c, e, f, g, i, n.

(b) There are fewer nonmetallic elements than metallic elements.

(d) Potassium belongs to the alkali metal family.

(h) An atom of aluminum (Group IIIA) has three electrons in its outer shell.

(j) The element [Kr] $5s^2$ is a metal.

(k) The element with Z = 12 forms compounds similar to the element with Z = 38.

(l) Nitrogen, fluorine, neon, and bromine are all nonmetals. Gallium is a metal.

(m) The atom having an outer shell electron structure of $5s^25p^2$ would be in period 5, Group IVA.

21. (a) H 1 proton

(b) B 5 protons

(c) Sc 21 protons

(d) U 92 protons

22. (a) F 9 protons

(b) Ag 47 protons

(c) Br 35 protons

(d) Sb 51 protons

23. (a) B $1s^22s^22p^1$

(b) Ti $1s^22s^22p^63s^23p^64s^23d^2$

(c) Zn $1s^22s^22p^63s^23p^64s^23d^{10}$

(d) Sr $1s^22s^22p^63s^23p^64s^23d^{10}4p^65s^2$

24. (a) Cl $1s^22s^22p^63s^23p^5$

(b) Ag $1s^22s^22p^63s^23p^64s^23d^{10}4p^65s^24d^9$

(c) Li $1s^22s^1$

(d) Fe $1s^22s^22p^63s^23p^64s^23d^6$

(e) I $1s^22s^22p^63s^23p^64s^23d^{10}4p^65s^24d^{10}5p^5$

25. The spectral lines of hydrogen are produced by energy emitted when an electron falls from a higher energy level to a lower energy level (closer to the nucleus).

26. Bohr said that a number of orbits were available, each corresponding to an energy level. When an electron falls from a higher energy orbit to a lower orbit, energy is given off as a specific frequency of light. Only those energies are seen and form the hydrogen spectrum. Each line corresponds to a change from one orbit to another.

27. 9 orbitals in the third energy level: 3s, $3p_x$, $3p_y$, $3p_z$ plus five d orbitals

28. 32 electrons in the fourth energy level

		Nucleus	Electrons	Symbol
29.	(a)	7p 7n	$2e^-$ $5e^-$	$^{14}_{7}N$
	(b)	17p 18n	$2e^-$ $8e^-$ $7e^-$	$^{35}_{17}Cl$
	(c)	30p 35n	$2e^-$ $8e^-$ $18e^-$ $2e^-$	$^{65}_{30}Zn$
	(d)	40p 51n	$2e^-$ $8e^-$ $18e^-$ $10e^-$ $2e^-$	$^{91}_{40}Zr$
	(e)	53p 74n	$2e^-$ $8e^-$ $18e^-$ $18e^-$ $7e^-$	$^{127}_{53}I$
30.	(a)	14p 14n	$2e^-$ $8e^-$ $4e^-$	$^{28}_{14}Si$
	(b)	16p 16n	$2e^-$ $8e^-$ $6e^-$	$^{32}_{16}S$
	(c)	18p 22n	$2e^-$ $8e^-$ $8e^-$	$^{40}_{18}Ar$
	(d)	23p 28n	$2e^-$ $8e^-$ $11e^-$ $2e^-$	$^{51}_{23}V$
	(e)	15p 16n	$2e^-$ $8e^-$ $5e^-$	$^{31}_{15}I$

31. (a) Mg

(b) Al

(c) Ni

(d) Mn

32. (a) Sc
(b) Zn
(c) Sn
(d) Cs

33.

	atomic #	electron structure
(a)	8	$1s^22s^22p^4$
(b)	11	$1s^22s^22p^63s^1$
(c)	17	$1s^22s^22p^63s^23p^5$
(d)	23	$1s^22s^22p^63s^23p^64s^23d^3$
(e)	28	$1s^22s^22p^63s^23p^64s^23d^8$
(f)	34	$1s^22s^22p^63s^23p^64s^23d^{10}4p^4$

34.

	atomic #	electron structure
(a)	9	$[He]2s^22p^5$
(b)	26	$[Ar]4s^23d^6$
(c)	31	$[Ar]4s^23d^{10}4p^1$
(d)	39	$[Kr]5s^24d^1$
(e)	52	$[Kr]5s^24d^{10}5p^4$
(f)	10	$[He]2s^22p^6$

35. (a) $^{32}_{16}S$ (b) $^{60}_{28}Ni$

36. (a) (13p 14n) $2e^-$ $8e^-$ $3e^-$ $^{27}_{13}Al$

(b) (22p 26n) $2e^-$ $8e^-$ $10e^-$ $2e^-$ $^{48}_{22}Ti$

37. The eleventh electron of sodium is located in the third energy level because the first and second levels are filled.

38. The last electron in potassium is located in the fourth energy level because the 4s orbital is at a lower energy level than the 3d orbital.

39. Noble gases all have filled s and p orbitals in the outermost energy level.

40. Noble gases each have filled s and p orbitals in the outermost energy level.

41. Moving from left to right in any period of elements, the atomic number increases by one from one element to the next and the atomic radius generally decreases. Each period (except period 1) begins with an alkali metal and ends with a noble gas. There is a trend in properties of the elements changing from metallic to nonmetallic from the beginning to the end of the period.

42. All the elements in a Group have the same number of outer shell electrons.

43. The outermost energy level contains one electron in an s orbital.

44. All of these elements have a s^2d^{10} electron configuration in their outermost energy levels.

45. (a) and (g)

(b) and (d)

(c) and (f) - wrong

46. (a) and (f)

(e) and (h)

47. 12, 38 since they are in the same group

48. 7, 33 since they are in the same group

49. (a) metal

(b) metal

(c) nonmetal

(d) metalloid

50. (a) nonmetal

(b) metal

(c) metal

(d) metalloid

51. Period 4 group IIIB contains the first element with an electron in a d orbital.

52. Period 6, lanthanide series, contains the first element with an electron in an f orbital.

53. Group IIIA contain 3 valence electrons
Group IIIB contain 2 electrons in the outermost level and one electron in an inner d orbital.
Group A elements are representative while Group B elements are transition elements.

54. Group VIIA contain 7 valence electrons.
Group VIIB contain 2 electrons in the outermost level and 5 electrons in an inner d orbital.
Group A elements are representative while Group B elements are transition elements.

55. $1s^3$ Li
$1s^3 2s^3 2p^9$ P
$1s^3 2s^3 2p^9 3s^3 3p^9$ Co

56. Nitrogen has more valence electrons on more energy levels. More varied electron transitions are possible.

57. (a) all 100

(b) all but H, He, Li, Be (96)

(c) 80 (all but those from H to Ca)

(d) the 44 elements beginning after Ba

58. (a) $\frac{2}{2} \times 100 = 100\%$

(b) $\frac{4}{4} \times 100 = 100\%$

(c) $\frac{10}{54} \times 100 = 19\%$

(d) $\frac{8}{34} \times 100 = 24\%$

(e) $\frac{11}{55} \times 100 = 20\%$

59. (a) :Ȯ: (one dot above, one below)   2 pair

(b) ·P· (two dots above, one below)   1 pair

(c) :I: (two dots above, one below)   3 pair

(d) :Xe: (two dots above, two below)   4 pairs

(e) Rb·   0 pairs

60. $\frac{1.5}{1.0 \times 10^{-8}} = 150{,}000{,}000 = \frac{1.5 \times 10^{8}}{1}$

61. (a) Ne

(b) Ge

(c) F

(d) N

62. In a scanning-probe microscope, a probe is placed near the surface of a sample and a parameter such as voltage is measured. The signals are translated electronically into a topographic image of the object. In a light microscope the image is formed by light reflecting from the object.

63. The object to be viewed is coated with a thin layer of metal and a probe is placed near the surface. Electrons jump from the metal to the probe producing the image by generating a varying current.

64. In the atomic force microscope the probe measures the electric forces between electrons in the molecule, while in scanning-tunneling the actual movement of electrons is measured.

65. The outermost electron structure for both sulfur and oxygen is s^2p^4.

66. Transition elements are found in Groups IB – VIIB and VIII.

67. In transition elements the last electron added is in a d or f orbital. The last electron added in a representative element is in an s or p orbital.

68. 8, 16, 34, 52, 84. All have 6 electrons in their outer shell.

69. The outermost energy level is the 7th. This element contains one electron in the 7s orbital.

70. If 36 is a noble gas, 35 would be in VIIA and 37 would be in 1A.

71. Answers will vary.

72. (a) $[Rn]7s^25f^{14}6d^{10}7p^5$

(b) 7 valence electrons

(c) F, Cl, Br, I, At

(d) halogen family, Group VIIA

73. (a) These elements are isotopes.

(b) These elements are adjacent to each other on the periodic table and happen to have equal atomic masses.

74. Most gases are located in the upper right part of the periodic table (H is an exception). They are nonmetals.

Liquids show no pattern. Neither do solids, except the vast majority of solids are metals.

CHAPTER 11

CHEMICAL BONDS: THE FORMATION OF COMPOUNDS FROM ATOMS

1. smallest Cl, Mg, Na, K, Rb largest.

2. More energy is required for neon because it has a very stable outer electron structure consisting of an octet of electrons in filled orbitals (noble gas electron structure). Sodium, an alkali metal, has a relatively unstable outer electron structure with a single electron in an unfilled orbital. The sodium electron is also farther away from the nucleus and is shielded by more inner electron shells than are neon outer shell electrons.

3. When a third electron is removed from beryllium, it must come from a very stable electron shell structure corresponding to that of the noble gas, helium. In addition, the third electron must be removed from a +2 beryllium ion, which increases the difficulty of removing it.

4. The first ionization energy decreases from top to bottom because in the successive alkali metals, the outermost electron is farther away from the nucleus and is shielded from the positive nucleus by additional electron shells.

5. The first ionization energy decreases from top to bottom because the outermost electrons in the successive noble gases are farther away from the nucleus and are shielded by additional inner electron shells.

6. Barium and beryllium are in the same family. The electron to be removed from barium is, however, located in an energy level father away from the nucleus than is the energy level holding the electron in beryllium. Hence, it requires less energy to remove the electron from barium than to remove the electron from beryllium. Barium, therefore, has a lower ionization energy than beryllium.

7. The first electron removed from a sodium atom is the one outer-shell electron, which is shielded from most of its nuclear charge by the electrons of lower levels. To remove a second electron from the sodium ion requires breaking into the noble gas structure. This requires much more energy than that necessary to remove the first electron, because the Na^+ is already positive.

8. (a) K > Na
(b) Na > Mg
(c) O > F
(d) I > Br
(e) Zr > Ti

9. The first element in each group has the smallest radius.

10. Atomic size increases down a column since each successive element has an additional energy level which contains electrons located farther from the nucleus.

11. Group IA IIA IIIA IVA VA VIA VIIA

IA	IIA	IIIA	IVA	VA	VIA	VIIA
E·	E:	E: (with one dot below)	·E: (with one dot below)	·E: (with one dot above and one below)	·E: (with one dot above and two below)	·E: (with two dots above and two below)

12. Lewis structures:

Cs· Ba: Tl: ·Pb: ·Po: ·At: :Rn:

Each of these is a representative element and has the same number of electrons in its outer shell as its periodic group number.

13. (a) Elements with the highest electronegativities are found in the upper right hand corner of the periodic table.

(b) Elements with the lowest electronegativities are found in the lower left of the periodic table.

14. Valence electrons are the electrons found in the outermost energy level of an atom.

15. By losing one electron, a potassium atom acquires a noble gas structure and becomes a K^+ ion. To become a K^{2+} ion requires loss of a second electron and breaking into the noble gas structure of the K^+ ion. This requires too much energy to generally occur.

16. An aluminum ion has a +3 charge because it has lost 3 electrons in acquiring a noble gas electron structure.

17. The correct statements are b, c, e, h, i, j, l, o, q, r, s, t, u, v, w, y

(a) If the formula for calcium iodide is CaI_2, then the formula for cesium iodide is CsI.

(d) Sodium and chlorine do not form molecules of NaCl.

(f) The noble gases have no tendency to lose electrons.

(g) The chemical bonds in a water molecule are covalent.

(k) Fluorine has the highest electronegativity of all the elements.

(m) A cation is smaller than its corresponding neutral atom.

(n) Cl_2 is more covalent in character than HCl.

(p) A nitrogen atom has five valence electrons.

(x) The simplest compound between oxygen and fluorine OF_2.

18. The correct statements are b, c, g, h, j, k, m, n, o

(a) The smaller the difference in electronegativity between two atoms, the more covalent the bond between them will be.

(d) The correct Lewis structure for CO_2 is $:\ddot{O}::C::\ddot{O}:$

(e) The Lewis structure for the noble gas helium is He:

(f) In period 5, the element having the highest ionization energy is Xe.

(i) The structures that show that H_2O is a dipole and that CO_2 is not a dipole are

```
   H
   |
  :O—H   and   :Ö=C=Ö:
   ··
```

(l) A molecule with a central atom having 3 bonding and one nonbonding electron pairs will have a tetrahedral electron structure and a pyramidal shape.

(p) The Lewis structure for potassium is K•

19. Magnesium atom is larger because it has electrons in the 3rd shell, while a magnesium ion does not. Also, the ion has 12 protons and 10 electrons, creating a charge imbalance and drawing the electrons in towards the nucleus more closely.

20. A bromine atom is smaller because it has one less electron than the bromine ion in its outer shell. Also, the ion has 35 protons and 36 electrons, creating a charge imbalance, resulting in a lessening of the attraction of the electrons towards the nucleus.

21.

	+	–
(a)	H	O
(b)	Na	F
(c)	H	N
(d)	Pb	S
(e)	N	O
(f)	H	C

22.

	+	–
(a)	H	Cl
(b)	Li	H
(c)	C	Cl
(d)	I	Br
(e)	Mg	H
(f)	O	F

23. (a) ionic
(b) covalent
(c) covalent
(d) ionic

24. (a) covalent
(b) ionic
(c) covalent
(d) ionic

25. Magnesium has an electron structure $1s^22s^22p^63s^2$, while the structure for chlorine is $1s^22s^22p^63s^23p^5$. When these two elements react with each other, each magnesium atom loses its $3s^2$ electrons, one to each of two chlorine atoms. The resulting structures for both magnesium and chlorine are noble gas configurations.

26. (a) $F + 1e^- \longrightarrow F^-$ (b) $Ca \longrightarrow Ca^{2+} + 2e^-$

27. (a) Mg: + ·F: + ·F: ⟶ MgF_2

(b) 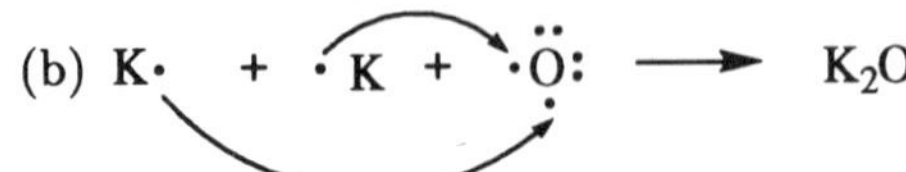

28. (a) Ca: + ·O: ⟶ CaO

(b) Na· + ·Br: ⟶ NaBr

29. Valence electrons: H (1) K (1) Mg (2) He (2) Al (3)

30. Valence electrons: Si (4) N (5) P (5) O (6) Cl (7)

31. Noble gas structures:

(a) Calcium atom, lose $2e^-$

(b) Sulfur atom, gain $2e^-$

(c) Helium, none

32. (a) Chloride ion, none

(b) Nitrogen atom, gain $3e^-$ or lose $5e^-$

(c) Potassium atom, lose $1e^-$

33. (a) A potassium atom is larger than a potassium ion because the atom has one more energy level containing an electron than the ion.

(b) A bromide ion is larger than a bromine atom because it has, within the same principle energy level, one more electron than a bromine atom.

34. (a) A magnesium ion, Mg^{2+}, is larger than an aluminum ion, Al^{3+}. Both ions have the same number of electrons (isoelectronic), with identical distribution of these electrons in their energy levels. The increased nuclear charge of the Al^{3+} ion pulls the electrons in the Al^{3+} ion closer to its nucleus, making it smaller than the Mg^{2+} ion.

(b) The Fe^{2+} ion is larger than the Fe^{3+} ion because the Fe^{2+} ion has one more electron than the Fe^{3+}ion. Both ions have the same nuclear charge.

35. (a) NaH, Na_2O

(b) CaH_2, CaO

(c) AlH_3, Al_2O_3

(d) SnH_4, SnO_2

36. (a) SbH_3, Sb_2O_3

(b) H_2Se, SeO_3

(c) HCl, Cl_2O_7

(d) CCl_4, CO_2

37. Li_2SO_4: lithium sulfate
K_2SO_4: potassium sulfate
Rb_2SO_4: rubidium sulfate
Cs_2SO_4: cesium sulfate
Fr_2SO_4: francium sulfate

38. $BeBr_2$, $MgBr_2$, $SrBr_2$, $BaBr_2$

39. Lewis structures:

(a) Na· (b) $[:\ddot{\underset{..}{Br}}:]^-$ (c) $[:\ddot{\underset{..}{O}}:]^{2-}$

40. (a) $\underset{\cdot}{Ga}$: (b) $[Ga]^{3+}$ (c) $[Ca]^{2+}$

41. (a) covalent
(b) ionic
(c) ionic
(d) covalent

42. (a) covalent
(b) ionic
(c) covalent
(d) covalent

43. (a) ionic
(b) covalent
(c) covalent

44. (a) covalent
(b) covalent
(c) ionic

45. (a) H : H (b) :N:::N: (c) :Cl:Cl:

46. (a) :O::O: (b) :Br:Br: (c) :I:I:

47. (a) :Cl : N : Cl:
:Cl:

(b) H:O:C::O:
:O:
H

(c) H H
H:C:C:H
H H

(d) [Na]$^{+}$[:O:N::O:]$^{-}$
:O:

48. (a) :S:H
H

(b) :S::C::S:

(c) H:N:H
H

(d) [H
H:N:H
H]$^{+}$ [:Cl:]$^{-}$

49. (a) $[Ba]^{2+}$

(b) $[Al]^{3+}$

(c) $\left[\begin{array}{c}:\ddot{\underset{..}{O}}:\ddot{\underset{..}{S}}:\ddot{\underset{..}{O}}: \\ :\underset{..}{O}:\end{array}\right]^{2-}$

(d) $[:C:::N:]^{-}$

(e) $\left[\begin{array}{c}:\ddot{O}::C:\ddot{\underset{..}{O}}: \\ :\ddot{\underset{..}{O}}: \\ H\end{array}\right]^{-}$

50. (a) $\left[:\ddot{\underset{..}{I}}:\right]^{-}$

(b) $\left[:\ddot{\underset{..}{S}}:\right]^{2-}$

(c) $\left[\begin{array}{c}:\ddot{\underset{..}{O}}:\underset{..}{C}::\ddot{O}: \\ :\underset{..}{O}:\end{array}\right]^{2-}$

(d) $\left[\begin{array}{c}:\ddot{\underset{..}{O}}:\ddot{\underset{..}{Cl}}:\ddot{\underset{..}{O}}: \\ :\underset{..}{O}:\end{array}\right]^{-}$

(e) $\left[\begin{array}{c}:\ddot{\underset{..}{O}}:\underset{..}{N}::\underset{..}{O}: \\ :\underset{..}{O}:\end{array}\right]^{-}$

51. (a) polar

(b) polar

(c) nonpolar

52. (a) nonpolar

(b) nonpolar

(c) polar

53. (a) 4 electron pairs, tetrahedral

(b) 4 electron pairs, tetrahedral

(c) 3 electron pairs, trigonal planar

54. (a) 2 electron pairs, linear

(b) 4 electron pairs, tetrahedral

(c) 4 electron pairs, tetrahedral

55. (a) tetrahedral
(b) pyramidal
(c) tetrahedral

56. (a) tetrahedral
(b) pyramidal
(c) tetrahedral

57. (a) tetrahedral
(b) pyramidal
(c) bent

58. (a) tetrahedral
(b) bent
(c) bent

59. Oxygen

60. Potassium

61. (a) Hg
(b) Be
(c) N
(d) Fr
(e) Zn

62. (a) Zn
(b) Be
(c) N

63. (1) Fluorine is missing one electron from its 2p level while neon has a full energy level.
(2) Fluorine's valence electrons are closer to the nucleus. The attraction between the electrons and the nucleus is greater.

64. Lithium has a +1 charge after the first electron is removed. It takes more energy to overcome that charge than to remove a single electron from an uncharged He atom.

65. Yes. These elements could lose an electron in the same way elements in group IA do. They would then form +1 ions and ionic compounds such as these.

66. $SnBr_2$, $GeBr_2$

67. The bond between sodium and chlorine is ionic. The electrons have been transferred from sodium to chlorine. The substance is composed of ions not molecules. Use of the word molecule implies covalent bonding.

68. A covalent bond results from the sharing of a pair of electrons between two atoms, while an ionic bond involves the transfer of one electron or more from one atom to another.

69. This structure is incorrect since the bond is ionic. It should be represented as:

$$[\text{Na}]^+[:\ddot{\underset{\cdot\cdot}{\text{O}}}:]^{2-}[\text{Na}]^+$$

70. The four most electronegative elements are N, O, F, Cl

71. highest F, O, S, H, Mg, Cs lowest

72. It is possible for a molecule to be nonpolar even though it contains polar bonds. If the molecule is symmetrical, the polarities of the bonds will cancel (in a manner similar to a positive and negative number of the same size) resulting in a nonpolar molecule. An example is CO_2 which is linear and nonpolar.

73. Both molecules contain polar bonds. CO_2 is symmetrical about the C atom, so the polarities cancel. In CO, there is only one polar bond, therefore the molecule is polar.

74. (a) 105°
(b) 107°
(c) 109.5°
(d) 109.5°

75. A superconductor is a substance in which the electrical resistance drops to zero at very low temperatures.

76. 77 K is the boiling point of liquid N_2. If a substance superconducts at this temperature, liquid N_2 can be used to cool the material. Since liquid N_2 is relatively inexpensive, superconductivity at this temperature would be more economical.

77. Chu realized that elements in the same family have similar chemistry. By substituting an element in the same family with a smaller radius, he brought the layers of the superconducting oxide closer together without changing the chemistry. This raised the critical temperature for superconductivity.

78. Current material used in superconductors is brittle, non-malleable and does not carry a high current per cross sectional area as conventional conductors.

79. Liquid crystals are used in mood rings, calculators, watches, thermometers, bullet resistant vests, canoes and the space shuttle.

80. Normally, the LCD acts as a mirror reflecting light. It has a series of layers, however. When the molecules at the top align with the lines etched on the first layer of glass and those at the bottom align with the grooves on the bottom layer of glass, the molecules in between form a twisted spiral (trying to align with those near them). If a current is applied to specific segments of the etched glass, the plates become charged and the spirals of molecules are attached to the charged plate destroying the arrangement. The pattern of reflected light is changed and a number appears.

81. Kevlar is a nematic liquid crystal which contains molecules all pointing in the same direction. It gets super strength by passing through a liquid crystal state in the manufacturing process. During this time the molecules are aligned and formed into fibers.

82. (a) Both use the p orbitals for bonding. B uses one s and two p orbitals while N uses three p orbitals for bonding.

(b) BF_3 is trigonal planar while NF_3 is pyramidal.

(c) BF_3 no lone pairs
NF_3 one lone pair

(d) BF_3 has 3 very polar covalent bonds. NF_3 has 3 covalent bonds

83. Fluorine's electronegativity is greater than any other element. Ionic bonds form between atoms of widely different electronegativities. Therefore Cs–F, Rb–F, K–F or Fr–F would be ionic substances with the greatest electronegativity difference.

84. Each element in a particular column has the same number of valence electrons and therefore the same Lewis structure.

85. S $\dfrac{1.409\text{ g}}{32.00\ \frac{\text{g}}{\text{mol}}} = 0.0438\text{ mol}$ $\qquad \dfrac{0.0438}{0.0438} = 1$

O $\dfrac{2.10\text{ g}}{16.00\ \frac{\text{g}}{\text{mol}}} = 0.131\text{ mol}$ $\qquad \dfrac{0.131}{0.0438} = 3$

Empirical formula is SO_3

```
     ..
    :O:
     |
 ..       ..
:O — S = O:
 ..
```

86. C $\dfrac{14.5\text{ g}}{12.01\ \frac{\text{g}}{\text{mol}}} = 1.21\text{ mol}$ $\qquad \dfrac{1.21}{1.21} = 1$

Cl $\dfrac{85.5\text{ g}}{35.45\ \frac{\text{g}}{\text{mol}}} = 2.41\text{ mol}$ $\qquad \dfrac{2.41}{1.21} = 2$

CCl_2 is the empirical formula

empirical mass = 1(12.01) + 2(35.45) = 83

$\frac{166}{83} = 2$

Therefore the molecular formula is $(CCl_2)_2 \longrightarrow C_2Cl_4$

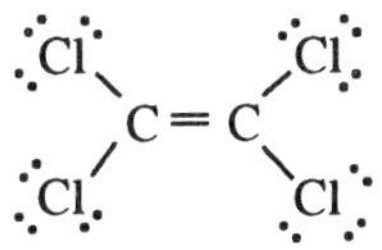

$$\begin{array}{ccccc} :\ddot{Cl}: & & & & :\ddot{Cl}: \\ & \diagdown & & \diagup & \\ & & C=C & & \\ & \diagup & & \diagdown & \\ :\ddot{Cl}: & & & & :\ddot{Cl}: \end{array}$$

CHAPTER 12

THE GASEOUS STATE OF MATTER

1. In Figure 12.1, color is the evidence of diffusion; bromine is colored and air is colorless. If hydrogen and oxygen had been the two gases, this would not work because both gases are colorless. Two ways could be used to show the diffusion. The change of density would be one method. Before diffusion the gas in the hydrogen flask would be much less dense. After diffusion the densities would be equal. A second method would require the introduction of spark gaps into both flasks. Before diffusion, neither gas would show a reaction when sparked. After diffusion, the gases in both flasks would explode because of the mixture of the two gases.

2. The air pressure inside the balloon is greater than the air pressure outside the balloon. The pressure inside must equal the sum of the outside air pressure plus the pressure exerted by the stretched rubber of the balloon.

3. The major components of dry air are nitrogen and oxygen.

4. 1 torr = 1 mm Hg

5. The molecules of H_2 at 100°C are moving faster. Temperature is a measure of average kinetic energy. At higher temperatures, the molecules will have more kinetic energy.

6. 1 atm corresponds to 4 L.

7. The pressure times the volume at any point on the curve is equal to the same value. This is an inverse relationship as is Boyle's law. ($PV = k$)

8. If $T_2 < T_1$, the volume of the cylinder would decrease (the piston would move downward).

9. The pressure inside the bottle is less than atmospheric pressure. We come to this conclusion because the water inside the bottle is higher than the water in the trough (outside the bottle).

10. The density of air is given as 1.29 g/L. Any gas with a greater density is listed below air on the table. Any five of these gases would be correct. (O_2, H_2S, HCl, F_2, CO_2).

11. Basic assumptions of Kinetic Molecular Theory include:

    (a) Gases consist of tiny particles.

    (b) The distance between particles is great compared to the size of the particles.

    (c) Gas particles move in straight lines. They collide with one another and with the walls of the container with no loss of energy.

(d) Gas particles have no attraction for each other.

(e) The average kinetic energy of all gases is the same at any given temperature. It varies directly with temperature.

12. The order of increasing molecular velocities is the order of decreasing molar masses.

molecular velocity increases ⟶

Rn, F_2, N_2, CH_4, He, H_2

⟵ molar mass increases

At the same temperature the kinetic energies of the gases are the same and equal to $\frac{1}{2}mv^2$. For the kinetic energies to be the same, the velocities must increase as the molar masses decrease.

13. Average kinetic energies of all these gases are the same, as the gases are all at the same temperature.

14. Gases are described by the following parameters:

(a) Pressure
(b) Volume
(c) Temperature
(d) Number of moles

15. An ideal gas is one which follows the described gas laws at all P, V, and T and whose behavior is described exactly by the Kinetic Molecular Theory.

16. A gas is least likely to behave ideally at low temperatures. Under this condition, the velocities of the molecules decrease and attractive forces between the molecules begin to play a significant role.

17. A gas is least likely to behave ideally at high pressures. Under this condition, the molecules are forced close enough to each other so that their volume is no longer small compared to the volume of the container. Attractive forces may also occur here and sooner or later, the gas will liquefy.

18. Equal volumes of H_2 and O_2 at the same T and P:

(a) have equal numbers of molecules (Avogadro's law)

(b) mass O_2 = 16 × mass H_2

(c) moles O_2 = moles H_2

(d) average kinetic energies are the same (T same)

(e) rate H_2 = 4 (rate O_2) (Graham's Law of Effusion)

(f) density O_2 = 16 (density H_2)

density $O_2 = \left(\frac{\text{mass } O_2}{\text{volume } O_2}\right)$ density $H_2 = \left(\frac{\text{mass } H_2}{\text{volume } H_2}\right)$

volume O_2 = volume H_2

$\left(\frac{\text{mass } O_2}{\text{den } O_2}\right) = \left(\frac{\text{mass } H_2}{\text{den } H_2}\right)$ density $O_2 = \left(\frac{\text{mass } O_2}{\text{mass } H_2}\right)(\text{den } H_2)$

density $O_2 = \left(\frac{32}{2}\right)(\text{density } H_2) = 16(\text{density } H_2)$

19. Behavior of gases as described by the Kinetic Molecular Theory.

(a) Boyle's law. Boyle's law states that the volume of a fixed mass of gas is inversely proportional to the pressure, at constant temperature. The Kinetic Molecular Theory assumes the volume occupied by gases is mostly empty space. Decreasing the volume of a gas by compressing it, increases the concentration of gas molecules, resulting in more collisions of the molecules and thus increased pressure upon the walls of the container.

(b) Charles' law. Charles' law states that the volume of a fixed mass of gas is directly proportional to the absolute temperature, at constant pressure. According to Kinetic Molecular Theory, the kinetic energies of gas molecules are proportional to the absolute temperature. Increasing the temperature of a gas causes the molecules to move faster, and in order for the pressure not to increase, the volume of the gas must increase.

(c) Dalton's law. Dalton's law states the pressure of a mixture of gases is the sum of the pressures exerted by the individual gases. According to the Kinetic Molecular Theory, there are no attractive forces between gas molecules; therefore, in a mixture of gases, each gas acts independently and the total pressure exerted will be the sum of the pressures exerted by the individual gases.

20. $N_2(g) + O_2(g) \longrightarrow 2\,NO(g)$

1 vol + 1 vol $\longrightarrow$ 2 vol

According to Avogadro's Law, equal volumes of nitrogen and oxygen at the same temperature and pressure contain the same number of molecules. In the reaction, nitrogen and oxygen molecules react in a 1:1 ratio. Since two volumes of nitrogen monoxide are produced, one molecule of nitrogen and one molecule of oxygen must produce two molecules of nitrogen monoxide. Therefore, each nitrogen and oxygen molecule must be made up of two atoms.

21. We refer gases to STP because some reference point is needed to relate volume to moles. A temperature and pressure must be specified to determine the moles of gas in a given volume, and 0°C and 760 torr are convenient reference points.

22. Conversion of oxygen to ozone is an endothermic reaction. Evidence for this statement is that energy (286 kJ/3 mol O_2) is required to convert O_2 to O_3.

23. Oxygen atom = O Oxygen molecule = O_2 Ozone molecule = O_3
An oxygen molecule contains 16 electrons.

24. Heating a mole of N_2 gas at constant pressure has the following effects:

(a) Density will decrease. Heating the gas at constant pressure will increase its volume. The mass does not change, so the increased volume results in a lower density.

(b) Mass does not change. Heating a substance does not change its mass.

(c) Average kinetic energy of the molecules increases. This is a basic assumption of the Kinetic Molecular Theory.

(d) Average velocity of the molecules will increase. Increasing the temperature increases the average kinetic energies of the molecules; hence, the average velocity of the molecules will increase also.

(e) Number of N_2 molecules remains unchanged. Heating does not alter the number of molecules present, except if extremely high temperatures were attained. Then, the N_2 molecules might dissociate into N atoms resulting in fewer N_2 molecules.

25. Henry's law states that the amount of gas which will dissolve in a liquid is directly proportional to the pressure above the liquid. This law is the basis for the "bends". As a diver descends the pressure increases allowing more gas to dissolve in the blood. As the diver ascends the pressure decreases and the solubility of a gas decreases as well. The "bends" are caused by a diver returning too rapidly to the surface. The quick reduction in pressure causes the dissolved gases to form bubbles in the blood.

26. The correct statements are b, d, f1, f4, h, i, n, o, p, q, t.

(a) The pressure exerted by a gas, at constant volume, is dependent on the temperature of the gas.

(c) At constant pressure, the volume of a gas is directly proportional to the absolute temperature.

(e) Compressing a gas at constant temperature will cause its density to increase. The mass will remain constant.

(f) 2. Equal volumes of CO_2 and CH_4 at the same temperature contain different masses.

3. Equal volumes of CO_2 and CH_4 at the same temperature have different densities.

5. Equal volumes of CO_2 and CH_4 at the same temperature contain different numbers of atoms.

(g) At constant temperature the average kinetic energy of O_2 molecules is the same at 100 atm and 200 atm pressure.

(j) The volume of a gas depends upon its temperature, pressure, and amount of gas present.

(k) In a mixture containing O_2 molecules and N_2 molecules, the O_2 molecules, on the average, are moving slower than the N_2 molecules.

(l) PV = k is a statement of Boyle's law.

(m) If the temperature of a sample of gas is increased from 25°C to 50°C, the volume of the gas will increase by about 8%. (The volume will not double.)

(r) According to the equation $2\ KClO_{3(s)} \xrightarrow{\Delta} 2\ KCl_{(s)} + 3\ O_{2(g)}$

1 mol of $KClO_3$ will produce 33.6 L of O_2 at STP

(s) PV = nRT is a statement of the ideal gas law.

27. (a) $(715\text{ mm Hg})\left(\frac{1\text{ atm}}{760\text{ mm Hg}}\right) = 0.941\text{ atm}$

(b) $(715\text{ mm Hg})\left(\frac{1\text{ in. Hg}}{25.4\text{ mm Hg}}\right) = 28.1\text{ in. Hg}$

(c) $(715\text{ mm Hg})\left(\frac{14.7\text{ lb/in.}^2}{760\text{ mm Hg}}\right) = 13.8\text{ lb/in.}^2$

28. (a) $(715\text{ mm Hg})\left(\frac{1\text{ torr}}{1\text{ mm Hg}}\right) = 715\text{ torr}$

(b) $(715\text{ mm Hg})\left(\frac{1013\text{ mbar}}{760\text{ mm Hg}}\right) = 953\text{ mbar}$

(c) $(715\text{ mm Hg})\left(\frac{101.325\text{ kPa}}{760\text{ mm Hg}}\right) = 95.3\text{ kPa}$

29. (a) $(28\text{ mm Hg})\left(\frac{1\text{ atm}}{760\text{ mm Hg}}\right) = 0.037\text{ atm}$

(b) $(6000.\text{ cm Hg})\left(\frac{1\text{ atm}}{76\text{ cm Hg}}\right) = 78.95\text{ atm}$

(c) $(795\text{ torr})\left(\frac{1\text{ atm}}{760\text{ torr}}\right) = 1.05\text{ atm}$

(d) $(5.00\text{ kPa})\left(\frac{1\text{ atm}}{101.325\text{ kPa}}\right) = 0.0493\text{ atm}$

30. (a) $(62\text{ mm Hg})\left(\frac{1\text{ atm}}{760\text{ mm Hg}}\right) = 0.082\text{ atm}$

(b) $(4250.\text{ cm Hg})\left(\frac{1\text{ atm}}{76\text{ cm Hg}}\right) = 55.92\text{ atm}$

(c) $(225\text{ torr})\left(\frac{1\text{ atm}}{760\text{ torr}}\right) = 0.296\text{ atm}$

(d) $(0.67\text{ kPa})\left(\frac{1\text{ atm}}{101.325\text{ kPa}}\right) = 0.0066\text{ atm}$

31. $P_1V_1 = P_2V_2$ or $V_2 = \left(\frac{P_1V_1}{P_2}\right)$.

(a) $\frac{(500.\text{ mm Hg})(400.\text{ mL})}{760\text{ mm Hg}} = 2.6 \times 10^2\text{ mL}$

(b) $\frac{(500.\text{ torr})(400.\text{ mL})}{250\text{ torr}} = 8.0 \times 10^2\text{ mL}$

32. $P_1V_1 = P_2V_2$ or $V_2 = \frac{P_1V_1}{P_2}$

(a) $\frac{(500.\text{ mm Hg})\ (1\text{ atm}/760\text{ mm Hg})(400.\text{ mL})}{2.00\text{ atm}} = 132\text{ mL}$

(b) $\frac{(500.\text{ mm Hg})(1\text{ torr}/1\text{ mm Hg})(400.\text{ mL})}{325\text{ torr}} = 615\text{ mL}$

33. $P_2 = \frac{P_1V_1}{V_2}$ $\quad \frac{(640.\text{ mm Hg})(500.\text{ mL})}{855\text{ mL}} = 374\text{ mm Hg}$

34. $P_2 = \frac{P_1V_1}{V_2}$ $\quad \frac{(640.\text{ mm Hg})(500.\text{ mL})}{450.\text{ mL}} = 711\text{ mm Hg}$

35. $\frac{V_1}{T_1} = \frac{V_2}{T_2}$ or $V_2 = \frac{V_1T_2}{T_1}$; Temperatures must be in Kelvin (°C + 273)

(a) $\frac{(6.00\text{ L})(273\text{ K})}{248\text{ K}} = 6.60\text{ L}$

(b) $\frac{(6.00\text{ L})(100.\text{ K})}{248\text{ K}} = 2.42\text{ L}$

36. $\frac{V_1}{T_1} = \frac{V_2}{T_2}$ or $V_2 = \frac{V_1T_2}{T_1}$; Temperatures must be in Kelvin (°C + 273)

0.0°F = −18°C

(a) $\frac{(6.00\text{ L})(255\text{ K})}{248\text{ K}} = 6.17\text{ L}$

(b) $\frac{(6.00\text{ L})(345\text{ K})}{248\text{ K}} = 8.35\text{ L}$

37. Use the combined gas law $\frac{P_1V_1}{T_1} = \frac{P_2V_2}{T_2}$ or $V_2 = \frac{P_1V_1T_2}{P_2T_1}$

$$V_2 = \frac{(740\text{ mm Hg})(410\text{ mL})(273\text{ K})}{(760\text{ mm Hg})(300.\text{ K})} = 3.6 \times 10^2\text{ mL}$$

38. Use the combined gas law $\frac{P_1V_1}{T_1} = \frac{P_2V_2}{T_2}$ or $V_2 = \frac{P_1V_1T_2}{P_2T_1}$

$$V_2 = \frac{(740\text{ mm Hg})(410\text{ mL})(523\text{ K})}{(680\text{ mm Hg})(300.\text{ K})} = 7.8 \times 10^2\text{ mL}$$

39. Use the combined gas law $\frac{P_1V_1}{T_1} = \frac{P_2V_2}{T_2}$ or $V_2 = \frac{P_1V_1T_2}{P_2T_1}$

$$V_2 = \frac{(0.950\text{ atm})(1400.\text{ mL})(275.0\text{ K})}{(4.0\text{ torr})(1\text{ atm}/760\text{ torr})(291\text{ K})} = 2.4 \times 10^5\text{ L}$$

40. Use the combined gas law $\frac{P_1V_1}{T_1} = \frac{P_2V_2}{T_2}$ or $V_2 = \frac{P_1V_1T_2}{P_2T_1}$

$$V_2 = \frac{(2.50\text{ atm})(22.4\text{ L})(268\text{ K})}{(1.50\text{ atm})(300.\text{ K})} = 33.4\text{ L}$$

41. $P_{total} = P_{N_2} + P_{H_2O\text{ vapor}} = 720.\text{ torr}$
$P_{H_2O\text{ vapor}} = 17.5\text{ torr}$
$P_{N_2} = 720.\text{ torr} - 17.5\text{ torr} = 703\text{ torr}$

42. $P_{total} = P_{N_2} + P_{H_2O\text{ vapor}} = 705\text{ torr}$
$P_{H_2O\text{ vapor}} = 23.8\text{ torr}$
$P_{N_2} = 705\text{ torr} - 23.8\text{ torr} = 681\text{ torr}$

43. $P_{total} = P_{N_2} + P_{H_2} + P_{O_2}$
$= 200.\text{ torr} + 600.\text{ torr} + 300.\text{ torr} = 1100.\text{ torr} = 1.100 \times 10^3\text{ torr}$

44. $P_{total} = P_{H_2} + P_{N_2} + P_{O_2}$
$= 325\text{ torr} + 475\text{ torr} + 650.\text{ torr} = 1450.\text{ torr} = 1.45 \times 10^3\text{ torr}$

45. $P_{total} = P_{CH_4} + P_{H_2O\text{ vapor}}$
$P_{H_2O\text{ vapor}} = 23.8\text{ torr}$
$P_{CH_4} = 720.\text{ torr} - 23.8\text{ torr} = 696\text{ torr}$
To calculate the volume of dry methane, note that the temperature is constant, so
$P_1V_1 = P_2V_2$

$$V_2 = \frac{P_1V_1}{P_2} = \frac{(696\text{ torr})(2.50\text{ L})}{(760.\text{ torr})} = 2.29\text{ L}$$

46. $P_{total} = P_{C_3H_8} + P_{H_2O}$ vapor

P_{H_2O} vapor = 21.9 torr

$P_{C_3H_8}$ = 745 torr – 21.9 torr = 723 torr

To calculate the volume of dry propane, note that the temperature is constant, so $P_1V_1 = P_2V_2$

$$V_2 = \frac{P_1V_1}{P_2} = \frac{(723 \text{ torr})(1.25 \text{ L})}{(760. \text{ torr})} = 1.19 \text{ L } C_3H_8$$

47. $(2.5 \text{ mol})\left(\frac{22.4 \text{ L}}{\text{mol}}\right) = 56 \text{ L } Cl_2$ (22.4 L is volume of 1 mol at STP)

48. $(1.25 \text{ mol})\left(\frac{22.4 \text{ L}}{\text{mol}}\right) = 28.0 \text{ L } N_2$

49. $(2500 \text{ mL})\left(\frac{1 \text{ L}}{1000 \text{ mL}}\right)\left(\frac{1 \text{ mol}}{22.4 \text{ L}}\right)\left(\frac{44.01 \text{ g } CO_2}{\text{mol}}\right) = 4.9 \text{ g } CO_2$

50. $(1.75 \text{ L})\left(\frac{1 \text{ mol}}{22.4 \text{ L}}\right)\left(\frac{17.03 \text{ g } NH_3}{\text{mol}}\right) = 1.33 \text{ g } NH_3$

51. (a) $(1.0 \text{ mol } NO_2)\left(\frac{22.4 \text{ L}}{\text{mol}}\right) = 22.4 \text{ L } NO_2$

(b) $(17.05 \text{ g } NO_2)\left(\frac{1 \text{ mol}}{46.01 \text{ g}}\right)\left(\frac{22.4 \text{ L}}{\text{mol}}\right) = 8.30 \text{ L } NO_2$

(c) $(1.20 \times 10^{24} \text{ molecules } NO_2)\left(\frac{1 \text{ mol}}{6.022 \times 10^{23} \text{ molecules}}\right)\left(\frac{22.4 \text{ L}}{\text{mol}}\right) = 44.6 \text{ L } NO_2$

52. (a) $(0.50 \text{ mol } H_2S)\left(\frac{22.4 \text{ L}}{\text{mol}}\right) = 11 \text{ mol } H_2S$

(b) $(22.41 \text{ g } H_2S)\left(\frac{1 \text{ mol}}{34.08 \text{ g}}\right)\left(\frac{22.4 \text{ L}}{\text{mol}}\right) = 14.7 \text{ L } H_2S$

(c) $(8.55 \times 10^{23} \text{ molecules } H_2S)\left(\frac{1 \text{ mol}}{6.022 \times 10^{23}}\right)\left(\frac{22.4 \text{ L}}{\text{mol}}\right) = 31.8 \text{ L } H_2S$

53. $(1.00 \text{ L } NH_3)\left(\frac{1 \text{ mol}}{22.4 \text{ L}}\right)\left(\frac{6.022 \times 10^{23} \text{ molecules}}{\text{mol}}\right) = 2.69 \times 10^{22} \text{ molecules } NH_3$

54. $(1.00 \text{ L } CH_4)\left(\frac{1 \text{ mol}}{22.4 \text{ L}}\right)\left(\frac{6.022 \times 10^{23} \text{ molecules}}{\text{mol}}\right) = 2.69 \times 10^{22} \text{ molecules } CH_4$

55. (a) $d = \left(\frac{83.80 \text{ g Kr}}{\text{mol}}\right)\left(\frac{1 \text{ mol}}{22.4 \text{ L}}\right) = 3.74 \text{ g/L Kr}$

(b) $d = \left(\frac{80.06 \text{ g } SO_2}{\text{mol}}\right)\left(\frac{1 \text{ mol}}{22.4 \text{ L}}\right) = 3.57 \text{ g/L } SO_3$

56. (a) $d = \left(\frac{4.003 \text{ g He}}{\text{mol}}\right)\left(\frac{1 \text{ mol}}{22.4 \text{ L}}\right) = 0.179 \text{ g/L He}$

(b) $d = \left(\frac{56.10 \text{ g } C_4H_8}{\text{mol}}\right)\left(\frac{1 \text{ mol}}{22.4 \text{ L}}\right) = 2.50 \text{ g/L } C_4H_8$

57. (a) $d = \left(\frac{38.00 \text{ g}}{\text{mol}}\right)\left(\frac{1 \text{ mol}}{22.4 \text{ L}}\right) = 1.70 \text{ g/L } F_2$

(b) Assume 1.00 mol of F_2 and determine the volume using the ideal gas equation, PV = nRT.

$$V = \frac{nRT}{P} = \frac{(1.00 \text{ mol})(0.0821 \text{ L atm/mol K})(300. \text{ K})}{1.00 \text{ atm}} = 24.6 \text{ L at } 27°\text{C and } 1.00 \text{ atm}$$

$$d = \frac{38.00\ g}{24.6\ L} = 1.54\ g/L\ F_2$$

58. (a) $d = \left(\frac{70.90 \text{ g } Cl_2}{\text{mol}}\right)\left(\frac{1 \text{ mol}}{22.4 \text{ L}}\right) = 3.17 \text{ g/L } Cl_2$

(b) Assume 1.00 mol of Cl_2 and determine the volume using the ideal gas equation, PV = nRT

$$V = \frac{nRT}{P} = \frac{(1.00 \text{ mol})(0.0821 \text{ L atm/mol K})(295 \text{ K})}{0.500 \text{ atm}} = 48.4 \text{ L at } 22°\text{C and } 0.500 \text{ atm}$$

$$d = \frac{70.90 \text{ g}}{48.4 \text{ L}} = 1.46 \text{ g/L } Cl_2$$

59. PV = nRT $\quad V = \frac{nRT}{P} \quad V = \frac{(2.3 \text{ mol})(0.0821 \text{ L atm/mol K})(300. \text{ K})}{\frac{750 \text{ torr}}{760 \frac{\text{torr}}{\text{atm}}}} = 57 \text{ L Ne}$

60. PV = nRT $\quad V = \frac{nRT}{P} \quad V = \frac{(0.75 \text{ mol})(0.0821 \text{ L atm/mol K})(298 \text{ K})}{\frac{725 \text{ torr}}{760 \frac{\text{torr}}{\text{atm}}}} = 19 \text{ L Kr}$

61. When working with gases, the identity of the gas does not affect the volume, as long as the number of moles are known.

Total moles = mol H_2 + mol CO_2 = 5.00 mol + 0.500 mol = 5.50 mol

$$V = (5.50 \text{ mol})\left(\frac{22.4 \text{ L}}{\text{mol}}\right) = 123 \text{ L}$$

62. When working with gases, the identity of the gas does not affect the volume, as long as the number of moles are known.

Total moles = mol N_2 + mol HCl = 2.50 mol + 0.750 mol = 3.25 mol

$$V = (3.25 \text{ mol})\left(\frac{22.4 \text{ L}}{\text{mol}}\right) = 72.8 \text{ L}$$

63. (a) $4\ NH_3(g) + 5\ O_2(g) \longrightarrow 4\ NO(g) + 6\ H_2O(g)$

$$(5.5 \text{ mol NO})\left(\frac{4 \text{ mol } NH_3}{4 \text{ mol NO}}\right) = 5.5 \text{ mol } NH_3$$

(b) Limiting reactant problem. Remember, volume - volume relationships are the same as mole - mole relationships when dealing with gases at constant T and P.

$$(12 \text{ L } O_2)\left(\frac{4 \text{ mol NO}}{5 \text{ mol } O_2}\right) = 9.6 \text{ L NO (from } O_2)$$

$$(10. \text{ L } NH_3)\left(\frac{4 \text{ mol NO}}{4 \text{ mol } NH_3}\right) = 10. \text{ L NO (from } NH_3)$$

Oxygen is the limiting reactant, 9.6 L NO is formed.

(c) Limiting reactant problem. Remember, volume - volume relationships are the same as mole - mole relationships when dealing with gases at constant T and P.

$$(3.0 \text{ L } NH_3)\left(\frac{4 \text{ mol NO}}{4 \text{ mol } NH_3}\right) = 3.0 \text{ L NO (from } NH_3)$$

$$(3.0 \text{ L } O_2)\left(\frac{4 \text{ mol NO}}{5 \text{ mol } O_2}\right) = 2.4 \text{ L NO (from } O_2)$$

Oxygen is the limiting reactant, 2.4 L NO is formed.

64. $4\ NH_3(g) + 5\ O_2(g) \longrightarrow 4\ NO(g) + 6\ H_2O(g)$

(a) $(7.0 \text{ mol } O_2)\left(\frac{4 \text{ mol } NH_3}{5 \text{ mol } O_2}\right) = 5.6 \text{ mol } NH_3$

(b) $(800. \text{ mL } O_2)\left(\frac{4 \text{ mol NO}}{5 \text{ mol } O_2}\right) = 640. \text{ mL NO} = 0.640 \text{ L NO}$

(c) $(60. \text{ L NO})\left(\frac{1 \text{ mol}}{22.4 \text{ L}}\right)\left(\frac{5 \text{ mol } O_2}{4 \text{ mol NO}}\right)\left(\frac{32.00 \text{ g}}{\text{mol}}\right) = 1.1 \times 10^2 \text{ g } O_2$

65. The balanced equation is

$$4\ FeS + 7\ O_2 \xrightarrow{\Delta} 2\ Fe_2O_3 + 4\ SO_2$$

$$(600. \text{ g FeS})\left(\frac{1 \text{ mol}}{87.91 \text{ g}}\right)\left(\frac{7 \text{ mol } O_2}{4 \text{ mol FeS}}\right)\left(\frac{22.4 \text{ L}}{\text{mol}}\right) = 268 \text{ L } O_2$$

66. $4\ FeS + 7\ O_2 \xrightarrow{\Delta} 2\ Fe_2O_3 + 4\ SO_2$

$$(600. \text{ g FeS})\left(\frac{1 \text{ mol}}{87.91 \text{ g}}\right)\left(\frac{4 \text{ mol } SO_2}{4 \text{ mol FeS}}\right)\left(\frac{22.4 \text{ L}}{\text{mol}}\right) = 153 \text{ L } SO_2$$

67. (a)

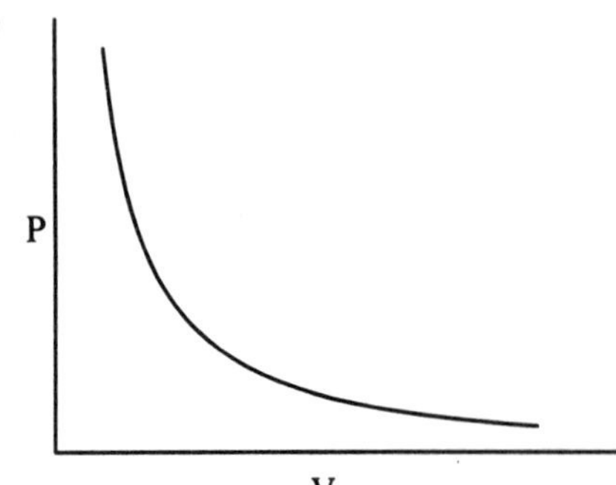

(b)

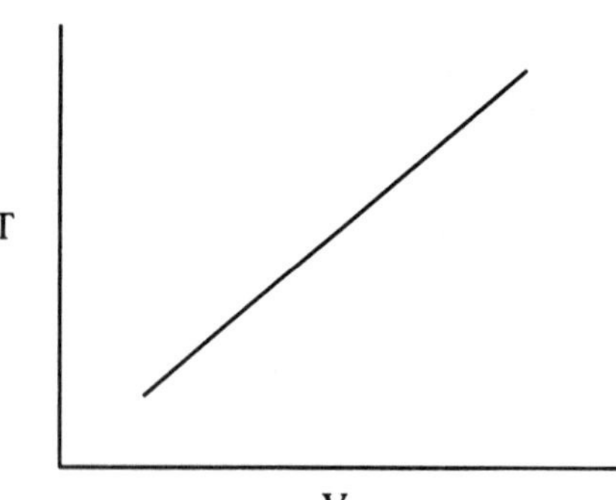

(c)

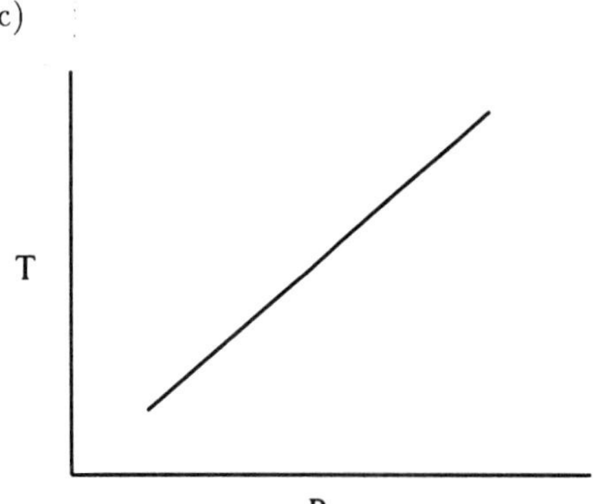

(d)

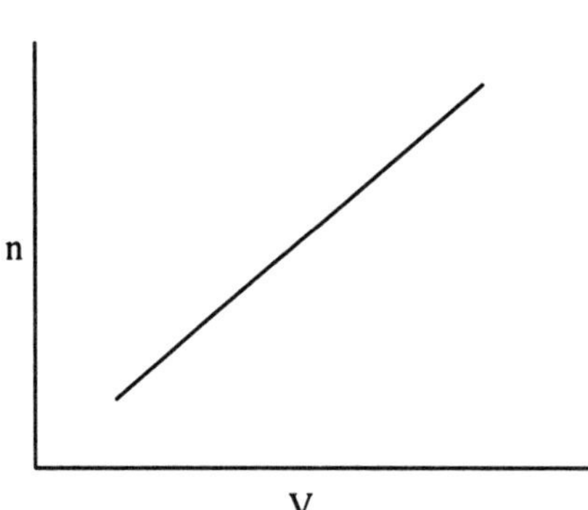

68. The can is a sealed unit and very likely still contains some of the aerosol. As the can is heated, pressure builds up in it eventually causing the can to explode and rupture with possible harm from flying debris.

69. One mole of an ideal gas occupies 22.4 liters at standard conditions. (0°C and 1 atm pressure)

$PV = nRT$

$(1.00 \text{ atm})(V) = (1.00 \text{ mol})(0.0821 \text{ L atm/mol K})(273 \text{ K})$

$V = 22.4 \text{ L}$

70. Solve for volume using $PV = nRT$

(a) $V = \dfrac{(0.2 \text{ mol } Cl_2)(0.0821 \text{ L atm/mol K})(321 \text{ K})}{(80 \text{ cm}/76 \text{ cm}) \text{ atm}} = 5 \text{ L } Cl_2$

(b) $V = \dfrac{(4.2 \text{ g } NH_3)\left(\dfrac{1 \text{ mol}}{17.03 \text{ g}}\right)\left(\dfrac{0.0821 \text{ L atm}}{\text{mol K}}\right)(161 \text{ K})}{0.65 \text{ atm}} = 5.0 \text{ L } NH_3$

(c) Assume 25°C for room temperature

$$V = \frac{(21 \text{ g } SO_3)\left(\frac{1 \text{ mol}}{80.06 \text{ g}}\right)\left(\frac{0.0821 \text{ L atm}}{\text{mol K}}\right)(298 \text{ K})}{\frac{110 \text{ kPa}}{101.3 \frac{\text{kPa}}{\text{atm}}}} = 5.9 \text{ L } SO_3$$

21 g SO_3 has the greatest volume

71. Assume 1 mol of each gas

(a) SF_6 = 146.1 g/mol

$$d = \left(\frac{146.1 \text{ g}}{\text{mol}}\right)\left(\frac{1 \text{ mol}}{22.4 \text{ L}}\right) = 6.52 \text{ g/L}$$

(b) Assume 25°C and 1 atm pressure

$$V(\text{at } 25°\text{C}) = (22.4 \text{ L})\left(\frac{298 \text{ K}}{273 \text{ K}}\right) = 24.5 \text{ L}$$

C_2H_6 = 30.07 g/mol

$$d = \left(\frac{30.07 \text{ g}}{\text{mol}}\right)\left(\frac{1 \text{ mol}}{24.5 \text{ L}}\right) = 1.23 \text{ g/L}$$

(c) He at –80°C and 2.15 atm

$$V = \frac{(1 \text{ mol})(0.0821 \text{ L atm/mol K})(193 \text{ K})}{2.15 \text{ atm}} = 7.37 \text{ L}$$

$$d = \left(\frac{4.003 \text{ g}}{\text{mol}}\right)\left(\frac{1 \text{ mol}}{7.37 \text{ L}}\right) = 0.543 \text{ g/L}$$

SF_6 has the greatest density

72. (a) Empirical formula. Assume 100. g starting material

$$\frac{80.0 \text{ g C}}{12.01 \text{ g/mol}} = 6.66 \text{ mol C} \qquad \frac{6.66}{6.66} = 1$$

$$\frac{20.0 \text{ g H}}{1.008 \text{ g/mol}} = 19.8 \text{ mol H} \qquad \frac{19.8}{6.66} = 2.97$$

Empirical formula = CH_3

Empirical mass = 12.01 g + 3.024 g = 15.03 g/mol

(b) Molecular formula. $\left(\frac{2.01 \text{ g}}{1.5 \text{ L}}\right)\left(\frac{22.4 \text{ L}}{\text{mol}}\right) = 30. \text{ g/mol}$ (molar mass)

$$\frac{30. \text{ g/mol}}{15.03 \text{ g/mol}} = 2;$$ Molecular formula is C_2H_6

(c) Valence electrons = 2(4) + 6 = 14

```
    H  H
    |  |
 H—C—C—H
    |  |
    H  H
```

Lewis structure

73. PV = nRT

(a) $\left(\frac{(790\text{ torr})(1\text{ atm})}{760\text{ torr}}\right)(2.0\text{L}) = (\text{n})(0.0821\text{ L atm/mol K})(298\text{ K})$

n = 0.085 mol (total moles)

(b) mol N_2 = total moles − mol O_2 − mol CO_2

$$= 0.085\text{ mol} - \frac{0.65\text{ g }O_2}{32.00\text{ g/mol}} - \frac{0.58\text{ g }CO_2}{44.01\text{ g/mol}}$$

mol N_2 = 0.085 mol − 0.020 mol O_2 − 0.013 mol CO_2 = 0.052 mol

$(0.052\text{ mol }N_2)\left(\frac{28.02\text{ g }N_2}{\text{mol}}\right) = 1.5\text{ g }N_2$

(c) $P_{O_2} = (790\text{ torr})\left(\frac{0.020\text{ mol }O_2}{0.085\text{ mol}}\right) = 1.9 \times 10^2\text{ torr}$

$P_{CO_2} = (790\text{ torr})\left(\frac{0.013\text{ mol }CO_2}{0.085\text{ mol}}\right) = 1.2 \times 10^2\text{ torr}$

$P_{N_2} = (790\text{ torr})\left(\frac{0.051\text{ mol }N_2}{0.085\text{ mol}}\right) = 4.7 \times 10^2\text{ torr}$

74. $2\ CO + O_2 \longrightarrow 2\ CO_2$

Calculate the moles of O_2 and CO_2 to find the limiting reactant.

PV = nRT

O_2: $(1.8\text{ atm})(0.500\text{ L }O_2) = (\text{n})(0.0821\text{ L atm/mol K})(288\text{ K})$

mol O_2 = 0.038 mol

CO: $\left(\frac{800\text{ torr} \times 1\text{ atm}}{760\text{ torr}}\right)(0.500\text{ L}) = (\text{n})(0.0821\text{ L atm/mol K})(333\text{ K})$

mol CO = 0.019 mole *Limiting Reactant*

0.038 mol O_2 will react with 0.076 mole CO.

$(0.019\text{ mol CO})\left(\frac{2\text{ mol }CO_2}{2\text{ mol CO}}\right)\left(\frac{22.4\text{ L}}{\text{mol}}\right) = 0.43\text{ L }CO_2 = 430\text{ mL }CO_2$

75. PV = nRT or $PV = \left(\frac{\text{g}}{\text{molar mass}}\right)RT$

$(1.4\text{ g/cm}^3 = 1.4 \times 10^3\text{ g/L}$

$$(1.3 \times 10^9 \text{ atm})(1.0 \text{ L}) = \left(\frac{1.4 \times 10^3 \text{ g}}{2.0 \text{ g/mol}}\right)(0.0821 \text{ L atm/mol K})(\text{T})$$

$$\text{T} = \frac{1.3 \times 10^9 \text{ atm})(1.0 \text{ L})(2.0 \text{ g/mol})}{(0.0821 \text{ L atm/mol K})(1.4 \times 10^3 \text{ g})} = 2.3 \times 10^7 \text{ K}$$

76. PV = nRT

(a) Assume atmospheric pressure of 14.7 lb/in.^{2} to begin with.
Total pressure in the ball = 14.7 lb/in.^{2} + 13 lb/in.^{2} = 28 lb/in.^{2}

$$(28 \text{ lb/in.}^2)\left(\frac{1 \text{ atm}}{14.7 \text{ lb/in.}^2}\right)(2.24 \text{ L}) = (\text{n})(0.0821 \text{ L atm/mol K})(293 \text{ K})$$

n = 0.18 mol air

(b) mass of air in the ball

$$\text{m} = (0.18 \text{ mol})\left(\frac{29 \text{ g}}{\text{mol}}\right) = 5.2 \text{ g air}$$

(c) Actually the pressure changes when the temperature changes. Since pressure is directly proportional to moles we can calculate the change in moles required to keep the pressure the same at 30°C as it was at 20°C.

$$(28 \text{ lb/in.}^2)\left(\frac{1 \text{ atm}}{14.7 \text{ lb/in.}^2}\right)(2.24 \text{ L}) = (\text{n})(0.0821 \text{ L atm/mol K})(303 \text{ K})$$

n = 0.17 mol of air required to keep the pressure the same at 30°C.

0.01 mol air (0.18 – 0.17) must be allowed to escape from the ball.

$$(0.01 \text{ mol air})\left(\frac{29 \text{ g}}{\text{mol}}\right) = 0.29 \text{ g or } 0.3 \text{ g air}$$

77. $\frac{P_1V_1}{T_1} = \frac{P_2V_2}{T_2}$

P_1 = 65 cm	P_2 = 1.00 atm (76 cm)
V_1 = 1.75 L	V_2 = 2.00 L
T_1 = 20°C (293K)	$T_2 = T_2$

$$T_2 = \frac{P_2V_2T_1}{P_1V_1} = \frac{(76 \text{ cm})(2.00 \text{ L})(293 \text{ K})}{(65 \text{ cm})(1.75 \text{ L})}$$

T_2 = 392 K (119°C)

78. $P_1V_1 = P_2V_2$ or $P_2 = \frac{P_1V_1}{V_2}$

$$P_2 = \frac{(1.0 \text{ atm})(2500 \text{ L})}{25 \text{ L}} = 1.0 \times 10^2 \text{ atm}$$

79. To double the volume of a gas, at constant pressure, the temperature (K) must be doubled.

$$\frac{V_1}{T_1} = \frac{V_2}{T_2} \qquad V_2 = 2\ V_1$$

$$\frac{V_1}{T_1} = \frac{2\ V_1}{T_2} \qquad T_2 = \frac{2\ V_1 T_1}{V_1} \qquad T_2 = 2\ T_1$$

$$T_2 = 2(300.\ \text{K}) = 600.\ \text{K} = 327°\text{C}$$

80. V = volume at 22°C and 740 torr

2 V = volume after change in temperature (P constant)

V = volume after change in pressure (T constant)

Since temperature is constant, $P_1V_1 = P_2V_2$ or $P_2 = \frac{P_1V_1}{V_2}$

$P_2 = (740\ \text{torr})\left(\frac{2\ V}{V}\right) = 1.5 \times 10^3$ torr (pressure to change 2 V to V)

81. Volume is constant, so $\frac{P_1}{T_1} = \frac{P_2}{T_2}$ or $T_2 = \frac{T_1P_2}{P_1}$;

$$T_2 = \frac{(500.\ \text{torr})(295\ \text{K})}{700.\ \text{torr}} = 211\ \text{K} = -62°\text{C}$$

82. The volume of the cylinder remains constant, so

$$\frac{P_1}{T_1} = \frac{P_2}{T_2} \text{ or } P_2 = \frac{P_1T_2}{T_1};$$

$$P_2 = \frac{(252\ \text{atm})(77\ \text{K})}{298\ \text{K}} = 65\ \text{atm}$$

83. The volume of the tires remain constant (until they burst), so

$$\frac{P_1}{T_1} = \frac{P_2}{T_2} \text{ or } T_2 = \frac{T_1P_2}{P_1};$$

71.0°F = 21.7°C = 295 K

$$T_2 = \frac{(44\ \text{psi})(295\ \text{K})}{30.\ \text{psi}} = 433\ \text{K} = 160°\text{C} = 320°\text{F}$$

84. $\frac{P_1V_1}{T_1} = \frac{P_2V_2}{T_2}$ or $V_2 = \frac{P_1V_1T_2}{P_2T_1}$

$$V_2 = \frac{(760\ \text{torr})(5.30\ \text{L})(343\ \text{K})}{(830\ \text{torr})(273\ \text{K})} = 6.1\ \text{L}$$

85. $\frac{P_1V_1}{T_1} = \frac{P_2V_2}{T_2}$ or $P_2 = \frac{P_1V_1T_2}{V_2T_1}$

$$P_2 = \frac{(1.00 \text{ atm})(800. \text{ mL})(303 \text{ K})}{(250. \text{ mL})(273 \text{ K})} = 3.55 \text{ atm}$$

86. Use the combined gas law $\frac{P_1V_1}{T_1} = \frac{P_2V_2}{T_2}$ or $V_2 = \frac{P_1V_1T_2}{P_2T_1}$

First calculate the volume at STP.

$$V_2 = \frac{(400. \text{ torr})(600. \text{ mL})(273 \text{ K})}{(760. \text{ torr})(313 \text{ K})} = 275 \text{ mL}$$

At STP, a mole of any gas has a volume of 22.4 L

$$(0.275 \text{ L})\left(\frac{1 \text{ mol}}{22.4 \text{ L}}\right)\left(\frac{6.022 \times 10^{23} \text{ molecules}}{1 \text{ mol}}\right) = 7.39 \times 10^{21} \text{ molecules}$$

Each molecule of N_2O contains 3 atoms, so:

$$(7.39 \times 10^{21} \text{ molecules})\left(\frac{3 \text{ atoms}}{1 \text{ molecule}}\right) = 2.22 \times 10^{22} \text{ atoms}$$

87. CO_2 $\quad \frac{(5.00 \text{ L})(500. \text{ torr})}{10.0 \text{ L}} = 250. \text{ torr}$

CH_4 $\quad \frac{(3.00 \text{ L})(400. \text{ torr})}{10.0 \text{ L}} = 120. \text{ torr}$

$$P_{total} = P_{CO_2} + P_{CH_4} = 250. \text{ torr} + 120. \text{ torr} = 370. \text{ torr}$$

88. The number of moles of a gas is proportional to pressure. (T and V constant)

$$\frac{P_1}{n_1} = \frac{P_2}{n_2}; \quad n_2 = \frac{n_1P_2}{P_1}$$

(a) $(60.0 \text{ mol } H_2)\left(\frac{850 \text{ lb/in.}^2}{1500 \text{ lb/in.}^2}\right) = 34 \text{ mol}$

(b) $(60.0 \text{ mol } H_2)\left(\frac{2.016 \text{ g}}{\text{mol}}\right) = 121 \text{ g } H_2$

89. $(0.560 \text{ L})\left(\frac{1\text{mol}}{22.4 \text{ L}}\right) = 0.0250 \text{ mol}$

$\frac{1.08 \text{ g}}{0.0250 \text{ mol}} = 43.2 \text{ g/mol (molar mass)}$

90. $(1.00 \text{ m}^3)\left(\frac{100 \text{ cm}}{1 \text{ m}}\right)^3\left(\frac{1 \text{ mL}}{1 \text{ cm}^3}\right)\left(\frac{1 \text{ L}}{1000 \text{ mL}}\right)\left(\frac{1 \text{ mol}}{22.4 \text{ L}}\right) = 44.6 \text{ mol } Cl_2$

91. $\left(\frac{1.78 \text{ g}}{\text{L}}\right)\left(\frac{22.4 \text{ L}}{\text{mol}}\right) = 39.9 \text{ g/mol (molar mass)}$

92. At STP 22.4 L of CH_4 has a mass of 16.04 g. $\frac{22.4\ L}{16.04\ g} = \frac{1.40\ L}{1.00\ g}$
Using 1.00 g as the mass of the sample:

$P_1 = 1$ atm $P_2 = 1$ atm
$V_1 = 1.40$ L $V_2 = 1.0$ L
$T_1 = 273$ K $T_2 = T_2$

Since the pressure is constant, $\frac{V_1}{T_1} = \frac{V_2}{T_2}$

$$T_2 = \frac{V_2 T_1}{V_1} = \frac{(1.00\ L)(273\ K)}{1.40\ L} = 195\ K = -78°C$$

93. (a) $V = \frac{nRT}{P} = \frac{(0.510\ mol)(0.0821\ L\ atm/mol\ K)(320.\ K)}{1.6\ atm} = 8.4\ L\ \ H_2$

(b) $n = \frac{PV}{RT} = \frac{(0.789\ atm)(16.0\ L)}{(0.0821\ L\ atm/mol\ K)(300.\ K)} = 0.513\ mol\ CH_4$

The molar mass for CH_4 is 16.04 g/mol

(16.04 g/mol)(0.513 mol) = 8.23 g CH_4

(c) $PV = nRT$, but $n = \frac{m}{M}$ where M is the molar mass and m is the mass of the gas. Thus,

$PV = \frac{mRT}{M}$. To determine density, $d = m/V$.

Solving $PV = \frac{mRT}{M}$ for $\frac{m}{V}$ produces $\frac{m}{V} = \frac{PM}{RT}$.

$$d = \frac{m}{V} = \frac{(4.00\ atm)(44.01\ g/mol)}{(0.0821\ L\ atm/mol\ K)(253\ K)} = 8.48\ g/L\ CO_2$$

(d) Since $d = \frac{m}{V} = \frac{PM}{RT}$ from part (c), solve for M (molar mass)

$$M = \frac{dRT}{P} = \frac{(2.58\ g/L)(0.0821\ L\ atm/mol\ K)(300.\ K)}{1.00\ atm} = 63.5\ g/mol\ \text{(molar mass)}$$

94. $n = \frac{PV}{RT} = \frac{(0.813\ atm)(0.215\ L)}{(0.0821\ L\ atm/mol\ K)(303\ K)} = 7.03 \times 10^{-3}$ mol

$$\text{molar mass} = \left(\frac{1.15\ g}{7.03 \times 10^{-3}\ mol}\right) = 164\ g/mol$$

95. $T = \frac{PV}{nR} = \frac{(4.15\ atm)(0.250\ L)}{(4.50\ mol)(0.0821\ L\ atm/mol\ K)} = 2.81$ K

96. $n = \frac{PV}{RT} = \frac{(0.500\ atm)(5.20\ L)}{(0.0821\ L\ atm/mol\ K)(250\ K)} = 0.13\ mol\ \ N_2$

97. $(8.30\ mol\ Al)\left(\frac{3\ mol\ H_2}{2\ mol\ Al}\right)\left(\frac{22.4\ L}{mol}\right) = 279\ L\ \ H_2$ at STP

98. $C_2H_2(g) + 2\ HF(g) \longrightarrow C_2H_4F_2(g)$

$1.0\ mol\ C_2H_2 \longrightarrow 1.0\ mol\ C_2H_4F_2$

$(5.0\ mol\ HF)\left(\frac{1\ mol\ C_2H_4F_2}{2\ mol\ HF}\right) = 2.5\ mol\ C_2H_4F_2$

C_2H_2 is the limiting reactant. 1.0 mol $C_2H_4F_2$ forms, no moles C_2H_2 remain. According to the equation, 2.0 mol HF yields 1.0 mol $C_2H_4F_2$. Therefore,

5.0 mol HF – 2.0 mol HF = 3.0 mol HF unreacted

The flask contains 1.0 mole $C_2H_4F_2$ and 3.0 mol HF when the reaction is complete.

$$P = \frac{nRT}{V} = \frac{(4.0\ mol)(0.0821\ L\ atm/mol\ K)(273\ K)}{10.0\ L} = 9.0\ atm$$

99. According to Grahams' Law of Effusion, the rates of effusion are inversely proportional to the molar mass.

$$\frac{rate\ He}{rate\ N_2} = \sqrt{\frac{molar\ mass\ N_2}{molar\ mass\ He}} = \sqrt{\frac{28.02}{4.003}} = \sqrt{7.000} = 2.646$$

Helium effuses 2.646 times faster than nitrogen.

100. (a) $\frac{rate\ CH_4}{rate\ He} = \sqrt{\frac{16.04}{4.003}} = \sqrt{4.007} = 2.002$

Helium effuses twice as fast as CH_4.

(b) x = distance He travels

$100 - x$ = distance CH_4 travels

$D_{He} = 2\ D_{CH_4}$ D = distance traveled

$x = 2(100 - x)$

$3x = 200$

$x = 66.7$ cm

The gases meet 66.7 cm from the helium end.

101. Assume 100. g of material to start with.

C $(85.7\ g)\left(\frac{1\ mol}{12.01\ g}\right) = 7.14\ mol$ $\frac{7.14}{7.14} = 1.00\ mol$

H $(14.3\ g)\left(\frac{1\ mol}{1.008\ g}\right) = 14.2\ mol$ $\frac{14.2}{7.14} = 1.99\ mol$

The empirical formula is CH_2. To determine the molecular formula, the molar mass must be known.

$\left(\frac{2.50\ g}{L}\right)\left(\frac{22.4\ L}{mol}\right) = 56.0\ g/mol$ (molar mass)

The empirical formula mass is 14.0. $\frac{56.0}{14.0} = 4$

Therefore, the molecular formula is $(CH_2)_4 = C_4H_8$

102. $2\ CO(g) + O_2(g) \longrightarrow 2\ CO_2(g)$

$(10.0 \text{ mol CO})\left(\frac{2 \text{ mol } CO_2}{2 \text{ mol CO}}\right) = 10.0 \text{ mol } CO_2 \text{ (from CO)}$

$(8.0 \text{ mol } O_2)\left(\frac{2 \text{ mol } CO_2}{1 \text{ mol } O_2}\right) = 16 \text{ mol } CO_2 \text{ (from } O_2)$

CO: the limiting reactant, O_2: in excess, 3.0 mole O_2 unreacted

(a) 10.0 mol CO react with 5.0 mol O_2

10.0 mol CO_2 and 3.0 mol O_2 are present, no CO will be present.

(b) $P = \frac{nRT}{V} = \frac{(13 \text{ mol})(0.0821 \text{ L atm/mol K})(273 \text{ K})}{10. \text{ L}} = 29 \text{ atm}$

103. $2\ KClO_3(s) \xrightarrow{\Delta} 2\ KCl(s) + 3\ O_2(g)$

$(0.25 \text{ L } O_2)\left(\frac{1 \text{ mol}}{22.4 \text{ L}}\right) = 0.011 \text{ mol } O_2$

$(0.011 \text{ mol } O_2)\left(\frac{2 \text{ mol } KClO_3}{3 \text{ mol } O_2}\right)\left(\frac{122.6 \text{ g}}{\text{mol}}\right) = 0.90 \text{ g } KClO_3$ in the sample

$\left(\frac{0.90 \text{ g}}{1.20 \text{ g}}\right)(100) = 75\%\ KClO_3$ in the mixture

104. Some ammonia gas dissolves in the water squirted into the flask, lowering the pressure inside the flask. The atmospheric pressure outside is greater than the pressure inside the flask and pushes water from the beaker up the tube and into the flask, filling the flask.

105. (a) The pressure of the helium is simply the difference between levels of Hg; 250 mm Hg (250 torr).

(b) The pressure of the oxygen is the difference between the pressure of the atmosphere and the difference in the levels of Hg.

$$\begin{aligned} P_{O_2} &= P_{atm} + 300 \text{ mm Hg} \\ &= 760 \text{ mm Hg} + 300 \text{ mm Hg} \\ &= 1060 \text{ mm Hg (1060 torr)} \end{aligned}$$

106. Each gas behaves as though it were alone in a 4.0 L system.

(a) After expansion: $P_1V_1 = P_2V_2$

For CO_2 $P_2 = \frac{P_1V_1}{V_2} = \frac{(150. \text{ torr})(3.0 \text{ L})}{4.0 \text{ L}} = 1.1 \times 10^2 \text{ torr}$

For H_2 $\qquad P_2 = \dfrac{P_1V_1}{V_2} = \dfrac{(50.\ \text{torr})(1.0\ \text{L})}{4.0\ \text{L}} = 13\ \text{torr}$

$P_{total} = P_{H_2} + P_{CO_2} = 110\ \text{torr} + 13\ \text{torr} = 120\ \text{torr (2 sig. figures)}$

107. Assume 1.00 L of air. The mass of the sample is 1.29 g.

$$\frac{P_1V_1}{T_1} = \frac{P_2V_2}{T_2}$$

$$V_2 = \frac{P_1V_1T_2}{P_2T_1} = \frac{(760\ \text{torr})(1.00\ \text{L})(290.\ \text{K})}{(450\ \text{torr})(273\ \text{K})} = 1.8\ \text{L}$$

$$d = \frac{m}{V} = \frac{1.29\ \text{g}}{1.8\ \text{L}} = 0.72\ \text{g/L}$$

108. Air enters the room. The volume is directly proportional to temperature.

Initial volume of the air in the room = $V_1 = (16\ \text{ft})(12\ \text{ft})(12\ \text{ft})$

$= 2300\ \text{ft}^3$

Final volume of the air in the room (after change in temperature)

$$V_2 = \frac{V_1T_2}{T_1} = \frac{(2300\ \text{ft}^3)(270\ \text{K})}{300\ \text{K}} = 2100\ \text{ft}^3$$

Since the volume of the air in the room is lower at the new temperature, air would enter the room.

109. $P_1 = 40.0\ \text{atm}$ $\qquad P_2 = P_2$

$V_1 = 50.0\ \text{L}$ $\qquad V_2 = 50.0\ \text{L}$

$T_1 = 25\ ^\circ\text{C} = 298\ \text{K}$ $\qquad T_2 = 25^\circ\text{C} + 152^\circ\text{C} = 177^\circ\text{C} = 450.\ \text{K}$

Gas cylinders have constant volume, so pressure and temperature vary directly.

$$P_2 = \frac{P_1T_2}{T_1} = \frac{(40.0\ \text{atm})(450.\ \text{K})}{298\ \text{K}} = 60.4\ \text{atm}$$

CHAPTER 13

WATER AND THE PROPERTIES OF LIQUIDS

1. The potential energy is greater in the liquid water than in the ice. The heat necessary to melt the ice increases the potential energy of the liquid, thus allowing the molecules greater freedom of motion.

2. At 0°C, all three substances, H_2S, H_2Se, and H_2Te, are gases, because they all have boiling points below 0°C.

3. The pressure of the atmosphere must be 1.00 atmosphere, otherwise the water would be boiling at some other temperature.

4. 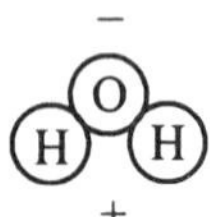

5. Melting point, boiling point, heat of fusion, heat of vapoization, density, and crystallization structure in the solid state are some of the physical properties of water that would be very different, if the the molecules were linear and nonpolar instead of bent and highly polar. For example, the boiling point, melting point, heat of fusion, and heat of vaporization would be lower because linear molecules have no dipole moment and the attraction among molecules would be much less.

6. Prefixes preceding the word hydrate are used in naming hydrates, indicating the number of molecules of water present in the formulas. The prefixes used are:

mono = 1	di = 2	tri = 3	tetra = 4	penta = 5
hexa = 6	hepta = 7	octa = 8	nona = 9	deca = 10

7. The distillation setup in Figure 13.10 would be satisfactory for separating salt and water, but not for separating ethyl alcohol and water. In the first case, the water is easily vaporized, the salt is not, so the water boils off and condenses to a pure liquid. In the second case, both substances are easily vaporized, so both would vaporize (though not to an equal degree) and the condensed liquid would contain both substances.

8. The thermometer would be at about 70°C. The liquid is boiling, which means its vapor pressure equals the confining pressure. From Table 13.1, we find that ethyl alcohol has a vapor pressure of 543 torr at 70°C.

9. The water in both containers would have the same vapor pressure, for it is a function of the temperature of the liquid.

10. In Figure 13.1, it would be case (b) in which the atmosphere would reach saturation. The pressure exerted by water vapor molecules would be less in the open container, because the opportunity to escape prevents the build-up of vapor pressure.

11. If ethyl ether and ethyl alcohol were both placed in a closed container, (a) both substances would be present in the vapor, for both are volatile liquids: (b) ethyl ether would have more molecules in the vapor because it has a higher vapor pressure at a given temperature.

12. The vapor pressure observed in (c) would remain unchanged. The presence of more water in (b) does not change the magnitude of the vapor pressure of the water. The temperature of the water determines the magnitude of the vapor pressure.

13. At 30 torr, H_2O would boil at approximately 29°C, ethyl alcohol at 14°C, and ethyl ether and ethyl chloride at some temperature below 0°C.

14. (a) At a pressure of 500 torr, water boils at 88°C.

(b) The normal boiling point of ethyl alcohol is 78°C.

(c) At a pressure of 0.50 atm (380 torr), ethyl ether boils at 16°C.

15. Based on Figure 13.5:

(a) Line BC is horizontal because the temperature remains constant during the entire process of melting. The energy input is absorbed in changing from the solid to the liquid state.

(b) During BC, both solid and liquid phases are present.

(c) The line DE represents the change from liquid water to steam (vapor) at the boiling temperature of water.

16. Physical properties of water:

(a) melting point, 0°C

(b) boiling point, 100°C (at 1 atm pressure)

(c) colorless

(d) odorless

(e) tasteless

(f) heat of fusion, 335 J/g (80 cal/g)

(g) heat of vaporization, 2.26 kJ/g (540 cal/g)

(h) density = 1.0 g/mL (at 4°C)

(i) specific heat = 4.184 J/g°C

17. For water, to have its maximum density, the temperature must be 4°C, and the pressure sufficient to keep it liquid. $d = 1.0$ g/mL

18. Apply heat to an ice-water mixture, the heat energy is absorbed to melt the ice, rather than warm the water, so the temperature remains constant until all the ice has melted.

19. Ice at 0°C contains less heat than water at 0°C. Heat must be added to convert ice to water, so the water will contain that much additional heat energy.

20. Ice floats in water because it is less dense than water. The density of ice at 0°C is 0.915 g/mL. Liquid water, however, has a density of 1.00 g/mL. Ice will sink in ethyl alcohol, which has a density of 0.789 g/mL.

21. The heat of vaporization of water would be lower if water molecules were linear instead of bent. If linear, the molecules of water would be nonpolar. The relatively high heat of vaporization of water is a result of the molecule being highly polar and having strong dipole-dipole hydrogen bonding attraction for other water molecules.

22. Ethyl alcohol exhibits hydrogen bonding; ethyl ether does not. This is indicated by the high heat of vaporization of ethyl alcohol, even though its molar mass is much less than the molar mass of ethyl ether.

23. A linear water molecule, being nonpolar, would exhibit less hydrogen bonding than the highly polar, bent, water molecule. The polar molecule has a greater intermolecular attractive force than a nonpolar molecule.

24. Ammonia exhibits hydrogen bonding; methane does not. The ammonia molecule is polar; the methane molecule is not. The nitrogen atom in ammonia is quite electronegative; the carbon atom in methane is much less electronegative.

25. Water, at 80°C, will have fewer hydrogen bonds than water at 40°C. At the higher temperature, the molecules of water are moving faster than at the lower temperature. This results in less hydrogen bonding at the higher temperature.

26. $H_2NCH_2CH_2NH_2$ has two polar NH_2 groups. It should, therefore, show more hydrogen bonding and a higher boiling point (117°C) versus 49°C for $CH_3CH_2CH_2NH_2$.

27. Rubbing alcohol feels cold when applied to the skin, because the evaporation of the alcohol absorbs heat from the skin. The alcohol has a fairly high vapor pressure (low boiling point) and evaporates quite rapidly. This produces the cooling effect.

28. (a) Order of increasing rate of evaporation: Mercury, acetic acid, water, toluene, benzene, carbon tetrachloride, methyl alcohol, bromine.

(b) Highest boiling point is mercury. Lowest boiling point is bromine.

29. Water boils when its vapor pressure equals the prevailing atmospheric pressure over the water. In order for water to boil at 50°C, the pressure over the water would need to be reduced to a point equal to the vapor pressure of the water (92.5 torr).

30. In a pressure cooker, the temperature at which the water boils increases above its normal boiling point, because the water vapor (steam) formed by boiling cannot escape. This results in an increased pressure over water and, consequently, an increased boiling temperature.

31. Vapor pressure varies with temperature. The temperature at which the vapor pressure of a liquid equals the prevailing pressure is the boiling point of the liquid.

32. As temperature increases, molecular velocities increase. At higher velocities, it becomes easier for molecules to break away from the attractive forces in the liquid.

33. Water has a relatively high boiling point because there is a high attraction between molecules due to hydrogen bonding.

34. Ammonia would have a higher vapor pressure than SO_2 at –40°C because it has a lower boiling point (NH_3 is more volatile then SO_2).

35. As the temperature of a liquid increases, the kinetic energy of the molecules as well as the vapor pressure of the liquid increases. When the vapor pressure of the liquid equals the external pressure, boiling begins with many of the molecules having enough energy to escape from the liquid. Bubbles of vapor are formed throughout the liquid and these bubbles rise to the surface, escaping as boiling continues.

36. HF has a higher boiling point that HCl because of the strong hydrogen bonding in HF (F is the most electronegative element). Neither F_2 nor Cl_2 will have hydrogen bonding, so the compound, F_2, with the smaller mass, has the lower boiling point.

37. The boiling liquid remains at constant temperature because the added heat energy is being used to convert the liquid to a gas, i.e., to supply the heat of vaporization for the liquid at its boiling point.

38. 34.6°C, the boiling point of ether. (See Table 13.2)

39. The lake freezes from the top down because, as the temperature drops to freezing or below, the water on the surface tends to cool faster than the water that lies deeper. As the surface water freezes, the ice formed floats because the ice is less dense than the liquid water below it.

40. If the lake is in an area where the temperature is below freezing for part of the year, the expected temperature would be 4°C at the bottom of the lake. This is because the surface water would cool to 4°C (maximum density) and sink.

41. Reactions of metals with water:

$2\ Al + 3\ H_2O$ (steam) $\longrightarrow 3\ H_{2(g)} + Al_2O_3$ requires steam

$Ca + 2\ H_2O \longrightarrow H_{2(g)} + Ca(OH)_2$ slowly at room temperature

$3\ Fe + 4\ H_2O$ (steam) $\longrightarrow 4\ H_{2(g)} + Fe_3O_4$ requires steam

$2\ Na + 2\ H_2O \longrightarrow H_{2(g)} + 2\ NaOH$ vigorously at room temp

$Zn + H_2O$ (steam) $\longrightarrow H_{2(g)} + ZnO$ requires steam

42. The formation of hydrogen and oxygen from water is an endothermic reaction, due to the following evidence:

(a) Energy must be continually be provided to the system for the reaction to proceed. The reaction will cease when the energy source is removed.

(b) The reverse reaction, burning hydrogen in oxygen, releases energy as heat.

43. (a) The word anhydride originates from the Greek, anhydrous, meaning waterless. An anhydride is a substance that will react with water to form an acid or base.

(b) An acid anhydride will be an oxide of a nonmetal.

(c) A basic anhydride will be an oxide of a metal.

44. The correct statements are a, b, c, f, h, l, m, o, p, s, t, w.

(d) The changing of ice into water is an endothermic process.

(e) Water and hydrogen fluoride are both polar molecules.

(g) $2\ H_2O_2 \longrightarrow 2\ H_2O + O_2$ represents a balanced equation for the decomposition of hydrogen peroxide.

(i) The density of water is dependent on temperature.

(j) Liquid A boils at a lower temperature than liquid B. This fact indicates that liquid A has a higher vapor pressure than liquid B at any particular temperature.

(k) Water boils at a lower temperature in the mountains than at sea level.

(n) The normal boiling temperature of water is 100°C.

(q) Calcium oxide reacts with water to form calcium hydroxide.

(r) Carbon dioxide is the anhydride of carbonic acid.

(u) Distillation is effective for softening water because the pure water is distilled, leaving the minerals behind.

(v) Disposal of toxic industrial wastes in toxic waste dumps has been found to be an unsatisfactory long-term solution to the problem.

(x) $CaCl_2 \cdot 2\ H_2O$ has a higher percentage of water than $BaCl_2 \cdot 2\ H_2O$.

45. Acid anyhdride: [H_2SO_3, SO_2] [H_2SO_4, SO_3] [HNO_3, N_2O_5]

46. Acid anhydride: [$HClO_4$, Cl_2O_7] [H_2CO_3, CO_2] [H_3PO_4, P_2O_5]

47. Basic anhydrides: [LiOH Li_2O] [NaOH, Na_2O] [$Mg(OH)_2$, MgO]

48. Basic anhydrides: [KOH, K_2O] [$Ba(OH)_2$, BaO] [$Ca(OH)_2$, CaO]

49. (a) $Ba(OH)_2 \xrightarrow{\Delta} BaO + H_2O$

(b) $2\ CH_3OH + 3\ O_2 \longrightarrow 2\ CO_2 + 4\ H_2O$

(c) $2\ Rb + 2\ H_2O \longrightarrow 2\ RbOH + H_2$

(d) $SnCl_2 \cdot 2\ H_2O \xrightarrow{\Delta} SnCl_2 + 2\ H_2O$

(e) $HNO_3 + NaOH \longrightarrow NaNO_3 + H_2O$

(f) $CO_2 + H_2O \longrightarrow H_2CO_3$

50. (a) $Li_2O + H_2O \longrightarrow 2\ LiOH$

(b) $2\ KOH \xrightarrow{\Delta} K_2O + H_2O$

(c) $Ba + 2\ H_2O \longrightarrow Ba(OH)_2 + H_2$

(d) $Cl_2 + H_2O \longrightarrow HCl + HClO$

(e) $SO_3 + H_2O \longrightarrow H_2SO_4$

(f) $H_2SO_3 + 2\ KOH \longrightarrow K_2SO_3 + 2\ H_2O$

51. (a) barium bromide dihydrate

(b) aluminum chloride hexahydrate

(c) iron(III) phosphate tetrahydrate

52. (a) magnesium ammonium phosphate hexahydrate

(b) iron(II) sulfate heptahydrate

(c) tin(IV) chloride pentahydrate

53. Deionized water is water from which the ions have been removed.

(a) Hard water contains dissolved calcium and magnesium salts.

(b) Soft water is free of ions that cause hardness (Ca^{2+} and Mg^{2+}) but it may contain other ions such as Na^+ and K^+.

54. Deionized water is water from which the ions have been removed.

(a) Distilled water has been vaporized by boiling and recondensed. It is free of nonvolatile impurities, but may still contain any volatile impurities that were initially present in the water.

(b) Natural waters are generally not pure, but contain dissolved minerals and suspended matter, and can even contain harmful bacteria.

55. $(100.\text{ g CoCl}_2 \cdot 6\text{ H}_2\text{O})\left(\frac{1\text{ mol}}{238.0\text{ g}}\right) = 0.420\text{ mol CoCl}_2 \cdot 6\text{ H}_2\text{O}$

56. $(100.\text{ g FeI}_2 \cdot 4\text{ H}_2\text{O})\left(\frac{1\text{ mol}}{381.7\text{ g}}\right) = 0.262\text{ mol FeI}_2 \cdot 4\text{ H}_2\text{O}$

57. $(100.\text{ g CoCl}_2 \cdot 6\text{ H}_2\text{O})\left(\frac{1\text{ mol}}{238.0\text{ g}}\right)\left(\frac{6\text{ mol H}_2\text{O}}{1\text{ mol CoCl}_2 \cdot 6\text{ H}_2\text{O}}\right) = 2.52\text{ mol H}_2\text{O}$

58. $(100.\text{ g FeI}_2 \cdot 4\text{ H}_2\text{O})\left(\frac{1\text{ mol}}{381.7\text{ g}}\right)\left(\frac{4\text{ mol H}_2\text{O}}{1\text{ mol FeI}_2 \cdot 4\text{ H}_2\text{O}}\right) = 1.05\text{ mol H}_2\text{O}$

59. Assume 1 mol of the compound.

$$\left(\frac{\text{g H}_2\text{O}}{\text{g MgSO}_4 \cdot 7\text{ H}_2\text{O}}\right)(100) = \left(\frac{(7)(18.02\text{ g})}{246.5\text{ g}}\right)(100) = 51.2\%\ \text{H}_2\text{O}$$

60. Assume 1 mol of hydrate.

$$\%\ \text{H}_2\text{O} = \frac{\text{g H}_2\text{O}}{\text{g Al}_2(\text{SO}_4)_3 \cdot 18\text{ H}_2\text{O}} = \left(\frac{(18)(18.02\text{ g})}{666.5\text{ g}}\right)(100) = 48.7\%\ \text{H}_2\text{O}$$

61. Assume 100. g of the compound.

$$(0.142)(100.\text{ g}) = 14.2\text{ g H}_2\text{O}$$
$$(0.858)(100.\text{ g}) = 85.8\text{ g Pb(C}_2\text{H}_3\text{O}_2)_2$$
$$(14.2\text{ g H}_2\text{O})\left(\frac{1\text{ mol}}{18.02\text{ g}}\right) = 0.788\text{ mol H}_2\text{O}$$
$$(85.8\text{ g Pb(C}_2\text{H}_3\text{O}_2)_2)\left(\frac{1\text{ mol}}{325.3\text{ g}}\right) = 0.264\text{ mol Pb(C}_2\text{H}_3\text{O}_2)_2$$

In the formula for the hydrate, there is one mole of $\text{Pb(C}_2\text{H}_3\text{O}_2)_2$, so divide each of the moles by 0.264.

$$\frac{0.264\text{ mol Pb(C}_2\text{H}_3\text{O}_2)_2}{0.264\text{ mol}} = 1\text{ Pb(C}_2\text{H}_3\text{O}_2)_2$$

$$\frac{0.788\text{ mol H}_2\text{O}}{0.264\text{ mol}} = 2.98\text{ H}_2\text{O}$$

Therefore, the formula is $\text{Pb(C}_2\text{H}_3\text{O}_2)_2 \cdot 3\text{ H}_2\text{O}$.

62. $25.0\text{ g hydrate} - 16.9\text{ g FePO}_4 = 8.1\text{ g H}_2\text{O}$

$(8.1\text{ g H}_2\text{O})\left(\frac{1\text{ mol}}{18.02\text{ g}}\right) = 0.45\text{ mol H}_2\text{O}$ $\qquad \frac{0.45}{0.112} = 4.0$

$(16.9\text{ g FePO}_4)\left(\frac{1\text{ mol}}{150.8\text{ g}}\right) = 0.112\text{ mol FePO}_4$ $\qquad \frac{0.112}{0.112} = 1.00$

The formula is $\text{FePO}_4 \cdot 4\text{ H}_2\text{O}$

63. (a) Heat water $20.^\circ C \longrightarrow 100.^\circ C$

$$E_a = (m)(\text{specific heat})(\Delta T) = (120.\ g)\left(\frac{4.184\ J}{g^\circ C}\right)(80.^\circ C) = 4.0 \times 10^4\ J$$

(b) Convert water to steam

$$E_b = (m)(\text{heat of vaporization}) = (120.\ g)(2.26 \times 10^3\ J/g) = 2.71 \times 10^5\ J$$

$$E_{total} = E_a + E_b = (4.0 \times 10^4\ J) + (2.71 \times 10^5\ J) = 3.11 \times 10^5\ J$$

64. (a) Cool water $24^\circ C \longrightarrow 0^\circ C$

$$E_a = (m)(\text{specific heat})(\Delta T) = (126\ g)\left(\frac{4.184\ J}{g\ ^\circ C}\right)(24^\circ C) = 1.3 \times 10^4\ J$$

(b) Convert water to ice

$$E_b = (m)(\text{heat of fusion}) = (126\ g)(335\ J/g) = 4.22 \times 10^4\ J$$

$$E_{total} = E_a + E_b = (1.3 \times 10^4\ J) + (4.22 \times 10^4\ J) = 5.5 \times 10^4\ J$$

65. Energy released in cooling water

$$E = (m)(\text{specific heat})(\Delta T) = (300.\ g)\left(\frac{1\ cal}{g\ ^\circ C}\right)(25^\circ C) = 7.5 \times 10^3\ cal$$

Energy required to melt ice

$$E = (m)(\text{heat of fusion}) = (100\ g)(80\ cal/g) = 8.0 \times 10^3\ cal$$

Less energy is released in cooling the water than is required to melt the ice. Ice will remain and the water will be at 0°C.

66. Energy to heat water = energy to condense steam

$$(300.\ g)\left(\frac{1\ cal}{g\ ^\circ C}\right)(100.^\circ C - 25^\circ C) = (m)(540\ cal/g)$$

42 g = m (grams of steam to heat the water to 100.°C)

42 g of steam are required to heat 300. g of water to 100.°C. Since only 35 g of steam are added to the system, the final temperature will be less than 100.°C. Not sufficient steam.

67. Energy lost by warm water = energy gained by the ice

x = final temperature

$$\text{mass}(H_2O) = (1.5\ L\ H_2O)\left(\frac{1000\ mL}{L}\right)\left(\frac{1.0\ g}{mL}\right) = 1500\ g$$

$$(1500\ g)\left(\frac{1\ cal}{g^\circ C}\right)(75^\circ C - x) = (75\ g)(80.\ \tfrac{cal}{g}) + (75\ g)\left(\frac{1\ cal}{g^\circ C}\right)(x - 0^\circ C)$$

$$(112{,}500\ cal) - (1500x\ cal/^\circ C) = 6.0 \times 10^3\ cal + 75x\ cal/^\circ C$$

$$106{,}500\ cal = 1575x\ cal/^\circ C$$

$$68^\circ C = x$$

68. E = (m)(heat of fusion)

(500. g)(335 J/g) = 167,000 J needed to melt the ice

9560 J < 167,500 J

Since 167,500 J are required to melt all the ice, and only 9560 J are available, the system will be at 0°C. It will be a mixture of ice and water.

69. (a) $2\ K + 2\ H_2O \longrightarrow 2\ KOH + H_2$

$$(1.00\ \text{mol K})\left(\frac{2\ \text{mol}\ H_2O}{2\ \text{mol K}}\right)\left(\frac{18.02\ \text{g}}{\text{mol}}\right) = 18.0\ \text{g}\ H_2O$$

(b) $Ca + 2\ H_2O \longrightarrow Ca(OH)_2 + H_2$

$$(1.00\ \text{mol Ca})\left(\frac{2\ \text{mol}\ H_2O}{1\ \text{mol Ca}}\right)\left(\frac{18.02\ \text{g}}{\text{mol}}\right) = 36.0\ \text{g}\ H_2O$$

(c) $SO_3 + H_2O \longrightarrow H_2SO_4$

$$(1.00\ \text{mol}\ SO_3)\left(\frac{1\ \text{mol}\ H_2O}{1\ \text{mol}\ SO_3}\right)\left(\frac{18.02\ \text{g}}{\text{mol}}\right) = 18.0\ \text{g}\ H_2O$$

70. (a) $2\ Na + 2\ H_2O \longrightarrow 2\ NaOH + H_2$

$$(1.00\ \text{g Na})\left(\frac{1\ \text{mol}}{22.99\ \text{g}}\right)\left(\frac{2\ \text{mol}\ H_2O}{2\ \text{mol Na}}\right)\left(\frac{18.02\ \text{g}}{\text{mol}}\right) = 0.784\ \text{g}\ H_2O$$

(b) $MgO + H_2O \longrightarrow Mg(OH)_2$

$$(1.00\ \text{g MgO})\left(\frac{1\ \text{mol}}{40.31\ \text{g}}\right)\left(\frac{1\ \text{mol}\ H_2O}{1\ \text{mol MgO}}\right)\left(\frac{18.02\ \text{g}}{\text{mol}}\right) = 0.447\ \text{g}\ H_2O$$

(c) $N_2O_5 + H_2O \longrightarrow 2\ HNO_3$

$$(1.00\ \text{g}\ N_2O_5)\left(\frac{1\ \text{mol}}{108.0\ \text{g}}\right)\left(\frac{1\ \text{mol}\ H_2O}{1\ \text{mol}\ N_2O_5}\right)\left(\frac{18.02\ \text{g}}{\text{mol}}\right) = 0.167\ \text{g}\ H_2O$$

71. Steam molecules will cause a more severe burn. Steam molecules contain more energy at 100°C than water molecules at 100°C due to the energy absorbed during the vaporization stage (heat of vaporization).

72. The alcohol has a higher vapor pressure than water and thus evaporates faster than water. When the alcohol evaporates it absorbs energy from the water, cooling the water. Eventually the water will lose enough energy to change from a liquid to a solid (freeze).

73. When one leaves the swimming pool, water starts to evaporate from the skin of the body. Part of the energy needed for the evaporation is absorbed from the skin, resulting in the cooling feeling.

74.

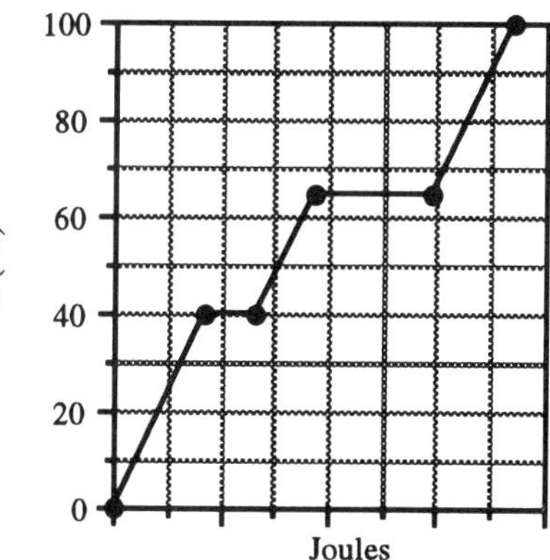

(a) From 0°C to 40.°C solid X warms until at 40.°C it begins to melt. The temperature remains at 40.°C until all of X is melted. After that, liquid X will warm steadily to 65°C where it will boil and remain at 65°C until all of the liquid becomes vapor. Beyond 65°C, the vapor will warm steadily until 100°C.

(b)

Joules needed (0°C to 40°C)	= (60. g)(3.5 J/g°C)(40. °C)	=	8400 J
Joules needed at 40°C	= (60. g)(80. J/g)	=	4800 J
Joules needed (40°C to 65°C)	= (60. g)(3.5 J/g°C)(25°C)	=	5300 J
Joules needed at 65°C	= (60. g)(190 J/g)	=	11,000 J
Joules needed (65°C to 100°C)	= (60. g)(3.5 J/g°C)(35°C)	=	7400 J
Total Joules needed			37,000 J

75. As the temperature of a liquid increases, the molecules gain kinetic energy thereby increasing their escaping tendency (vapor pressure).

76. Since boiling occurs when vapor pressure equals atmospheric pressure, the graph in Figure 13.4 indicates that water will boil at about 75°C at 270 torr pressure.

77. $CuSO_4$ (anhydrous) is gray white. When exposed to moisture, it turns bright blue ($CuSO_4 \cdot 5\ H_2O$). The color change is an indicator of moisture in the environment.

78. $MgSO_4 \cdot 7\ H_2O$ $\quad$ $Na_2HPO_4 \cdot 12\ H_2O$

79. Soap can soften hard water by forming a precipitate with, and thus removing, the calcium and magnesium ions. This precipitate is a greasy scum and is very objectional, so it is a poor way to soften water.

80. Chlorine is commonly used to destroy bacteria in water. Ozone and ultraviolet radiation are also used in some places.

81. Ozone, O_3

82. When organic pollutants in water are oxidized by dissolved oxygen, there may not be sufficient dissolved oxygen to sustain marine life, such as fish. Most marine life forms depend on dissolved oxygen for cellular respiration.

83. Liquids that are stored in ceramic containers should never be drunk, for they are likely to have dissolved some of the lead from the ceramic. If the ceramic is glazed, the liquid is less apt to dissolve lead from the ceramic.

84. Na_2 zeolite*(s)* + Mg^{2+}*(aq)* $\longrightarrow$ Mg zeolite*(s)* + 2 Na^+*(aq)*

85. Softening of hard water using sodium carbonate:

$CaCl_2(aq) + Na_2CO_3(aq) \longrightarrow CaCO_3(s) + 2\ NaCl(aq)$

86. Humectants are polar compounds while emollients are nonpolar compounds.

87. Humectants and emollients soften the skin by increasing its water content. Humectants attract water vapor from the air, while emollients cover the skin with a layer of material which is immiscible with water.

88. The triangle theory of sweetness states that molecules which are sweet contain 3 specific sites which produce the proper structure to attach to the taste buds and trigger the "sweet" response.

89. A good sweetener is:

1. as sweet or sweeter than sugar (sucrose)
2. nontoxic
3. quick to register sweet on the taste buds
4. easy to release
5. noncaloric
6. stable when cooked or dissolved
7. inexpensive

90. (a) Melt ice: E_a = (m)(heat of fusion) = (225 g)(80. cal/g) = 18,000 cal

(b) Warm the water: E_b = (m)(specific heat)(ΔT) = $(225\ g)\left(\frac{1\ cal}{g\ °C}\right)(100.°C)$ = 22,500 cal

(c) Vaporize water: E_c = (m)(heat of vaporization) = (225 g)(540 cal/g) = 121,500 cal

$E_{total} = E_a + E_b + E_c = 1.6 \times 10^5$ cal

91. $(2.26\ kJ/g)\left(\frac{18.02\ g}{mol}\right) = 40.7\ kJ/mol$

92. $E = (m)(\text{specific heat})(\Delta T) = (250.\ g)\left(\frac{0.096\ cal}{g\ °C}\right)(150. - 20.0°C) = 3.1 \times 10^3\ cal$ (3.1 kcal)

93. Heat lost by warm water = heat gained by ice

m = grams of ice to lower temperature of water to 0.0°C.

$$(120.\ g)\left(\frac{1\ cal}{g\ °C}\right)(45°C - 0.0°C) = (m)(80\ cal/g)$$

68 g = m (grams of ice melted)

68 g of ice melted. Therefore, 150. g – 68 g = 82 g ice remains.

94. Energy liberated when steam at 100.0°C condenses to water at 100.0°C

$$(50.0 \text{ mol steam})\left(\frac{18.02 \text{ g}}{\text{mol}}\right)\left(\frac{2.26 \text{ kJ}}{\text{g}}\right)\left(\frac{1000 \text{ J}}{\text{kJ}}\right) = 2.04 \times 10^6 \text{ J}$$

Energy liberated in cooling water from 100.0°C to 30.0°C

$$(50.0 \text{ mol } H_2O)\left(\frac{18.02 \text{ g}}{\text{mol}}\right)\left(\frac{4.184 \text{ J}}{\text{g°C}}\right)(100.0°\text{C} - 30.0°\text{C}) = 2.64 \times 10^5 \text{ J}$$

Total energy liberated

$$2.04 \times 10^6 \text{ J} + 2.64 \times 10^5 \text{ J} = 2.30 \times 10^6 \text{ J}$$

95. Energy to warm the ice from −10.0°C to 0°C

$$(100.\text{ g})\left(\frac{2.01 \text{ J}}{\text{g °C}}\right)(10.0°\text{C}) = 2010 \text{ J}$$

Energy to melt the ice at 0°C

$$(100.\text{ g})(335 \text{ J/g}) = 33{,}500 \text{ J}$$

Energy to heat the water from 0°C to 20.0°C

$$(100.\text{ g})\left(\frac{4.184 \text{ J}}{\text{g °C}}\right)(20.0°\text{C}) = 8370 \text{ J}$$

$$E_{total} = 2010 \text{ J} + 33{,}500 \text{ J} + 8370 \text{ J} = 4.39 \text{ x } 10^4 \text{ J} = 43.9 \text{ kJ}$$

96. $2\ H_2O \longrightarrow 2\ H_2 + O_2$

$$(25.0 \text{ L } O_2)\left(\frac{1 \text{ mol}}{22.4 \text{ L}}\right)\left(\frac{2 \text{ mol } H_2O}{1 \text{ mol } O_2}\right)\left(\frac{18.02 \text{ g}}{\text{mol}}\right) = 40.2 \text{ g } H_2O$$

97. Liquid water has a density of 1.00 g/mL.

$$d = \frac{m}{V} \qquad V = \frac{m}{d} = \frac{18.02 \text{ g}}{1.00 \text{ g/mL}} = 18.0 \text{ mL}$$

1.00 mol of water vapor at STP has a volume of 22.4 L (gas)

98. $\left(\frac{1.00 \text{ mol } H_2O}{\text{day}}\right)\left(\frac{6.022 \text{ x } 10^{23} \text{ molecules}}{\text{mol}}\right)\left(\frac{1.00 \text{ day}}{24 \text{ hr}}\right)\left(\frac{1 \text{ hr}}{60 \text{ min}}\right)\left(\frac{1 \text{ min}}{60 \text{ s}}\right) = 6.97 \times 10^{18} \text{ molecules/s}$

99. Mass solution − mass H_2O = mass H_2SO_4

$$(122 \text{ mL})(1.26 \text{ g/mL}) - (100.\text{ mL})(1.00 \text{ g/mL}) = 54 \text{ g } H_2SO_4$$

100. $2\ H_2 + O_2 \longrightarrow 2\ H_2O$

(a) $(80.0 \text{ mL } H_2)\left(\frac{1 \text{ mL } O_2}{2 \text{ mL } H_2}\right) = 40.0 \text{ mL } O_2$

Since 60.0 mL of O_2 are available, some oxygen remains unreacted.

(b) 60.0 mL − 40.0 mL = 20.0 mL O_2 unreacted.

101. Energy absorbed by the student when steam at 100.°C changes to water at 100.°C

$(1.5 \text{ g steam})\left(\frac{2.26 \text{ kJ}}{\text{g}}\right) = 3.4 \text{ kJ } (3.4 \times 10^3 \text{ J})$

Energy absorbed when water cools from 100.°C to 20.0°C

E = (m)(specific heat(ΔT)

$(1.5 \text{ g})\left(\frac{4.184 \text{ J}}{\text{g}^\circ\text{C}}\right)(100.^\circ\text{C} - 20.0^\circ\text{C}) = 5.0 \times 10^2 \text{ J}$

$\text{Total} = 3.4 \times 10^3 \text{ J} + 5.0 \times 10^2 \text{ J} = 3.9 \times 10^3 \text{ J}$

CHAPTER 14

SOLUTIONS

1. (a)

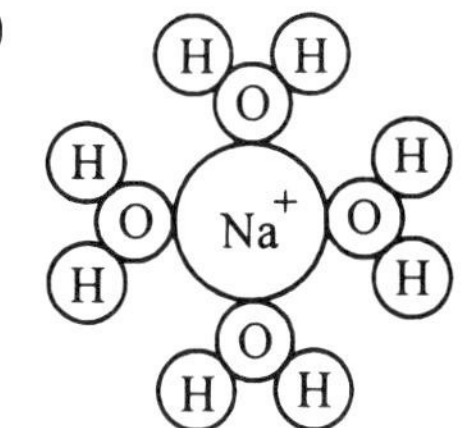

(b)

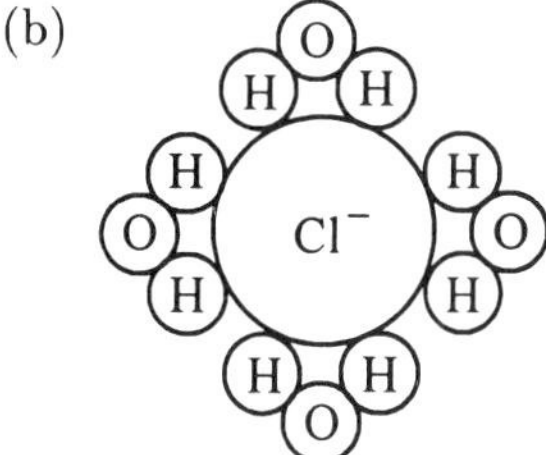

These diagrams are intended to illustrate the orientation of the water molecules about the ions, not the number of water molecules.

2. From Table 14.3, approximately 4.5 g of NaF would be soluble in 100 g of water at 50°C.

3. From Figure 14.3, solubilities in water at 25°C are:

   (a) KCl 35 g/100 g H_2O
   (b) $KClO_3$ 9 g/100 g H_2O
   (c) KNO_3 39 g/100 g H_2O

4. Potassium fluoride has a relatively high solubility when compared to lithium or sodium fluoride. For lithium and sodium halides, the order of solubility (in order of increasing solubilities) is:

   F^- Cl^- Br^- I^-

   For potassium halides, the order of increasing solubilities is:

   Cl^- Br^- F^- I^-

5. (a) $KClO_3$ at 60°C, 25 g
   (b) HCl at 20°C, 72 g
   (c) Li_2SO_4 at 80°C, 31 g
   (d) KNO_3 at 0°C, 14 g

6. KNO_3

7. A one molal solution in camphor will show a greater freezing point depression than a 2 molal solution in benzene.

$$\Delta t_f = \left(\frac{1 \text{ mol solute}}{\text{kg camphor}}\right)\left(\frac{40^\circ\text{C kg camphor}}{\text{mol solute}}\right) = 40^\circ\text{C (freezing point depression)}$$

$$\Delta t_f = \left(\frac{2 \text{ mol solute}}{\text{kg benzene}}\right)\left(\frac{5.1^\circ\text{C kg benzene}}{\text{mol solute}}\right) = 10.2^\circ\text{C (freezing point depression)}$$

8.

Cube	1 cm	0.01 cm
Volume	1 cm^3	1 x 10^{-6} cm^3
Number/1 cm cube	1	10^6 (1 cm^3/1 × 10^{-6} cm^3 = 10^6 cubes)
Area of face	1 cm^2	1 × 10^{-4} cm^2
Total surface area	6 cm^2	6 × 10^2 cm^2

(1 × 10^6 cubes)(6 faces/cube)(1 × 10^{-4} cm^2/face) = 6 × 10^2 cm^2

9. $\dfrac{63 \text{ g } NH_4Cl}{150 \text{ g } H_2O} = \dfrac{42 \text{ g } NH_4Cl}{100 \text{ g } H_2O}$ From Figure 14.3, the solubility of NH_4Cl in water is approximately 42 g/100 g H_2O at 30°C, 46 g/100 g H_2O at 40°C. Therefore, the solution of 63 g/150 g of water would be saturated at 10°C, 20°C, and 30°C. The solution would be unsaturated at 40°C and 50°C.

10. The dissolving process involves solvent molecules attaching to the solute ions or molecules. This rate decreases as more of the solvent molecules are already attached to solute molecules. As the solution becomes more saturated, the number of unused solvent molecules decreases. Also, the rate of recrystallization increases as the concentration of dissolved solute increases.

11. Yes. A volumetric flask is satisfactory for preparing normal solutions:

(a) Determine the number of equivalents of solute required for the volume of solution.

(b) Measure out the corresponding number of grams of solute.

(c) Place the solute in the volumetric flask.

(d) Add water to dissolve the solute.

(e) Dilute to the appropriate calibration mark with water and mix.

12. Because the concentration of water is greater in the thistle tube, the water will flow through the membrane from the thistle tube to the urea solution in the beaker. The solution level in the thistle tube will fall.

13. The two components of a solution are the solute and the solvent. The solute is dissolved into the solvent or is the least abundant component. The solvent is the dissolving agent or the most abundant component.

14. It is not always apparent which component in a solution is the solute. For example, in a solution composed of equal volumes of two liquids, the designation of solute and solvent would be simply a matter of preference on the part of the person making the designation.

15. The ions or molecules of a dissolved solute do not settle out because the individual particles are so small that the force of molecular collisions is large compared to the force of gravity.

16. Yes. It is possible to have one solid dissolved in another solid. Metal alloys are of this type. Atoms of one metal are dissolved among atoms of another metal.

17. Orange. The three reference solutions are KCl, $KMnO_4$, and $K_2Cr_2O_7$. They all contain K^+ ions in solution. The different colors must result from the different anions dissolved in the solutions: MnO_4^- (purple) and $Cr_2O_7^{2-}$ (orange). Therefore, it is predictable that the $Cr_2O_7^{2-}$ ion present in an aqueous solution of $Na_2Cr_2O_7$ will impart an orange color to the solution.

18. Hexane and benzene are both nonpolar molecules. There are no strong intermolecular forces between molecules of either substance or with each other, so they are miscible. Sodium chloride consists of ions strongly attracted to each other by electrical attractions. The hexane molecules, being nonpolar, have no strong forces to pull the ions apart, so sodium chloride is insoluble in hexane.

19. Coca Cola has two main characteristics, taste and fizz (carbonation). The carbonation is due to a dissolved gas, carbon dioxide. Since dissolved gases become less soluble as temperature increases, warm Coca Cola would be flat, with little to no carbonation. It would, therefore, be unappealing to most people.

20. Air is considered to be a solution because it is a homogeneous mixture of several gaseous substances and does not have a fixed composition.

21. A teaspoon of sugar would definitely dissolve more rapidly in 200 mL of hot coffee than in 200 mL of iced tea. The much greater thermal agitation of the hot coffee will help break the sugar molecules away from the undissolved solid and disperse them throughout the solution. Other solutes in coffee and tea would have no significant effect. The temperature difference is the critical factor.

22. The solubility of gases in liquids is greatly affected by the pressure of a gas above the liquid. The greater the pressure, the more soluble the gas. There is very little effect of pressure regarding the dissolution of solids in liquids.

23. For a given mass of solute, the smaller the particles, the faster the dissolution of the solute. This is due to the smaller particles having a greater surface area exposed to the dissolving action of the solvent.

24. In a saturated solution, the net rate of dissolution is zero. There is no further increase in the amount of dissolved solute, even though undissolved solute is continuously dissolving, because dissolved solute is continuously coming out of solution, crystallizing at a rate equal to the rate of dissolving.

25. When crystals of $AgNO_3$ and NaCl are mixed, the contact between the individual ions is not intimate enough for the double displacement reaction to occur. When solutions of the two chemicals are mixed, the ions are free to move and come into intimate contact with each other, allowing the reaction to occur easily. The AgCl formed is insoluble.

26. A 16 molar solution of nitric acid is a solution where the ratio of moles of solute (HNO_3) to liters of solution is 16:1. 16 moles HNO_3 per liter of solution.

27. The two solutions contain the same number of chloride ions. One liter of 1 M NaCl contains 1 mole of NaCl, therefore 1 mole of chloride ions. 0.5 liter of 1 M $MgCl_2$ contains 0.5 mole of $MgCl_2$ and 1 mole of chloride ions.

$$(0.5\ \text{L})\left(\frac{1\ \text{mol MgCl}_2}{\text{L}}\right)\left(\frac{2\ \text{mol Cl}^-}{1\ \text{mol MgCl}_2}\right) = 1\ \text{mol Cl}^-$$

28. The champagne would spray out of the bottle all over the place. The rise in temperature and the increase in kinetic energy of the molecules by shaking both act to decrease the solubility of gas within the liquid. The pressure inside the bottle would be great. As the cork is popped, much of the gas would escape from the liquid very rapidly, causing the champagne to spray.

29. A supersaturated solution of $NaC_2H_3O_2$ may be prepared in the following sequence:

(a) Determine the mass of $NaC_2H_3O_2$ necessary to saturate a specific amount of water at room temperature.

(b) Place a bit more $NaC_2H_3O_2$ in the water than the amount needed to saturate the solution.

(c) Heat the solution until all the solid dissolves.

(d) Cover the container and allow it to cool undisturbed. The cool solution, which should contain no solid Na $C_2H_3O_2$, is supersaturated.

To test for supersaturation, add one small crystal of $NaC_2H_3O_2$ to the solution. Immediate crystallization is an indication that the solution was supersaturated.

30. A semipermeable membrane will allow water molecules to pass through in both directions. If it has pure water on one side and 10% sugar solutions on the other side of the membrane, there is a higher concentration of water molecules on the pure water side. More water molecules will pass from the pure water to the sugar solution.

31. The urea solution will have the greater osmotic pressure because it has 1.67 moles solute/kg H_2O, while the glucose solution has only 0.83 mole solute/kg H_2O.

32. A lettuce leaf immersed in salad dressing containing salt and vinegar will become limp and wilted as a result of osmosis. As the water inside the leaf flows into the dressing where the solute concentration is higher the leaf becomes limp from fluid loss. In water, osmosis proceeds in the opposite direction flowing into the lettuce leaf maintaining a high fluid content and crisp leaf.

33. The concentration of solutes (such as salts) is higher in seawater than in body fluids. The survivors who drank seawater suffered further dehydration from the transfer of water by osmosis from body tissues to the intestinal tract.

34. The correct statements are a, b, f, h, j, k, l, n, p, r, s, t, v, x, y, z

(c) A solute cannot be removed from a solution by filtration.

(d) Saturated solutions may or may not be concentrated solutions.

(e) If a solution of sugar in water is allowed to stand undisturbed for a lengthy period, the sugar will remain dispersed in solution.

(g) Gases are usually more soluble in cold H_2O than in hot H_2O.

(i) A solution that is 10% NaCl by mass always contains 10 g NaCl for each 100 g solution.

(m) Dissolving 1 mole of NaCl in 1 liter (1000 g) of water will give a 1 molal solution.

(o) When 100 mL of 0.200 M HCl is diluted to 200 mL volume by the addition of water, the resulting solution is 0.100 M and contains the same number of moles of HCl as were in the original solution.

(q) 100 mL of 0.1 N H_2SO_4 will neutralize the same volume of 0.1 M NaOH as 100 mL of 0.1 M HCl.

(u) An aqueous solution that freezes below 0°C will have a normal boiling point above 100°C.

(w) A solution of 1.00 mol of a nonionizable solute and 1000 g of water will freeze at −1.86°C and will boil at 100.5°C at atmospheric pressure.

35. "Dilute" and "concentrated" are rather general, qualitative terms, which are not very useful, because they are so indefinite. A solution that is termed dilute in one situation might well be considered concentrated under different circumstances. Neither term provides data on "how much" solute is dissolved in a given solution.

36. A solution of H_2SO_4 that is 18 molar contains 18 moles of H_2SO_4 per liter of solution. Since there can be 2 equivalents of H_2SO_4 per mole of H_2SO_4, a solution of 18 molar would contain 36 equivalents of H_2SO_4 per liter of solution. Such a solution would be 36 normal.

37. The number of grams of NaCl in 750 mL of 5 molar solution is

$$(0.75\ \text{L})\left(\frac{5\ \text{mol NaCl}}{\text{L}}\right)\left(\frac{58.44\ \text{g}}{1\ \text{mol}}\right) = 200\ \text{g NaCl}$$

Dissolve the 200 g of NaCl in a minimum amount of water, then dilute the resulting solution to a final volume of 750 mL.

38. Ranking of the specified bases in descending order of the volume of each required to react with 1 liter of 1 M HCl. The volume of each required to yield 1 mole of OH^- ion is shown.

(a) 1 M NaOH 1 liter

(d) 0.6 M $Ba(OH)_2$ 0.83 liter

(c) 2 M KOH 0.50 liter

(b) 1.5 M $Ca(OH)_2$ 0.33 liter

39. The boiling point of a liquid or solution is the temperature at which the vapor pressure of the liquid equals the pressure of the atmosphere. Since a solution containing a nonvolatile solute has a lower vapor pressure than the pure solvent, the boiling point of the solution must be at a higher temperature than for the pure solvent. This will result in the vapor pressure of the solution equaling the atmospheric pressure.

40. The freezing point is the temperature at which a liquid changes to a solid. The vapor pressure of a solution is lower than that of a pure solute. Therefore, the vapor pressure curve of the solution intersects the vapor pressure curve of the pure solid, at a temperature lower than the freezing point of the pure solvent. (See Figure 14.8.) At this point of intersection, the vapor pressure of the solution equals the vapor pressure of the pure solvent.

41. A glass filled with Seven-Up and crushed ice would be colder than a glass of water and crushed ice. The ice will keep the water at its freezing point. The Seven-Up will have a lower freezing point because it contains dissolved solutes.

42. Water and ice are different phases of the same substance in equilibrium at the freezing point of water, 0°C. The presence of the methanol lowers the vapor pressure and hence the freezing point of the water. If the ratio of alcohol to water is high, the freezing point can be lowered an much as 10°C or more.

43. Effectiveness in lowering the freezing point of 500. g water:

(a) 100. g (2.17 mol) of ethyl alcohol is more effective than 100. g (0.292 mol) of sucrose.

(b) 20.0 g (0.435 mol) of ethyl alcohol is more effective than 100. g (0.292 mol) of sucrose.

(c) 20.0 g (0.625 mol) of methyl alcohol is more effective than 20.0 g (0.435 mol) of ethyl alcohol.

44. 5 molal NaCl = 5 mol NaCl/kg H_2O; 5 molar NaCl = 5 mol NaCl/L of solution. The volume of the 5 molal solution will be larger than 1 liter (1 L H_2O + 5 mol NaCl). The volume of the 5 molar solution is exactly 1 L (5 mol NaCl + sufficient H_2O to produce 1 L of solution). The molarity of a 5 molal solution is therefore, less than 5 molar.

45. Microencapsulation is a technique in which reactive chemicals are sealed in tiny capsules. The chemicals are released at the appropriate time for the reaction to occur.

46. When the paper on a scratch and sniff label is scratched or pulled open the encapsulated fragrance is released into the air.

47. Timed released microencapsulated products are used for perfumes, pesticides, medications, neutralizer for contact lenses, and other purposes.

48. Three types of microencapulation systems are:

1) water diffuses through capsule forming a solution which diffuses out again. This can be used in medications to release the drug over a period of time.

2) mechanical. This type of microcapsule can be disturbed by pressure as in carbonless paper receipts, or adhesive label.

3) thermal. This type of system is found in flavorings in prepackaged mixes and release contents during cooking.

49. Reasonably soluble: (a) KOH (b) $NiCl_2$ (d) $AgC_2H_3O_2$ (e) Na_2CrO_4
Insoluble: (c) ZnS

50. Reasonable soluble: (c) $CaCl_2$ (d) $Fe(NO_3)_3$
Insoluble: (a) PbI_2 (b) $MgCO_3$ (e) $BaSO_4$

51. Mass percent calculations.

(a) 25.0 g NaBr + 100. g H_2O = 125 g solution

$$\left(\frac{25.0\text{ g NaBr}}{125\text{ g solution}}\right)(100) = 20.0\%\text{ NaBr}$$

(b) 1.20 g K_2SO_4 + 10.0 g H_2O = 11.2 g solution

$$\left(\frac{1.20\text{ g K}_2\text{SO}_4}{11.2\text{ g solution}}\right)(100) = 10.7\%\text{ K}_2\text{SO}_4$$

52. (a) 40.0 g $Mg(NO_3)_2$ + 500. g H_2O = 540. g solution

$$\left(\frac{40.0\text{ g Mg(NO}_3)_2}{540.\text{ g solution}}\right)(100) = 7.41\%\text{ Mg(NO}_3)_2$$

(b) 17.5 g $NaNO_3$ + 250. g H_2O = 268 g solution

$$\left(\frac{17.5\text{ g NaNO}_3}{268\text{ g solution}}\right)(100) = 6.53\%\text{ NaNO}_3$$

53. A 12.5% $AgNO_3$ solution contains 12.5 g $AgNO_3$ per 100. g solution

$$(30.0\text{ g AgNO}_3)\left(\frac{100.\text{ g solution}}{12.5\text{ g AgNO}_3}\right) = 240.\text{ g solution}$$

54. A 12.5% $AgNO_3$ solution contains 12.5 g $AgNO_3$ per 100. g solution

$$(0.400\text{ mol AgNO}_3)\left(\frac{169.9\text{ g}}{\text{mol}}\right)\left(\frac{100.\text{ g solution}}{12.5\text{ g AgNO}_3}\right) = 544\text{ g solution}$$

55. Mass percent calculations.

(a) 60.0 g NaCl + 200.0 g H_2O = 260.0 g solution

$$\left(\frac{60.0 \text{ g NaCl}}{260.0 \text{ g solution}}\right)(100) = 23.1\% \text{ NaCl}$$

(b) $(0.25 \text{ mol } HC_2H_3O_2)\left(\frac{60.03 \text{ g}}{\text{mol}}\right) = 15 \text{ g } HC_2H_3O_2$

$$(3.0 \text{ mol } H_2O)\left(\frac{18.02 \text{ g}}{\text{mol}}\right) = 54 \text{ g } H_2O$$

$$\left(\frac{15 \text{ g } HC_2H_3O_2}{69 \text{ g solution}}\right)(100) = 22\% \ HC_2H_3O_2$$

56. Mass percent calculation.

(a) 145.0 g NaOH + 1500 g H_2O = 1645 g solution

$$\left(\frac{145.0 \text{ g NaOH}}{1645 \text{ g solution}}\right)(100) = 8.815\% \text{ NaOH}$$

(b) 1.0 molal solution of $C_6H_{12}O_6 = \left(\frac{1 \text{ mol } C_6H_{12}O_6}{1000. \text{ g } H_2O}\right)$

$$(1.0 \text{ mol } C_6H_{12}O_6)\left(\frac{180.2 \text{ g}}{\text{mol}}\right) = 180 \text{ g } C_6H_{12}O_6$$

1000. g H_2O + 180 g $C_6H_{12}O_6$ = 1180 g solution

$$\left(\frac{180 \text{ g } C_6H_{12}O_6}{1180 \text{ g solution}}\right)(100) = 15\% \ C_6H_{12}O_6$$

57. $(65 \text{ g solution})\left(\frac{5.0 \text{ g KCl}}{100. \text{ g solution}}\right) = 3.3 \text{ g KCl}$

58. $(250. \text{ g solution})\left(\frac{15.0 \text{ g } K_2CrO_4}{100. \text{ g solution}}\right) = 37.5 \text{ g } K_2CrO_4$

59. Mass/volume percent.

$$\left(\frac{22.0 \text{ g } CH_3OH}{100. \text{ mL solution}}\right)(100) = 22.0\% \ CH_3OH$$

60. Mass/volume percent.

$$\left(\frac{4.20 \text{ g NaCl}}{12.5 \text{ mL solution}}\right)(100) = 33.6\ \% \text{ NaCl}$$

61. Volume percent.

$$\left(\frac{10.0 \text{ mL } CH_3OH}{40.0 \text{ mL solution}}\right)(100) = 25.0\% \ CH_3OH$$

62. Volume percent.

$$\left(\frac{2.0\ \text{mL}\ C_6H_{14}}{9.0\ \text{mL solution}}\right)(100) = 22\%\ C_6H_{14}$$

63. Molarity problems ($M = \frac{\text{mol}}{\text{L}}$)

(a) $\left(\frac{0.10\ \text{mol}}{250\ \text{mL}}\right)\left(\frac{1000\ \text{mL}}{\text{L}}\right) = 0.40\ \text{M}$

(b) $\left(\frac{2.5\ \text{mol NaCl}}{0.650\ \text{L}}\right) = 3.8\ \text{M NaCl}$

(c) $\left(\frac{53.0\ \text{g}\ Na_2CrO_4}{1.00\ \text{L}}\right)\left(\frac{1\ \text{mol}}{162.0\ \text{g}}\right) = 0.327\ \text{M}\ Na_2CrO_4$

(d) $\left(\frac{260\ \text{g}\ C_6H_{12}O_6}{800.\ \text{mL}}\right)\left(\frac{1000\ \text{mL}}{\text{L}}\right)\left(\frac{1\ \text{mol}}{180.2\ \text{g}}\right) = 1.8\ \text{M}\ C_6H_{12}O_6$

64. (a) $\left(\frac{0.025\ \text{mol HCl}}{10.\ \text{mL}}\right)\left(\frac{1000\ \text{mL}}{\text{L}}\right) = 2.5\ \text{M HCl}$

(b) $\left(\frac{0.35\ \text{mol}\ BaCl_2 \cdot 2\ H_2O}{593\ \text{mL}}\right)\left(\frac{1000\ \text{mL}}{\text{L}}\right) = 0.59\ \text{M}\ BaCl_2 \cdot 2\ H_2O$

(c) $\left(\frac{1.50\ \text{g}\ Al_2(SO_4)_3}{2.00\ \text{L}}\right)\left(\frac{1\ \text{mol}}{342.1\ \text{g}}\right) = 2.19 \times 10^{-3}\ \text{M}\ \ Al_2(SO_4)_3$

(d) $\left(\frac{0.0282\ \text{g}\ Ca(NO_3)_2}{1.00\ \text{mL}}\right)\left(\frac{1000\ \text{mL}}{\text{L}}\right)\left(\frac{1\ \text{mol}}{164.1\ \text{g}}\right) = 0.172\ \text{M}\ Ca(NO_3)_2$

65. Molarity $= \frac{\text{mol solute}}{\text{L solution}}$ or mol solute = (L solution)(Molarity)

(a) $(40.0\ \text{L})\left(\frac{1.0\ \text{mol LiCl}}{1\ \text{L}}\right) = 40.\ \text{mol LiCl}$

(b) $(25.0\ \text{mL})\left(\frac{1\ \text{L}}{1000\ \text{mL}}\right)\left(\frac{3.00\ \text{mol}\ H_2SO_4}{\text{L}}\right) = 0.0750\ \text{mol}\ H_2SO_4$

66. Molarity $= \frac{\text{mol solute}}{\text{L solution}}$ or mol solute = (L solution)(Molarity)

(a) $(349\ \text{mL})\left(\frac{1\ \text{L}}{1000\ \text{mL}}\right)\left(\frac{0.0010\ \text{mol NaOH}}{\text{L}}\right) = 3.5 \times 10^{-4}\ \text{mol NaOH}$

(b) $(5000.\ \text{mL})\left(\frac{1\ \text{L}}{1000\ \text{mL}}\right)\left(\frac{3.1\ \text{mol}\ CoCl_2}{\text{L}}\right) = 16\ \text{mol}\ CoCl_2$

67. (a) $(150\ \text{L})\left(\frac{1.0\ \text{mol NaCl}}{\text{L}}\right)\left(\frac{58.44\ \text{g}}{\text{mol}}\right) = 8.8 \times 10^3\ \text{g NaCl}$

(b) $(260\ \text{mL})\left(\frac{18\ \text{mol}\ H_2SO_4}{1000\ \text{mL}}\right)\left(\frac{98.08\ \text{g}}{\text{mol}}\right) = 4.6 \times 10^2\ \text{g}\ H_2SO_4$

68. (a) $(0.035\ \text{L})\left(\frac{10.0\ \text{mol HCl}}{\text{L}}\right)\left(\frac{36.46\ \text{g}}{\text{mol}}\right) = 13\ \text{g HCl}$

(b) $(8.00\ \text{mL})\left(\frac{1\ \text{L}}{1000\ \text{mL}}\right)\left(\frac{8.00\ \text{mol}\ Na_2C_2O_4}{\text{L}}\right)\left(\frac{134.0\ \text{g}}{\text{mol}}\right) = 8.58\ \text{g}\ Na_2C_2O_4$

69. (a) $(0.430\ \text{mol})\left(\frac{1\ \text{L}}{0.256\ \text{mol}}\right)\left(\frac{1000\ \text{mL}}{\text{L}}\right) = 1.68 \times 10^3\ \text{mL}$

(b) $(20.0\ \text{g KCl})\left(\frac{1\ \text{mol}}{74.55\ \text{g}}\right)\left(\frac{1\ \text{L}}{0.256\ \text{mol}}\right)\left(\frac{1000\ \text{mL}}{\text{L}}\right) = 1.05 \times 10^3\ \text{mL}$

70. (a) $(10.0\ \text{mol})\left(\frac{1\ \text{L}}{0.256\ \text{mol}}\right)\left(\frac{1000\ \text{mL}}{\text{L}}\right) = 3.91 \times 10^4\ \text{mL}$

(b) $(71.0\ \text{g}\ Cl^-)\left(\frac{1\ \text{mol}}{35.45\ \text{g}}\right)\left(\frac{1\ \text{mol KCl}}{1\ \text{mol}\ Cl^-}\right)\left(\frac{1\ \text{L}}{0.256\ \text{mol KCl}}\right)\left(\frac{1000\ \text{mL}}{\text{L}}\right) = 7.82 \times 10^3\ \text{mL}$

71. Dilution problem
$V_1M_1 = V_2M_2$

(a) $V_1 = 200.\ \text{mL}$ $\quad V_2 = 400.\ \text{mL}$
$M_1 = 12\ \text{M}$ $\quad M_2 = M_2$
$(200.\ \text{mL})(12\ \text{M}) = (400.\ \text{mL})(M_2)$

$$M_2 = \frac{(200.\ \text{mL})(12\ \text{M})}{400.\ \text{mL}} = 6.0\ \text{M HCl}$$

(b) $V_1 = 60.0\ \text{mL}$ $\quad V_2 = 560.\ \text{mL}$
$M_1 = 0.60\ \text{M}$ $\quad M_2 = M_2$
$(60.0\ \text{mL})(0.60\ \text{M}) = (560.\ \text{mL})(M_2)$

$$M_2 = \frac{(60.0\ \text{mL})(0.60\ \text{M})}{560.\ \text{mL}} = 0.064\ \text{M}\ ZnSO_4$$

72. (a) First calculate the moles of HCl in each solution. Then calculate the molarity.

$(100.\ \text{mL})\left(\frac{1\ \text{L}}{1000\ \text{mL}}\right)\left(\frac{1.0\ \text{mol}}{\text{L}}\right) = 0.10\ \text{mol HCl}$

$(150.\ \text{mL})\left(\frac{1\ \text{L}}{1000\ \text{mL}}\right)\left(\frac{2.0\ \text{mol}}{\text{L}}\right) = 0.30\ \text{mol HCl}$

Total mol $=$ 0.40 mol HCl

Total volume $=$ 100. mL + 150. mL $=$ 250. mL (0.250 L)

$$\frac{0.40 \text{ mol HCl}}{0.250 \text{ L}} = 1.6 \text{ M HCl}$$

(b) First calculate the moles of NaCl in each solution. Then calculate the molarity.

$$(25.0 \text{ mL})\left(\frac{1 \text{ L}}{1000 \text{ mL}}\right)\left(\frac{1.25 \text{ mol}}{\text{L}}\right) = 0.0313 \text{ mol NaCl}$$

$$(75.0 \text{ mL})\left(\frac{1 \text{ L}}{1000 \text{ mL}}\right)\left(\frac{2.00 \text{ mol}}{\text{L}}\right) = 0.150 \text{ mol NaCl}$$

Total mol = 0.181 mol NaCl

Total volume = 25.0 mL + 75.0 mL = 100. mL = 0.100 L

$$\frac{0.181 \text{ mol NaCl}}{0.100 \text{ L}} = 1.81 \text{ M NaCl}$$

73. $V_1M_1 = V_2M_2$

(a) $(V_1)(12 \text{ M}) = (400. \text{ mL})(6.0 \text{ M})$

$$V_1 = \frac{(400. \text{ mL})(6.0 \text{ M})}{12 \text{ M}} = 2.0 \times 10^2 \text{ mL 12 M HCl}$$

(b) 2.5 N HNO_3 = 2.5 M HNO_3

$(V_1)(16 \text{ M}) = (100. \text{ mL})(2.5 \text{ M})$

$$V_1 = \frac{(100. \text{ mL})(2.5 \text{ M})}{16 \text{ M}} = 16 \text{ mL 16 M } HNO_3$$

74. (a) $(V_1)(15 \text{ M}) = (50. \text{ mL})(6.0 \text{ M})$

$$V_1 = \frac{(50. \text{ mL})(6.0 \text{ M})}{15 \text{ M}} = 20. \text{ mL 15 M } NH_3$$

(b) $10.0 \text{ N } H_2SO_4 = \left(\frac{10.0 \text{ equivalents } H_2SO_4}{\text{L}}\right)\left(\frac{1 \text{ mol}}{2 \text{ equivalents}}\right) = 5.00 \text{ M } H_2SO_4$

$(V_1)(18 \text{ M}) = (250 \text{ mL})(5.00 \text{ M})$

$$V_1 = \frac{(250 \text{ mL})(5.00 \text{ M})}{18 \text{ M}} = 69 \text{ mL 18 M } H_2SO_4$$

75. $(0.250 \text{ L})\left(\frac{0.75 \text{ mol}}{\text{L}}\right) = 0.19 \text{ mol } H_2SO_4$

(a) Final volume after mixing

250. mL + 150. mL $=$ 400. mL $=$ 0.400 L

$$\frac{0.19 \text{ mol } H_2SO_4}{0.400 \text{ L}} = 0.48 \text{ M } H_2SO_4$$

(b) $(250.\ \text{mL})\left(\frac{1\ \text{L}}{1000\ \text{mL}}\right)\left(\frac{0.70\ \text{mol}\ H_2SO_4}{\text{L}}\right) = 0.18\ \text{mol}\ H_2SO_4$

Total moles $= 0.19\ \text{mol} + 0.18\ \text{mol} = 0.37\ \text{mol}\ H_2SO_4$

Final volume $= 250.\ \text{mL} + 250.\ \text{mL} = 500.\ \text{mL} = 0.500\ \text{L}$

$\frac{0.37\ \text{mol}\ H_2SO_4}{0.500\ \text{L}} = 0.74\ \text{M}\ H_2SO_4$

76. $(0.250\ \text{L})\left(\frac{0.75\ \text{mol}}{\text{L}}\right) = 0.19\ \text{mol}\ H_2SO_4$

(a) $(400.\ \text{mL})\left(\frac{1\ \text{L}}{1000\ \text{mL}}\right)\left(\frac{2.50\ \text{mol}\ H_2SO_4}{\text{L}}\right) = 1.00\ \text{mol}\ H_2SO_4$

Total moles $= 0.19\ \text{mol} + 1.00\ \text{mol} = 1.19\ \text{mol}\ H_2SO_4$

Final volume $= 250.\ \text{mL} + 400.\ \text{mL} = 650.\ \text{mL} = 0.650\ \text{L}$

$\frac{1.19\ \text{mol}\ H_2SO_4}{0.650\ \text{L}} = 1.83\ \text{M}\ H_2SO_4$

(b) Final volume after mixing

$250.\ \text{mL} + 375\ \text{mL} = 625\ \text{mL} = 0.625\ \text{L}$

$\frac{0.19\ \text{mol}\ H_2SO_4}{0.625\ \text{L}} = 0.30\ \text{M}\ H_2SO_4$

77. $BaCl_2 + K_2CrO_4 \longrightarrow BaCrO_4 + 2\ KCl$

mL $BaCl_2$ $\longrightarrow$ mol $BaCl_2$ $\longrightarrow$ mol $BaCrO_4$ $\longrightarrow$ g $BaCrO_4$

(a) $(100.\ \text{mL}\ BaCl_2)\left(\frac{0.300\ \text{mol}}{1000\ \text{mL}}\right)\left(\frac{1\ \text{mol}\ BaCrO_4}{1\ \text{mol}\ BaCl_2}\right)\left(\frac{253.3\ \text{g}}{\text{mol}}\right) = 7.60\ \text{g}\ BaCrO_4$

(b) mL K_2CrO_4 $\longrightarrow$ mol K_2CrO_4 $\longrightarrow$ mol $BaCl_2$ $\longrightarrow$ mL $BaCl_2$

$(50.0\ \text{mL}\ K_2CrO_4)\left(\frac{0.300\ \text{mol}}{1000\ \text{mL}}\right)\left(\frac{1\ \text{mol}\ BaCl_2}{1\ \text{mol}\ K_2CrO_4}\right)\left(\frac{1000\ \text{mL}}{1.0\ \text{mol}}\right) = 15\ \text{mL of}\ 1.0\ \text{mol}\ BaCl_2$

78. $3\ MgCl_2 + 2\ Na_3PO_4 \longrightarrow Mg_3(PO_4)_2 + 6\ NaCl$

(a) mL $MgCl_2$ $\longrightarrow$ mol $MgCl_2$ $\longrightarrow$ mol Na_3PO_4 $\longrightarrow$ mL Na_3PO_4

$(50.0\ \text{mL}\ MgCl_2)\left(\frac{0.250\ \text{mol}}{1000\ \text{mL}}\right)\left(\frac{2\ \text{mol}\ Na_3PO_4}{3\ \text{mol}\ MgCl_2}\right)\left(\frac{1000\ \text{mL}}{0.250\ \text{mol}}\right) = 33.3\ \text{mL of}\ 0.250\ \text{M}\ Na_3PO_4$

(b) mL $MgCl_2$ $\longrightarrow$ mol $MgCl_2$ $\longrightarrow$ mol $Mg_3(PO_4)_2$ $\longrightarrow$ g $Mg_3(PO_4)_2$

$(50.0\ \text{mL}\ MgCl_2)\left(\frac{0.250\ \text{mol}}{1000\ \text{mL}}\right)\left(\frac{1\ \text{mol}\ Mg_3(PO_4)_2}{3\ \text{mol}\ MgCl_2}\right)\left(\frac{262.9\ \text{g}}{\text{mol}}\right) = 1.10\ \text{g}\ Mg_3(PO_4)_2$

79. The balanced equation is

$$6\ FeCl_2 + K_2Cr_2O_7 + 14\ HCl \longrightarrow 6\ FeCl_3 + 2\ CrCl_3 + 2\ KCl + 7H_2O$$

(a) $(2.0 \text{ mol } FeCl_2)\left(\frac{2 \text{ mol } KCl}{6 \text{ mol } FeCl_2}\right) = 0.67 \text{ mol } KCl$

(b) $(1.0 \text{ mol } FeCl_2)\left(\frac{2 \text{ mol } CrCl_3}{6 \text{ mol } FeCl_2}\right) = 0.33 \text{ mol } CrCl_3$

(c) $(0.050 \text{ mol } K_2Cr_2O_7)\left(\frac{6 \text{ mol } FeCl_2}{1 \text{ mol } K_2Cr_2O_7}\right) = 0.30 \text{ mol } FeCl_2$

(d) $(0.025 \text{ mol } FeCl_2)\left(\frac{1 \text{ mol } K_2Cr_2O_7}{6 \text{ mol } FeCl_2}\right)\left(\frac{1000 \text{ mL}}{0.060 \text{ mol}}\right) = 69 \text{ mL of } 0.060 \text{ M } K_2Cr_2O_7$

(e) $(15.0 \text{ mL } FeCl_2)\left(\frac{6.0 \text{ mol}}{1000 \text{ mL}}\right)\left(\frac{14 \text{ mol } HCl}{6 \text{ mol } FeCl_2}\right)\left(\frac{1000 \text{ mL}}{6.0 \text{ mol}}\right) = 35 \text{ mL of } 6.0 \text{ M } HCl$

80. $2\ KMnO_4 + 16\ HCl \longrightarrow 2\ MnCl_2 + 5\ Cl_2 + 8\ H_2O + 2\ KCl$

(a) $(0.050 \text{ mol } KMnO_4)\left(\frac{5 \text{ mol } Cl_2}{2 \text{ mol } KMnO_4}\right) = 0.13 \text{ mol } Cl_2$

(b) $(1.0 \text{ L } KMnO_4)\left(\frac{2.0 \text{ mol}}{\text{L}}\right)\left(\frac{16 \text{ mol } HCl}{2 \text{ mol } KMnO_4}\right) = 16 \text{ mol } HCl$

(c) $(200. \text{ mL } KMnO_4)\left(\frac{0.50 \text{ mol}}{1000 \text{ mL}}\right)\left(\frac{16 \text{ mol } HCl}{2 \text{ mol } KMnO_4}\right)\left(\frac{1000 \text{ mL}}{6.0 \text{ mol}}\right) = 1.3 \times 10^2 \text{ mL of } 6 \text{ M } HCl$

(d) $(75.0 \text{ mL } HCl)\left(\frac{6.0 \text{ mol}}{1000 \text{ mL}}\right)\left(\frac{5 \text{ mol } Cl_2}{16 \text{ mol } HCl}\right)\left(\frac{22.4 \text{ L}}{\text{mol}}\right) = 3.2 \text{ L } Cl_2$

81. For an acid, the number of hydrogen ions reacting per formula unit of acid equals the number of equivalents per mole of acid. For a base, the number of hydroxide ions reacting per formula unit of base equals the number of equivalents per mole of base.

(a) HCl $\left(\frac{1 \text{ mol } HCl}{\text{eq}}\right)\left(\frac{36.46 \text{ g}}{\text{mol}}\right) = \left(\frac{36.46 \text{ g } HCl}{\text{eq}}\right)$

NaOH $\left(\frac{1 \text{ mol } NaOH}{\text{eq}}\right)\left(\frac{40.00 \text{ g}}{\text{mol}}\right) = \left(\frac{40.00 \text{ g } NaOH}{\text{eq}}\right)$

(b) HCl $\left(\frac{1 \text{ mol HCl}}{\text{eq}}\right)\left(\frac{36.46 \text{ g}}{\text{mol}}\right) = \left(\frac{36.46 \text{ g HCl}}{\text{eq}}\right)$

$Ba(OH)_2$ $\left(\frac{1 \text{ mol Ba(OH)}_2}{2 \text{ eq}}\right)\left(\frac{171.3 \text{ g}}{\text{mol}}\right) = \left(\frac{85.65 \text{ g Ba(OH)}_2}{\text{eq}}\right)$

(c) H_2SO_4 $\left(\frac{1 \text{ mol H}_2\text{SO}_4}{2 \text{ eq}}\right)\left(\frac{98.08 \text{ g}}{\text{mol}}\right) = \left(\frac{49.04 \text{ g H}_2\text{SO}_4}{\text{eq}}\right)$

$Ca(OH)_2$ $\left(\frac{1 \text{ mol Ca(OH)}_2}{2 \text{ eq}}\right)\left(\frac{74.10 \text{ g}}{\text{mol}}\right) = \left(\frac{37.05 \text{ g Ca(OH)}_2}{\text{eq}}\right)$

82. (a) H_2SO_4 $\left(\frac{1 \text{ mol H}_2\text{SO}_4}{1 \text{ eq}}\right)\left(\frac{98.08 \text{ g}}{\text{mol}}\right) = \left(\frac{98.08 \text{ g H}_2\text{SO}_4}{\text{eq}}\right)$

KOH $\left(\frac{1 \text{ mol KOH}}{1 \text{ eq}}\right)\left(\frac{56.11 \text{ g}}{\text{mol}}\right) = \left(\frac{56.11 \text{ g KOH}}{\text{eq}}\right)$

(b) H_3PO_4 $\left(\frac{1 \text{ mol H}_3\text{PO}_4}{2 \text{ eq}}\right)\left(\frac{97.99 \text{ g}}{\text{mol}}\right) = \left(\frac{49.00 \text{ g H}_3\text{PO}_4}{\text{eq}}\right)$

LiOH $\left(\frac{1 \text{ mol LiOH}}{1 \text{ eq}}\right)\left(\frac{23.95 \text{ g}}{\text{mol}}\right) = \left(\frac{23.95 \text{ g LiOH}}{\text{eq}}\right)$

(c) HNO_3 $\left(\frac{1 \text{ mol HNO}_3}{1 \text{ eq}}\right)\left(\frac{63.02 \text{ g}}{\text{mol}}\right) = \left(\frac{63.02 \text{ g HNO}_3}{\text{eq}}\right)$

NaOH $\left(\frac{1 \text{ mol NaOH}}{1 \text{ eq}}\right)\left(\frac{40.00 \text{ g}}{\text{mol}}\right) = \left(\frac{40.00 \text{ g NaOH}}{\text{eq}}\right)$

83. Normality

(a) 4.0 M HCl $\left(\frac{4.0 \text{ mol}}{\text{L}}\right)\left(\frac{1 \text{ eq}}{\text{mol}}\right) = \left(\frac{4.0 \text{ eq}}{\text{L}}\right) = 4.0 \text{ N HCl}$

(b) 0.243 M HNO_3 $\left(\frac{0.243 \text{ mol}}{\text{L}}\right)\left(\frac{1 \text{ eq}}{\text{mol}}\right) = \left(\frac{0.243 \text{ eq}}{\text{L}}\right) = 0.243 \text{ N HNO}_3$

(c) 3.0 M H_2SO_4 $\left(\frac{3.0 \text{ mol}}{\text{L}}\right)\left(\frac{2 \text{ eq}}{\text{mol}}\right) = \left(\frac{6.0 \text{ eq}}{\text{L}}\right) = 6.0 \text{ N H}_2\text{SO}_4$

84. (a) 1.85 M H_3PO_4 $\left(\frac{1.85 \text{ mol}}{\text{L}}\right)\left(\frac{3 \text{ eq}}{\text{mol}}\right) = \left(\frac{5.55 \text{ eq}}{\text{L}}\right) = 5.55 \text{ N H}_3\text{PO}_4$

(b) 0.250 M $HC_2H_3O_2$ $\left(\frac{0.250 \text{ mol}}{\text{L}}\right)\left(\frac{1 \text{ eq}}{\text{mol}}\right) = \left(\frac{0.250 \text{ eq}}{\text{L}}\right) = 0.250 \text{ N HC}_2\text{H}_3\text{O}_2$

(c) 1.25 M NaOH $\left(\frac{1.25 \text{ mol}}{\text{L}}\right)\left(\frac{1 \text{ eq}}{\text{mol}}\right) = \left(\frac{1.25 \text{ eq}}{\text{L}}\right) = 1.25 \text{ N NaOH}$

85. $N_AV_A = N_BV_B$

(a) $(0.1254\ N)(20.22\ mL) = (0.2250\ N)(V_B)$

$$V_B = \frac{(0.1254\ N)(20.22\ mL)}{0.2550\ N} = 9.943\ mL\ \ NaOH$$

(b) $(0.1246\ N)(14.86\ mL) = (0.2550\ N)(V_B)$

$$V_B = \frac{(0.1246\ N)(14.86\ mL)}{0.2550\ N} = 7.261\ mL\ \ NaOH$$

86. $N_AV_A = N_BV_B$

(a) $(0.1430\ N)(21.30\ mL) = (0.2550\ N)(V_B)$

$$V_B = \frac{(0.1430\ N)(21.30\ mL)}{0.2550\ N} = 11.94\ mL\ NaOH$$

(b) $(0.1430\ N)(18.00\ mL) = (0.2550\ N)(V_B)$

$$V_B = \frac{(0.1430\ N)(18.00\ mL)}{0.2550\ N} = 10.09\ mL\ NaOH$$

87. Molality $= m = \dfrac{\text{mol solute}}{\text{kg solvent}}$

(a) $\left(\dfrac{14.0\ g\ CH_3OH}{100.\ g\ \ H_2O}\right)\left(\dfrac{1000\ g}{kg}\right)\left(\dfrac{1\ mol}{32.04\ g}\right) = \left(\dfrac{4.37\ mol\ CH_3OH}{kg\ H_2O}\right) = 4.37\ m\ \ CH_3OH$

(b) $\left(\dfrac{2.50\ mol\ C_6H_6}{250\ g\ C_6H_{14}}\right)\left(\dfrac{1000\ g}{kg}\right) = \left(\dfrac{10.\ mol\ C_6H_6}{kg\ C_6H_{14}}\right) = 10.\ m\ C_6H_6$

88. (a) $\left(\dfrac{1.0\ g\ C_6H_{12}O_6}{1.0\ g\ H_2O}\right)\left(\dfrac{1000\ g}{kg}\right)\left(\dfrac{1\ mol}{180.2\ g}\right) = \left(\dfrac{5.5\ mol\ C_6H_{12}O_6}{kg\ H_2O}\right) = 5.5\ m\ C_6H_{12}O_6$

(b) Molality $= m = \dfrac{\text{mol solute}}{\text{kg solvent}}$

$\left(\dfrac{0.250\ mol\ I_2}{1.0\ kg\ H_2O}\right) = 0.25\ m\ I_2$

89. (a) $\left(\dfrac{100.0\ g\ C_2H_6O_2}{150.0\ g\ H_2O}\right)\left(\dfrac{1\ mol}{62.07\ g}\right)\left(\dfrac{1000\ g}{kg}\right) = 10.74\ m$

(b) $\Delta t_b = mK_b = (10.74\ m)\left(\dfrac{0.512^\circ C}{m}\right) = 5.50^\circ C$

Boiling point $= 100.00^\circ C + 5.50^\circ C = 105.50^\circ C$

(c) $\Delta t_f = mK_f = (10.74\ m)\left(\dfrac{1.86^\circ C}{m}\right) = 20.0^\circ C$

Freezing point $= 0.00^\circ C - 20.0^\circ C = -20.0^\circ C$

90. (a) $\left(\frac{2.68 \text{ g } C_{10}H_8}{38.4 \text{ g } C_6H_6}\right)\left(\frac{1 \text{ mol}}{128.2 \text{ g}}\right)\left(\frac{1000 \text{ g}}{\text{kg}}\right) = 0.544\ m$

(b) K_f (for benzene) $= \frac{5.1°C}{m}$ Freezing point of benzene $= 5.5°C$

$\Delta t_f = (0.544\ m)\left(\frac{5.1°C}{m}\right) = 2.8°C$

Freezing point of solution $= 5.5°C - 2.8°C = 2.7°C$

(c) K_b (for benzene) $= \frac{2.53°C}{m}$ Boiling point of benzene $= 80.1°C$

$\Delta t_b = (0.544 \text{ m})\left(\frac{2.53°C}{m}\right) = 1.38°C$

Boiling point of solution $= 80.1°C + 1.38°C = 81.5°C$

91. Freezing point of acetic acid is 16.6°C K_f acetic acid $= \frac{3.90°C}{m}$

$\Delta t_f = 16.6°C - 13.2°C = 3.4°C$

$\Delta t_f = mK_f$

$m = \frac{3.4°C}{3.90°C\ /\text{m}} = 0.87\ m$

Convert 8.00 g unknown/60.0 g $HC_2H_3O_2$ to g/mol (molar mass)

$\left(\frac{8.00 \text{ g unknown}}{60.0 \text{ g } HC_2H_3O_2}\right)\left(\frac{1000 \text{ g}}{\text{kg}}\right)\left(\frac{1 \text{ kg } HC_2H_3O_2}{0.87 \text{ mol unknown}}\right) = 153 \text{ g/mol}$

92. $\Delta t_f = 2.50°C$ K_f (for H_2O) $= \frac{1.86°C}{m}$

$\Delta t_f = mK_f$

$m = \frac{2.50°C}{1.86°C/m} = 1.34\ m$

Convert 4.80 g unknown/22.0 g H_2O to g/mol (molar mass)

$\left(\frac{4.80 \text{ g unknown}}{22.0 \text{ g } H_2O}\right)\left(\frac{1000 \text{ g}}{\text{kg}}\right)\left(\frac{1 \text{ kg } H_2O}{1.34 \text{ mol unknown}}\right) = 163 \text{ g/mol}$

93. $NaOH + HCl \longrightarrow NaCl + H_2O$

$(0.15 \text{ L HCl})\left(\frac{1.0 \text{ mol}}{\text{L}}\right)\left(\frac{1 \text{ mol NaOH}}{1 \text{ mol HCl}}\right)\left(\frac{40.00 \text{ g}}{\text{mol}}\right) = 6.0 \text{ g NaOH}$

$\frac{6.0 \text{ g NaOH}}{x} = \frac{10.0 \text{ g NaOH}}{100.0 \text{ g } 10.0\% \text{ NaOH solution}}$

$x = 60. \text{ g } 10.0\% \text{ NaOH solution}$

94. $NaOH + HCl \longrightarrow NaCl + H_2O$

$$1.0\ m\ HCl = \frac{1\ mol\ HCl}{1\ kg\ H_2O} = \frac{36.46\ g\ HCl}{1000\ g\ H_2O}$$

Total mass of solution $= 1000\ g + 36.46\ g = 1036.46\ g$

$$\text{Therefore, } 1.0\ m\ HCl = \frac{1 mol\ HCl}{1036.46\ g\ HCl\ solution}$$

$$(250.\ g\ solution)\left(\frac{1\ mol\ HCl}{1036.46\ solution}\right)\left(\frac{1\ mol\ NaOH}{1\ mol\ HCl}\right)\left(\frac{40.00\ g}{mol}\right) = 9.65\ g\ NaOH$$

$$\frac{9.65\ g\ NaOH}{x} = \frac{10.\ g\ NaOH}{100.0\ g\ 10\%\ NaOH\ solution}$$

$x = 97\ g\ 10.\%$ NaOH solution

95. (a) $(1.0\ L\ syrup)\left(\frac{1000\ mL}{L}\right)\left(\frac{1.06\ g}{mL}\right)\left(\frac{15.0\ g\ sugar}{100.\ g\ syrup}\right) = 1.6 \times 10^2\ g\ sugar$

(b) $\left(\frac{1.6 \times 10^2\ g\ C_{12}H_{22}O_{11}}{L}\right)\left(\frac{1\ mol}{342.3\ g}\right) = 0.47\ M$

(c) $m = \frac{mol\ sugar}{kg\ H_2O}$

$$\left(\frac{15.0\ g\ C_{12}H_{22}O_{11}}{85.0\ g\ H_2O}\right)\left(\frac{1000\ g\ H_2O}{1\ kg\ H_2O}\right)\left(\frac{1\ mol}{342.3\ g}\right) = 0.516\ m$$

96. $K_f = \frac{5.1°C}{m} \qquad \Delta t_f = 0.614°C$

$$\left(\frac{3.84\ g\ C_4H_2N}{250.\ g\ C_6H_6}\right)\left(\frac{1000\ g}{kg}\right) = \frac{15.4\ g\ C_4H_2N}{kg\ C_6H_6}$$

$\Delta t_f = mK_f$

$$m = \frac{0.614°C}{5.1°C/m} = 0.12\ m = \frac{0.12\ mol\ C_4H_2N}{kg\ C_6H_6}$$

$$\left(\frac{15.4\ g\ C_4H_2N}{kg\ C_6H_6}\right)\left(\frac{1\ kg\ C_6H_6}{0.12\ mol\ C_4H_2N}\right) = 1.3 \times 10^2\ g/mol$$

Empirical mass $= 64.07\ g$

$\frac{130\ g}{64.07\ g} = 2.0$ number of empirical formulas per molecular formula.

Therefore, the molecular formula is twice the empirical formula, or $C_8H_4N_2$.

97. $(12.0 \text{ mol HCl})\left(\frac{36.46 \text{ g}}{\text{mol}}\right) = 438 \text{ g HCl in } 1.00 \text{ L solution}$

$$(1.00 \text{ L})\left(\frac{1.18 \text{ g solution}}{\text{mL}}\right)\left(\frac{1000 \text{ mL}}{\text{L}}\right) = 1180 \text{ g solution}$$

$$1180 \text{ g solution} - 438 \text{ g HCl} = 742 \text{ g } H_2O$$

$$\text{Since molality} = \frac{\text{mol HCl}}{\text{kg } H_2O} = \frac{12.0 \text{ mol HCl}}{0.742 \text{ kg } H_2O} = 16.2\ m \text{ HCl}$$

98. $\left(\frac{5.5 \text{ mg } K^+}{\text{mL}}\right)\left(\frac{1 \text{ g}}{1000 \text{ mg}}\right)\left(\frac{101.1 \text{ g } KNO_3}{39.10 \text{ g } K^+}\right)(450 \text{ mL}) = 6.4 \text{ g } KNO_3$

$$(6.4 \text{ g } KNO_3)\left(\frac{1 \text{ mol}}{101.1 \text{ g}}\right) = 0.063 \text{ mol } KNO_3$$

$$\frac{0.063 \text{ mol } KNO_3}{0.450 \text{ L}} = 0.14 \text{ M}$$

99. $(25.0 \text{ g KCl})\left(\frac{100. \text{ g solution}}{5.50 \text{ g KCl}}\right) = 455 \text{ g solution}$

Alternate solution:

$$\left(\frac{25.0 \text{ g KCl}}{x}\right) = \left(\frac{5.50 \text{ g KCl}}{100. \text{ g solution}}\right)$$

$x = 455$ g solution

100. (a) $(500. \text{ mL solution})\left(\frac{0.90 \text{ g NaCl}}{100. \text{ mL solution}}\right) = 4.5 \text{ g NaCl}$

(b) $\left(\frac{4.5 \text{ g NaCl}}{\text{X mL}}\right)(100) = 9.0\%$ X = volume of 9.0% solution

$$\text{X} = \frac{450 \text{ g NaCl}}{9.0\%} = 50. \text{ mL (4.5 g NaCl in solution)}$$

$500. \text{ mL} - 50. \text{ mL} = 450. \text{ mL } H_2O$ must be evaporated

101. From Figure 14.4, the solubility of KNO_3 in H_2O at 20°C is 32 g per 100 g H_2O.

$(50.0 \text{ g } KNO_3)\left(\frac{100. \text{ g } H_2O}{32.0 \text{ g } KNO_3}\right) = 156 \text{ g } H_2O$ to produce a saturated solution.

$175 \text{ g } H_2O - 156 \text{ g } H_2O = 19 \text{ g } H_2O$ must be evaporated.

102. $(150 \text{ mL alcohol})\left(\frac{100. \text{ mL solution}}{70.0 \text{ mL alcohol}}\right) = 210 \text{ mL solution}$

103. (a) $(1.00 \text{ L solution})\left(\frac{1000 \text{ mL solution}}{\text{L solution}}\right)\left(\frac{1.21 \text{ g}}{\text{mL}}\right)\left(\frac{35.0 \text{ g } HNO_3}{100. \text{ g solution}}\right) = 424 \text{ g } HNO_3$

(b) $(500. \text{ g } HNO_3)\left(\frac{1000 \text{ mL solution}}{424 \text{ g } HNO_3}\right)\left(\frac{1.00 \text{ L}}{1000 \text{ mL}}\right) = 1.18 \text{ L solution}$

104. Assume 1.000 L of solution

$$\left(\frac{1000.\ \text{mL}}{\text{L}}\right)\left(\frac{1.21\ \text{g solution}}{\text{mL}}\right)\left(\frac{35.0\ \text{g}\ HNO_3}{100.\ \text{g solution}}\right)\left(\frac{1\ \text{mol}}{63.02\ \text{g}}\right) = 6.72\ \text{M}\ HNO_3$$

105. First calculate the molarity of the solution

$$\left(\frac{80.0\ \text{g}\ H_2SO_4}{500.\ \text{mL}}\right)\left(\frac{1000\ \text{mL}}{\text{L}}\right)\left(\frac{1\ \text{mol}}{98.08\ \text{g}}\right) = 1.63\ \text{M}\ H_2SO_4$$

$$M_1V_1 = M_2V_2$$

$$(1.63\ \text{M})(500.\ \text{mL}) = (0.10\ \text{M})(V_2)$$

$$V_2 = \frac{(1.63\ \text{M})(500.\ \text{mL})}{0.10\ \text{M}} = 8.2 \times 10^3\ \text{mL} = 8.2\ \text{L}$$

106. Note that the problem asks for the volume of water to be added, not the final volume of the solution.

$$(300.\ \text{mL})(1.40\ \text{M}) = (V_2)(0.500\ \text{M})$$

$$V_2 = \frac{(300.\ \text{mL})(1.40\ \text{M})}{0.500\ \text{M}} = 840.\ \text{mL}\ \ \text{(final volume)}$$

840. mL – 300. mL = 540. mL water to be added

107. $(10.0\ \text{mL})(16\ \text{M}) = (500.\ \text{mL})(M_2)$

$$M_2 = \frac{(10.0\ \text{mL})(16\ \text{M})}{500.\ \text{mL}} = 0.32\ \text{M}\ HNO_3$$

108. $(V_1)(5.00\ \text{M}) = (250\ \text{mL})(0.625\ \text{M})$

$$V_1 = \frac{(250\ \text{mL})(0.625\ \text{M})}{5.00\ \text{M}} = 31\ \text{mL}\ 5.00\ \text{M KOH}$$

To make 250. mL of 0.625 M KOH, take 31.3 mL of 5.00 M KOH and dilute with water to a volume of 250. mL.

109. $Mg + 2\ HCl \longrightarrow MgCl_2 + H_{2(g)}$

(a) mL HCl ⟶ mol HCl ⟶ mol H_2

$$(200.\ \text{mL HCl})\left(\frac{3.00\ \text{mol}}{1000\ \text{mL}}\right)\left(\frac{1\ \text{mol}\ H_2}{2\ \text{mol HCl}}\right) = 0.300\ \text{mol}\ H_2$$

(b) $PV = nRT$

$$P = (720\ \text{torr})\left(\frac{1\ \text{atm}}{760\ \text{torr}}\right) = 0.95\ \text{atm}$$

$$T = 27^\circ\text{C} = 300.\ \text{K}$$

$$n = 0.300\ \text{mol}$$

$$V = \frac{nRT}{P} = \frac{(0.300\ \text{mol})(0.0821\ \text{L atm/mol K})(300.\ \text{K})}{0.95\ \text{atm}} = 7.8\ \text{L}\ H_2$$

110. $Mg + 2\ HCl \longrightarrow MgCl_2 + H_{2(g)}$

$L\ H_2 \longrightarrow mol\ H_2 \longrightarrow mol\ HCl \longrightarrow M\ HCl$

$$(3.50\ L\ H_2)\left(\frac{1\ mol}{22.4\ L}\right)\left(\frac{2\ mol\ HCl}{1\ mol\ H_2}\right)\left(\frac{1}{0.150\ L}\right) = 2.08\ M\ \ HCl$$

111. Equivalent mass of $Ca(OH)_2 = \left(\frac{74.10\ g}{mol}\right)\left(\frac{1\ mol}{2\ eq}\right) = \left(\frac{37.05\ g}{eq}\right)$

Equivalents acid = equivalents base

$$(N_A)(0.03626\ L) = (2.50\ g\ Ca(OH)_2)\left(\frac{1\ eq}{37.05\ g}\right)$$

$$N_A = \left(\frac{1.86\ g}{eq}\right) = 1.86\ N\ \ H_2SO_4$$

112. $(12.0\ g\ Mg(OH)_2)\left(\frac{1\ eq}{29.16\ g\ Mg(OH)_2}\right) = 0.412\ eq\ \ Mg(OH)_2$

$$(10.0\ g\ Al(OH)_3)\left(\frac{1\ eq}{26.00\ g\ Al(OH)_3}\right) = 0.385\ eq\ \ Al(OH)_3$$

1 equivalent of base will neutralize 1 equivalent of acid. Therefore, 12.0 g $Mg(OH)_2$ will neutralize more stomach acid than 10.0 g $Al(OH)_3$.

113. (a) With equal masses of CH_3OH and C_2H_5OH, the substance with the lower molar mass will represent more moles of solute in solution. Therefore, the CH_3OH will be more effective than C_2H_5OH as an antifreeze.

(b) Equal molal solutions will lower the freezing point of the solution by the same amount.

114. Calculate molarity and molality. Assume 1000 mL of solution to calculate the amounts of H_2SO_4 and H_2O in the solution.

$$(1000\ mL\ solution)\left(\frac{1.29\ g}{mL}\right) = 1.29 \times 10^3\ g\ \ solution$$

$$(1.29 \times 10^3\ g\ solution)\left(\frac{38\ g\ H_2SO_4}{100\ g\ solution}\right) = 4.9 \times 10^2\ g\ \ H_2SO_4$$

$$1.29 \times 10^3\ g\ solution - 4.9 \times 10^2\ g\ H_2SO_4 = 8.0 \times 10^2\ g\ H_2O \text{ in the solution}$$

$$m = \left(\frac{490\ g\ H_2SO_4}{8.0 \times 10^2\ g\ H_2O}\right)\left(\frac{1000\ g}{kg}\right)\left(\frac{1\ mol}{98.08\ g}\right) = 6.2\ m\ \ H_2SO_4$$

$$M = \left(\frac{4.9 \times 10^2\ g\ H_2SO_4}{L}\right)\left(\frac{1\ mol}{98.08\ g}\right) = 5.0\ M\ \ H_2SO_4$$

115. 1.00 lb = 454 g sugar ($C_{12}H_{22}O_{11}$)

$$(4.00 \text{ lb } H_2O)\left(\frac{453.6 \text{ g}}{\text{lb}}\right) = 1.82 \times 10^3 \text{ g } H_2O \ (1.82 \text{ kg } H_2O)$$

$$(453.6 \text{ g } C_{12}H_{22}O_{11})\left(\frac{1 \text{ mol}}{342.3 \text{ g}}\right) = 1.33 \text{ mol } C_{12}H_{22}O_{11}$$

K_f (for H_2O) = 1.86°C kg solvent/mol solute

$$\Delta t_f = mK_f = \left(\frac{1.33 \text{ mol } C_{12}H_{22}O_{11}}{1.82 \text{ kg } H_2O}\right)\left(\frac{1.86 \text{ °C kg } H_2O}{\text{mol } C_{12}H_{22}O_{11}}\right) = 1.36\text{°C}$$

Freezing point of solution = 0°C – 1.36°C = –1.36°C = 29.6°F

If the sugar solution is placed outside, where the temperature is 20°F, the solution will freeze.

116. Freezing point depression is 5.4°C

(a) $\Delta t_f = mK_f$

$$m = \frac{\Delta t_f}{K_f} = \frac{5.4 \text{ °C}}{1.86\text{°C kg solvent/mol solute}} = 2.9 \ m$$

(b) $$K_b \text{ (for } H_2O) = \frac{0.512\text{°C kg solvent}}{\text{mol solute}} = \frac{0.512\text{°C}}{m}$$

$$\Delta t_b = mK_b = (2.9 \text{ m})\left(\frac{0.512\text{°C}}{m}\right) = 1.5\text{°C}$$

Boiling point = 100.0°C + 1.5°C = 101.5°C

117. Freezing point depression = 0.372°C $K_f = \frac{1.86\text{°C}}{m}$

$\Delta t_f = mK_f$

$$m = \frac{0.373\text{°C}}{1.86\text{°C/mol}} = 0.200 \ m$$

$$(6.20 \text{ g } C_2H_6O_2)\left(\frac{1 \text{ mol}}{62.07 \text{ g}}\right) = 0.100 \text{ mol } C_2H_6O_2$$

$$(0.100 \text{ mol } C_2H_6O_2)\left(\frac{1 \text{ kg } H_2O}{0.200 \text{ mol } C_2H_6O_2}\right)\left(\frac{1000 \text{ g } H_2O}{\text{kg } H_2O}\right) = 500. \text{ g } H_2O$$

118. (a) Freezing point depression = 20.0°C

$$12.0 \text{ L } H_2O\left(\frac{1000 \text{ mL}}{\text{L}}\right)\left(\frac{1.00 \text{ g}}{\text{mL}}\right) = 1.20 \times 10^4 \text{ g } H_2O$$

$\Delta t_f = mK_f$

$$m = \frac{20.0\text{°C}}{1.86\text{°C}/m} = 10.8 \ m$$

$$(1.20 \times 10^4 \text{ g } H_2O)\left(\frac{10.8 \text{ mol } C_2H_6O_2}{1000 \text{ g } H_2O}\right)\left(\frac{62.07 \text{ g}}{\text{mol}}\right) = 8.04 \times 10^3 \text{ g } C_2H_6O_2$$

(b) $$(8.04 \times 10^3 \text{ g } C_2H_6O_2)\left(\frac{1.00 \text{ mL}}{1.11 \text{ g}}\right) = 7.24 \times 10^3 \text{ mL } C_2H_6O_2$$

(c) $1.8(-20.0) + 32 = -4.0\ ^\circ F$

119. Yes, a saturated solution can also be a dilute solution. For example, the solutibility of AgCl in water at 25°C is 1.3×10^{-5} mol/L. Thus, the solution formed by dissolving AgCl in water is both saturated and very dilute.

120. $\underset{1\ mol}{HCl} + \underset{1\ mol}{NaOH} \longrightarrow NaCl + H_2O$

g NaOH $\longrightarrow$ mol NaOH $\longrightarrow$ mol HCl $\longrightarrow$ L HCl

$$(12\ g\ NaOH)\left(\frac{1\ mol}{40.00\ g}\right)\left(\frac{1\ mol\ HCl}{1\ mol\ NaOH}\right)\left(\frac{1\ L\ HCl}{0.65\ mol\ HCl}\right) = 0.46\ L\ HCl\ (460\ mL)$$

121. $HNO_3 + NaHCO_3 \longrightarrow NaNO_3 + H_2O + CO_2$

First calculate the grams of $NaHCO_3$ in the sample.

mL HNO_3 $\longrightarrow$ L HNO_3 $\longrightarrow$ mol HNO_3 $\longrightarrow$ mol $NaHCO_3$ $\longrightarrow$ g $NaHCO_3$

$$(150\ mL\ HNO_3)\left(\frac{1\ L}{1000\ mL}\right)\left(\frac{0.055\ mol}{L}\right)\left(\frac{1\ mol\ NaHCO_3}{1\ mol\ HNO_3}\right)\left(\frac{84.01\ g}{mol}\right)$$

$= 0.69$ g $NaHCO_3$ in the sample

$$\left(\frac{0.69\ g}{1.48\ g}\right)(100) = 47\%\ NaHCO_3$$

122. (a) Dilution problem: $M_1V_1 = M_2V_2$

$(1.5\ M)(8.4\ L) = (17.8\ M)(V_2)$

$$V_2 = \frac{(1.5\ M)(8.4\ L)}{17.8\ M} = 0.71\ L$$

0.71 L of 17.8 M H_2SO_4 is to be diluted to 8.4 L.

$8.4\ L - 0.71\ L = 7.7\ L\ H_2O$ must be added

(b) $\left(\frac{17.8\ mol}{1000.\ mL}\right)(1.00\ mL) = 0.0178\ mol$

(c) $\left(\frac{1.5\ mol}{1000.\ mL}\right)(1.00\ mL) = 0.0015\ mol$

123. Freezing point depresion is 3.6°C

$\Delta t_f = mk_f$

$$m = \frac{\Delta t_f}{k_f} = \frac{3.6^\circ C}{1.86^\circ C\ kg\ solvent/mole\ solute} = 1.9\ m\ solution$$

$$\Delta t_b = mk_b = (1.9\ m)\left(\frac{0.512^\circ C}{m}\right) = 0.97^\circ C$$

Boiling point = 100.00°C + 0.97°C = 100.97°C

124. moles HNO_3 total = moles HNO_3 from 3.00 M + moles HNO_3 from 12.0 M

$M_T V_T = M_{3.00\ M} V_{3.00\ M} + M_{12.0\ M} V_{12.0\ M}$

Assume preparation of 1000. mL of 6 M solution

Let Y = volume of 3.00 M solution; volume of 12.0 M = 1000. mL – Y

(6.00 M)(1000. mL) = (3.00 M)(Y) + (12.0 M)(1000. mL – Y)

6000. mL = 3.00 Y mL + 12,000 mL – 12.0 Y

6000. mL = 9.00 Y $\qquad Y = \frac{6000.\ \text{mL}}{9.00} = 667$ mL 3 M

1000. mL – 667 mL = 333 mL 12 M

Mix together 667 mL 3.00 M HNO_3 and 333 mL 12.0 M HNO_3 to get 1000. mL of 6.00 M HNO_3

125. $HBr + NaOH \longrightarrow NaBr + H_2O$

First calculate the molarity of the diluted HBr solution.

The reaction is 1 mol HBr to 1 mol NaOH, so

$M_A V_A = M_B V_B$

$M_A(100.0\ \text{mL}) = (0.37\ \text{M})(88.4\ \text{mL})$

$$M_A = \frac{(0.37\ \text{M})(88.4\ \text{mL})}{100.0\ \text{mL}} = 0.33\ \text{M HBr}$$

Now calculate the molarity of the HBr before dilution

$M_1 V_1 = M_2 V_2$

$(M_1)(20.0\ \text{mL}) = (0.33\ \text{M})(240.\ \text{mL})$

$$M_1 = \frac{(0.33\ \text{M})(240.\ \text{mL})}{20.0\ \text{mL}} = 4.0\ \text{M HBr}$$

126. $Ba(NO_3)_2 + 2\ KOH \longrightarrow Ba(OH)_2 + 2\ KNO_3$

This is a limiting reactant problem. First calculate the moles of each reactant and determine the limiting reactant.

$$M \times L = \left(\frac{\text{moles}}{L}\right)(L) = \text{moles}$$

$$\left(\frac{0.642\ \text{mol}}{L}\right)(0.0805\ L) = 0.0517\ \text{mol}\ Ba(NO_3)_2$$

$$\left(\frac{0.743\ \text{mol}}{L}\right)(0.0445\ L) = 0.0331\ \text{mol KOH}$$

According to the equation, twice as many moles of KOH as $Ba(NO_3)_2$ are needed, so KOH is the limiting reactant.

$$(0.0331\ \text{mol KOH})\left(\frac{1\ \text{mol}\ Ba(OH)_2}{2\ \text{mol KOH}}\right)\left(\frac{171.3\ g}{\text{mol}}\right) = 2.84\ g\ Ba(OH)_2 \text{ is formed}$$

127. $(300.\text{ g solution})\left(\dfrac{5.0\text{ g sucrose}}{100.\text{ g solution}}\right) = 15\text{ g sucrose}$

$(\text{Y g 2.0\% solution})\left(\dfrac{2.0\text{ g sucrose}}{100\text{ g solution}}\right) = 15\text{ g sucrose}$

$\text{Y} = \left(\dfrac{100\text{ g solution}}{2.0\text{ g sucrose}}\right)(15\text{ g sucrose}) = 750\text{ g 2\% solution}$

128. (a) $\left(\dfrac{0.25\text{ mol}}{\text{L}}\right)(0.0458\text{ L}) = 0.011\text{ mol } Li_2CO_3$

(b) $\left(\dfrac{0.25\text{ mol}}{\text{L}}\right)(0.75\text{ L})\left(\dfrac{73.89\text{ g}}{\text{mol}}\right) = 14\text{ g } Li_2CO_3$

(c) $6.0\text{ g } Li_2CO_3\left(\dfrac{1\text{ mol}}{73.89\text{ g}}\right)\left(\dfrac{1000.\text{ mL}}{0.25\text{ mol}}\right) = 3.2 \times 10^2\text{ mL solution}$

(d) Assume 1000. mL solution

$\left(\dfrac{1.22\text{ g}}{\text{mL}}\right)(1000.\text{ mL}) = 1220\text{ g solution}$

$\left(\dfrac{0.25\text{ mol}}{\text{L}}\right)\left(\dfrac{73.89\text{ g } Li_2CO_3}{\text{mol}}\right) = 18\text{ g } Li_2CO_3 \text{ per L solution}$

$\% = \left(\dfrac{\text{g solute}}{\text{g solution}}\right)(100) = \left(\dfrac{18\text{ g}}{1220\text{ g}}\right)(100) = 1.5\%$

129. Calculate the total moles of HCl and divide by the total volume.

$\left(\dfrac{0.35\text{ mol}}{\text{L}}\right)(0.4000\text{ L}) = 0.14\text{ mol HCl}$

$\left(\dfrac{0.65\text{ mol}}{\text{L}}\right)(1.1\text{ L}) = \dfrac{0.72\text{ mol HCl}}{0.86\text{ mol HCl (total mol HCl)}}$

Total volume = 1.1 L + 0.40 L = 1.5 L

$\text{Molarity} = \dfrac{0.86\text{ mol}}{1.5\text{ L}} = 0.57\text{ M HCl}$

130. Use $M_1V_1 = M_2V_2$

$1\text{ drop} = \dfrac{1}{20.}\text{mL} = 0.050\text{ mL}$

100. mL + 0.050 mL = 100.050 mL

(17.8 M)(0.050 mL) = (M)(100.050 mL)

$\text{M} = \dfrac{(17.8\text{ M})(0.050\text{ mL})}{100.050\text{ mL}} = 8.9 \times 10^{-3}\text{ M}$

CHAPTER 15

ACIDS, BASES, AND SALTS

1. The Arrhenius definition is restricted to aqueous solutions, while the Bronsted-Lowry definition is not.

2. An electrolyte must be present in the solution for the bulb to glow.

3. Electrolytes include acids, bases, and salts.

4. First, the orientation of the polar water molecules about the Na^+ and Cl^- is different. The positive end (hydrogen) of the water molecule is directed towards Cl^-, while the negative end (oxygen) of the water molecule is directed towards the Na^+. Second, more water molecules will fit around Cl^-, since it is larger than the Na^+ ion.

5. The pH for a solution with a hydrogen ion concentration of 0.003 M will be between 2 and 3.

6. Tomato juice is more acidic than blood, since its pH is lower.

7. By the Arrhenius theory, an acid is a substance that produces hydrogen ions in aqueous solution. A base is a substance that produces hydroxide ions in aqueous solution. By the Bronsted-Lowry theory, an acid is a proton donor, while a base accepts protons. Since a proton is a hydrogen ion, then the two theories are very similar for acids, but not for bases. A chloride ion can accept a proton (producing HCl), so it is a Bronsted-Lowry base, but would not be a base by the Arrhenius theory, since it does not produce hydroxide ions.

   By the Lewis theory, an acid is an electron pair acceptor, and a base is an electron pair donor. Many individual substances would be similarly classified as bases by Bronsted-Lowry or Lewis theories, since a substance with an electron pair to donate, can accept a proton. But, the Lewis definition is almost exclusively applied to reactions where the acid and base combine into a single molecule. The Bronsted-Lowry definition is usually applied to reactions that involve a transfer of a proton from the acid to the base. The Arrhenius definition is most often applied to individual substances, not to reactions. According to the Arrhenius theory, neutralization involves the reaction between a hydrogen ion and a hydroxide ion to form water. Neutralization, according to the Bronsted-Lowry theory, involves the transfer of a proton to a negative ion. The formation of a coordinate-covalent bond constitues a Lewis neutralization.

8. Neutralization reactions:

Arrhenius: $HCl + NaOH \longrightarrow NaCl + H_2O \quad (H^+ + OH^- \longrightarrow H_2O)$

Bronsted-Lowry: $HCl + KCN \longrightarrow HCN + KCl \ (H^+ + CN^- \longrightarrow HCN)$

Lewis: $AlCl_3 + NaCl \longrightarrow AlCl_4^- + Na^+$

$$\mathrm{:\ddot{Cl}:\ddot{Al}:\ddot{Cl}:\ddot{Cl}:} \ + \ [:\ddot{\underset{..}{Cl}}:]^- \longrightarrow [:\ddot{Cl}:\ddot{Al}:\ddot{Cl}: \ldots]^-$$

9. (a) $[:\ddot{\underset{..}{Br}}:]^-$ (b) $[:\ddot{\underset{..}{O}}:H]^-$ (c) $[:C:::N:]^-$

These ions are considered to be bases according to the Brønsted-Lowry theory, because they can accept a proton at any of their unshared pairs of electrons. They are considered to be bases according to the Lewis acid-base theory, because they can donate an electron pair.

10. The classes of compounds containing electrolytes are acids, bases, and salts.

11. Names of the compounds in Table 15.3

H_2SO_4	sulfuric acid	$HC_2H_3O_2$	acetic acid
HNO_3	nitric acid	H_2CO_3	carbonic acid
HCl	hydrochloric acid	HNO_2	nitrous acid
HBr	hydrobromic acid	H_2SO_3	sulfurous acid
$HClO_4$	perchloric acid	H_2S	hydrosulfuric acid
$NaOH$	sodium hydroxide	$H_2C_2O_4$	oxalic acid
KOH	potassium hydroxide	H_3BO_3	boric acid
$Ca(OH)_2$	calcium hydroxide	$HClO$	hypochlorous acid
$Ba(OH)_2$	barium hydroxide	NH_3	ammonia
		HF	hydrofluoric acid

12. Hydrogen chloride dissolved in water conducts an electric current. HCl reacts with polar water molecules to produce H_3O^+ and Cl^- ions, which conduct electric current. Benzene is a nonpolar solvent, so it cannot pull the HCl molecules apart. Since there are no ions in the benzene solution, it does not conduct an electric current. HCl does not ionize in benzene.

13. In their crystalline structure, salts exist as positive and negative ions in definite geometric arrangement to each other, held together by the attraction of the opposite charges. When dissolved in water, the salt dissociates as the ions are pulled away from each other by the polar water molecules.

14. Testing the electrical conductivity of the solutions shows that CH_3OH is a nonelectrolyte, while NaOH is an electrolyte. This indicates that the OH group in CH_3OH must be covalently bonded to the CH_3 group.

15. Molten NaCl conducts electricity because the ions are free to move. In the solid state, however, the ions are immobile and do not conduct electricity.

16. Dissociation is the separation of already existing ions in an ionic compound. Ionization is the formation of ions from molecules. The dissolving of NaCl is a dissociation, since the ions already existed. The dissolving of HCl in water is an ionization process, because ions are formed from HCl molecules.

17. Strong electrolytes are those which are essentially 100% ionized or dissociated in water. Weak electrolytes are those which are only slightly ionized in water.

18. Ions are hydrated in solution because there is an electrical attraction between the charged ions and the polar water molecules.

19. The main distinction between water solutions of strong and weak electrolytes is the degree of ionization of the electrolyte. A solution of an electrolyte contains many more ions than a solution of a nonelectrolyte. Strong electrolytes are essentially 100% ionized. Weak electrolytes are only slightly ionized in water.

20. (a) In a neutral solution, the concentration of H^+ and OH^- are equal.

(b) In an acid solution, the concentration of H^+ is greater than the concentration of OH^-.

(c) In a basic solution, the concentration of OH^- is greater than the concentration of H^+.

21. The net ionic equation for an acid-base reaction in aqueous solution is $H^+ + OH^- \longrightarrow H_2O$.

22. The HCl molecule is polar and, consequently, is much more soluble in the polar solvent, water, than in the nonpolar solvent, benzene. There is also a chemical reaction between HCl and H_2O molecules. $HCl + H_2O \longrightarrow H_3O^+ + Cl^-$

23. Pure water is neutral because the concentrations of acid and base ions are equal.

24. The correct statements are a, c, f, i, j, k, l, n, p, r, t, u

(b) The Brønsted-Lowry theory of acids and bases is not restricted to aqueous solutions.

(d) Some, but not all, substances that are acids according to the Lewis theory will also be acids according to the Brønsted-Lowry theory.

(e) An electron pair acceptor is a Lewis acid.

(g) When an ionic compound dissolves in water, the ions separate. This process is called dissociation.

(h) In autoionization of water, the H_2O and OH^- and the H_3O^+ and H_2O constitute conjugate acid-base pairs.

(m) The terms dissociation and ionization are not synonymous. Dissociation is the separation of an ionic compound into its ions. Ionization is the formation of ions from molecular compounds such as HCl.

(o) The terms strong acid and strong base refer to solutions in which ionization approaches 100%. Weak acid and weak base, however, refer to solutions in which ionization is very low.

(q) Ionic reactions may be represented by net ionic equations.

(s) It is possible to boil seawater at a higher temperature than that required to boil pure water (both at the same atmospheric pressure).

(v) The Tyndall effect is observable in colloidal dispersions.

25. The fundamental difference between a colloidal dispersion and a true solution lies in the size of the particles. In a true solution particles are usually ions or hydrated molecules. In colloidals the particles are aggregates of ions or molecules.

26. Colloids are prepared by two methods:

1) dispersion in which larger particles are reduced to colloidal size. This is usually accomplished by mechanical means.

2) condensation in which small particles coalesce into larger ones of colloidal size. This is often accomplished by precipitation in a dilute solution.

27. The Tyndall effect is observed when a narrow beam of light is passed through a colloidal suspension. The light is reflected from the colloidal particles effectively illuminating the path of the light through the liquid. In a true solution the light path cannot be seen because the dissolved particles are too small to reflect light.

28. Adsorption refers to the adhesion of particles to a surface (like adsorption of gas bubbles on a metal surface) while absorption refers to the taking in of one material by another (like absorption of water by a sponge).

29. The particles in a colloid remain dispered because:

1) they are subjected to constant bombardment by the dispersing phase which keeps the particles in motion and prevents settling out.

2) since colloid particles have the same kind of charge, they repel each other preventing coalescing to larger particles and settling out.

30. Dialysis is the process of removing dissolved solutes from colloidal dispersion by use of a dialyzing membrane. The dissolved solutes pass through the membrane leaving the colloidal dispersion behind. Dialysis is used in artificial kidneys to remove soluble waste products from the blood.

31.

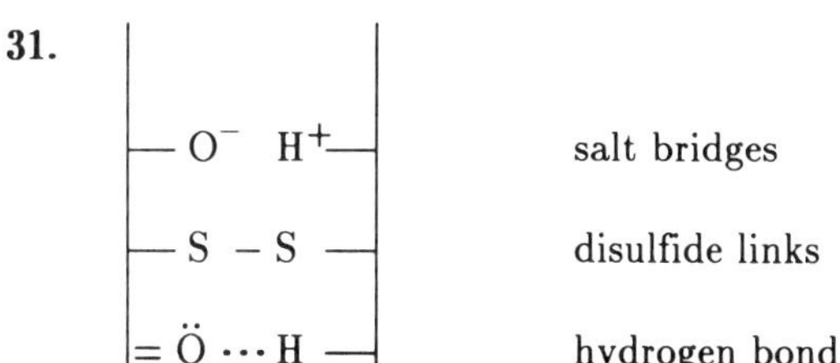

32. Acidic shampoo breaks hydrogen bonds and salt bridges in the hair leaving only disulfide bonds. In a depilatory a basic solution (pH about 12) is used which breaks all the types of bonds (H, salt bridges, disulfide) and the hair dissolves.

33. Conjugate acid-base pairs:

(a) HCl - Cl^- ; NH_4^+ - NH_3

(b) HCO_3^- - CO_3^{2-} ; H_2O - OH^-

(c) H_3O^+ - H_2O ; H_2CO_3 - HCO_3^-

(d) $HC_2H_3O_2$ - $C_2H_3O_2^-$; H_3O^+ - H_2O

34. (a) H_2SO_4 - HSO_4^- ; $H_2C_2H_3O_2^+$ - $HC_2H_3O_2$

(b) step 1: H_2SO_4 - HSO_4^- ; H_3O^+ - H_2O

step 2: HSO_4^- - SO_4^{2-}; H_3O^+ - H_2O

(c) $HClO_4$ - ClO_4^- ; H_3O^+ - H_2O

(d) H_3O^+ - H_2O ; CH_3OH - CH_3O^-

35. Balancing equations

(a) $Mg(s) + 2\ HCl(aq) \longrightarrow MgCl_2(aq) + H_2(g)$

(b) $BaO(s) + 2\ HBr(aq) \longrightarrow BaBr_2(aq) + H_2O(l)$

(c) $2\ Al(s) + 3\ H_2SO_4(aq) \longrightarrow Al_2(SO_4)_3(aq) + 3\ H_2(g)$

(d) $Na_2CO_3(aq) + 2\ HCl(aq) \longrightarrow 2\ NaCl(aq) + H_2O(l) + CO_2(g)$

(e) $Fe_2O_3(s) + 6\ HBr(aq) \longrightarrow 2\ FeBr_3(aq) + 3\ H_2O(l)$

(f) $Ca(OH)_2(aq) + H_2CO_3(aq) \longrightarrow CaCO_3(s) + 2\ H_2O(l)$

36. (a) $NaOH(aq) + HBr(aq) \longrightarrow NaBr(aq) + H_2O(l)$

(b) $KOH(aq) + HCl(aq) \longrightarrow KCl(aq) + H_2O(l)$

(c) $Ca(OH)_2(aq) + 2\ HI(aq) \longrightarrow CaI_2(aq) + 2\ H_2O(l)$

(d) $Al(OH)_3(s) + 3\ HBr(aq) \longrightarrow AlBr_3(aq) + 3\ H_2O(l)$

(e) $Na_2O(s) + 2\ HClO_4(aq) \longrightarrow 2\ NaClO_4(aq) + H_2O(l)$

(f) $3\ LiOH(aq) + FeCl_3(aq) \longrightarrow Fe(OH)_3(s) + 3\ LiCl(aq)$

37. The following compounds are electrolytes:

(a) HCl

(b) CO_2

(c) $CaCl_2$

38. The following compounds are electrolytes in **aqueous** solution:

(a) $NaHCO_3$

(c) $AgNO_3$

(d) $HCOOH$

(e) $RbOH$

(f) K_2CrO_4

39. Calculation of molarity of ions.

(a) $$(0.015\ M\ NaCl)\left(\frac{1\ mol\ Na^+}{1\ mol\ NaCl}\right) = 0.015\ M\ \ Na^+$$

$$(0.015\ M\ NaCl)\left(\frac{1\ mol\ Cl^-}{1\ mol\ NaCl}\right) = 0.015\ M\ \ Cl^-$$

(b) $$(4.25\ M\ NaKSO_4)\left(\frac{1\ mol\ Na^+}{1\ mol\ NaKSO_4}\right) = 4.25\ M\ \ Na^+$$

$$(4.25\ M\ NaKSO_4)\left(\frac{1\ mol\ K^+}{1\ mol\ NaKSO_4}\right) = 4.25\ M\ \ K^+$$

$$(4.25\ M\ NaKSO_4)\left(\frac{1\ mol\ SO_4^{\ 2-}}{1\ mol\ NaKSO_4}\right) = 4.25\ M\ \ SO_4^{\ 2-}$$

(c) $$(0.20\ M\ CaCl_2)\left(\frac{1\ mol\ Ca^{2+}}{1\ mol\ CaCl_2}\right) = 0.20\ M\ \ Ca^{2+}$$

$$(0.20\ M\ CaCl_2)\left(\frac{2\ mol\ Cl^-}{1\ mol\ CaCl_2}\right) = 0.40\ M\ \ Cl^-$$

(d) $\left(\frac{22.0 \text{ g KI}}{500. \text{ mL}}\right)\left(\frac{1 \text{ mol}}{166.0 \text{ g}}\right)\left(\frac{1000 \text{ mL}}{\text{L}}\right) = 0.265 \text{ M KI}$

$(0.265 \text{ M KI})\left(\frac{1 \text{ mol K}^+}{1 \text{ mol KI}}\right) = 0.265 \text{ M K}^+$

$(0.265 \text{ M KI})\left(\frac{1 \text{ mol I}^-}{1 \text{ mol KI}}\right) = 0.265 \text{ M I}^-$

40. (a) $(0.75 \text{ M ZnBr}_2)\left(\frac{1 \text{ mol Zn}^{2+}}{1 \text{ mol ZnBr}_2}\right) = 0.75 \text{ M Zn}^{2+}$

$(0.75 \text{ M ZnBr}_2)\left(\frac{2 \text{ mol Br}^-}{1 \text{ mol ZnBr}_2}\right) = 1.5 \text{ M Br}^-$

(b) $(1.65 \text{ M Al}_2(\text{SO}_4)_3)\left(\frac{3 \text{ mol SO}_4{}^{2-}}{1 \text{ mol Al}_2(\text{SO}_4)_3}\right) = 4.95 \text{ M SO}_4{}^{2-}$

$(1.65 \text{ M Al}_2(\text{SO}_4)_3)\left(\frac{2 \text{ mol Al}^{3+}}{1 \text{ mol Al}_2(\text{SO}_4)_3}\right) = 3.30 \text{ M Al}^{3+}$

(c) $\left(\frac{900. \text{ g (NH}_4)_2\text{SO}_4}{20.0 \text{ L}}\right)\left(\frac{1 \text{ mol}}{132.1 \text{ g}}\right) = 0.341 \text{ M (NH}_4)_2\text{SO}_4$

$(0.341 \text{ M (NH}_4)_2\text{SO}_4)\left(\frac{2 \text{ mol NH}_4{}^+}{1 \text{ mol (NH}_4)_2\text{SO}_4}\right) = 0.682 \text{ M NH}_4{}^+$

$(0.341 \text{ M (NH}_4)_2\text{SO}_4)\left(\frac{1 \text{ mol SO}_4{}^{2-}}{1 \text{ mol (NH}_4)_2\text{SO}_4}\right) = 0.341 \text{ M SO}_4{}^{2-}$

(d) $\left(\frac{0.0120 \text{ g Mg(ClO}_3)_2}{0.00100 \text{ L}}\right)\left(\frac{1 \text{ mol}}{191.2 \text{ g}}\right) = 0.0628 \text{ M Mg(ClO}_3)_2$

$(0.0628 \text{ M Mg(ClO}_3)_2)\left(\frac{1 \text{ mol Mg}^{2+}}{1 \text{ mol Mg(ClO}_3)_2}\right) = 0.0628 \text{ M Mg}^{2+}$

$(0.0628 \text{ M Mg(ClO}_3)_2)\left(\frac{2 \text{ mol ClO}_3{}^-}{1 \text{ mol Mg(ClO}_3)_2}\right) = 0.126 \text{ M ClO}_3{}^-$

41. The molarities of each ion, as calculated in Excercise 39 will be used to calculate the mass of each ion present in 100. mL of solution.

(a) $(0.100 \text{ L})\left(\frac{0.015 \text{ mol Na}^+}{\text{L}}\right)\left(\frac{22.99 \text{ g}}{\text{mol}}\right) = 0.034 \text{ g Na}^+$

$(0.100 \text{ L})\left(\frac{0.015 \text{ mol Cl}^-}{\text{L}}\right)\left(\frac{35.45 \text{ g}}{\text{mol}}\right) = 0.053 \text{ g Cl}^-$

(b) $(0.100\ \text{L})\left(\frac{4.25\ \text{mol Na}^+}{\text{L}}\right)\left(\frac{22.99\ \text{g}}{\text{mol}}\right) = 9.77\ \text{g Na}^+$

$(0.100\ \text{L})\left(\frac{4.25\ \text{mol K}^+}{\text{L}}\right)\left(\frac{39.10\ \text{g}}{\text{mol}}\right) = 16.6\ \text{g K}^+$

$(0.100\ \text{L})\left(\frac{4.25\ \text{mol SO}_4^{2-}}{\text{L}}\right)\left(\frac{96.06\ \text{g}}{\text{mol}}\right) = 40.8\ \text{g SO}_4^{2-}$

(c) $(0.100\ \text{L})\left(\frac{0.20\ \text{mol Ca}^{2+}}{\text{L}}\right)\left(\frac{40.08\ \text{g}}{\text{mol}}\right) = 0.80\ \text{g Ca}^{2+}$

$(0.100\ \text{L})\left(\frac{0.40\ \text{mol Cl}^-}{\text{L}}\right)\left(\frac{35.45\ \text{g}}{\text{mol}}\right) = 1.4\ \text{g Cl}^-$

(d) $(0.100\ \text{L})\left(\frac{0.265\ \text{mol K}^+}{\text{L}}\right)\left(\frac{39.10\ \text{g}}{\text{mol}}\right) = 1.04\ \text{g K}^+$

$(0.100\ \text{L})\left(\frac{0.265\ \text{mol I}^-}{\text{L}}\right)\left(\frac{126.9\ \text{g}}{\text{mol}}\right) = 3.36\ \text{g I}^-$

42. (a) $(0.100\ \text{L})\left(\frac{0.75\ \text{mol Zn}^{2+}}{\text{L}}\right)\left(\frac{65.38\ \text{g}}{\text{mol}}\right) = 4.9\ \text{g Zn}^{2+}$

$(0.100\ \text{L})\left(\frac{1.5\ \text{mol Br}^-}{\text{L}}\right)\left(\frac{79.90\ \text{g}}{\text{mol}}\right) = 12\ \text{g Br}^-$

(b) $(0.100\ \text{L})\left(\frac{3.30\ \text{mol Al}^{3+}}{\text{L}}\right)\left(\frac{26.98\ \text{g}}{\text{mol}}\right) = 8.90\ \text{g Al}^{3+}$

$(0.100\ \text{L})\left(\frac{4.95\ \text{mol SO}_4^{2-}}{\text{L}}\right)\left(\frac{96.06\ \text{g}}{\text{mol}}\right) = 47.5\ \text{g SO}_4^{2-}$

(c) $(0.100\ \text{L})\left(\frac{0.682\ \text{mol NH}_4^+}{\text{L}}\right)\left(\frac{18.04\ \text{g}}{\text{mol}}\right) = 1.23\ \text{g NH}_4^+$

$(0.100\ \text{L})\left(\frac{0.341\ \text{mol SO}_4^{2-}}{\text{L}}\right)\left(\frac{96.06\ \text{g}}{\text{mol}}\right) = 3.28\ \text{g SO}_4^{2-}$

(d) $(0.100\ \text{L})\left(\frac{0.0628\ \text{mol Mg}^{2+}}{\text{L}}\right)\left(\frac{24.31\ \text{g}}{\text{mol}}\right) = 0.153\ \text{g Mg}^{2+}$

$(0.100\ \text{L})\left(\frac{0.126\ \text{mol ClO}_3^-}{\text{L}}\right)\left(\frac{83.45\ \text{g}}{\text{mol}}\right) = 1.05\ \text{g ClO}_3^-$

43. (a) $(30.0\ \text{mL})\left(\frac{1.0\ \text{mol NaCl}}{1000\ \text{mL}}\right) = 0.030\ \text{mol NaCl}$

$(40.0\ \text{mL})\left(\frac{1.0\ \text{mol NaCl}}{1000\ \text{mL}}\right) = 0.040\ \text{mol NaCl}$

Total mol NaCl = 0.030 mol + 0.040 mol = 0.070 mol NaCl

$\frac{0.070\ \text{mol NaCl}}{0.070\ \text{L}} = 1.0\ \text{M NaCl}$

$(1.0\ \text{M NaCl})\left(\frac{1\ \text{mol Na}^+}{1\ \text{mol NaCl}}\right) = 1.0\ \text{M Na}^+$

$(1.0\ \text{M NaCl})\left(\frac{1\ \text{mol Cl}^-}{1\ \text{mol NaCl}}\right) = 1.0\ \text{M Cl}^-$

(b) $HCl + NaOH \longrightarrow NaCl + H_2O$

$$(30.0 \text{ mL HCl})\left(\frac{1 \text{ L}}{1000 \text{ mL}}\right)\left(\frac{1.0 \text{ mol}}{\text{L}}\right) = 0.030 \text{ mol HCl}$$

$$(30.0 \text{ mL NaOH})\left(\frac{1 \text{ L}}{1000 \text{ mL}}\right)\left(\frac{1.0 \text{ mol}}{\text{L}}\right) = 0.030 \text{ mol NaOH}$$

0.030 mol HCl reacts with 0.030 mol NaOH and produces 0.030 mol NaCl. The final volume is 0.060 L. 0.030 mol NaCl/0.060 L = 0.50 M NaCl. Since there is one mole each of sodium and chloride ions per mole of NaCl, the molar concentrations of Na^+ and Cl^- will be 0.50 M Na^+, and 0.50 M Cl^-.

(c) $KOH + HCl \longrightarrow KCl + H_2O$

$$(100.0 \text{ mL})\left(\frac{1 \text{ L}}{1000 \text{ mL}}\right)\left(\frac{0.40 \text{ mol KOH}}{\text{L}}\right) = 0.040 \text{ mol KOH}$$

$$(100.0 \text{ mL})\left(\frac{1 \text{ L}}{1000 \text{ mL}}\right)\left(\frac{0.80 \text{ mol HCl}}{\text{L}}\right) = 0.080 \text{ mol HCl}$$

0.040 mol KOH reacts with 0.080 mol HCl. 0.040 mol HCl remains and 0.040 mol KCl is produced. The final volume is 200.0 mL and contains 0.040 mol HCl and 0.040 mol KCl. Moles of ions are: 0.040 mol H^+, 0.040 mol K^+, and 0.080 mol Cl^-.
Concentrations of ions are:

$$\frac{0.040 \text{ mol H}^+}{0.200 \text{ L}} = 0.20 \text{ M H}^+ \qquad \text{molarity K}^+ = \text{molarity H}^+$$

$$\frac{0.080 \text{ mol Cl}^-}{0.200 \text{ L}} = 0.40 \text{ M Cl}^-$$

44. (a) 100.0 mL of 2.0 M KCl and 100.0 mL of 1.0 M $CaCl_2$ are mixed, giving a final volume of 200.0 mL and concentrations of 1.0 M KCl and 0.5 M $CaCl_2$. The concentration of K^+ will be 1.0 M and the concentration of Ca^{2+} will be 0.5 M. The chloride ion concentration will be 2.0 M(1.0 M from the KCl and 2(0.5 M) from the $CaCl_2$).

(b) $$(35.0 \text{ mL})\left(\frac{1 \text{ L}}{1000 \text{ mL}}\right)\left(\frac{0.20 \text{ mol Ba(OH)}_2}{\text{L}}\right) = 0.0070 \text{ mol Ba(OH)}_2$$

$$(35.0 \text{ mL})\left(\frac{1 \text{ L}}{1000 \text{ mL}}\right)\left(\frac{0.20 \text{ mol H}_2\text{SO}_4}{\text{L}}\right) = 0.0070 \text{ mol H}_2\text{SO}_4$$

Final volume = 35.0 mL + 35.0 mL = 70.0 mL

H_2SO_4	+	$Ba(OH)_2$	$\longrightarrow$	$BaSO_{4(s)}$	+	$2\ H_2O$
0.0070 mol		0.0070 mol		0.0070 mol		0.014 mol

The H_2SO_4 and the $Ba(OH)_2$ react completely producing insoluble $BaSO_4$ and H_2O. No ions are present in solution.

(c) $(0.500\ \text{L NaCl})\left(\frac{2.0\ \text{mol}}{\text{L}}\right) = 1.0\ \text{mol NaCl}$

$(1.00\ \text{L AgNO}_3)\left(\frac{1.00\ \text{mol}}{\text{L}}\right) = 1.0\ \text{mol AgNO}_3$

$NaCl(aq)$ +	$AgNO_3(aq) \longrightarrow$	$AgCl(s)$ +	$NaNO_3(aq)$
1.0 mol	1.0 mol	1.0 mol	1.0 mol

The AgCl is insoluble and produces no ions. The 1.0 mol $NaNO_3$ will produce 1.0 mol Na^+ ions and 1.0 mol NO_3^- ions. The final volume of the solution is 1.5 L.
The concentration of ions are:

$$\frac{1.0\ \text{mol Na}^+}{1.5\ \text{L}} = 0.67\ \text{M Na}^+ \qquad \frac{1.0\ \text{mol NO}_3^-}{1.5\ \text{L}} = 0.67\ \text{M NO}_3^-$$

45. The reaction of HCl and NaOH occurs on a 1:1 mole ratio.

$$HCl + NaOH \longrightarrow NaCl + H_2O$$

At the endpoint in these titration reactions, equal moles of HCl and NaOH will have reacted. Moles = (molarity)(volume). At the endpoint, mol HCl = mol NaOH. Therefore, at the endpoint,

$M_AV_A = M_BV_B$

(a) $(37.70\ \text{mL})(0.728\ \text{M}) = (40.13\ \text{mL})(\text{M HCl})$

$$\text{M HCl} = \frac{(37.70\ \text{mL})(0.728\ \text{M})}{40.13\ \text{mL}} = 0.684\ \text{M HCl}$$

(b) $\frac{(33.66\ \text{mL})(0.306\ \text{M})}{19.00\ \text{mL}} = 0.542\ \text{M HCl}$

(c) $\frac{(18.00\ \text{mL})(0.555\ \text{M})}{27.25\ \text{mL}} = 0.367\ \text{M HCl}$

46. The reaction of HCl and NaOH occurs on a 1:1 mole ratio.

$$HCl + NaOH \longrightarrow NaCl + H_2O$$

At the endpoint in these titration reactions, equal moles of HCl and NaOH will have reacted. Moles = (molarity)(volume). At the endpoint, mol HCl = mol NaOH. Therefore, at the endpoint,

$M_AV_A = M_BV_B$

(a) $\frac{(37.19\ \text{mL})(0.126\ \text{M})}{31.91\ \text{mL}} = 0.147\ \text{M NaOH}$

(b) $\frac{(48.04\ \text{mL})(0.482\ \text{M})}{24.02\ \text{mL}} = 0.964\ \text{M NaOH}$

(c) $\frac{(13.13 \text{ mL})(1.425 \text{ M})}{39.39} = 0.4750 \text{ M NaOH}$

47. (a) $SO_4^{2-}(aq) + Ba^{2+}(aq) \longrightarrow BaSO_4(s)$

(b) $CaCO_3(s) + 2\ H^+(aq) \longrightarrow Ca^{2+}(aq) + CO_2(g) + H_2O(l)$

(c) $Mg(s) + 2\ HC_2H_3O_2(aq) \longrightarrow Mg^{2+}(aq) + H_2(g) + 2\ C_2H_3O_2^-(aq)$

48. (a) $H_2S(g) + Cd^{2+}(aq) \longrightarrow CdS(s) + 2\ H^+(aq)$

(b) $Zn(s) + 2\ H^+(aq) \longrightarrow Zn^{2+}(aq) + H_2(g)$

(c) $Al^{3+}(aq) + PO_4^{3-}(aq) \longrightarrow AlPO_4(s)$

49. The more acidic solution is listed followed by an explanation.

(a) 1 molar H_2SO_4. The concentration of H^+ in 1 M H_2SO_4 is greater than 1 M, since there are two ionizable hydrogens per mole of H_2SO_4. In HCl the concentration of H^+ will be 1 M, since there is only one ionizable hydrogen per mole of HCl.

(b) 1 molar HCl. HCl is a strong electrolyte, while $HC_2H_3O_2$ is a weak electrolyte.

50. The more acidic solution is listed followed by an explanation.

(a) 2 molar HCl. 2 M HCl will yield 2 M H^+ concentration. 1 M HCl will yield 1 M H^+ concentration.

(b) 1 molar H_2SO_4. 1 M H_2SO_4 is equivalent to 2 N H_2SO_4 and will therefore have a greater concentration of H^+ than 1 N H_2SO_4.

51. $2\ HCl + Ca(OH)_2 \longrightarrow CaCl_2 + 2\ H_2O$

M $Ca(OH)_2 \longrightarrow$ mol $Ca(OH)_2 \longrightarrow$ mol HCl $\longrightarrow$ mL HCl

$$(0.0500 \text{ L } Ca(OH)_2)\left(\frac{0.100 \text{ mol}}{\text{L}}\right)\left(\frac{2 \text{ mol HCl}}{1 \text{ mol } Ca(OH)_2}\right)\left(\frac{1000 \text{ mL}}{0.245 \text{ mol}}\right) = 40.8 \text{ mL of } 0.245 \text{ M HCl}$$

52. $3\ HCl + Al(OH)_3 \longrightarrow AlCl_3 + 3\ H_2O$

g $Al(OH)_3 \longrightarrow$ mol $Al(OH)_3 \longrightarrow$ mol HCl $\longrightarrow$ mL HCl

$$(10.0 \text{ g } Al(OH)_3)\left(\frac{1 \text{ mol}}{78.01 \text{ g}}\right)\left(\frac{3 \text{ mol HCl}}{1 \text{ mol } Al(OH)_3}\right)\left(\frac{1000 \text{ ml}}{0.245 \text{ mol}}\right) = 1.57 \times 10^3 \text{ mL of } 0.245 \text{ M HCl}$$

53. $NaOH + HCl \longrightarrow NaCl + H_2O$

L HCl $\longrightarrow$ mol HCl $\longrightarrow$ mol NaOH $\longrightarrow$ g NaOH

$$(0.01825 \text{ L HCl})\left(\frac{0.2406 \text{ mol}}{\text{L}}\right)\left(\frac{1 \text{ mol NaOH mol}}{1 \text{ mol HCl}}\right)\left(\frac{40.00 \text{ g}}{\text{mol}}\right) = 0.1756 \text{ g NaOH}$$

$$\left(\frac{0.1756 \text{ g NaOH}}{0.200 \text{ g sample}}\right)(100) = 87.8\% \text{ NaOH}$$

54. $NaOH + HCl \longrightarrow NaCl + H_2O$

L HCl $\longrightarrow$ mol HCl $\longrightarrow$ mol NaOH $\longrightarrow$ g NaOH

$$(0.04990\ \text{L HCl})\left(\frac{0.466\ \text{mol}}{\text{L}}\right)\left(\frac{1\ \text{mol NaOH}}{1\ \text{mol HCl}}\right)\left(\frac{40.00\ \text{g}}{\text{mol}}\right) = 0.930\ \text{g NaOH}$$

1.00 g sample − 0.930 g NaOH = 0.070 g NaCl in sample

$$\left(\frac{0.070\ \text{g NaCl}}{1.00\ \text{g sample}}\right)(100) = 7.0\%\ \text{NaCl in sample}$$

55. $Zn + 2\ HCl \longrightarrow ZnCl_2 + H_2$

This is a limiting reactant problem. First find the moles of Zn and HCl from the given data and then identify the limiting reactant.

g Zn $\longrightarrow$ mol Zn

$$(5.00\ \text{g Zn})\left(\frac{1\ \text{mol}}{65.38\ \text{g}}\right) = 0.0765\ \text{mol Zn}$$

$$(0.100\ \text{L HCl})\left(\frac{0.350\ \text{mol}}{\text{L}}\right) = 0.0350\ \text{mol HCl}$$

Therefore Zn is in excess and HCl is the limiting reactant.

$$(0.0350\ \text{mol HCl})\left(\frac{1\ \text{mol}\ H_2}{2\ \text{mol HCl}}\right) = 0.0175\ \text{mol}\ H_2$$

$$T = 27°C = 300.\ K$$

$$P = (700.\ \text{torr})\left(\frac{1\ \text{atm}}{760\ \text{torr}}\right) = 0.921\ \text{atm}$$

$$PV = nRT$$

$$V = \frac{nRT}{P} = \frac{(0.0175\ \text{mol})(0.0821\ \text{L atm/mol K})(300.\ \text{K})}{0.921\ \text{atm}} = 0.468\ \text{L}\ H_2$$

56. $Zn + 2\ HCl \longrightarrow ZnCl_2 + H_2$

This is a limiting reactant problem. First find the moles of Zn and HCl from the given data and then identify the limiting reactant.

g Zn $\longrightarrow$ mol Zn

$$(5.00\ \text{g Zn})\left(\frac{1\ \text{mol}}{65.38\ \text{g}}\right) = 0.0765\ \text{mol Zn}$$

$$(0.200\ \text{L HCl})\left(\frac{0.350\ \text{mol}}{\text{L}}\right) = 0.0700\ \text{mol HCl}$$

Zn is in excess and HCl is the limiting reactant.

$$(0.0700\ \text{mol HCl})\left(\frac{1\ \text{mol}\ H_2}{2\ \text{mol HCl}}\right) = 0.0350\ \text{mol}\ H_2$$

$$T = 27°C = 300.\ K$$

$P = (700.\ \text{torr})\left(\frac{1\ \text{atm}}{760\ \text{torr}}\right) = 0.921\ \text{atm}$

$PV = nRT$

$V = \frac{nRT}{P} = \frac{(0.0350\ \text{mol})(0.0821\ \text{L atm/mol K})(300.\ \text{K})}{0.921\ \text{atm}} = 0.936\ \text{L}\ H_2$

57. Calculation of the pH solutions:

(a) $H^+ = 0.01\ M = 1 \times 10^{-2}\ M$; $pH = -\log 1 \times 10^{-2} = 2.0$

(b) $H^+ = 1.0\ M$; $pH = -\log 1.0 = 0$

(c) $H^+ = 6.5 \times 10^{-9}\ M$; $pH = -\log (6.5 \times 10^{-9}) = 8.19$

58. (a) $H^+ = 1 \times 10^{-7}\ M$; $pH = -\log (1 \times 10^{-7}) = 7.0$

(b) $H^+ = 0.50\ M$; $pH = -\log (5.0 \times 10^{-1}) = 0.30$

(c) $H^+ = 0.00010\ M = 1.0 \times 10^{-4}\ M$; $pH = -\log (1.0 \times 10^{-4}) = 4.00$

59. (a) Orange juice $= 3.7 \times 10^{-4}\ M\ H^+$

$pH = -\log (3.7 \times 10^{-4}) = 3.43$

(b) Vinegar $= 2.8 \times 10^{-3}\ M\ H^+$

$pH = -\log (2.8 \times 10^{-3}) = 2.55$

60. (a) Black coffee $= 5.0 \times 10^{-5}\ M\ H^+$

$pH = -\log (5.0 \times 10^{-5}) = 4.30$

(b) Limewater $= 3.4 \times 10^{-11}\ M\ H^+$

$pH = -\log (3.4 \times 10^{-11}) = 10.47$

61. $V_A N_A = V_B N_B \qquad V_A = \frac{V_B N_B}{N_A} = \frac{(32.8\ \text{mL})(0.225\ \text{N})}{0.325\ \text{N}} = 22.7\ \text{mL of } 0.325\ \text{N}\ HNO_3$

62. $V_A N_A = V_B N_B \qquad V_A = \frac{V_B N_B}{N_A} = \frac{(32.8\ \text{mL})(0.225\ \text{N})}{0.325\ \text{N}} = 22.7\ \text{mL of } 0.325\ \text{N}\ H_2SO_4$

63. Equivalents of acid $=$ equivalents of base

$= N_B V_B$

$= (0.10\ \text{eq/L})(0.025\ \text{L})$

$= 2.5 \times 10^{-3}\ \text{eq}$

Equivalent mass $= \frac{0.305\ \text{g}}{2.5 \times 10^{-3}\ \text{eq}} = 1.2 \times 10^2\ \text{g/eq}$ benzoic acid

64. Equivalents of acid $= \text{equivalents of base}$

$$= N_B V_B$$

$$= (0.10 \text{ eq/L})(0.125 \text{ L})$$

$$= 1.3 \times 10^{-2} \text{ eq}$$

Equivalent mass $= \dfrac{0.738 \text{ g}}{1.3 \times 10^{-2} \text{ eq}} = 57 \text{ g/eq succinic acid}$

65. $CaI_2 \longrightarrow Ca^{2+} + 2\ I^-$

$$\left(\frac{0.520 \text{ mol } I^-}{L}\right)\left(\frac{1 \text{ mol } Ca^{2+}}{2 \text{ mol I}}\right) = \left(\frac{0.260 \text{ mol } Ca^{2+}}{L}\right) = 0.260 \text{ M } Ca^{2+}$$

66. Dilution problem

$V_1M_1 = V_2M_2$

$(100. \text{ mL})(12 \text{ M}) = (V_2)(0.40 \text{ M})$

$$V_2 = \frac{(100. \text{ mL})(12 \text{ M})}{0.40 \text{ M}} = 3.0 \times 10^3 \text{ mL}$$

67. $Ba(OH)_2 + 2\ HCl \longrightarrow BaCl_2 + 2\ H_2O$

$\text{M HCl} \longrightarrow \text{mol HCl} \longrightarrow \text{mol } Ba(OH)_2 \longrightarrow \text{M } Ba(OH)_2$

$$\left(\frac{0.430 \text{ mol HCl}}{L}\right)\left(\frac{1 \text{ L}}{1000 \text{ mL}}\right)(29.26 \text{ mL}) = 0.0126 \text{ mol HCl}$$

$$(0.0126 \text{ mol HCl})\left(\frac{1 \text{ mol } Ba(OH)_2}{2 \text{ mol HCl}}\right) = 0.00630 \text{ mol } Ba(OH)_2$$

$$\frac{0.00630 \text{ mol } Ba(OH)_2}{0.02040 \text{ L}} = 0.309 \text{ M } Ba(OH)_2$$

68. The acetic acid solution freezes at a lower temperature than the alcohol solution. The acetic acid ionizes slightly, while the alcohol does not. The ionization of the acetic acid increases its particle concentration above that of the alcohol solution, resulting in a lower freezing point for the acetic acid solution.

69. It is more economical to purchase CH_3OH at the same cost per pound as C_2H_5OH. Because CH_3OH has a lower molar mass than C_2H_5OH, the CH_3OH solution will contain more particles per pound in a given solution and therefore, have a greater effect on the freezing point of the radiator solution.

70. A hydronium ion is a hydrated hydrogen ion.

$$H^+ + H_2O \longrightarrow H_3O^+$$

(hydrogen ion) (hydronium ion)

71. Freezing point depression is directly related to the concentration of particles in the solution.

$$\begin{array}{lllllll} C_{12}H_{22}O_{11} & > & HC_2H_3O_2 > & HCl & > & CaCl_2 & \\ 1\text{ mol} & > & 1+\text{ mol} & > 2\text{ mol} & > & 3\text{ mol} & \text{(particles in solution)} \end{array}$$

72. (a) 100°C pH $= -\log(1 \times 10^{-6}) = 6.0$ pH of H_2O is greater at 25°C
25°C pH $= -\log(1 \times 10^{-7}) = 7.0$

(b) $1 \times 10^{-6} > 1 \times 10^{-7}$ so, H^+ concentration is higher at 100°C.

(c) The water is neutral at both temperatures, because the H_2O ionizes into equal concentrations of H^+ and OH^- at any temperature.

73. As the pH changes by 1 unit, the concentration of H^+ in solution changes by a factor of 10. For example, the pH of 0.10 M HCl is 1.00, while the pH of 0.010 M HCl is 2.00.

74. $Na_2CO_3 + 2\ HCl \longrightarrow 2\ NaCl + CO_2 + H_2O$

g $Na_2CO_3 \longrightarrow$ mol $Na_2CO_3 \longrightarrow$ mol HCl $\longrightarrow$ M HCl

$$(0.452\text{ g }Na_2CO_3)\left(\frac{1\text{ mol}}{106.0\text{ g}}\right)\left(\frac{2\text{ mol HCl}}{1\text{ mol }Na_2CO_3}\right) = 0.00853\text{ mol HCl}$$

$$\frac{0.00853\text{ mol}}{0.0424\text{ L}} = 0.201\text{ M HCl}$$

75. $2\ HCl + Ca(OH)_2 \longrightarrow CaCl_2 + 2\ H_2O$

g $Ca(OH)_2 \longrightarrow$ mol $Ca(OH)_2 \longrightarrow$ mol HCl $\longrightarrow$ mL HCl

$$(2.00\text{ g }Ca(OH)_2)\left(\frac{1\text{ mol}}{74.10\text{ g}}\right)\left(\frac{2\text{ mol HCl}}{1\text{ mol }Ca(OH)_2}\right)\left(\frac{1000\text{ mL}}{0.1234\text{ mol}}\right) = 437\text{ mL of }0.1234\text{ M HCl}$$

76. $KOH + HNO_3 \longrightarrow KNO_3 + H_2O$

L $HNO_3 \longrightarrow$ mol $HNO_3 \longrightarrow$ mol KOH $\longrightarrow$ g KOH

$$(0.05000\text{ L }HNO_3)\left(\frac{0.240\text{ mol}}{\text{L}}\right)\left(\frac{1\text{ mol KOH}}{1\text{ mol }HNO_3}\right)\left(\frac{56.11\text{ g}}{\text{mol}}\right) = 0.673\text{ g KOH}$$

77. pH of 1.0 L solution containing 0.1 mL of 1.0 M HCl

$$(0.1\text{ mL})\left(\frac{1.0\text{ L}}{1000\text{ mL}}\right)\left(\frac{1.0\text{ mol HCl}}{\text{L}}\right) = 1 \times 10^{-4}\text{ mol HCl}$$

$$\frac{1 \times 10^{-4}\text{ mol HCl}}{1.0\text{ L}} = 1 \times 10^{-4}\text{ mol HCl}$$

1×10^{-4} M HCl produces 1×10^{-4} M H^+

pH $= -\log(1 \times 10^{-4}) = 4.0$

78. Dilution problem

$$V_1M_1 = V_2M_2 \qquad V_1 = \frac{V_2M_2}{M_1} = \frac{(50.0\text{ L})(5.00\text{ M})}{18.0\text{ M}} = 13.9\text{ L of }18.0\text{ M }H_2SO_4$$

79. $NaOH + HCl \longrightarrow NaCl + H_2O$

$$(3.0\text{ g NaOH})\left(\frac{1\text{ mol}}{40.00\text{ g}}\right) = 0.075\text{ mol NaOH}$$

$$(500\text{ mL HCl})\left(\frac{1\text{ L}}{1000\text{ mL}}\right)\left(\frac{0.10\text{ mol}}{\text{L}}\right) = 0.050\text{ mol HCl}$$

The solution is basic. The NaOH will neutralize the HCl with an excess of 0.025 mol of NaOH remaining.

80. $$V_AN_A = V_BN_B \qquad V_B = \frac{V_AN_A}{V_B} = \frac{(28.92\text{ mL})(0.1240\text{ N})}{10.00\text{ mL}} = 0.3586\text{ N NaOH}$$

81. $$V_AN_A = V_BN_B \qquad N_A = \frac{V_BN_B}{V_A} = \frac{(22.68\text{ mL})(0.5000\text{ N})}{25.00\text{ mL}} = 0.4536\text{ N }H_3PO_4$$

$$\left(\frac{0.4536\text{ eq }H_3PO_4}{\text{L}}\right)\left(\frac{1\text{ mol}}{3\text{ eq}}\right) = 0.1512\text{ M }H_3PO_4$$

82. $2\ NaOH + H_2SO_4 \longrightarrow H_2O + 2\ Na_2SO_4$

$$N_AV_A = N_BV_B$$
$$(0.20\ N_A)(60.\text{ mL}) = (0.10\ N_B)(V_B)$$
$$1.2 \times 10^2\text{ mL} = V_B$$

83. (a) $N_AV_A = N_BV_B$

$$N_A(25\text{ mL}) = (0.20\ N_B)(40.\text{ mL})$$
$$N_A = 0.32\text{ N }H_2SO_4$$

(b) $$\left(\frac{0.32\text{ eq}}{\text{L}}\right)\left(\frac{1\text{ mol}}{2\text{ eq}}\right)(0.025\text{ L})\left(\frac{98.08\text{ g }H_2SO_4}{\text{mol}}\right) = 0.39\text{ g }H_2SO_4$$

84. $HCl + NaOH \longrightarrow NaCl + H_2O$

The residue is NaCl.

g NaCl $\longrightarrow$ mol NaCl $\longrightarrow$ mol HCl or mol NaOH

$$(0.117\text{ g NaCl})\left(\frac{1\text{ mol}}{58.44\text{ g}}\right)\left(\frac{1\text{ mol HCl}}{1\text{ mol NaCl}}\right) = 0.00200\text{ mol HCl or }0.00200\text{ mol NaOH}$$

Molarity and normality are the same for HCl and NaOH, since each has 1 eq/ mol.

$$\frac{0.00200\text{ eq HCl}}{0.0400\text{ L}} = 0.0500\text{ N HCl}$$

$$\frac{0.00200\text{ eq NaOH}}{0.0200\text{ L}} = 0.100\text{ N NaOH}$$

85. A concentrated solution of a weak electrolyte would contain many dissolved molecules, but only relatively few ions. The dilute solution of a strong electrolyte will contain essentially all of its particles as ions, the number of ions depending on the concentration.

86. The molarity and normality of an acid solution will be the same when the acid has one ionizable hydrogen. The molarity and normality of a base solution will be the same when the base has one ioniziable hydroxide. The normality and molarity of a salt (ionic compound) solution will be the same when the salt contains one ionizable cation with a charge of +1.

87. Dilution problem

$N_1V_1 = N_2V_2$ (applies for diluting normal solutions)

$(1.000\ \text{N})(85.00\ \text{mL}) = (0.6500\ \text{N})(V_2)$

$V_2 = 130.8\ \text{mL}$ (total volume of 0.6500 N solution)

$130.8\ \text{mL} - 85.00\ \text{mL} = 45.8\ \text{mL}\ H_2O$ to be added

88. $Ba(OH)_2(aq) + 2\ HCl(aq) \longrightarrow BaCl_2(aq) + 2\ H_2O(l)$

$(0.380\ \text{L}\ Ba(OH)_2)\left(\frac{0.35\ \text{mol}}{\text{L}}\right) = 0.13\ \text{mol}\ Ba(OH)_2$

$0.13\ \text{mol}\ Ba(OH)_2 \longrightarrow 0.26\ \text{mol}\ OH^-$

$(0.5000\ \text{L}\ HCl)\left(\frac{0.65\ \text{mol}}{\text{L}}\right) = 0.33\ \text{mol}\ HCl$

$0.33\ \text{mol}\ HCl \longrightarrow 0.33\ \text{mol}\ H^+$

0.33 mol H^+ will neutralize 0.26 mol OH^- and leave 0.07 mol H^+ (0.33 – 0.26) remaining in solution.

Total volume = 500.0 mL + 380 mL = 880 mL (0.88 L)

$[H^+]$ in solution $= \frac{0.07\ \text{mol}}{0.88\ \text{L}} = 0.08\ \text{M}$

$\text{pH} = -\log [H^+] = -\log (8 \times 10^{-2}) = 1.1$

89. $(0.05000\ \text{L}\ HCl)\left(\frac{0.2000\ \text{mol}}{\text{L}}\right) = 0.01000\ \text{mol}\ HCl = 0.01000\ \text{mol}\ H^+$ in 50.00 mL HCl

(a) no base added: $\text{pH} = -\log (0.2000) = 0.700$

(b) 10.00 mL base added: $(0.01000\ \text{L})\left(\frac{0.2000\ \text{mol}}{\text{L}}\right) = 0.002000\ \text{mol NaOH} = 0.002000\ \text{mol}\ OH^-$

$(0.01000\ \text{mol}\ H^+ - 0.002000\ \text{mol}\ OH^- = 0.00800\ \text{mol}\ H^+$ in 60.00 mL solution

$[H^+] = \frac{0.00800\ \text{mol}}{0.06000\ \text{L}}$ $\qquad \text{pH} = -\log\left(\frac{0.00800}{0.06000}\right) = 0.880$

(c) 25.00 mL base added: $(0.02500\ \text{L})\left(\frac{0.2000\ \text{mol}}{\text{L}}\right) = 0.005000\ \text{mol NaOH} = \text{mol}\ OH^-$

$(0.01000\ \text{mol}\ H^+ - 0.005000\ \text{mol}\ OH^-) = 0.00500\ \text{mol}\ H^+$ in 75.00 mL solution

$[H^+] = \frac{0.00500\ \text{mol}}{0.07500\ \text{L}}$ $\qquad \text{pH} = -\log\left(\frac{0.00500}{0.07500}\right) = 1.2$

(d) 49.00 mL base added: $(0.04900\ \text{L})\left(\frac{0.2000\ \text{mol}}{\text{L}}\right) = 0.009800$ mol NaOH = mol OH^-

0.01000 mol H^+ – 0.009800 mol OH^- = 0.00020 mol H^+ in 99.00 mL solution

$[H^+] = \frac{0.00020\ \text{mL}}{0.09900\ \text{L}}$ $\quad$ $\text{pH} = -\log\left(\frac{0.00020}{0.09900}\right) = 2.69$

(e) 49.90 mL base added: $(0.04990\ \text{L})\left(\frac{0.2000\ \text{mol}}{\text{L}}\right) = 0.009980$ mol NaOH = mol OH^-

0.01000 mol H^+ – 0.009980 mol OH^- = 2×10^{-5} mol H^+ in 99.9 mL solution

$[H^+] = \frac{2 \times 10^{-5}}{0.0999\ \text{L}}$ $\quad$ $\text{pH} = -\log\frac{2 \times 10^{-5}}{0.0999} = 3.7$

(f) 49.99 mL base added: $(0.04999\ \text{L})\left(\frac{0.2000\ \text{mol}}{\text{L}}\right) = 0.009998$ mol NaOH = mol OH^-

0.01000 mol HCl – 0.009998 mol OH^- = 2×10^{-6} mol H^+ in 99.99 mL solution

$[H^+] = \frac{2 \times 10^{-6}\ \text{mol}}{0.09999\ \text{L}}$ $\quad$ $\text{pH} = -\log\frac{2 \times 10^{-6}}{9.999 \times 10^{-2}} = 4.7$

(g) 50.00 mL of 0.2000 M NaOH neutralizes 50.00 mL of 0.2000 M HCl. No excess acid or base is in the solution. Therefore, the solution is neutral with a pH = 7.0

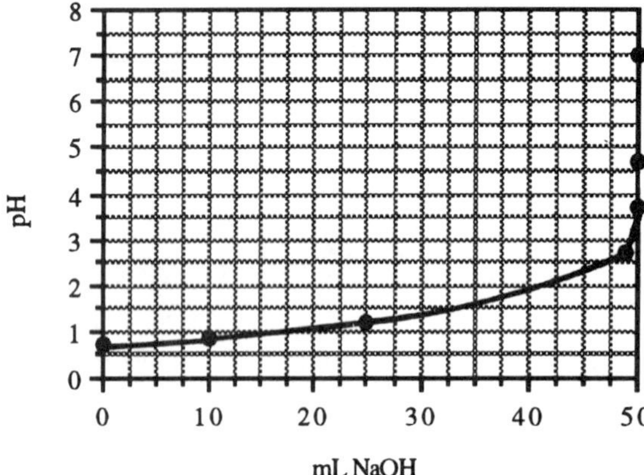

90. (a) $2\ NaOH(aq) + H_2SO_4(aq) \longrightarrow Na_2SO_4(aq) + 2\ H_2O(l)$

(b) mol H_2SO_4 ⟶ mol NaOH ⟶ mL NaOH

$$(0.0050\ \text{mol}\ H_2SO_4)\left(\frac{2\ \text{mol NaOH}}{1\ \text{mol}\ H_2SO_4}\right)\left(\frac{1000\ \text{mL}}{0.10\ \text{mol}}\right) = 1.0 \times 10^2\ \text{mL NaOH}$$

(c) $(0.0050\ \text{mol}\ H_2SO_4)\left(\frac{1\ \text{mol}\ Na_2SO_4}{1\ \text{mol}\ H_2SO_4}\right)\left(\frac{142.0\ \text{g}}{\text{mol}}\right) = 0.71\ \text{g}\ Na_2SO_4$

91. mol acid = mol base (lactic acid has one acidic H)

$$\frac{1.0\ \text{g Acid}}{\text{molar mass}} = (0.017\ \text{L})\left(\frac{0.65\ \text{mol}}{\text{L}}\right) = 90.\ \text{g/mol}$$

mass of empirical formula ($HC_3H_5O_3$) = 90.17 g/mol

molar mass = mass of empirical formula

Therefore the molecular formula is $HC_3H_5O_3$

92. $HNO_3 + KOH \longrightarrow KNO_3 + H_2O$

$M_AV_B = M_BV_B$

$(M_A)(25 \text{ mL}) = (0.60 \text{ M})(50.0 \text{ mL})$

$M_A = 1.2 \text{ M}$ (diluted solution)

Dilution problem $M_1V_1 = M_2V_2$

$(M_1)(10.0 \text{ mL}) = (1.2 \text{ M})(100.00 \text{ mL})$

$M_1 = 12 \text{ M } HNO_3$ (original solution)

93. Yes, adding water changes the concentration of the acid, which changes the concentration of the $[H^+]$, and changes the pH.

No, the solution theoretically will never reach a pH of 7, but it will approach pH 7 as water is added.

94. $pH = -\log [H^+]$ $\quad pH = 2, [H^+] = 1 \times 10^{-2}$
$\quad pH = 4, [H^+] = 1 \times 10^{-4}$

statement (c) is correct. The $[H^+]$ of X is 100 times that of Y

95. (a) The solution is neutral, neither acidic nor basic (pH = 7.0).

(b) The solution is basic, $[OH^-] > 10^{-7}$; pH = 12.

(c) The solution is acidic

(d) Cannot determine whether the solution is acidic or basic.

CHAPTER 16

CHEMICAL EQUILIBRIUM

1. At 25°C both tubes would appear the same and contain more molecules in the gaseous state than the tube at 0°C, and less molecules in the gaseous state than the tube at 80°C.

2. The reaction is endothermic because the increased temperature increases the concentration of product present at equilibrium.

3. At equilibrium, the rate of the forward reaction equals the rate of the reverse reaction.

4. The yield of NH_3 would be greater if the reaction were carried out in a one liter vessel. The pressure would be greater in the 1 L vessel. Increased pressure favors more product because the reaction of four moles of reactants (3 moles H_2 and 1 mole N_2) produces 2 moles of product, NH_3.

5. The sum of the pH and the pOH is 14. A solution whose pH is -1 would have a pOH of 15.

6. Acids stronger than acetic acid are: benzoic, cyanic, formic, hydrofluoric, and nitrous acids (all equilibrium constants are greater than the equilibrium constant for acetic acid). Acids weaker than acetic acid are: carbolic, hydrocyanic, and hypochlorous acids (all have equilibrium constants smaller than the equilibrium constant for acetic acid). All have one ionizable hydrogen atom.

7. The order of solubility will correspond to the order of the values of the solubility product constants of the salts being compared. This occurs because each salt in the comparison produces the same number of ions (two in this case) for each formula unit of salt that dissolves. This type of comparison would not necessarily be valid if the salts being compared gave different numbers of ions per formula unit of salt dissolving. The order is: $AgC_2H_3O_2$, $PbSO_4$, $BaSO_4$, AgCl, $BaCrO_4$, AgBr, AgI, PbS.

8. (a) K_{sp} $Mn(OH)_2$ $= 2.0 \times 10^{-13}$; K_{sp} Ag_2CrO_4 $= 1.9 \times 10^{-12}$.

   Each salt gives 3 ions per formula units of salt dissolving. Therefore, the salt with the largest K_{sp} (in this case Ag_2CrO_4) is more soluble.

   (b) K_{sp} $BaCrO_4$ $= 8.5 \times 10^{-11}$; K_{sp} Ag_2CrO_4 $= 1.9 \times 10^{-12}$. Ag_2CrO_4 has a greater molar solubility than $BaCrO_4$, even though its K_{sp} is smaller, because the Ag_2CrO_4 produces more ions per formula unit of salt dissolving than $BaCrO_4$.

   $BaCrO_{4(s)} \rightleftarrows Ba^{2+} + CrO_4^{2-}$ $\qquad$ $K_{sp} = [Ba^{2+}]\,[CrO_4^{2-}]$

   Let y = molar solubility of $BaCrO_4$

$K_{sp} = [y]\,[y] = 8.5 \times 10^{-11}$

$y = \sqrt{8.5 \times 10^{-11}} = 9.2 \times 10^{-6}$ mol $BaCrO_4$/L

$Ag_2CrO_4(s) \rightleftarrows 2\,Ag^+ + CrO_4^{\,2-}$ $\qquad K_{sp} = [Ag^+]^2[CrO_4^{\,2-}]$

Let y = molar solubility of Ag_2CrO_4

$K_{sp} = [2y]^2[y] = 1.9 \times 10^{-12}$

$y = \sqrt[3]{\frac{1.9 \times 10^{-12}}{4}} = 7.8 \times 10^{-5}$ mol Ag_2CrO_4/L

9. $HC_2H_3O_2 \rightleftarrows H^+ + C_2H_3O_2^-$

Inital Concentrations		Added	Concentration After Equilibrium Shifts
$HC_2H_3O_2$	1.00 M	------	1.01 M
H^+	1.8×10^{-5} M	0.010 M	1.9×10^{-5}
$C_2H_3O_2^-$	1.00 M	------	0.99 M

The initial concentration of H^+ in the buffer solution is very low (1.8×10^{-5} M) because of the large excess of acetate ions. 0.010 mole of HCl is added to one liter of the buffer solution. This will supply 0.010 M H^+. The added H^+ creates a stress on the right side of the equation. The equilibrium shifts to the left, using up almost all the added H^+, reducing the acetate ion by approximately 0.010 M, and increasing the acetic acid by approximately 0.010 M. The concentration of H^+ will not increase significantly and the pH is maintained relatively constant.

10. If the reaction shown in Figure 16.4 were exothermic, the figure should be modified to show the energy level of the product as lower than the energy level of the reactants.

11. In a saturated sodium chloride solution, the equilibrium is

$Na^+(aq) + Cl^-(aq) \rightleftarrows NaCl(s)$

Bubbling in HCl gas increases the concentration of Cl^-, creating a stress, which will cause the equilibrium to shift to the right, precipitating solid NaCl.

12. The rate of a reaction increases when the concentration of one of the reactants increases. The increase in concentration causes the number of collisions between the reactants to increase. The rate of a reaction, being proportional to the frequency of such collisions, as a result will increase.

13. If pure HI is placed in a vessel at 700 K, some of it will decompose. The mechanism would be for two molecules to collide, forming the activated complex, which would, in turn, decompose to form H_2 and I_2. $2\,HI \rightleftarrows [H_2I_2] \rightleftarrows H_2 + I_2$

14. An increase in temperature causes the rate of reaction to increase, because it speeds up the motion of the molecules. Faster moving molecules increase the number and effectiveness of the collisions, resulting in enough energy transfer to cause a reaction.

15. $A + B \rightleftarrows C + D$

When A and B are initially mixed, the rate of the forward reaction to produce C and D is at its maximum. As the reaction proceeds, the rate of production of C and D decreases because the concentrations of A and B decrease. As C and D are produced, some of the collisions between C and D will result in the reverse reactions, forming A and B. Finally, an equilibrium is achieved in which the forward rate exactly equals the reverse rate.

16. $HC_2H_3O_2 + H_2O \leftrightarrows H_3O^+ + C_2H_3O_2^-$
As water is added (diluting the solution from 1.0 M to 0.10 M), the equilibrium shifts to the right, yielding a higher percent ionization.

17. The statement does not contradict Le Chatelier's Principle. The previous question deals with the case of dilution. If pure acetic acid is added to a dilute solution, the reaction will shift to the right, producing more ions in accordance with Le Chatelier's Principle. But, the concentration of the un-ionized acetic acid will increase faster than the concentration of the ions, thus yielding a smaller percent ionization.

18. At different temperatures, the degree of ionization of water varies, being higher at higher temperatures. Consequently, the pH of water can be different at different temperatures.

19. In pure water, H^+ and OH^- are produced in equal quantities by the ionization of the water molecules, $H_2O \leftrightarrows H^+ + OH^-$. Since pH = –log $[H^+]$, and pOH = –log $[OH^-]$, they will always be identical for pure water. At 25°C, they each have the value of 7, but at higher temperatures, the degree of ionization is greater, so the pH and pOH would both be less than 7, but still equal.

20. In water, the silver acetate dissociates until the equilibrium concentration of ions is reached. In nitric acid solution, the acetate ions will react with hydrogen ions to form acetic acid molecules. Since the acetate ion concentration is kept low, more silver acetate can dissolve. If HCl is used, a precipitate of silver chloride would be formed, since silver chloride is less soluble than silver acetate. Thus, more silver acetate would dissolve in HCl than in pure water.

$$AgC_2H_3O_2(s) \leftrightarrows Ag^+(aq) + C_2H_3O_2^-(aq)$$

21. When the salt, sodium acetate, is dissolved in water, the solution becomes basic. The dissolving reaction is

$$NaC_2H_3O_2(s) \xrightarrow{H_2O} Na^+(aq) + C_2H_3O_2^-(aq)$$

The acetate ion reacts with water (hydrolysis). The reaction does not go to completion, but some OH^- ions are produced and at equilibrium the solution is basic.

$$C_2H_3O_2^-(aq) + H_2O(l) \leftrightarrows OH^-(aq) + HC_2H_3O_2(aq)$$

22. A buffer solution contains a weak acid or base plus a salt of that weak acid or base, such as dilute acetic acid and sodium acetate.

$$HC_2H_3O_2(aq) \leftrightarrows H^+(aq) + C_2H_3O_2^-(aq)$$
$$NaC_2H_3O_2(aq) \leftrightarrows Na^+(aq) + C_2H_3O_2^-(aq)$$

When a small amount of a strong acid (H^+) is added to this buffer solution, the H^+ reacts with the acetate ions to form un-ionized acetic acid, thus neutralizing the added acid. When a strong base, OH^-, is added it reacts with un-ionized acetic acid to neutralize the added base. As a result, in both cases, the approximate pH of the solution is maintained.

23. The equilibrium reaction is

$$Hb + O_2 \rightleftarrows HbO_2$$

In the lungs the concentration of O_2 is high and favors binding O_2 to hemoglobin. From there it is transported to the tissues.

24. Both molecules bind oxygen. The myoglobin molecule binds oxygen differently than hemoglobin. Since the affintiy between oxygen and myoglobin is higher than that between oxygen and hemoglobin, the hemoglobin will release oxygen to myoglobin for storage.

25. Hemoglobin depleted of oxygen has the potential to carry CO_2. The CO_2 binds at one end of the hemoglobin protein chain at a different site than O_2. When CO_2 from the tissues dissolve in water the following reaction occurs:

$$CO_2 + H_2O \rightleftarrows HCO_3^- + H^+$$

To facilitate removal of CO_2 the hemoglobin binds H^+ shifting the equilibrium towards the right liberating, CO_2. In the lungs the whole process is reversed, liberating CO_2.

26. The correct statements are c, d, g, i, j, k, l, n, p, q, s, t, u, v, z

(a) In a reaction, at equilibrium, the concentration of reactants and products usually are not equal.

(b) A catalyst has no effect on the concentration of the products at equilibrium.

(e) If an increase in temperature causes an increase in the concentration of products present at equilibrium, the reaction is endothermic.

(f) The magnitude of an equilibrium constant is dependent upon temperature.

(h) The amount of product obtained at equilibrium is independent of the time required to reach equilibrium.

(m) The reaction shown is exothermic.

(o) Increasing the temperature will decrease the magnitude of the equilibrium constant, K_{eq}.

(r) High temperatures lead to decreased yields of NO_2.

(w) Addition of solid $BaSO_4$ to a saturated solution of $BaSO_4$ has no effect on the magnitude of K_{sp}.

(x) A solution of pOH 10 is acidic.

(y) The pH decreases as the $[H^+]$ increases.

27. Reversible systems

(a) $H_2O(s) \overset{0^\circ C}{\leftrightarrows} H_2O(l)$

(b) $Na_2SO_4(s) \leftrightarrows 2\ Na^+(aq) + SO_4^{2-}(aq)$

28. Reversible systems

(a) $H_2O(l) \overset{100^\circ C}{\leftrightarrows} H_2O(g)$

(b) $SO_2(l) \leftrightarrows SO_2(g)$

29. Equilibrium system

$$4\ NH_3(g) + 3\ O_2(g) \leftrightarrows 2\ N_2(g) + 6\ H_2O(g) + 1531\ kJ$$

(a) The reaction is exothermic with heat being evolved.

(b) The addition of O_2 will shift the reaction to the right until equilibrium is reestablished. The concentration of N_2, H_2O, and O_2 will be increased. The concentration of the NH_3 will be decreased.

30. Equilibrium system

$$4\ NH_3(g) + 3\ O_2(g) \leftrightarrows 2\ N_2(g) + 6\ H_2O(l) + 1531\ kJ$$

(a) The addition of N_2 will shift the reaction to the left until equilibrium is reestablished. The concentration of NH_3, O_2, and N_2 will be increased. The concentration of H_2O will be decreased.

(b) The addition of heat will shift the reaction to the left. This shift will use up the heat added.

31. $N_2(g) + 3\ H_2(g) \leftrightarrows 2\ NH_3(g) + 92.5\ kJ$

Change or stress imposed on the system at equilibrium	Direction of reaction, left or right, to re-establish equilibrium	Change in number of moles N_2	H_2	NH_3
(a) Add N_2	right	I	D	I
(b) Remove H_2	left	I	D	D
(c) Decrease volume of reaction vessel	right	D	D	I
(d) Increase temperature	left	I	I	D

I = Increase; D = Decrease; N = No Change;
? = insufficient information to determine

32. $N_2(g) + 3\,H_2(g) \rightleftarrows 2\,NH_3(g) + 92.5$ kJ

	Change or stress imposed on the system at equilibrium	Direction of reaction, left or right, to re-establish equilibrium	Change in number of moles N_2	H_2	NH_3
(a)	Add NH_3	left	I	I	D
(b)	Increase volume of reaction vessel	left	I	I	D
(c)	Add a catalyst	no change	N	N	N
(d)	Add H_2 and NH_3	?	?	I	I

I = Increase; D = Decrease; N = No Change;
? = insufficient information to determine

33. Direction of shift in equilibrium:

Reaction	Increase Temperature	Increased Pressure (Volume Decrease)	Add Catalyst
(a)	right	right	no change
(b)	left	no change	no change
(c)	left	right	no change

34. Direction of shift in equilibrium:

Reaction	Increase Temperature	Increased Pressure (Volume Decrease)	Add Catalyst
(a)	right	left	no change
(b)	left	left	no change
(c)	left	left	no change

35. (a) right
(b) left
(c) none

36. (a) left
(b) right
(c) right

37. (a) $K_{eq} = \dfrac{[Cl_2]^2\,[H_2O]^2}{[HCl]^4\,[O_2]}$

(b) $K_{eq} = \dfrac{[NH_3]^2}{[N_2]\,[H_2]^3}$

(c) $K_{eq} = \dfrac{[PCl_3]\,[Cl_2]}{[PCl_5]}$

38. (a) $K_{eq} = \dfrac{[H^+]\,[ClO_2^-]}{[HClO_2]}$

(b) $K_{eq} = \dfrac{[H^+]\,[C_2H_3O_2^-]}{[HC_2H_3O_2]}$

(c) $K_{eq} = \dfrac{[NO]^4\,[H_2O]^6}{[NH_3]^4\,[O_2]^5}$

39. (a) $K_{sp} = [Cu^{2+}]\,[S^{2-}]$

(b) $K_{sp} = [Ba^{2+}]\,[SO_4^{2-}]$

(c) $K_{sp} = [Pb^{2+}]\,[Br^-]^2$

(d) $K_{sp} = [Ag^+]^3\,[AsO_4^{3-}]$

40. (a) $K_{sp} = [Fe^{3+}]\,[OH^-]^3$

(b) $K_{sp} = [Sb^{5+}]^2\,[S^{2-}]^5$

(c) $K_{sp} = [Ca^{2+}]\,[F^-]^2$

(d) $K_{sp} = [Ba^{2+}]^3\,[PO_4^{3-}]^2$

41. If the H^+ ion concentration is decreased:

(a) pH is increased

(b) pOH is decreased

(c) $[OH^-]$ is increased

(d) K_w remains the same. K_w is a constant at a given temperature.

42. If the H^+ ion concentration is increased:

(a) pH is decreased (pH of 1 is more acidic than that of 4)
(b) pOH is increased
(c) OH^- is decreased
(d) K_w remains unchanged. K_w is a constant at a given temperature.

43. The basis for deciding if a salt dissolved in water produces an acidic, a basic, or a neutral solution, is whether or not the salt reacts with water (hydrolysis). Salts that contain an ion derived from a weak acid or base will hydrolyze to produce an acidic or a basic solution.

(a) KCl, neutral

(b) Na_2CO_3, basic

(c) K_2SO_4, neutral

(d) $(NH_4)_2SO_4$, acidic

44. The basis for deciding if a salt dissolved in water produces an acidic, a basic, or a neutral solution, is whether or not the salt reacts with water (hydrolysis). Salts that contain an ion derived from a weak acid or base will hydrolyze to produce an acidic or a basic solution.

(a) $Ca(CN)_2$, basic

(b) $BaBr_2$, neutral

(c) $NaNO_2$, basic

(d) NaF, basic

45. (a) $NO_2^-(aq) + H_2O(l) \rightleftarrows OH^-(aq) + HNO_2(aq)$

(b) $C_2H_3O_2^-(aq) + H_2O(l) \rightleftarrows OH^-(aq) + HC_2H_3O_2(aq)$

46. (a) $NH_4^+(aq) + H_2O(l) \rightleftarrows H_3O^+(aq) + NH_3(aq)$

(b) $SO_3^{2-}(aq) + H_2O(l) \rightleftarrows OH^-(aq) + HSO_3^-(aq)$

47. (a) $HCO_3^-(aq) + H_2O(l) \rightleftarrows OH^-(aq) + H_2CO_3(aq)$

(b) $NH_4^+(aq) + H_2O(l) \rightleftarrows H_3O^+(aq) + NH_3(aq)$

48. (a) $OCl^-(aq) + H_2O(l) \rightleftarrows OH^-(aq) + HOCl(aq)$

(b) $ClO_2^-(aq) + H_2O(l) \rightleftarrows OH^-(aq) + HClO_2(aq)$

49. When excess acid (H^+) gets into the blood stream it reacts with HCO_3^- to form un-ionized H_2CO_3, thus neutralizing the acid and maintaining the approximate pH of the solution.

50. When excess base gets into the blood stream it reacts with H^+ to form water. Then H_2CO_3 ionizies to replace H^+, thus maintaining the approximate pH of the solution.

51. (a) $HC_2H_3O_2 \rightleftarrows H^+ + C_2H_3O_2^-$

Let x = molarity of $HC_2H_3O_2$ ionizing to establish equilibrium. Equilibrium concentrations are:

$[H^+] = [C_2H_3O_2^-] = x$

$[HC_2H_3O_2] = 0.25 - x = 0.25$ (since x is small)

$$K_a = \frac{[H^+][C_2H_3O_2^-]}{[HC_2H_3O_2]} = \frac{x^2}{0.25} = 1.8 \times 10^{-5}$$

$x^2 = (0.25)(1.8 \times 10^{-5})$

$x = \sqrt{(0.25)(1.8 \times 10^{-5})} = 2.1 \times 10^{-3}\ M = [H^+]$

(b) $pH = -\log [H^+] = -\log (2.1 \times 10^{-3}) = 2.68$

(c) Percent ionization

$$\frac{[H^+]}{[HC_2H_3O_2]}(100) = \left(\frac{2.1 \times 10^{-3}}{0.25}\right)(100) = 0.84\%$$

52. (a) $HC_6H_5O(aq) \rightleftarrows H^+(aq) + C_6H_5O^-(aq)$

Let x = molarity of HC_6H_5O ionizing to establish equilibrium. Equilibrium concentrations are:

$[H^+] = [C_6H_5O^-] = x$

$[HC_6H_5O] = 0.25 - x = 0.25$ (since x is small)

$$K_a = \frac{[H^+]\,[C_6H_5O^-]}{[HC_6H_5O]} = \frac{x^2}{0.25} = 1.3 \times 10^{-10}$$

$x^2 = (0.25)(1.3 \times 10^{-10})$

$x = \sqrt{(0.25)(1.3 \times 10^{-10})} = 5.7 \times 10^{-6} = [H^+]$

(b) $pH = -\log [H^+] = -\log [5.7 \times 10^{-6}] = 5.24$

(c) Percent ionization

$$\left(\frac{[H^+]}{[HC_6H_5O]}\right)(100) = \left(\frac{5.7 \times 10^{-6}}{0.25}\right)(100) = 2.3 \times 10^{-3}\ \%$$

53. $HA \rightleftarrows H^+ + A^-$ $\quad 0.52\% = 0.0052$

$[H^+] = [A^-] = (1.000\ M)(0.0052) = 5.2 \times 10^{-3}\ M$

$[HA] = 1.000\ M - 0.0052\ M = 0.9948\ M$

$$K_a = \frac{[H^+]\,[A^-]}{[HA]} = \frac{(5.2 \times 10^{-3})^2}{0.9948} = 2.7 \times 10^{-5}$$

54. $HA \rightleftarrows H^+ + A^-$ $\quad pH = 5 = -\log [H^+] \quad [H^+] = 1 \times 10^{-5}\ M$

$[H^+] = [A^-] = 1 \times 10^{-5}\ M$

$[HA] = 0.15\ M - 1 \times 10^{-5}\ M = 0.15\ M$

$$K_a = \frac{[H^+]\,[A^-]}{[HA]} = \frac{(1 \times 10^{-5})^2}{0.15} = 7 \times 10^{-10}$$

55. $HC_2H_3O_2 \rightleftarrows H^+ + C_2H_3O_2^-$

$$K_a = \frac{[H^+][C_2H_3O_2^-]}{[HC_2H_3O_2]} = 1.8 \times 10^{-5}$$

Let x = molarity of $HC_2H_3O_2$, which is ionized, to establish equilibrium. Equilibrium concentrations are:

$[H^+] = [C_2H_3O_2^-] = x$

$[HC_2H_3O_2]$ = intitial concentration $-$ x = initial concentration

Since K_a is small, the degree of ionization is small. Therefore, the approximation, initial concentration – x = initial concentration, is valid.

(a) $[H^+] = [C_2H_3O_2^-] = x$ $\qquad [HC_2H_3O_2] = 1.0\ M$

$$\frac{(x)(x)}{1.0} = 1.8 \times 10^{-5}$$

$$x^2 = (1.0)(1.8 \times 10^{-5})$$

$$x = \sqrt{1.8 \times 10^{-5}} = 4.2 \times 10^{-3}\ M$$

$$\left(\frac{4.2 \times 10^{-3}\ M}{1.0\ M}\right)(100) = 0.42\%\ \text{ionized}$$

$$pH = -\log\ (4.2 \times 10^{-3}) = 2.38$$

(b) $[HC_2H_3O_2] = 0.10\ M$

$$\frac{(x)(x)}{0.10} = 1.8 \times 10^{-5}$$

$$x^2 = (0.10)(1.8 \times 10^{-5}) = 1.8 \times 10^{-6}$$

$$x = \sqrt{1.8 \times 10^{-6}} = 1.3 \times 10^{-3}\ M$$

$$\left(\frac{1.3 \times 10^{-3}\ M}{0.10\ M}\right)(100) = 1.3\%\ \text{ionized}$$

$$pH = -\log\ (1.3 \times 10^{-3}) = 2.89$$

(c) $[HC_2H_3O_2] = 0.010\ M$

$$\frac{(x)(x)}{0.010} = 1.8 \times 10^{-5}$$

$$x^2 = (0.010)(1.8 \times 10^{-5}) = 1.8 \times 10^{-7}$$

$$x = \sqrt{1.8 \times 10^{-7}} = 4.2 \times 10^{-4}\ M$$

$$\left(\frac{4.2 \times 10^{-4}\ M}{0.010\ M}\right)(100) = 4.2\%\ \text{ionized}$$

$$pH = -\log\ (4.2 \times 10^{-4}) = 3.38$$

56. $HClO \rightleftarrows H^+ + ClO^-$

$$K_a = \frac{[H^+]\,[ClO^-]}{[HClO]} = 3.5 \times 10^{-8}$$

Let x = molarity of HClO, which is ionized, to establish equilibrium. Equilibrium concentrations are:

$[H^+] = [ClO^-] = x$

$[HClO]$ = initial concentration $- x$ = initial concentration

Since K_a is small, the degree of ionization is small. Therefore, the approximation, initial concentration $- x$ = initial concentration

(a) $[H^+] = [ClO^-] = x$ $\qquad$ $[HClO] = 1.0$ M

$\frac{(x)(x)}{1.0} = 3.5 \times 10^{-8}$ $\qquad$ $x^2 = (3.5 \times 10^{-8})(1.0)$

$x = \sqrt{3.5 \times 10^{-8}} = 1.9 \times 10^{-4}$ M

$\left(\frac{1.9 \times 10^{-7}\text{ M}}{1.0\text{ M}}\right)(100) = 1.9 \times 10^{-2}\%$ ionized

$pH = -\log(1.9 \times 10^{-4}) = 3.72$

(b) $[HClO] = 0.10$ M

$\frac{(x)(x)}{0.10} = 3.5 \times 10^{-8}$ $\qquad$ $x^2 = (0.10)(3.5 \times 10^{-8})$

$x = \sqrt{3.5 \times 10^{-9}} = 5.9 \times 10^{-5}$ M

$\left(\frac{5.9 \times 10^{-5}\text{ M}}{0.10\text{ M}}\right)(100) = 0.059\%$ ionized

$pH = -\log(5.9 \times 10^{-8}) = 4.23$

(c) $[HClO] = 0.10$ M

$\frac{(x)(x)}{0.10} = 3.5 \times 10^{-8}$ $\qquad$ $x^2 = (0.010)(3.5 \times 10^{-8})$

$x = \sqrt{3.5 \times 10^{-10}} = 1.9 \times 10^{-5}$ M

$\left(\frac{1.9 \times 10^{-5}\text{ M}}{0.010\text{ M}}\right)(100) = 0.19\%$ ionized

$pH = -\log(1.9 \times 10^{-5}) = 4.72$

57. HA $\leftrightarrows$ $H^+ + A^-$ $\quad K_a = \dfrac{[H^+][A^-]}{[HA]}$

First, find the $[H^+]$. This is calculated from the pH expression, pH = $-\log [H^+]$ = 3.7. Enter –3.7 into the calculator and push the 10^x key.
This yields the $[H^+] = 2 \times 10^{-4}$.

$[H^+] = [A^-] = 2 \times 10^{-4}$ $\qquad [HA] = 0.37$

$$K_a = \frac{[H^+][A^-]}{[HA]} = \frac{(2 \times 10^{-4})(2 \times 10^{-4})}{0.37} = 1 \times 10^{-7}$$

58. See problem 57 for a discussion of calculating $[H^+]$ from pH.

HA $\leftrightarrows$ $H^+ + A^-$ $\quad K_a = \dfrac{[H^+][A^-]}{[HA]}$

$-\log [H^+] = 2.89$ $\quad [H^+] = 1.3 \times 10^{-3}$ $\quad$ pH = 2.89

$[H^+] = 1.3 \times 10^{-3} = [A^-]$ $\qquad [HA] = 0.23$

$$K_a = \frac{[H^+][A^-]}{[HA]} = \frac{(1.3 \times 10^{-3})(1.3 \times 10^{-3})}{0.23} = 7.3 \times 10^{-6}$$

59. 6.0 M HCl yields $[H^+]$ = 6.0 M $\quad$ (100% ionized)

pH = $-\log 6.0$ = –0.78

pOH = 14 – pH = 14 – (–0.78) = 14.78

$$[OH^-] = \frac{K_w}{[H^+]} = \frac{1 \times 10^{-14}}{6.0} = 1.7 \times 10^{-15}$$

60. 1.0 M NaOH yields $[OH^-]$ = 1.0 M $\quad$ (100% ionized)

pOH = $-\log 1.0$ = 0.00

pH = 14 – pOH = 14.00

$$[H^+] = \frac{K_w}{OH^-} = \frac{1.0 \times 10^{-14}}{1.0} = 1 \times 10^{-14}$$

61. pH + pOH = 14.0 $\qquad$ pOH = 14.0 – pH

(a) 0.00010 M HCl $\quad [H^+]$ = 0.00010 M = 1.0×10^{-4} M

pH = $-\log (1.0 \times 10^{-4})$ = 4.00

pOH = 14.0 – 4.00 = 10.0

(b) 0.010 M NaOH $\quad [OH^-]$ = 0.010 M = 1.0×10^{-2} M

pOH = $-\log (1.0 \times 10^{-2})$ = 2.00

pH = 14.0 – 2.00 = 12.0

62. pH + pOH = 14.0

(a) 0.0025 M NaOH $[OH^-] = 2.5 \times 10^{-3}$ M

$pOH = -\log(2.5 \times 10^{-3}) = 2.60$

$pH = 14.0 - 2.60 = 11.4$

(b) $HClO \rightleftarrows H^+ + ClO^-$

0.10 M x x

$$K_a = \frac{[H^+][ClO^-]}{[HClO]} = 3.5 \times 10^{-8}$$

$$\frac{(x)(x)}{0.10} = 3.5 \times 10^{-8}$$

$x^2 = (0.10)(3.5 \times 10^{-8})$ $x = \sqrt{3.5 \times 10^{-9}}$

$x = 5.9 \times 10^{-5} = [H^+]$

$pH = -\log(5.9 \times 10^{-5}) = 4.23$

$pOH = 14.0 - 4.23 = 9.8$

(c) $Fe(OH)_{2(s)} \rightleftarrows Fe^{2+} + 2\ OH^-$

x x $2x$

$K_{sp} = [Fe^{2+}][OH^-]^2 = (x)(2x)^2 = 8.0 \times 10^{-16}$

$4x^3 = 8.0 \times 10^{-16}$

$$x = \sqrt[3]{\frac{8.0 \times 10^{-16}}{4}} = 5.8 \times 10^{-6}$$

$[OH^-] = 2x = 2(5.8 \times 10^{-6}) = 1.2 \times 10^{-5}$

$pOH = -\log(1.2 \times 10^{-5}) = 4.92$

$pH = 14.0 - 4.92 = 9.1$

63. Calculate the $[OH^-]$. $[OH^-] = \dfrac{K_w}{[H^+]}$

(a) $[H^+] = 1.0 \times 10^{-4}$ $[OH^-] = \dfrac{1.0 \times 10^{-14}}{1.0 \times 10^{-4}} = 1.0 \times 10^{-10}$

(b) $[H^+] = 2.8 \times 10^{-6}$ $[OH^-] = \dfrac{1.0 \times 10^{-14}}{2.8 \times 10^{-6}} = 3.6 \times 10^{-9}$

64. Calculate the $[OH^-]$. $[OH^-] = \dfrac{K_w}{[H^+]}$

(a) $[H^+] = 4.0 \times 10^{-9}$ $[OH^-] = \dfrac{1.0 \times 10^{-14}}{4.0 \times 10^{-9}} = 2.5 \times 10^{-6}$

(b) $[H^+] = 8.9 \times 10^{-2}$ $\quad [OH^-] = \dfrac{1.0 \times 10^{-14}}{8.9 \times 10^{-2}} = 1.1 \times 10^{-13}$

65. Calculate $[H^+]$. $[H^+] = \dfrac{K_w}{[OH^-]}$

(a) $[OH^-] = 6.0 \times 10^{-7}$ $\quad [H^+] = \dfrac{1.0 \times 10^{-14}}{6.0 \times 10^{-7}} = 1.7 \times 10^{-8}$

(b) $[OH^-] = 1 \times 10^{-8}$ $\quad [H^+] = \dfrac{1 \times 10^{-14}}{1 \times 10^{-8}} = 1 \times 10^{-6}$

66. Calculate $[H^+]$. $[H^+] = \dfrac{K_w}{[OH^-]}$

(a) $[OH^-] = 4.5 \times 10^{-6}$ $\quad [H^+] = \dfrac{1.0 \times 10^{-14}}{4.5 \times 10^{-6}} = 2.2 \times 10^{-9}$

(b) $[OH^-] = 7.3 \times 10^{-4}$ $\quad [OH^-] = \dfrac{1.0 \times 10^{-14}}{7.3 \times 10^{-4}} = 1.4 \times 10^{-11}$

67. The molar solubilities of the salts and their ions are indicated below the formulas in the equilibrium equations.

(a) $BaSO_{4}(s) \rightleftarrows Ba^{2+} + SO_4^{2-}$

$3.9 \times 10^{-5} \qquad 3.9 \times 10^{-5} \qquad 3.9 \times 10^{-5}$

$K_{sp} = [Ba^{2+}][SO_4^{2-}] = (3.9 \times 10^{-5})^2 = 1.5 \times 10^{-9}$

(b) $Ag_2CrO_{4}(s) \rightleftarrows 2\ Ag^+ + CrO_4^{2-}$

$7.8 \times 10^{-5} \qquad 2(7.8 \times 10^{-5}) \qquad 7.8 \times 10^{-5}$

$K_{sp} = [Ag^+]^2[CrO_4^{2-}] = (15.6 \times 10^{-5})^2(7.8 \times 10^{-5}) = 1.9 \times 10^{-12}$

(c) $\left(\dfrac{0.67\text{ g } CaSO_4}{L}\right)\left(\dfrac{1\text{ mol}}{136.1\text{ g}}\right) = 4.9 \times 10^{-3}\text{ M } CaSO_4$

$CaSO_{4}(s) \rightleftarrows Ca^{2+} + SO_4^{2-}$

$4.9 \times 10^{-3} \qquad 4.9 \times 10^{-3} \qquad 4.9 \times 10^{-3}$

$K_{sp} = [Ca^{2+}][SO_4^{2-}] = (4.9 \times 10^{-3})^2 = 2.4 \times 10^{-5}$

(d) $\left(\dfrac{0.0019\text{ g AgCl}}{L}\right)\left(\dfrac{1\text{ mol}}{143.4\text{ g}}\right) = 1.3 \times 10^{-5}\text{ M AgCl}$

$AgCl(s) \rightleftarrows Ag^+ + Cl^-$

$1.3 \times 10^{-5} \qquad 1.3 \times 10^{-5} \qquad 1.3 \times 10^{-5}$

$K_{sp} = [Ag^+][Cl^-] = (1.3 \times 10^{-5})^2 = 1.7 \times 10^{-10}$

68. The molar solubilities of the salts and their ions are indicated below the formulas in the equilibrium equations.

(a) $ZnS_{(s)} \rightleftarrows Zn^{2+} + S^{2-}$

3.5×10^{-12} $\quad$ 3.5×10^{-12} $\quad$ 3.5×10^{-12}

$K_{sp} = [Zn^{2+}][S^{2-}] = (3.5 \times 10^{-12})^2 = 1.2 \times 10^{-23}$

(b) $Pb(IO_3)_{2(s)} \rightleftarrows Pb^{2+} + 2\ IO_3^-$

$4.0 \times 10^{-5}2$ $\quad$ 4.0×10^{-5} $\quad$ (4.0×10^{-5})

$K_{sp} = [Pb^{2+}][IO_3^-]^2 = (4.0 \times 10^{-5})(8.0 \times 10^{-5})^2 = 2.6 \times 10^{-13}$

(c) $\left(\dfrac{6.73 \times 10^{-3}\text{ g } Ag_3PO_4}{\text{L}}\right)\left(\dfrac{1\text{ mol}}{418.7\text{ g}}\right) = 1.61 \times 10^{-5}\text{ M } Ag_3PO_4$

$Ag_3PO_{4(s)} \rightleftarrows 3\ Ag^+ + PO_4^{3-}$

1.61×10^{-5} $\quad$ $3(1.61 \times 10^{-5})$ $\quad$ 1.61×10^{-5}

$K_{sp} = [Ag^+]^3[PO_4^{3-}] = (4.83 \times 10^{-5})^3(1.61 \times 10^{-5}) = 1.81 \times 10^{-18}$

(d) $\left(\dfrac{2.33 \times 10^{-4}\text{ g } Zn(OH)_2}{\text{L}}\right)\left(\dfrac{1\text{ mol}}{99.40\text{ g}}\right) = 2.34 \times 10^{-6}\text{ M } Zn(OH)_2$

$Zn(OH)_{2(s)} \rightleftarrows Zn^{2+} + 2\ OH^-$

2.34×10^{-6} $\quad$ 2.34×10^{-6} $\quad$ $2(2.34 \times 10^{-6})$

$K_{sp} = [Zn^{2+}][OH^-]^2 = (4.68 \times 10^{-6})(4.68 \times 10^{-6})^2 = 5.13 \times 10^{-17}$

69. The molar solubilities of the salts and their ions will be represented in terms of x below their formulas in the equilibrium equations.

(a) $BaCO_{3(s)} \rightleftarrows Ba^{2+} + CO_3^{2-}$

x $\quad$ x $\quad$ x

$K_{sp} = [Ba^{2+}][CO_3^{2-}] = x^2 = 2.0 \times 10^{-9}$

$x = \sqrt{2.0 \times 10^{-9}} = 4.5 \times 10^{-5}\text{ M}$

(b) $AlPO_{4}(s) \rightleftarrows Al^{3+} + PO_4^{3-}$

$x \qquad x \qquad x$

$K_{sp} = [Al^{3+}][PO_4^{3-}] = x^2 = 5.8 \times 10^{-19}$

$x = \sqrt{5.8 \times 10^{-19}} = 7.6 \times 10^{-10}$ M

70. (a) $Ag_2SO_{4}(s) \rightleftarrows 2\,Ag^+ + SO_4^{2-}$

$x \qquad 2x \qquad x$

$K_{sp} = [Ag^+]^2[SO_4^{2-}] = (2x)^2(x) = 4x^3 = 1.5 \times 10^{-5}$

$x = \sqrt[3]{\dfrac{1.5 \times 10^{-5}}{4}} = 1.6 \times 10^{-2}$ M

(b) $Mg(OH)_{2}(s) \rightleftarrows Mg^{2+} + 2\,OH^-$

$x \qquad x \qquad 2x$

$K_{sp} = [Mg^{2+}][OH^-]^2 = (x)(2x)^2 = 4x^3 = 7.1 \times 10^{-12}$

$x = \sqrt[3]{\dfrac{7.1 \times 10^{-12}}{4}} = 1.2 \times 10^{-4}$ M

71. (a) $\left(\dfrac{4.5 \times 10^{-5}\text{ mol }BaCO_3}{L}\right)(0.100\text{ L})\left(\dfrac{197.3\text{ g}}{\text{mol}}\right) = 8.9 \times 10^{-4}\text{ g }BaCO_3$

(b) $\left(\dfrac{7.6 \times 10^{-10}\text{ mol }AlPO_4}{L}\right)(0.100\text{ L})\left(\dfrac{122.0\text{ g}}{\text{mol}}\right) = 9.3 \times 10^{-9}\text{ g }AlPO_4$

72. (a) $\left(\dfrac{1.6 \times 10^{-2}\text{mol }Ag_2SO_4}{L}\right)(0.100\text{ L})\left(\dfrac{311.9\text{ g}}{\text{mol}}\right) = 0.50\text{ g }Ag_2SO_4$

(b) $\left(\dfrac{1.2 \times 10^{-4}\text{ mol }Mg(OH)_2}{L}\right)(0.100\text{ L})\left(\dfrac{58.33\text{ g}}{\text{mol}}\right) = 7.0 \times 10^{-4}\text{ g }Mg(OH)_2$

73. The molar concentrations of ions, after mixing, are calculated and these concentrations are substituted into the equilibrium expression. The value obtained is compared to the K_{sp} of the salt. If the value is greater than the K_{sp}, precipitation occurs. If the value is less than the K_{sp}, no precipitation occurs.

100. mL 0.010 M $Na_2SO_4 \longrightarrow$ 100. mL 0.010 M SO_4^{2-}

100. mL 0.001 M $Pb(NO_3)_2 \longrightarrow$ 100. mL 0.001 M Pb^{2+}

Volume after mixing = 200. mL

Concentrations after mixing: SO_4^{2-} = 0.0050 M $\qquad Pb^{2+}$ = 0.0005 M

$[Pb^{2+}][SO_4^{2-}] = (5 \times 10^{-3})(5 \times 10^{-4}) = 2.5 \times 10^{-6}$

$K_{sp} = 1.3 \times 10^{-8}$ which is less than 2.5×10^{-6}, therefore, precipitation occurs.

74. The molar concentrations of ions, after mixing, are calculated and these concentrations are substituted into the equilibrium expression. The value obtained is compared to the K_{sp} of the salt. If the value is greater than the K_{sp}, precipitation occurs. If the value is less than the K_{sp}, no precipitation occurs.

$50.0 \text{ mL } 1.0 \times 10^{-4} \text{ M AgNO}_3 \longrightarrow 50.0 \text{ mL } 1.0 \times 10^{-4} \text{ M Ag}^+$

$100. \text{ mL } 1.0 \times 10^{-4} \text{ M NaCl} \longrightarrow 100. \text{ mL } 1.0 \times 10^{-4} \text{ M Cl}^-$

Volume after mixing = 150. mL

Concentrations after mixing:

$$(1.0 \times 10^{-4} \text{ M Ag}^+)\left(\frac{50.0 \text{ mL}}{150. \text{ mL}}\right) = 3.3 \times 10^{-5} \text{ M Ag}^+$$

$$(1.0 \times 10^{-4} \text{ M Cl}^-)\left(\frac{100. \text{ mL}}{150. \text{ mL}}\right) = 6.7 \times 10^{-5} \text{ M Cl}^-$$

$$[\text{Ag}^+][\text{Cl}^-] = (3.3 \times 10^{-5})(6.7 \times 10^{-5}) = 2.2 \times 10^{-9}$$

$K_{sp} = 1.7 \times 10^{-10}$ which is less than 2.2×10^{-9}, therefore precipitation occurs.

75. The concentration of $\text{Br}^- = 0.10$ M in 1.0 L of 0.10 M NaBr.

$$K_{sp} = [\text{Ag}^+][\text{Br}^-] = 5.0 \times 10^{-13}$$

$$[\text{Ag}^+] = \frac{5.0 \times 10^{-13}}{[\text{Br}^-]} = \frac{5.0 \times 10^{-13}}{0.10} = 5.0 \times 10^{-12} \text{ M}$$

$$\left(\frac{5.0 \times 10^{-12} \text{ mol Ag}^+}{\text{L}}\right)\left(\frac{1 \text{ mol AgBr}}{1 \text{ mol Ag}^+}\right)(1.0 \text{ L}) = 5.0 \times 10^{-12} \text{ mol AgBr will dissolve}$$

76. $\left(\frac{0.10 \text{ mol MgBr}_2}{\text{L}}\right)\left(\frac{2 \text{ mol Br}^-}{1 \text{ mol MgBr}_2}\right) = \left(\frac{0.20 \text{ mol Br}^-}{\text{L}}\right) = 0.20 \text{ M Br}^-$ in solution

$$[\text{Ag}^+] = \frac{5.0 \times 10^{-13}}{[\text{Br}^-]} = \frac{5.0 \times 10^{-13}}{0.20} = 2.5 \times 10^{-12} \text{ M}$$

$$\left(\frac{2.5 \times 10^{-12} \text{ mol Ag}^+}{\text{L}}\right)\left(\frac{1 \text{ mol AgBr}}{1 \text{ mol Ag}^+}\right)(1.0 \text{ L}) = 2.5 \times 10^{-12} \text{ mol AgBr will dissolve}$$

77. $HC_2H_3O_2 \rightleftarrows H^+ + C_2H_3O_2^-$

$$K_a = \frac{[H^+][C_2H_3O_2^-]}{[HC_2H_3O_2]} = 1.8 \times 10^{-5}$$

$[H^+] = K_a\left(\frac{[HC_2H_3O_2]}{[C_2H_3O_2^-]}\right)$ $\qquad [HC_2H_3O_2] = 0.20$ M

$[C_2H_3O_2^-] = 0.10$ M $\qquad [H^+] = (1.8 \times 10^{-5})\left(\frac{0.20}{0.10}\right) = 3.6 \times 10^{-5}$ M

$\text{pH} = -\log(3.6 \times 10^{-5}) = 4.44$

78. $HC_2H_3O_2 \rightleftarrows H^+ + C_2H_3O_2^-$

$$K_a = \frac{[H^+][C_2H_3O_2^-]}{[HC_2H_3O_2]} = 1.8 \times 10^{-5}$$

$[H^+] = K_a\left(\frac{[HC_2H_3O_2]}{[C_2H_3O_2^-]}\right)$ $\quad$ $[HC_2H_3O_2] = 0.20$ M

$[C_2H_3O_2^-] = 0.20$ M $\quad$ $[H^+] = (1.8 \times 10^{-5})\left(\frac{0.20}{0.20}\right) = 1.8 \times 10^{-5}$ M

$pH = -\log(1.8 \times 10^{-5}) = 4.74$

79. Initially, the solution of NaCl is neutral. $[H^+] = 1 \times 10^{-7}$

$pH = -\log(1 \times 10^{-7}) = 7.0$

Final $H^+ = 2.0 \times 10^{-2}$ M

$pH = -\log(2.0 \times 10^{-2}) = 1.70$

Change in pH $= 7.0 - 1.70 = 5.3$ units

80. Initially, $[H^+] = 1.8 \times 10^{-5}$

$pH = -\log(1.8 \times 10^{-5}) = 4.74$

Final $[H^+] = 1.9 \times 10^{-5}$

$pH = -\log(1.9 \times 10^{-5}) = 4.72$

Change in pH $= 4.74 - 4.72 = 0.02$ units in the buffered solution

81. $H_2 + I_2 \rightleftarrows 2\ HI$
The reaction is on a 1 to 1 mole ratio of hydrogen to iodine. The data given indicates that hydrogen is the limiting reactant.

$$(2.10 \text{ mol } H_2)\left(\frac{2 \text{ mol HI}}{1 \text{ mol } H_2}\right) = 4.20 \text{ mol HI}$$

82. $H_2 + I_2 \rightleftarrows 2\ HI$

(a) $(2.00 \text{ mol } H_2)\left(\frac{2 \text{ mol HI}}{1 \text{ mol } H_2}\right)\left(\frac{0.79 \text{ mol}}{1.00 \text{ mol}}\right) = 3.16 \text{ mol HI}$

(b) The addition of 0.27 mol I_2 makes the iodine present in excess and the 2.00 mol H_2 the limiting reactant. The yield increases to 85%.

$(2.00 \text{ mol } H_2)\left(\frac{2 \text{ mol HI}}{1 \text{ mol } H_2}\right)\left(\frac{0.85 \text{ mol}}{1.00 \text{ mol}}\right) = 3.4 \text{ mol HI}$

There will be 15% unreacted H_2 and I_2.

$(0.15)(2.0 \text{ mol } H_2) = 0.30 \text{ mol } H_2$ present.

In addition to the 0.30 mol of unreacted I_2, will be the 0.27 mol I_2 added.

$0.27 \text{ mol} + 0.30 \text{ mol} = 0.57 \text{ mol } I_2$ present.

(c) $K = \dfrac{[HI]^2}{[H_2]\,[I_2]}$

The formation of 3.16 mol HI required the reaction of 1.58 mol I_2 and 1.58 mol H_2. At equilibrium, the concentrations are:

3.16 mol HI; $2.00 - 1.58 = 0.42$ mol H_2 = 0.42 mol I_2

$$K_{eq} = \frac{(3.16)^2}{(0.42)(0.42)} = 57$$

In the calculation of the equilibrium constant, the actual number of moles of reactants and products present at equilibrium can be used in the calculation in place of molar concentrations. This occurs because the reaction is gaseous and the liters of HI produced equals the sum of the liters of H_2 and I_2 reacting. In the equilibrium expression, the volumes will cancel.

83. $H_2 + I_2 \rightleftarrows 2\,HI$

$(64.0\text{ g HI})\left(\frac{1\text{ mol}}{127.9\text{ g}}\right) = 0.500$ mol HI present

$(0.500\text{ mol HI})\left(\frac{1\text{ mol }I_2}{2\text{ mol HI}}\right) = 0.250$ mol I_2 reacted

$(0.500\text{ mol HI})\left(\frac{1\text{ mol }H_2}{2\text{ mol HI}}\right) = 0.250$ mol H_2 reacted

$(6.00\text{ g }H_2)\left(\frac{1\text{ mol}}{2.016\text{ g}}\right) = 2.98$ mol H_2 initially present

$(200.\text{ g }I_2)\left(\frac{1\text{ mol}}{253.8\text{ g}}\right) = 0.788$ mol I_2 initially present

At equilibrium, moles present are:

0.500 mol HI; $2.98 - 0.250 = 2.73$ mol H_2

$0.788 - 0.250 = 0.538$ mol I_2

84. $PCl_{3(g)} + Cl_{2(g)} \rightleftarrows PCl_{5(g)}$

$$K_{eq} = \frac{[PCl_5]}{[PCl_3]\,[Cl_2]}$$

The concentrations are:

$$PCl_5 = \frac{0.22 \text{ mol}}{20. \text{ L}} = 0.011 \text{ M}$$

$$PCl_3 = \frac{0.10 \text{ mol}}{20. \text{ L}} = 0.0050 \text{ M}$$

$$Cl_2 = \frac{1.50 \text{ mol}}{20. \text{ L}} = 0.075 \text{ M}$$

$$K_{eq} = \frac{0.011}{(0.0050)(0.075)} = 29$$

85. 100°C – 30°C = 70°C temperature increase. This increase is equal to seven 10°C increments. The reaction rate will be increased $2^7 = 128$ times.

86. Hypochlorous acid $HOCl \rightleftarrows H^+ + OCl^-$

Equilibrium concentrations:

$[H^+] = [OCl^-] = 5.9 \times 10^{-5}$ M

$[HOCl] = 0.10 - 5.9 \times 10^{-5} = 0.10$ M (neglecting 5.9×10^{-3})

$$K_a = \frac{[H^+]\,[OCl^-]}{[HOCl]} = \frac{(5.9 \times 10^{-5})(5.9 \times 10^{-5})}{0.10} = 3.5 \times 10^{-8}$$

Propanoic acid $HC_3H_5O_2 \rightleftarrows H^+ + C_3H_5O_2^-$

Equilibrium concentrations:

$[H^+] = [C_3H_5O_2^-] = 1.4 \times 10^{-3}$ M

$[HC_3H_5O_2] = 0.15 - 1.4 \times 10^{-3} = 0.15$ M (neglecting 1.4×10^{-3})

$$K_a = \frac{[H^+]\,[C_3H_5O_2^-]}{[HC_3H_5O_2]} = \frac{(1.4 \times 10^{-3})(1.4 \times 10^{-3})}{0.15} = 1.3 \times 10^{-3}$$

Hydrocyanic acid $HCN \rightleftarrows H^+ + CN^-$

Equilibrium concentrations:

$[H^+] = [CN^-] = 8.9 \times 10^{-6}$ M

$[HCN] = 0.20 - 8.9 \times 10^{-6} = 0.20$ M (neglecting 8.9×10^{-6})

$$K_a = \frac{[H^+]\,[CN^-]}{[HCN]} = \frac{(8.9 \times 10^{-6})^2}{0.20} = 4.0 \times 10^{-10}$$

87. Let y = M CaF_2 dissolving

$$CaF_{2(s)} \rightleftarrows Ca^{2+} + 2\ F^-$$
$$y \qquad\qquad y \qquad\qquad 2y$$

(a) $K_{sp} = [Ca^{2+}][F^-]^2 = (y)(2y)^2 = 4y^3 = 3.9 \times 10^{-11}$

$y = \sqrt[3]{\dfrac{3.9 \times 10^{-11}}{4}} = 2.1 \times 10^{-4}$ M (CaF_2 dissolved)

$$\left(\frac{2.1 \times 10^{-4}\ \text{mol } CaF_2}{L}\right)\left(\frac{1\ \text{mol } Ca^{2+}}{1\ \text{mol } CaF_2}\right) = 2.1 \times 10^{-4}\ M\ Ca^{2+}$$

$$\left(\frac{2.1 \times 10^{-4}\ \text{mol } CaF_2}{L}\right)\left(\frac{2\ \text{mol } F^-}{1\ \text{mol } CaF_2}\right) = 4.2 \times 10^{-4}\ M\ F^-$$

(b) $\left(\dfrac{2.1 \times 10^{-4}\ \text{mol } CaF_2}{L}\right)(0.500\ L)\left(\dfrac{78.08\ g}{mol}\right) = 8.2 \times 10^{-3}\ g\ CaF_2$

88. The molar concentrations of ions, after mixing, are calculated and these concentrations are substituted into the equilibrium expression. The value obtained is compared to the K_{sp} of the salt. If the value is greater than the K_{sp}, precipitation occurs. If the value is less than the K_{sp}, no precipitation occurs.

(a) 100. mL 0.010 M Na_2SO_4 ⟶ 100. mL 0.010 M SO_4^{2-}

100. mL 0.001 M $Pb(NO_3)_2$ ⟶ 100. mL 0.001 M Pb^{2+}

Volume after mixing = 200. mL

Concentrations after mixing:

SO_4^{2-} = 0.0050 M $\qquad$ Pb^{2+} = 0.0005 M

$[Pb^{2+}][SO_4^{2-}] = (5.0 \times 10^{-3})(5 \times 10^{-4}) = 3 \times 10^{-6}$

$K_{sp} = 1.3 \times 10^{-8}$ which is less than 3×10^{-6}, therefore, precipitation occurs.

(b) 50.0 mL 1.0×10^{-4} M $AgNO_3$ ⟶ 50.0 mL 1.0×10^{-4} M Ag^+

100. mL 1.0×10^{-4} M NaCl ⟶ 100. mL 1.0×10^{-4} M Cl^-

Volume after mixing = 150. mL

Concentrations after mixing:

$(1.0 \times 10^{-4}\ M\ Ag^+)\left(\dfrac{50.0\ mL}{150.\ mL}\right) = 3.3 \times 10^{-5}\ M\ Ag^+$

$(1.0 \times 10^{-4}\ M\ Cl^-)\left(\dfrac{100.\ mL}{150.\ mL}\right) = 6.7 \times 10^{-5}\ M\ Cl^-$

$[Ag^+][Cl^-] = (3.3 \times 10^{-5})(6.7 \times 10^{-5}) = 2.2 \times 10^{-9}$

$K_{sp} = 1.7 \times 10^{-10}$ which is less than 2.2×10^{-9}, therefore precipitation occurs.

(c) $\left(\frac{1.0 \text{ g } Ca(NO_3)_2}{0.150 \text{ L}}\right)\left(\frac{1 \text{ mol}}{164.1 \text{ g}}\right)\left(\frac{1 \text{ mol } Ca^{2+}}{1 \text{ mol } Ca(NO_3)_2}\right) = 0.041 \text{ M } Ca^{2+}$

250 mL 0.01 M NaOH $\longrightarrow$ 250 mL 0.01 M OH^-

Final volume $= 4.0 \times 10^2$ mL

Concentrations after mixing:

$(0.041 \text{ M } Ca^{2+})\left(\frac{150 \text{ mL}}{4.0 \times 10^2 \text{ mL}}\right) = 0.015 \text{ M } Ca^{2+}$

$(0.01 \text{ M } OH^-)\left(\frac{250 \text{ mL}}{4.0 \times 10^2 \text{ mL}}\right) = 0.0063 \text{ M } OH^-$

$[Ca^{2+}][OH^-]^2 = (0.015)(0.0063)^2 = 6.0 \times 10^{-7}$

$K_{sp} = 1.3 \times 10^{-6}$ which is greater than 6.0×10^{-7}, therefore, no precipitation occurs.

89. With a known Ba^{2+} concentration, the SO_4^{2-} concentration can be calculated using the K_{sp} value. $BaSO_{4}(s) \rightleftarrows Ba^{2+} + SO_4^{2-}$

$K_{sp} = [Ba^{2+}][SO_4^{2-}] = 1.5 \times 10^{-9}$ $Ba^{2+} = 0.050$ M

(a) $[SO_4^{2-}] = \frac{K_{sp}}{[Ba^{2+}]} = \frac{1.5 \times 10^{-9}}{0.050} = 3.0 \times 10^{-8} \text{ M } SO_4^{2-}$ in solution

(b) M SO_4^{2-} = M $BaSO_4$ in solution

$\left(\frac{3.0 \times 10^{-8} \text{ mol } BaSO_4}{\text{L}}\right)(0.100 \text{ L})\left(\frac{233.4 \text{ g}}{\text{mol}}\right) = 7.0 \times 10^{-7}$ g $BaSO_4$ remain in solution

90. $[Ba^{2+}][SO_4^{2-}] = 1.5 \times 10^{-9}$ $[Sr^{2+}][SO_4^{2-}] = 3.5 \times 10^{-7}$

Both cations are present in equal concentrations (0.10 M). Therefore, as SO_4^{2-} is added, the K_{sp} of $BaSO_4$ will be exceeded before that of $SrSO_4$. $BaSO_4$ precipitates first.

91. If $[Pb^{2+}][Cl^-]^2$ exceeds the K_{sp}, precipitation will occur.

$K_{sp} = [Pb^{2+}][Cl^-]^2 = 2.0 \times 10^{-5}$

0.050 M $Pb(NO_3)_2 \longrightarrow 0.050$ M Pb^{2+}

0.010 M NaCl $\longrightarrow$ 0.010 M Cl^-

$(0.050)(0.010)^2 = 5.0 \times 10^{-6}$

$[Pb^{2+}][Cl^-]^2$ is smaller than the K_{sp} value. Therefore, no precipitate of $PbCl_2$ will form.

92. $2\ SO_2(g) + O_2(g) \rightleftarrows 2\ SO_3(g)$

$$K_{eq} = \frac{[SO_3]^2}{[SO_2]^2[O_2]} = \frac{(11.0)^2}{(4.20)^2(0.60 \times 10^{-3})} = 1.1 \times 10^4$$

93. $(0.048\text{ g }BaF_2)\left(\frac{1\text{ mol}}{175.3\text{ g}}\right) = 2.7 \times 10^{-4}\text{ mol }BaF_2$

$\left(\frac{2.7 \times 10^{-4}\text{ mol}}{0.015\text{ L}}\right) = 1.8 \times 10^{-2}$ M BaF_2 dissolved

$BaF_2(s) \rightleftarrows Ba^{2+} + 2\ F^-$

1.8×10^{-2} 1.8×10^{-2} $2(1.8 \times 10^{-2})$ (molar concentration)

$K_{sp} = [Ba^{2+}][F^-]^2 = (1.8 \times 10^{-2})(3.6 \times 10^{-2})^2 = 2.3 \times 10^{-5}$

94. $N_2 + 3\ H_2 \rightleftarrows 2\ NH_3$

$K_{eq} = \frac{[NH_3]^2}{[N_2][H_2]^3} = 4.0$ Let $y = [NH_3]$

$4.0 = \frac{y^2}{(2.0)(2.0)^3}$ $y^2 = 64$ $y = \sqrt{64}$

$y = 8.0\text{ M} = [NH_3]$

95. Total volume of mixture = 40.0 mL (0.0400 L)

$K_{sp} = [Sr^{2+}]\,[SO_4{}^{2-}] = 7.6 \times 10^{-7}$

$$[Sr^{2+}] = \frac{(1.0 \times 10^{-3}\text{ M})(0.0250\text{ L})}{0.0400\text{ L}} = 6.3 \times 10^{-4}\text{ M}$$

$$[SO_4{}^{2-}] = \frac{(2.0 \times 10^{-3}\text{ M})(0.0150\text{ L})}{0.0400\text{ L}} = 7.5 \times 10^{-4}\text{ M}$$

$[Sr^{2+}][SO_4{}^{2-}] = (6.3 \times 10^{-4})(7.5 \times 10^{-4}) = 4.7 \times 10^{-7}$

$4.7 \times 10^{-7} < 7.6 \times 10^{-7}$ no precipitation should occur.

96. $\left(\frac{3.04 \times 10^{-7}\text{ g }Hg_2I_2}{\text{L}}\right)\left(\frac{1\text{ mol}}{655.0\text{ g}}\right) = 4.64 \times 10^{-10}\text{ M }Hg_2I_2$ (molar solubility)

$Hg_2I_2 \rightleftarrows Hg_2{}^{2+} + 2\ I^-$

$K_{sp} = [Hg_2{}^{2+}][I^-]^2 = (4.64 \times 10^{-10})(9.28 \times 10^{-10})^2 = 4.00 \times 10^{-28}$

97. $3\ O_2(g) + \text{heat} \rightleftarrows 2\ O_3(g)$

Three ways to increase ozone

(a) increase heat

(b) increase amount of O_2

(c) increase pressure

(d) remove O_3 as it is made

98. $H_2O(l) \rightleftarrows H_2O(g)$

Conditions on the second day

(a) the temperature could have been cooler

(b) the humidity in the air could have been higher

(c) the air pressure could have been greater

99. Treat this as an equilibrium where W = whole nuts, S = shell halves, and K = kernals

W $\rightleftarrows$	2 S	+ K	
144	0	0	amount before cracking
$144 - x$	$2x$	x	x = number of kernels after cracking

$144 - x + 2x + x = 194$ total pieces

$144 + 2x = 194; \quad 2x = 50$

$x = 25$ kernels; 50 shell halves; 119 whole nuts left

$$K_{eq} = \frac{(2x)^2(x)}{144 - x} = \frac{(50)^2(25)}{119} = 5.3 \times 10^2$$

100. $CO(g) + H_2O(g) \rightleftarrows CO_2(g) + H_2(g)$

(c) is the correct answer

$$K_{eq} = \frac{[CO_2]\,[H_2]}{[CO]\,[H_2O]} = 1$$

With equal concentations of products and reactants, the K_{eq} value will equal 1.

101. (a) $K_{eq} = \dfrac{[O_3]^2}{[O_2]^3}$ (b) $K_{eq} = \dfrac{[H_2O(l)]}{[H_2O(g)]}$

(c) $K_{eq} = \dfrac{[MgO]\,[CO_2]}{[MgCO_3]}$ (d) $K_{eq} = \dfrac{[Bi_2S_3]\,[H^+]^6}{[Bi^{3+}]^2[H_2S]^3}$

102. $2A \quad + \quad B \quad \rightleftarrows \quad C$

1.0 M	1.0 M	0	Initial conditions
1.0 – 2(0.30)	1.0 – 0.30	0.30	Equilibrium concentrations
0.4 M	0.7 M	0.30 M	

$$K_{eq} = \frac{[C]}{[A]^2[B]} = \frac{0.30}{(0.4)^2(0.7)} = 3$$

103. Since the second reaction is the reverse of the first, the K_{eq} value of the second reaction will be the reciprical of the K_{eq} value of the first reaction.

$$K_{eq} = \frac{[I_2]\,[Cl_2]}{[ICl]^2} = 2.2 \times 10^{-3} \quad \text{(first reaction)}$$

$$K_{eq} = \frac{[ICl]^2}{[I_2][Cl_2]} \qquad K_{eq} = \frac{1}{2.2 \times 10^{-3}} = 450$$

104. $HNO_{2}(aq) \rightleftarrows H^{+}(aq) + NO_2{}^{-}(aq)$

OH^- reacts with H^+ and equilibrium shifts to the right.

(a) After an initial increase, $[OH^-]$ will be neutralized and equilibrium shifts to the right.

(b) $[H^+]$ will be reduced (reacts with OH^-]. Equilibrium shifts to the right.

(c) $[NO_2{}^-]$ increases as equilibrium shifts to the right.

(d) $[HNO_2]$ decreases as equilbrium shifts to the right.

105. $SO_{2}(g) \quad + \quad NO_{2}(g) \quad \rightleftarrows \quad SO_{3}(g) \quad + \quad NO(g)$

0.50 M	0.50 M	0	0	Initial concentrations
$0.50 - x$	$0.50 - x$	x	x	Equilibrium concentrations

$$K_{eq} = \frac{[SO_3]\,[NO]}{[SO_2]\,[NO_2]} = \frac{x^2}{(0.50 - x)^2} = 90.$$

Take the square root of both sides

$$\frac{x}{0.50 - x} = 9.5 \qquad x = 0.45 \text{ M}$$

$[SO_3] = [NO] = 0.45$ M

$[SO_2] = [NO_2] = 0.05$ M

106. $CaSO_4(s) \rightleftarrows Ca^{2+}(aq) + SO_4^{2-}(aq)$

$K_{sp} = [Ca^{2+}]\,[SO_4^{2-}] = 2.0 \times 10^{-4}$

Let x = moles $CaSO_4$ that dissolve/L = $[Ca^{2+}] = [SO_4^{2-}]$

$(x)(x) = 2.0 \times 10^{-4}$

$x = 0.014$ M $CaSO_4$

M $\longrightarrow$ moles $\longrightarrow$ grams

$$\left(\frac{0.014 \text{ mol } CaSO_4}{\text{L}}\right)(0.600 \text{ L})\left(\frac{136.1 \text{ g}}{\text{mol}}\right) = 1.1 \text{ g } CaSO_4$$

107. $PbF_2(s) \rightleftarrows Pb^{2+} + 2\ F^-$

$$\left(\frac{0.098 \text{ g } PbF_2}{0.400 \text{ L}}\right)\left(\frac{1 \text{ mol}}{245.2 \text{ g}}\right) = 1.0 \times 10^{-3} \text{ mol/L} = 1.0 \times 10^{-3} \text{ M } PbF_2$$

$K_{sp} = (Pb^{2+})(F^-)^2$

$[Pb^{2+}] = 1.0 \times 10^{-3}$; $[F^-] = 2(1.0 \times 10^{-3}) = 2.0 \times 10^{-3}$

$K_{sp} = (1.0 \times 10^{-3})(2.0 \times 10^{-3})^2 = 4.0 \times 10^{-9}$

CHAPTER 17

OXIDATION-REDUCTION

1. (a) Iodine is oxidized. Its oxidation number increases from 0 to +5.

   (b) Chlorine is reduced. Its oxidation number decreases from 0 to −1.

2. The higher metal on the list is more reactive.

   (a) Al (b) Ba (c) Ni

3. If the free element is higher on the list than the ion with which it is paired, the reaction occurs.

   (a) Yes. $Zn(s) + Cu^{2+}(aq) \longrightarrow Zn^{2+}(aq) + Cu(s)$

   (b) No

   (c) Yes. $Sn(s) + 2\ Ag^{+}(aq) \longrightarrow Sn^{2+}(aq) + 2\ Ag(s)$

   (d) No

   (e) Yes. $Ba(s) + FeCl_2(aq) \longrightarrow BaCl_2(aq) + Fe(s)$

   (f) No

   (g) Yes. $Ni(s) + Hg(NO_3)_2(aq) \longrightarrow Ni(NO_3)_2(aq) + Hg(l)$

   (h) Yes. $2\ Al(s) + 3\ CuSO_4(aq) \longrightarrow Al_2(SO_4)_3(aq) + 3\ Cu(s)$

4. (a) $2\ Al + Fe_2O_3 \longrightarrow Al_2O_3 + 2\ Fe + \text{Heat}$

   (b) Al is above Fe in the activity series, which indicates Al is more active than Fe.

   (c) No. Iron is less active than aluminum and will not displace aluminum from its compounds.

   (d) Yes. Aluminum is above chromium in the activity series and will displace Cr^{3+} from its compounds.

5. (a) $2\ Al(s) + 6\ HCl(aq) \longrightarrow 2\ AlCl_3(aq) + 3\ H_2(g)$

   $2\ Al(s) + 3\ H_2SO_4(aq) \longrightarrow Al_2(SO_4)_3(aq) + 3\ H_2(g)$

   (b) $2\ Cr(s) + 6\ HCl(aq) \longrightarrow 2\ CrCl_3(aq) + 3\ H_2(g)$

   $2\ Cr(s) + 3\ H_2SO_4(aq) \longrightarrow Cr_2(SO_4)_3(aq) + 3\ H_2(g)$

   (c) $Au(s) + HCl(aq) \longrightarrow$ no reaction

   $Au(s) + H_2SO_4(aq) \longrightarrow$ no reaction

(d) $Fe(s) + 2\ HCl(aq) \longrightarrow FeCl_2(aq) + H_2(g)$

$Fe(s) + H_2SO_4(aq) \longrightarrow FeSO_4(aq) + H_2(g)$

(e) $Cu(s) + HCl(aq) \longrightarrow$ no reaction

$Cu(s) + H_2SO_4(aq) \longrightarrow$ no reaction

(f) $Mg(s) + 2\ HCl(aq) \longrightarrow MgCl_2(aq) + H_2(g)$

$Mg(s) + H_2SO_4(aq) \longrightarrow MgSO_4(aq) + H_2(g)$

(g) $Hg(l) + HCl(aq) \longrightarrow$ no reaction

$Hg(l) + H_2SO_4(aq) \longrightarrow$ no reaction

(h) $Zn(s) + 2\ HCl(aq) \longrightarrow ZnCl_2(aq) + H_2(g)$

$Zn(s) + H_2SO_4(aq) \longrightarrow ZnSO_4(aq) + H_2(g)$

6. (a) Oxidation occurs at the anode. The reaction is

$2\ Cl^-(aq) \longrightarrow Cl_2(g) + 2\ e^-$

(b) Reduction occurs at the cathode. The reaction is

$Ni^{2+}(aq) + 2\ e^- \longrightarrow Ni(s)$

(c) The net chemical reaction is

$$Ni^{2+}(aq) + 2\ Cl^-(aq) \xrightarrow[\text{energy}]{\text{electrical}} Ni(s) + Cl_2(g)$$

7. In Figure 17.3, electrical energy is causing chemical reactions to occur. In Figure 17.4, chemical reactions are used to produce electrical energy.

8. (a) It would not be possible to monitor the voltage produced, but the reactions in the cell would still occur.

(b) If the salt bridge were removed, the reaction would stop. Ions must be mobile to maintain an electrical neutrality of ions in solution. The two solutions would be isolated with no complete electrical circuit.

9. Oxidation and reduction are complementary processes because one does not occur without the other. The loss of e^- in oxidation is accompanied by a gain of e^- in reduction.

10. $Ca^{2+} + 2\ e^- \longrightarrow Ca$ cathode reaction, reduction

$2\ Br^- \longrightarrow Br_2 + 2\ e^-$ anode reaction, oxidation

11. During electroplating of metals, the metal is plated by reducing the positive ions of the metal in the solution. The plating will occur at the cathode, the source of electrons. With an alternating current, the polarity of the electrode would be constantly changing, so at one instant the metal would be plating and the next instant the metal would be dissolving.

12. Since lead dioxide and lead(II) sulfate are insoluble, it is unnecessary to have salt bridges in the cells of a lead storage battery.

13. The electolyte in a lead storage battery is dilute sulfuric acid. In the discharge cycle, SO_4^{2-}, is removed from solution as it reacts with PbO_2 and H^+ to form $PbSO_{4(s)}$ and H_2O. Therefore, the electrolyte solution contains less H_2SO_4 and becomes less dense.

14. If Hg^{2+} ions are reduced to metallic mercury, this would occur at the cathode, because reduction takes place at the cathode.

15. In both electrolytic and voltaic cells, oxidation and reduction reactions occur. In an electrolytic cell an electric current is forced through the cell causing a chemical change to occur. In voltaic cells, spontaneous chemical changes occur, generating an electric current.

16. In some voltaic cells, the reactants at the electrodes are in solution. For the cell to function, these reactants must be kept separated. A salt bridge permits movement of ions in the cell. This keeps the solution neutral with respect to the charged particles (ions) in the solution.

17. Iron will corrode (oxidize) in moist air. If the iron is in contact with copper and sea water (an electrolyte), the corrision rate increases dramatically. This is what occurred in the deterioration of the Statue of Liberty.

18. Patina is a blue-green product of the atmospheric corrosion of copper. It forms a coating on the surface which can protect the copper from further oxidation.

19. Photochromic glass is made from tetrahedrons of silicon and oxygen in a disorderly array with crystals of AgCl in between the tetrahedrons. Visible light passes through the glass but UV light triggers an oxidation reduction reaction between Ag^+ and Cl^- ($Ag^+ + Cl^- \xrightarrow{\text{UV Light}} Ag^0 + Cl^0$). Cu^+ ions react with the Cl^0 atoms while the Ag^0 atoms move to the surface of the crystal and form clusters of Ag metal. When the glass is removed from the light, Cu^{2+} ions migrate to the surface and react with the Ag^0 reforming Ag^+ and Cu^+ ($Cu^{2+} + Ag^0 \longrightarrow Cu^+ + Ag^+$), cleaning the glass.

20. The correct statements are a, c, e, g, j, k, m, p, q, r, s

(b) The oxidation number of molybdenum in Na_2MoO_4 is +6.

(d) The process in which an atom or an ion loses electrons is called oxidation.

(f) In the reaction $2\ Al + 3\ CuCl_2 \longrightarrow 2\ AlCl_3 + 3\ Cu$ aluminum is the reducing agent.

(h) $Cu^0 \longrightarrow Cu^{2+} + 2\ e^-$ is a balanced oxidation half–reaction.

(i) In the electrolysis of sodium chloride (brine) solution, Cl_2 gas is formed at the anode.

(l) The reaction $Zn + MgCl_2 \longrightarrow Mg + ZnCl_2$ is not a spontaneous reaction.

(n) Silver metal will not react with acids to liberate hydrogen gas.

(o) Zinc is a better reducing agent than iron.

(t) In an electrolytic cell, electrical energy is converted to chemical energy.

21. The oxidation number of the underlined element is indicated by the number following the formula.

(a) $\underline{Na}Cl$ +1
(b) $Fe\underline{Cl}_3$ −1
(c) $\underline{Pb}O_2$ +4
(d) $Na\underline{N}O_3$ +5
(e) $H_2\underline{S}O_3$ +4
(f) $\underline{N}H_4Cl$ −3

22. The oxidation number of the underlined element is indicated by the number following the formula.

(a) $K\underline{Mn}O_4$ +7
(b) $\underline{I}_2$ 0
(c) $\underline{N}H_3$ −3
(d) $K\underline{Cl}O_3$ +5
(e) $K_2\underline{Cr}O_4$ +6
(f) $K_2\underline{Cr}_2O_7$ +6

23. The oxidation number of the underlined element is indicated by the number following the formula.

(a) $\underline{S}^{2-}$ −2
(b) $\underline{N}O_2^-$ +3
(c) $Na_2\underline{O}_2$ −1
(d) Bi^{3+} +3

24. The oxidation number of the underlined element is indicated by the number following the formula.

(a) $\underline{O}_2$ 0
(b) $\underline{As}O_4^{3-}$ +5
(c) $Fe(\underline{O}H)_3$ −2
(d) $\underline{I}O_3^-$ +5

25. Half–reactions

Balanced half–reaction	Changing Element	Type of reaction
(a) $Zn^{2+} + 2\,e^- \longrightarrow Zn$	Zn	reduction
(b) $2\,Br^- \longrightarrow Br_2 + 2\,e^-$	Br	oxidation
(c) $MnO_4^- + 8\,H^+ + 5\,e^- \longrightarrow Mn^{2+} + 4\,H_2O$	Mn	reduction
(d) $Ni \longrightarrow Ni^{2+} + 2\,e^-$	Ni	oxidation

26. Half–reactions

Balanced half–reaction	Changing Element	Type of reaction
(a) $SO_3^{2-} + H_2O \longrightarrow SO_4^{2-} + 2\,H^+ + 2\,e^-$	S	oxidation
(b) $NO_3^- + 4\,H^+ + 3\,e^- \longrightarrow NO + 2\,H_2O$	N	reduction
(c) $S_2O_4^{2-} + 2\,H_2O \longrightarrow 2\,SO_3^{2-} + 4\,H^+ + 2\,e^-$	S	oxidation
(d) $Fe^{2+} \longrightarrow Fe^{3+} + 1\,e^-$	Fe	oxidation

27. (1) $Cr + HCl \longrightarrow CrCl_3 + H_2$

(a) Cr is oxidized, H is reduced

(b) HCl is the oxidizing agent, Cr the reducing agent

(2) $SO_4^{2-} + I^- + H^+ \longrightarrow H_2S + I_2 + H_2O$

(a) I is oxidized, S is reduced

(b) SO_4^{2-} is the oxidizing agent, I^- the reducing agent

28. (1) $AsH_3 + Ag^+ + H_2O \longrightarrow H_3AsO_4 + Ag + H^+$

(a) As is oxidized, Ag is reduced

(b) Ag^+ is the oxidizing agent, AsH_3 the reducing agent

(2) $Cl_2 + NaBr \longrightarrow NaCl + Br_2$

(a) Br is oxidized, Cl is reduced

(b) Cl_2 is the oxidizing agent, NaBr the reducing agent

29. Balancing oxidation–reduction equations

(a) $Zn + S \longrightarrow ZnS$

ox $Zn^0 \longrightarrow Zn^{2+} + 2\,e^-$ — Add half reactions

red $S^0 + 2\,e^- \longrightarrow S^{2-}$ — the $2e^-$ cancel

$Zn + S \longrightarrow ZnS$

(b) $AgNO_3 + Pb \longrightarrow Pb(NO_3)_2 + Ag$

ox $Pb^0 \longrightarrow Pb^{2+} + 2\,e^-$

red $Ag^+ + 1\,e^- \longrightarrow Ag^0$ — Multiply by 2, add the half

$Pb + 2\,Ag^+ \longrightarrow Pb^{2+} + 2\,Ag$ — reactions, the $2\,e^-$ cancel

Transfer the coefficients to the original equation and complete the balancing by inspection.

$2\,AgNO_3 + Pb \longrightarrow Pb(NO_3)_2 + 2\,Ag$

(c) $Fe_2O_3 + CO \longrightarrow Fe + CO_2$

ox $C^{2+} \longrightarrow C^{4+} + 2\ e^-$ Multiply by 3

red $Fe^{3+} + 3\ e^- \longrightarrow Fe^0$ Multiply by 2, add, the 6 e^- cancel

$3\ C^{2+} + 2\ Fe^{3+} \longrightarrow 3\ C^{4+} + 2\ Fe$

Transfer the coefficients to the original equation (the coefficient 2 in front of the Fe^{3+} becomes the subscript 2 in Fe_2O_3). Complete the balancing by inspection.

$Fe_2O_3 + 3\ CO \longrightarrow 2\ Fe + 3\ CO_2$

(d) $H_2S + HNO_3 \longrightarrow S + NO + H_2O$

$S^{2-} \longrightarrow S^0 + 2\ e^-$ Multiply by 3

$N^{5+} + 3\ e^- \longrightarrow N^{2+}$ Multiply by 2, add, the 6 e^- cancel

$3\ S^{2-} + 2\ N^{5+} \longrightarrow 3\ S + 2\ N^{2+}$

Transfer the coefficients to the original equations and complete the balancing by inspection.

$3\ H_2S + 2\ HNO_3 \longrightarrow 3\ S + 2\ NO + 4\ H_2O$

(e) $MnO_2 + HBr \longrightarrow MnBr_2 + Br_2 + H_2O$

$Br^- \longrightarrow Br^0 + 1\ e^-$ Multiply by 2

$Mn^{4+} + 2\ e^- \longrightarrow Mn^{2+}$ Add equations and the 2 e^- cancel

$Mn^{4+} + 2\ Br^- \longrightarrow Mn^{2+} + 2\ Br$

Transfer the coefficients to the original equation. The coefficient 2 in front of the Br^- becomes the subscript 2 in Br_2. Also, 2 more Br^- ions are required to account for the 2 Br^- ions that do not change oxidation numbers. These 2 are part of the compound $MnBr_2$.

$MnO_2 + 4\ HBr \longrightarrow MnBr_2 + Br_2 + 2\ H_2O$

30. (a) Balancing oxidation-reduction equations

$Cl_2 + KOH \longrightarrow KCl + KClO_3 + H_2O$

$Cl^0 \longrightarrow Cl^{5+} + 5\ e^-$

$Cl^0 + e^- \longrightarrow Cl^-$ Multiply by 5, add, the 5 e^- cancel

$3\ Cl_2 \longrightarrow Cl^{5+} + 5\ Cl^-$ 6 Cl^0 becomes 3 Cl_2

Tranfer the coefficients to the original equations and complete the balancing by inspection.

$3\ Cl_2 + 6\ KOH \longrightarrow KClO_3 + 5\ KCl + 3\ H_2O$

(b) $Ag + HNO_3 \longrightarrow AgNO_3 + NO + H_2O$

$Ag^0 \longrightarrow Ag^+ + e^-$ Multiply by 3, add, the 3 e^- cancel

$N^{5+} + 3\ e^- \longrightarrow N^{2+}$

$3\ Ag + N^{5+} \longrightarrow 3\ Ag^+ + N^{2+}$

Transfer the coefficients to the original equation and complete the balancing by inspection.

$3\ Ag + 4\ HNO_3 \longrightarrow 3\ AgNO_3 + NO + 2\ H_2O$

(c) $CuO + NH_3 \longrightarrow N_2 + Cu + H_2O$

$N^{3-} \longrightarrow N^0 + 3\ e^-$ Multiply by 2

$Cu^{2+} + 2\ e^- \longrightarrow Cu^0$ Multiply by 3, add, the 6 e^- cancel

$2\ N^{3-} + 3\ Cu^{2+} \longrightarrow N_2 + 3\ Cu$

Transfer the coefficients to the original equation and complete the balancing by inspection.

$3\ CuO + 2\ NH_3 \longrightarrow N_2 + 3\ Cu + 3\ H_2O$

(d) $PbO_2 + Sb + NaOH \longrightarrow PbO + NaSbO_2 + H_2O$

$Sb^0 \longrightarrow Sb^{3+} + 3\ e^-$ Mutiply by 2

$Pb^{4+} + 2\ e^- \longrightarrow 2\ Pb^{2+}$ Multiply by 3, add, the 6 e^- cancel

$2\ Sb + 3\ Pb^{4+} \longrightarrow 2\ Sb^{3+} + 3\ Pb^{2+}$

Transfer the coefficients to the original equation and complete the balancing by inspection.

$3\ PbO_2 + 2\ Sb + 2\ NaOH \longrightarrow 3\ PbO + 2\ NaSbO_2 + H_2O$

(e) $H_2O_2 + KMnO_4 + H_2SO_4 \longrightarrow O_2 + MnSO_4 + K_2SO_4 + H_2O$

${O_2}^{2-} \longrightarrow {O_2}^0 + 2\ e^-$ Multiply by 5

$Mn^{7+} + 5\ e^- \longrightarrow Mn^{2+}$ Multiply by 2, add, the 10 e^- cancel

$5\ {O_2}^{2-} + 2\ Mn^{7+} \longrightarrow 5\ O_2 + 2\ Mn^{2+}$

Transfer the coefficients to the original equation and complete the balancing by inspection.

$5\ H_2O_2 + 2\ KMnO_4 + 3\ H_2SO_4 \longrightarrow 5\ O_2 + 2\ MnSO_4 + K_2SO_4 + 8\ H_2O$

31. (a) $Zn + {NO_3}^- \longrightarrow Zn^{2+} + {NH_4}^+$ (acidic solution)

Step 1 Write half-reaction equations. Balance except H and O.

$Zn \longrightarrow Zn^{2+}$

${NO_3}^- \longrightarrow {NH_4}^+$

Step 2 Balance H and O using H_2O and H^+

$Zn \longrightarrow Zn^{2+}$

$10\ H^+ + {NO_3}^- \longrightarrow {NH_4}^+ + 3\ H_2O$

Step 3 Balance electrically with electrons

$Zn \longrightarrow Zn^{2+} + 2\ e^-$

$10\ H^+ + {NO_3}^- + 8\ e^- \longrightarrow {NH_4}^+ + 3\ H_2O$

Step 4 Equalize the loss and gain of electrons

$4\ (Zn \longrightarrow Zn^{2+} + 2\ e^-)$

$10\ H^+ + {NO_3}^- + 8\ e^- \longrightarrow {NH_4}^+ + H_2O$

Step 5 Add the half reactions — electrons cancel

$$10\ H^+ + 4\ Zn + NO_3^- \longrightarrow 4\ Zn^{2+} + NH_4^+ + 3\ H_2O$$

(b) $NO_3^- + S \longrightarrow NO_2 + SO_4^{2-}$ (acidic solution)

Step 1 Write half-reactions equations. Balance except H and O.

$$S \longrightarrow SO_4^{2-}$$

$$NO_3^- \longrightarrow NO_2$$

Step 2 Balance H and O using H_2O and H^+

$$4\ H_2O + S \longrightarrow SO_4^{2-} + 8\ H^+$$

$$2\ H^+ + NO_3^- \longrightarrow NO_2 + H_2O$$

Step 3 Balance electrically with electrons

$$4\ H_2O + S \longrightarrow SO_4^{2-} + 8\ H^+ + 6\ e^-$$

$$2\ H^+ + NO_3^- + e^- \longrightarrow NO_2 + H_2O$$

Steps 4 and 5 Equalize loss and gain of electrons; add half-reactions

$$4\ H_2O + S \longrightarrow SO_4^{2-} + 8\ H^+ + 6\ e^-$$

$$\underline{6\ (2\ H^+ + NO_3^- + e^- \longrightarrow NO_2 + H_2O)}$$

$$4\ H^+ + S + 6\ NO_3^- \longrightarrow 6\ NO_2 + SO_4^{2-} + 2\ H_2O$$

(c) $PH_3 + I_2 \longrightarrow H_3PO_2 + I^-$ (acidic solution)

Step 1 Write half-reaction equations. Balance except H and O.

$$PH_3 \longrightarrow H_3PO_2$$

$$I_2 \longrightarrow 2\ I^-$$

Step 2 Balance H and O using H_2O and H^+

$$2\ H_2O + PH_3 \longrightarrow H_3PO_2 + 4\ H^+$$

$$I_2 \longrightarrow 2\ I^-$$

Step 3 Balance electrically with electrons

$$2\ H_2O + PH_3 \longrightarrow H_3PO_2 + 4\ H^+ + 4\ e^-$$

$$I_2 + 2\ e^- \longrightarrow 2\ I^-$$

Steps 4 and 5 Equalize the loss and gain of electrons; add half-reactions

$$2\ H_2O + PH_3 \longrightarrow H_3PO_2 + 4\ H^+ + 4\ e^-$$

$$\underline{2\ (I_2 + 2\ e^- \longrightarrow 2\ I^-)}$$

$$PH_3 + 2\ H_2O + 2\ I_2 \longrightarrow H_3PO_2 + 4\ I^- + 4\ H^+$$

(d) $Cu + NO_3^- \longrightarrow Cu^{2+} + NO$ (acidic solution)

Step 1 Write half–reaction equations. Balance except H and O.

$Cu \longrightarrow Cu^{2+}$

$NO_3^- \longrightarrow NO$

Step 2 Balance H and O using H_2O and H^+

$Cu \longrightarrow Cu^{2+}$

$4\ H^+ + NO_3^- \longrightarrow NO + 2\ H_2O$

Step 3 Balance electrically with electrons

$Cu \longrightarrow Cu^{2+} + 2\ e^-$

$4\ H^+ + NO_3^- + 3\ e^- \longrightarrow NO + 2\ H_2O$

Steps 4 and 5 Equalize the loss and gain of electrons; add half-reactions

$3\ (Cu \longrightarrow Cu^{2+} + 2\ e^-)$

$2\ (4\ H^+ + NO_3^- + 3\ e^- \longrightarrow NO + 2\ H_2O)$

$3\ Cu + 8\ H^+ + 2\ NO_3^- \longrightarrow 3\ Cu^{2+} + 2\ NO + 4\ H_2O$

(e) $ClO_3^- + Cl^- \longrightarrow Cl_2$ (acidic solution)

Step 1 Write half reaction equations. Balance, except H and O.

$Cl^- \longrightarrow Cl^0$

$ClO_3^- \longrightarrow Cl^0$

Step 2 Balance H and O using H_2O and H^+

$Cl^- \longrightarrow Cl^0$

$6\ H^+ + ClO_3^- \longrightarrow Cl^0 + 3\ H_2O$

Step 3 Balance electrically with electrons

$Cl^- \longrightarrow Cl^0 + e^-$

$6\ H^+ + ClO_3^- + 5\ e^- \longrightarrow Cl^0 + 3\ H_2O$

Steps 4 and 5 Equalize the loss and gain of electrons; add half-reactions

$5\ (Cl^- \longrightarrow Cl^0 + e^-)$

$6\ H^+ + ClO_3^- + 5\ e^- \longrightarrow Cl^0 + 3\ H_2O$

$6\ H^+ + ClO_3^- + 5\ Cl^- \longrightarrow 3\ Cl_2 + 3\ H_2O$

32. (a) $ClO_3^- + I^- \longrightarrow I_2 + Cl^-$ (acidic solution)

Step 1 Write half-reaction equations. Balance except H and O.

$2\,I^- \longrightarrow I_2$

$ClO_3^- \longrightarrow Cl^-$

Step 2 Balance H and O using H_2O and H^+

$2\,I^- \longrightarrow I_2$

$6\,H^+ + ClO_3^- \longrightarrow Cl^- + 3\,H_2O$

Step 3 Balance electrically with electrons

$2\,I^- \longrightarrow I_2 + 2\,e^-$

$6\,H^+ + ClO_3^- + 6\,e^- \longrightarrow Cl^- + 3\,H_2O$

Steps 4 and 5 Equalize the loss and gain of electrons; add half-reactions

$3\,(2\,I^- \longrightarrow I_2 + 2\,e^-)$

$6\,H^+ + ClO_3^- + 6\,e^- \longrightarrow Cl^- + 3\,H_2O$

$6\,H^+ + ClO_3^- + 6\,I^- \longrightarrow 3\,I_2 + Cl^- + 3\,H_2O$

(b) $Cr_2O_7^{2-} + Fe^{2+} \longrightarrow Cr^{3+} + Fe^{3+}$ (acidic solution)

Step 1 Write half-reaction equations. Balance except H and O.

$Fe^{2+} \longrightarrow Fe^{3+}$

$Cr_2O_7^{2-} \longrightarrow 2\,Cr^{3+}$

Step 2 Balance H and O using H_2O and H^+

$Fe^{2+} \longrightarrow Fe^{3+}$

$14\,H^+ + Cr_2O_7^{2-} \longrightarrow 2\,Cr^{3+} + 7\,H_2O$

Step 3 Balance electrically with electrons

$Fe^{2+} \longrightarrow Fe^{3+} + e^-$

$14\,H^+ + Cr_2O_7^{2-} + 6\,e^- \longrightarrow 2\,Cr^{3+} + 7\,H_2O$

Steps 4 and 5 Equalize the loss and gain of electrons; add half-reactions

$6\,(Fe^{2+} \longrightarrow Fe^{3+} + e^-)$

$14\,H^+ + Cr_2O_7^{2-} + 6\,e^- \longrightarrow 2\,Cr^{3+} + 7\,H_2O$

$14\,H^+ + Cr_2O_7^{2-} + 6\,Fe^{2+} \longrightarrow 2\,Cr^{3+} + 6\,Fe^{3+} + 7\,H_2O$

(c) $MnO_4^- + SO_2 \longrightarrow Mn^{2+} + SO_4^{2-}$ (acidic solution)

Step 1 Write half–reaction equations. Balance except H and O.

$$SO_2 \longrightarrow SO_4^{2-}$$
$$MnO_4^- \longrightarrow Mn^{2+}$$

Step 2 Balance H and O using H_2O and H^+

$$2\ H_2O + SO_2 \longrightarrow SO_4^{2-} + 4\ H^+$$
$$8\ H^+ + MnO_4^- \longrightarrow Mn^{2+} + 4\ H_2O$$

Step 3 Balance electrically with electrons

$$2\ H_2O + SO_2 \longrightarrow SO_4^{2-} + 4\ H^+ + 2\ e^-$$
$$8\ H^+ + MnO_4^- + 5\ e^- \longrightarrow Mn^{2+} + 4\ H_2O$$

Steps 4 and 5 Equalize the loss and gain of electrons; add half-reactions

$$5\ (2\ H_2O + SO_2 \longrightarrow SO_4^{2-} + 4\ H^+ + 2\ e^-)$$
$$\underline{2\ (8\ H^+ + MnO_4^- + 5\ e^- \longrightarrow Mn^{2+} + 4\ H_2O)}$$

$$2\ H_2O + 2\ MnO_4^- + 5\ SO_2 \longrightarrow 4\ H^+ + 2\ Mn^{2+} + 5\ SO_4^{2-}$$

(d) $H_3AsO_3 + MnO_4^- \longrightarrow H_3AsO_4 + Mn^{2+}$ (acidic solution)

Step 1 Write half–reaction equations. Balance except H and O.

$$H_3AsO_3 \longrightarrow H_3AsO_4$$
$$MnO_4^- \longrightarrow Mn^{2+}$$

Step 2 Balance H and O using H_2O and H^+

$$H_2O + H_3AsO_3 \longrightarrow 2\ H^+ + H_3AsO_4$$
$$8\ H^+ + MnO_4^- \longrightarrow Mn^{2+} + 4\ H_2O$$

Step 3 Balance electrically with electrons

$$H_2O + H_3AsO_3 \longrightarrow 2\ H^+ + H_3AsO_4 + 2\ e^-$$
$$8\ H^+ + MnO_4^- + 5\ e^- \longrightarrow Mn^{2+} + 4\ H_2O$$

Steps 4 and 5 Equalize the loss and gain of electrons; add half-reactions

$$5\ (H_2O + H_3AsO_3 \longrightarrow 2\ H^+ + H_3AsO_4 + 2\ e^-)$$
$$\underline{2\ (8\ H^+ + MnO_4^- + 5\ e^- \longrightarrow Mn^{2+} + 4\ H_2O)}$$

$$6\ H^+ + 5\ H_3AsO_4 + 2\ MnO_4^- \longrightarrow 5\ H_3AsO_4 + 2\ Mn^{2+} + 3\ H_2O$$

(e) $Cr_2O_7^{2-} + H_3AsO_3 \longrightarrow Cr^{3+} + H_3AsO_4$ (acidic solution)

Step 1 Write half–reaction equations. Balance, except H and O.

$H_3AsO_3 \longrightarrow H_3AsO_4$

$Cr_2O_7^{2-} \longrightarrow 2\ Cr^{3+}$

Step 2 Balance H and O using H_2O and H^+

$H_2O + H_3AsO_3 \longrightarrow 2\ H^+ + H_3AsO_4$

$14\ H^+ + Cr_2O_7^{2-} \longrightarrow 2\ Cr^{3+} + 7\ H_2O$

Step 3 Balance electrically with electrons

$H_2O + H_3AsO_3 \longrightarrow 2\ H^+ + H_3AsO_4 + 2\ e^-$

$14\ H^+ + Cr_2O_7^{2-} + 6\ e^- \longrightarrow 2\ Cr^{3+} + 7\ H_2O$

Steps 4 and 5 Equalize the loss and gain of electrons; add half-reactions

$3\ (H_2O + H_3AsO_3 \longrightarrow 2\ H^+ + H_3AsO_4 + 2\ e^-)$

$14\ H^+ + Cr_2O_7^{2-} + 6\ e^- \longrightarrow 2\ Cr^{3+} + 7\ H_2O$

$8\ H^+ + Cr_2O_7^{2-} + 3\ H_3AsO_3 \longrightarrow 2\ Cr^{3+} + 3\ H_3AsO_4 + 4\ H_2O$

33. (a) $Cl_2 + IO_3^- \longrightarrow Cl^- + IO_4^-$ (basic solution)

Step 1 Write half–reaction equations. Balance, except H and O.

$IO_3^- \longrightarrow IO_4^-$

$Cl_2 \longrightarrow 2\ Cl^-$

Step 2 Balance H and O using H_2O and H^+

$H_2O + IO_3^- \longrightarrow IO_4^- + 2\ H^+$

$Cl_2 \longrightarrow 2\ Cl^-$

Step 3 Add OH^- ions to both sides (same number as H^+ ions)

$2\ OH^- + H_2O + IO_3^- \longrightarrow IO_4^- + 2\ H^+ + 2\ OH^-$

$Cl_2 \longrightarrow 2\ Cl^-$

Step 4 Combine H^+ and OH^- to form H_2O; cancel H_2O where possible

$2\ OH^- + H_2O + IO_3^- \longrightarrow IO_4^- + 2\ H_2O$

$Cl_2 \longrightarrow 2\ Cl^-$

$2\ OH^- + IO_3^- \longrightarrow IO_4^- + H_2O$

$Cl_2 \longrightarrow 2\ Cl^-$

Step 5 Balance electrically with electrons

$$2\ OH^- + IO_3^- \longrightarrow IO_4^- + H_2O + 2\ e^-$$

$$Cl_2 + 2\ e^- \longrightarrow 2\ Cl^-$$

Step 6 Electron loss and gain is balanced

Step 7 Add half-reactions

$$2\ OH^- + IO_3^- + Cl_2 \longrightarrow IO_4^- + 2\ Cl^- + H_2O$$

(b) $MnO_4^- + ClO_2^- \longrightarrow MnO_2 + ClO_4^-$ (basic solution)

Step 1 Write half-reaction equations. Balance, except H and O.

$$ClO_2^- \longrightarrow ClO_4^-$$

$$MnO_4^- \longrightarrow MnO_2$$

Step 2 Balance H and O using H_2O and H^+

$$2\ H_2O + ClO_2^- \longrightarrow ClO_2^- + 4\ H^+$$

$$MnO_4^- + 4\ H^+ \longrightarrow MnO_2 + 2\ H_2O$$

Step 3 Add OH^- ions to both sides (same number as H^+ ions)

$$4\ OH^- + 2\ H_2O + ClO_2^- \longrightarrow ClO_4^- + 4\ H^+ + 4\ OH^-$$

$$4\ OH^- + MnO_4^- + 4\ H^+ \longrightarrow MnO_2 + 2\ H_2O + 4\ OH^-$$

Step 4 Combine H^+ and OH^- to form H_2O; cancel H_2O where possible

$$4\ OH^- + 2\ H_2O + ClO_2^- \longrightarrow ClO_4^- + 4\ H_2O$$

$$4\ H_2O + MnO_4^- \longrightarrow MnO_2 + 2\ H_2O + 4\ OH^-$$

$$4OH^- + ClO_2^- \longrightarrow ClO_4^- + 2\ H_2O$$

$$2\ H_2O + MnO_4^- \longrightarrow MnO_2 + 4\ OH^-$$

Step 5 Balance electrically with electrons

$$4\ OH^- + ClO_2^- \longrightarrow ClO_4^- + 2\ H_2O + 4\ e^-$$

$$2\ H_2O + MnO_4^- + 3\ e^- \longrightarrow MnO_2 + 4\ OH^-$$

Steps 6 and 7 Equalize gain and loss of electrons; add half-reactions

$$3\ (4\ OH^- + ClO_2^- \longrightarrow ClO_4^- + 2\ H_2O + 4\ e^-)$$

$$\underline{4\ (2\ H_2O + MnO_4^- + 3\ e^- \longrightarrow MnO_2 + 4\ OH^-)}$$

$$2\ H_2O + 4\ MnO_4^- + 3\ ClO_2^- \longrightarrow 4\ MnO_2 + 3\ ClO_4^- + 4\ OH^-$$

(c) $Se \longrightarrow SeO_3^{2-} + Se^{2-}$ (basic solution)

Step 1 Write half-reaction equations. Balance, except H,O.

$Se \longrightarrow SeO_3^{2-}$

$Se \longrightarrow Se^{2-}$

Step 2 Balance H and O using H_2O and H^+

$3\ H_2O + Se \longrightarrow SeO_3^{2-} + 6\ H^+$

$Se \longrightarrow Se^{2-}$

Step 3 Add OH^- ions to both sides (same number as H^+ ions)

$6\ OH^- + 3\ H_2O + Se \longrightarrow SeO_3^{2-} + 6\ H^+ + 6\ OH^-$

$Se \longrightarrow Se^{2-}$

Step 4 Combine H^+ and OH^- to form H_2O; cancel H_2O where possible

$6\ OH^- + 3\ H_2O + Se \longrightarrow SeO_3^{2-} + 6\ H_2O$

$Se \longrightarrow Se^{2-}$

$6\ OH^- + Se \longrightarrow SeO_3^{2-} + 3\ H_2O$

Step 5 Balance electrically with electrons

$6\ OH^- + Se \longrightarrow SeO_3^{2-} + 3\ H_2O + 4\ e^-$

$Se + 2\ e^- \longrightarrow Se^{2-}$

Steps 6 and 7 Equalize loss and gain of electrons; add half-reactions

$6\ OH^- + Se \longrightarrow SeO_3^{2-} + 3\ H_2O + 4\ e^-$

$2\ (Se + 2\ e^- \longrightarrow Se^{2-})$

$6\ OH^- + 3\ Se \longrightarrow SeO_3^{2-} + 2\ Se^{2-} + 3\ H_2O$

(d) $Fe_3O_4 + MnO_4^- \longrightarrow Fe_2O_3 + MnO_2$ (basic solution)

Step 1 Write half-reaction equations. Balance, except H and O.

$2\ Fe_3O_4 \longrightarrow 3\ Fe_2O_3$

$MnO_4^- \longrightarrow MnO_2$

Step 2 Balance H and O using H_2O and H^+

$H_2O + 2\ Fe_3O_4 \longrightarrow 3\ Fe_2O_3 + 2\ H^+$

$4\ H^+ + MnO_4^- \longrightarrow MnO_2 + 2\ H_2O$

Step 3 Add OH^- ions to both sides (same number as H^+ ions)

$2\ OH^- + H_2O + 2\ Fe_3O_4 \longrightarrow 3\ Fe_2O_3 + 2\ H^+ + 2\ OH^-$

$4\ OH^- + 4\ H^+ + MnO_4^- \longrightarrow MnO_2 + 2\ H_2O + 4\ OH^-$

Step 4 Combine H^+ and OH^- to form H_2O; cancel H_2O where possible

$2\ OH^- + H_2O + 2\ Fe_3O_4 \longrightarrow 3\ Fe_2O_3 + 2\ H_2O$

$4\ H_2O + MnO_4^- \longrightarrow MnO_2 + 2\ H_2O + 4\ OH^-$

$2\ OH^- + 2\ Fe_3O_4 \longrightarrow 3\ Fe_2O_3 + H_2O$

$2\ H_2O + MnO_4^- \longrightarrow MnO_2 + 4\ OH^-$

Step 5 Balance electrically with electrons

$2\ OH^- + 2\ Fe_3O_4 \longrightarrow 3\ Fe_2O_3 + H_2O + 2\ e^-$

$2\ H_2O + MnO_4^- + 3\ e^- \longrightarrow MnO_2 + 4\ OH^-$

Steps 6 and 7 Equalize gain and loss of electrons; add half-reactions

$3\ (2\ OH^- + 2\ Fe_3O_4 \longrightarrow 3\ Fe_2O_3 + H_2O + 2\ e^-)$

$2\ (2\ H_2O + MnO_4^- + 3\ e^- \longrightarrow MnO_2 + 4\ OH^-)$

$H_2O + 6\ Fe_3O_4 + 2\ MnO_4^- \longrightarrow 9\ Fe_2O_3 + 2\ MnO_2 + 2\ OH^-$

(e) $BrO^- + Cr(OH)_4^- \longrightarrow Br^- + CrO_4^{2-}$ (basic solution)

Step 1 Write half-reaction equations. Balance, except H, O

$Cr(OH)_4^- \longrightarrow CrO_4^{2-}$

$BrO^- \longrightarrow Br^-$

Step 2 Balance H and O using H_2O and H^+

$Cr(OH)_4^- \longrightarrow CrO_4^{2-} + 4\ H^+$

$2\ H^+ + BrO^- \longrightarrow Br^- + H_2O$

Step 3 Add OH^- ions to both sides (same number as H^+ ions)

$4\ OH^- + Cr(OH)_4^- \longrightarrow CrO_4^{2-} + 4\ H^+ + 4\ OH^-$

$2\ OH^- + 2\ H^+ + BrO^- \longrightarrow Br^- + H_2O + 2\ OH^-$

Step 4 Combine H^+ and OH^- to form H_2O; cancel H_2O where possible

$4\ OH^- + Cr(OH)_4^- \longrightarrow CrO_4^{2-} + 4\ H_2O$

$2\ H_2O + BrO^- \longrightarrow Br^- + H_2O + 2\ OH^-$

$H_2O + BrO^- \longrightarrow Br^- + 2\ OH^-$

Step 5 Balance electrically with electrons

$4\ OH^- + Cr(OH)_4^- \longrightarrow CrO_4^{2-} + 4\ H_2O + 3\ e^-$

$H_2O + BrO^- + 2\ e^- \longrightarrow Br^- + 2\ OH^-$

Steps 6 and 7 Equalize gain and loss of electrons; add half-reactions

$2\ (4\ OH^- + Cr(OH)_4^- \longrightarrow CrO_4^{2-} + 4\ H_2O + 3\ e^-)$

$3\ (H_2O + BrO^- + 2\ e^- \longrightarrow Br^- + 2\ OH^-)$

$2\ OH^- + 3\ BrO^- + 2\ Cr(OH)_4^- \longrightarrow 3\ Br^- + 2\ CrO_4^{2-} + 5\ H_2O$

34. (a) $MnO_4^- + SO_3^{2-} \longrightarrow MnO_2 + SO_4^{2-}$ (basic solution)

Step 1 Write half-reaction equations. Balance except H and O.

$SO_3^{2-} \longrightarrow SO_4^{2-}$

$MnO_4^- \longrightarrow MnO_2$

Step 2 Balance H and O using H_2O and H^+

$H_2O + SO_3^{2-} \longrightarrow SO_4^{2-} + 2\ H^+$

$MnO_4^- + 4\ H^+ \longrightarrow MnO_2 + 2\ H_2O$

Step 3 Add OH^- ions to both sides (same number as H^+ ions)

$2\ OH^- + H_2O + SO_3^{2-} \longrightarrow SO_4^{2-} + 2\ H^+ + 2\ OH^-$

$4\ OH^- + MnO_4^- + 4\ H^+ \longrightarrow MnO_2 + 3\ H_2O + 4\ OH^-$

Step 4 Combine H^+ and OH^- to form H_2O; cancel H_2O where possible

$2\ OH^- + H_2O + SO_3^{2-} \longrightarrow SO_4^{2-} + 2\ H_2O$

$MnO_4^- + 4\ H_2O \longrightarrow MnO_2 + 2\ H_2O + 4\ OH^-$

$2\ OH^- + SO_3^{2-} \longrightarrow SO_4^{2-} + H_2O$

$MnO_4^- + 2\ H_2O \longrightarrow MnO_2 + 4\ OH^-$

Step 5 Balance electrically with electrons

$2\ OH^- + SO_3^{2-} \longrightarrow SO_4^{2-} + H_2O + 2\ e^-$

$3\ e^- + MnO_4^- + 2\ H_2O \longrightarrow MnO_2 + 4\ OH^-$

Steps 6 and 7 Equalize gain and loss of electrons; add half-reactions

$3\ (2\ OH^- + SO_3^{2-} \longrightarrow SO_4^{2-} + H_2O + 2\ e^-)$

$2\ (MnO_4^- + 2\ H_2O + 3\ e^- \longrightarrow MnO_2 + 4\ OH^-)$

$H_2O + 2\ MnO_4^- + 3\ SO_3^{2-} \longrightarrow 2\ MnO_2 + 3\ SO_4^{2-} + 2\ OH^-$

(b) $ClO_2 + SbO_2^- \longrightarrow ClO_2^- + Sb(OH)_6^-$ (basic solution)

Step 1 Write half-reaction equations. Balance, except H and O.

$SbO_2^- \longrightarrow Sb(OH)_6^-$

$ClO_2 \longrightarrow ClO_2^-$

Step 2 Balance H and O using H_2O and H^+

$4\ H_2O + SbO_2^- \longrightarrow Sb(OH)_6^- + 2\ H^+$

$ClO_2 \longrightarrow ClO_2^-$

Step 3 Add OH^- ions to both sides (same number as H^+ ions)

$2\ OH^- + 4\ H_2O + SbO_2^- \longrightarrow Sb(OH)_6^- + 2\ H^+ + 2\ OH^-$

$ClO_2 \longrightarrow ClO_2^-$

Step 4 Combine H^+ and OH^- to form H_2O; cancel H_2O where possible

$2\ OH^- + 4\ H_2O + SbO_2^- \longrightarrow Sb(OH)_6^- + 2\ H_2O$

$ClO_2 \longrightarrow ClO_2^-$

$2\ OH^- + 2\ H_2O + SbO_2^- \longrightarrow Sb(OH)_6^-$

Step 5 Balance electrically with electrons

$2\ OH^- + 2\ H_2O + SbO_2^- \longrightarrow Sb(OH)_6^- + 2\ e^-$

$ClO_2 + e^- \longrightarrow ClO_2^-$

Steps 6 and 7 Equalize gain and loss of electrons; add half-reactions

$2\ H_2O + 2\ OH^- + SbO_2^- \longrightarrow Sb(OH)_6^- + 2\ e^-$

$2\ (ClO_2 + e^- \longrightarrow ClO_2^-)$

$2\ H_2O + 2\ ClO_2 + 2\ OH^- + SbO_2^- \longrightarrow 2\ ClO_2^- + Sb(OH)_6^-$

(c) $Al + NO_3^- \longrightarrow NH_3 + Al(OH)_4^-$ (basic solution)

Step 1 Write half-reaction equation. Balance except H and O.

$Al \longrightarrow Al(OH)_4^-$

$NO_3^- \longrightarrow NH_3$

Step 2 Balance H and O using H_2O and H^+

$4\ H_2O + Al \longrightarrow Al(OH)_4^- + 4\ H^+$

$9\ H^+ + NO_3^- \longrightarrow NH_3 + 3\ H_2O$

Step 3 Add OH^- ions to both sides (same number as H^+ ions)

$4\ OH^- + 4\ H_2O + Al \longrightarrow Al(OH)_4^- + 4\ H^+ + 4\ OH^-$

$9\ OH^- + 9\ H^+ + NO_3^- \longrightarrow NH_3 + 3\ H_2O + 9\ OH^-$

Step 4 Combine H^+ and OH^- ions to form H_2O; cancel H_2O where possible

$4\ OH^- + 4\ H_2O + Al \longrightarrow Al(OH)_4^- + 4\ H_2O$

$9\ H_2O + NO_3^- \longrightarrow NH_3 + 3\ H_2O + 9\ OH^-$

$4\ OH^- + Al \longrightarrow Al(OH)_4^-$

$6\ H_2O + NO_3^- \longrightarrow NH_3 + 9\ OH^-$

Step 5 Balance electrically with electrons

$4\ OH^- + Al \longrightarrow Al(OH)_4^- + 3\ e^-$

$6\ H_2O + NO_3^- + 8\ e^- \longrightarrow NH_3 + 9\ OH^-$

Steps 6 and 7 Equalize gain and loss of electrons; add half-reactions

$8\ (4\ OH^- + Al \longrightarrow Al(OH)_4^- + 3\ e^-)$

$\underline{3\ (6\ H_2O + NO_3^- + 8\ e^- \longrightarrow NH_3 + 9\ OH^-)}$

$8\ Al + 3\ NO_3^- + 18\ H_2O + 5\ OH^- \longrightarrow 3\ NH_3 + 8\ Al(OH)_4^-$

(d) $P_4 \longrightarrow HPO_3^{2-} + PH_3$ (basic solution)

Step 1 Write half-reaction equations. Balence except H and O.

$P_4 \longrightarrow 4\ HPO_3^{2-}$

$P_4 \longrightarrow 4\ PH_3$

Step 2 Balance H and O using H_2O and H^+

$12\ H_2O + P_4 \longrightarrow 4\ HPO_3^{2-} + 20\ H^+$

$12\ H^+ + P_4 \longrightarrow 4\ PH_3$

Step 3 Add OH^- ions to both sides (same number as H^+ ions)

$20\ OH^- + 12\ H_2O + P_4 \longrightarrow 4\ HPO_3^{2-} + 20\ H^+ + 20\ OH^-$

$12\ OH^- + 12\ H^+ + P_4 \longrightarrow 4\ PH_3 + 12\ OH^-$

Step 4 Combine H^+ and OH^- to form H_2O; cancel H_2O where possible

$20\ OH^- + 12\ H_2O + P_4 \longrightarrow 4\ HPO_3^{2-} + 20\ H_2O$

$12\ H_2O + P_4 \longrightarrow 4\ PH_3 + 12\ OH^-$

$20\ OH^- + P_4 \longrightarrow 4\ HPO_3^{2-} + 8\ H_2O$

Step 5 Balance electrically with electrons

$20\ OH^- + P_4 \longrightarrow 4\ HPO_3^{2-} + 8\ H_2O + 12\ e^-$

$12\ H_2O + P_4 + 12\ e^- \longrightarrow 4\ PH_3 + 12\ OH^-$

Steps 6 and 7 Loss and gain of electrons are equal; add half-reaction

$$8\ OH^- + 4\ H_2O + 2\ P_4 \longrightarrow 4\ HPO_3^{2-} + 4\ PH_3$$

Divide equation by 2

$$4\ OH^- + 2\ H_2O + P_4 \longrightarrow 2\ HPO_3^{2-} + 2\ PH_3$$

(e) $Al + OH^- \longrightarrow Al(OH)_4^- + H_2$ (basic solution)

Step 1 Write half–reaction equations. Balance, except H and O.

$$Al \longrightarrow Al(OH)_4^-$$

$$OH^- \longrightarrow H_2$$

Step 2 Balance H and O using H_2O and H^+

$$4\ H_2O + Al \longrightarrow Al(OH)_4^- + 4\ H^+$$

$$3\ H^+ + OH^- \longrightarrow H_2 + H_2O$$

Step 3 Add OH^- ions to both sides (same number as H^+ ions)

$$4\ OH^- + 4\ H_2O + Al \longrightarrow Al(OH)_4^- + 4\ H^+ + 4\ OH^-$$

$$3\ OH^- + 3\ H^+ + OH^- \longrightarrow H_2 + H_2O + 3\ OH^-$$

Step 4 Combine H^+ and OH^- to form H_2O; cancel H_2O where possible

$$4\ OH^- + 4\ H_2O + Al \longrightarrow Al(OH)_4^- + 4\ H_2O$$

$$3\ H_2O + OH^- \longrightarrow H_2 + H_2O + 3\ OH^-$$

$$4\ OH^- + Al \longrightarrow Al(OH)_4^-$$

$$2\ H_2O + OH^- \longrightarrow H_2 + 3\ OH^-$$

Step 5 Balance electrically with electrons

$$4\ OH^- + Al \longrightarrow Al(OH)_4^- + 3\ e^-$$

$$2\ H_2O + OH^- + 2\ e^- \longrightarrow H_2 + 3\ OH^-$$

Steps 6 and 7 Equalize gain and loss of electrons; add half-reactions

$$2\ (4\ OH^- + Al \longrightarrow Al(OH)_4^- + 3\ e^-)$$

$$3\ (2\ H_2O + OH^- + 2\ e^- \longrightarrow H_2 + 3\ OH^-)$$

$$2\ Al + 6\ H_2O + 2\ OH^- \longrightarrow 2\ Al(OH)_4^- + 3\ H_2$$

35. (a) $Pb + SO_4^{2-} \longrightarrow PbSO_4 + 2\ e^-$

$$PbO_2 + SO_4^{2-} + 4\ H^+ + 2\ e^- \longrightarrow PbSO_4 + 2\ H_2O$$

(b) The first reaction is oxidation (Pb^0 is oxidized to Pb^{2+}).
The second reaction is reduction (Pb^{4+} is reduced to Pb^{2+})

(c) The first reaction (oxidation) occurs at the anode of the battery.

36. (a) The oxidizing agent is $KMnO_4$.

(b) The reducing agent is HCl.

(c) 5 moles of electrons $5\ e^- + Mn^{7+} \longrightarrow Mn^{2+}$

$$\left(\frac{5 \text{ mol } e^-}{\text{mol } KMnO_4}\right)\left(\frac{6.022 \times 10^{23}\ e^-}{\text{mol } e^-}\right) = 3.01 \times 10^{24}\ \frac{\text{electrons}}{\text{mol } KMnO_4}$$

37. $3\ Ag + 4\ HNO_3 \longrightarrow 3\ AgNO_3 + NO + 2\ H_2O$

g Ag $\longrightarrow$ mol Ag $\longrightarrow$ mol NO

$$(25.0 \text{ g Ag})\left(\frac{1 \text{ mol}}{107.9 \text{ g}}\right)\left(\frac{1 \text{ mol NO}}{3 \text{ mol Ag}}\right) = 0.0772 \text{ mol NO}$$

38. $3\ Cl_2 + 6\ KOH \longrightarrow KClO_3 + 5\ KCl + 3\ H_2O$

mol $KClO_3$ $\longrightarrow$ mol Cl_2 $\longrightarrow$ L Cl_2

$$(0.300 \text{ mol } KClO_3)\left(\frac{3 \text{ mol } Cl_2}{1 \text{ mol } KClO_3}\right)\left(\frac{22.4 \text{ L}}{1 \text{ mol}}\right) = 20.2 \text{ L } Cl_2$$

39. $5\ H_2O_2 + 2\ KMnO_4 + 3\ H_2SO_4 \longrightarrow 5\ O_2 + 2\ MnSO_4 + K_2SO_4 + 8\ H_2O$

mL H_2O_2 $\longrightarrow$ g H_2O_2 $\longrightarrow$ mol H_2O_2 $\longrightarrow$ mol $KMnO_4$ $\longrightarrow$ g $KMnO_4$

$$(100 \text{ mL } H_2O_2 \text{ solution})\left(\frac{1.031 \text{ g}}{\text{mL}}\right)\left(\frac{9.0 \text{ g } H_2O_2}{100. \text{ g } H_2O_2 \text{ solution}}\right)\left(\frac{1 \text{ mol}}{34.02 \text{ g}}\right)\left(\frac{2 \text{ mol } KMnO_4}{5 \text{ mol } H_2O_2}\right)\left(\frac{158.0 \text{ g}}{\text{mol}}\right)$$

$= 17$ g $KMnO_4$

40. $Cr_2O_7{}^{2-} + 3\ H_3AsO_3 + 8\ H^+ \longrightarrow 2\ Cr^{3+} + 3\ H_3AsO_4 + 4\ H_2O$

g H_3AsO_3 $\longrightarrow$ mol H_3AsO_3 $\longrightarrow$ mol $Cr_2O_7{}^{2-}$ $\longrightarrow$ mL $Cr_2O_7{}^{2-}$

$$(5.00 \text{ g } H_3AsO_4)\left(\frac{1 \text{ mol}}{125.9 \text{ g}}\right)\left(\frac{1 \text{ mol } Cr_2O_7{}^{2-}}{3 \text{ mol } H_3AsO_3}\right)\left(\frac{1000 \text{ mL}}{0.200 \text{ mol}}\right) = 66.2 \text{ mL of } 0.200 \text{ M } K_2Cr_2O_7$$

41. $Cr_2O_7{}^{2-} + 6\ Fe^{2+} + 14\ H^+ \longrightarrow 2\ Cr^{3+} + 6\ Fe^{3+} + 7\ H_2O$

mL $FeSO_4$ $\longrightarrow$ mol $FeSO_4$ $\longrightarrow$ mol $Cr_2O_7{}^{2-}$ $\longrightarrow$ mL $Cr_2O_7{}^{2-}$

$$(60.0 \text{ mL } FeSO_4)\left(\frac{0.200 \text{ mol}}{1000 \text{ mL}}\right)\left(\frac{1 \text{ mol } Cr_2O_7{}^{2-}}{6 \text{ mol } FeSO_4}\right)\left(\frac{1000 \text{ mL}}{0.200 \text{ mol}}\right) = 10.0 \text{ mL } 0.200 \text{ M } K_2Cr_2O_7$$

42. $8\ KI + 5\ H_2SO_4 \longrightarrow 4\ I_2 + H_2S + 4\ K_2SO_4 + 4\ H_2O$

g I_2 $\longrightarrow$ mol I_2 $\longrightarrow$ mol KI $\longrightarrow$ g KI

$$(2.79 \text{ g } I_2)\left(\frac{1 \text{ mol}}{253.8 \text{ g}}\right)\left(\frac{8 \text{ mol KI}}{4 \text{ mol } I_2}\right)\left(\frac{166.0 \text{ g}}{\text{mol}}\right) = 3.65 \text{ g KI in sample}$$

$$\left(\frac{3.65 \text{ g KI}}{4.00 \text{ g sample}}\right)(100) = 91.3\% \text{ KI}$$

43. $3\ Ag + 4\ HNO_3 \longrightarrow 3\ AgNO_3 + NO + 2\ H_2O$

mol Ag $\longrightarrow$ mol NO

$(0.500 \text{ mol Ag})\left(\frac{1 \text{ mol NO}}{3 \text{ mol Ag}}\right) = 0.167 \text{ mol NO}$

$PV = nRT \quad V = nRT/P$

$P = (744 \text{ torr})\left(\frac{1 \text{ atm}}{760. \text{ torr}}\right) = 0.979 \text{ atm}$

$T = 301 \text{ K}$

$$V = \frac{(0.167 \text{ mol NO})(0.0821 \text{ L atm/mol K})(301 \text{ K})}{(0.979 \text{ atm})} = 4.22 \text{ L NO}$$

44. $2\ Al + 2\ OH^- + 6\ H_2O \longrightarrow 2\ Al(OH)_4^- + 3\ H_2$

g Al $\longrightarrow$ mol Al $\longrightarrow$ mol H_2

$(100. \text{ g Al})\left(\frac{1 \text{ mol Al}}{26.98 \text{ g}}\right)\left(\frac{3 \text{ mol } H_2}{2 \text{ mol Al}}\right) = 5.56 \text{ mol } H_2$

45. (a) $Cu^+ \longrightarrow Cu^{2+}$ is an oxidation, but when electrons are gained reduction should occur.

$Cu^+ + e^- \longrightarrow Cu^0$

(b) When Pb^{2+} is reduced, it requires two individual electrons. $Pb^{2+} + 2\ e^- \longrightarrow Pb^0$. An electron has only a single negative charge (e^-).

46. The electrons lost by the species undergoing oxidation must be gained (or attracted) by another species which then undergoes reduction.

47. $A + B^{2+} \longrightarrow NR$ — B^{2+} cannot take e^- from A

$A + C^+ \longrightarrow NR$ — C^+ cannot take e^- from A

$D + 2\ C^+ \longrightarrow 2\ C + D^{2+}$ — C^+ takes e^- from D

$B + D^{2+} \longrightarrow D + B^{2+}$ — D^{2+} takes e^- from B

Therefore, B^{2+} is least able to attract e^-, then D^{2+}, then C^+, then A^+

48. Sn^{4+} can only be an oxidizing agent.
$Sn^{4+} + 2\ e^- \longrightarrow Sn^{2+}$
$Sn^{4+} + 4\ e^- \longrightarrow Sn^0$

Sn^0 can only be a reducing agent.
$Sn^0 \longrightarrow Sn^{2+} + 2\ e^-$
$Sn^0 \longrightarrow Sn^{4+} + 4\ e^-$

Sn^{2+} can be both oxidizing and reducing.
$Sn^{2+} + 2\ e^- \longrightarrow Sn^0$ (oxidizing)
$Sn^{2+} \longrightarrow Sn^{4+} + 2\ e^-$ (reducing)

49. $Mn(OH)_2$ +2
MnF_3 +3
MnO_2 +4
K_2MnO_4 +6
$KMnO_4$ +7

$KMnO_4$ is the best oxidizing agent of the group, since its great positive charge (+7) makes it very attractive to electrons.

50. Equations (a) and (b) represent oxidation

(a) $Mg \longrightarrow Mg^{2+} + 2\ e^-$

(b) $SO_2 \longrightarrow SO_3$; $(S^{4+} \longrightarrow S^{6+} + 2\ e^-)$

51. (a) $MnO_2 + 2\ Br^- + 4\ H^+ \longrightarrow Mn^{2+} + Br_2 + 2\ H_2O$

(b) mL $Mn^{2+} \longrightarrow$ mol $Mn^{2+} \longrightarrow$ mol $MnO_2 \longrightarrow$ g MnO_2

$$(100.0\ \text{mL}\ Mn^{2+})\left(\frac{0.05\ \text{mol}}{1000\ \text{mL}}\right)\left(\frac{1\ \text{mol}\ MnO_2}{1\ \text{mol}\ Mn^{2+}}\right)\left(\frac{86.94\ \text{g}}{\text{mol}}\right) = 0.4\ \text{g}\ MnO_2$$

(c) $$(100\ \text{mL}\ Mn^{2+})\left(\frac{0.05\ \text{mol}}{1000\ \text{mL}}\right)\left(\frac{1\ \text{mol}\ Br_2}{1\ \text{mol}\ Mn^{2+}}\right) = 0.005\ \text{mol}\ Br_2$$

$$PV = nRT \qquad V = \frac{nRT}{P}$$

$$V = \left(\frac{0.005\ \text{mol}}{1.4\ \text{atm}}\right)\left(\frac{0.0821\ \text{L atm}}{\text{mol K}}\right)(323\ \text{K}) = 0.09\ \text{L}\ Br_2\ \text{vapor}$$

52. (a) $F_2 + 2\ Cl^- \longrightarrow 2\ F^- + Cl_2$

(b) $Br_2 + Cl^- \longrightarrow$ NR

(c) $I_2 + Cl^- \longrightarrow$ NR

(d) $Br_2 + 2\ I^- \longrightarrow 2\ Br^- + I_2$

53. $Mn + 2\ HCl \longrightarrow Mn^{2+} + H_2 + 2\ Cl^-$

54. $4\ Zn + NO_3^- + 10\ H^+ \longrightarrow 4\ Zn^{2+} + NH_4^+ + 3\ H_2O$
See Exercise 31(a).

55.

(1)	(2)	(3)	(4)	(5)
a) C oxidized	a) S oxidized	a) N oxidized	a) S oxidized	a) O_2^{2-} oxidized
b) O_2 reduced	b) N reduced	b) Cu reduced	b) O reduced	b) O_2^{2-} reduced
c) O_2 O.A.	c) HNO_3, O.A.	c) CuO, O.A.	c) H_2O_2, O.A.	c) H_2O_2, O.A.
d) C_3H_8 R.A.	d) H_2S, R.A.	d) NH_3, R.A.	d) Na_2SO_3, R.A.	d) H_2O_2, O.A.
e) $2\frac{2}{3} \longrightarrow 4$	e) $S^{2-} \longrightarrow S^0$	e) $N^{3-} \longrightarrow N_2^0$	e) $S^{4+} \longrightarrow S^{6+}$	e) $O_2^{2-} \longrightarrow O_2^0$
f) $0 \longrightarrow -2$	f) $N^{5+} \longrightarrow N^{2+}$	f) $Cu^{2+} \longrightarrow Cu^0$	f) $O_2^{2-} \longrightarrow O^{2-}$	f) $O_2^{2-} \longrightarrow O^{2-}$

O.A. = oxidizing agent
R.A. = Reducing agent

56.

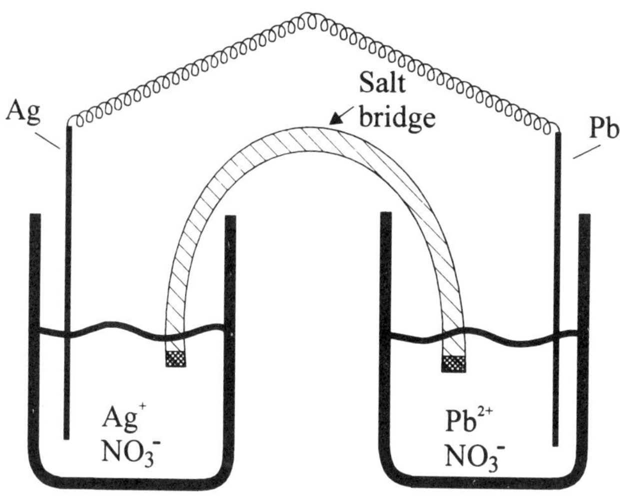

$Pb + 2\ Ag^{+} \longrightarrow 2\ Ag + Pb^{2+}$

(a) Pb is the anode

(b) Ag is the cathode

(c) Oxidation occurs at Pb (anode)

(d) Reduction occurs at Ag (cathode)

(e) Electrons flow from the lead through the wire to the silver

(f) Positive ions flow through the salt solution towards the negatively charged strip of silver; negative ions flow toward the positively charged strip of lead.

CHAPTER 18

NUCLEAR CHEMISTRY

1. (a) Gamma radiation requires the most shielding.

   (b) Alpha radiation requires the least shielding.

2. Alpha particles are deflected less than beta particles while passing through a magnetic field, because they are much heavier (more than 7,000 times heavier) than beta particles.

3. Pairs of nuclides that would be found in the fission reaction of U-235. Any two nuclides, whose atomic numbers add up to 92 and mass numbers (in the range of 70-160) add up to 230-234. Examples include:

   $^{90}_{38}Sr$ and $^{141}_{56}Xe$ $\quad$ $^{139}_{56}Ba$ and $^{94}_{36}Kr$ $\quad$ $^{101}_{42}Mo$ and $^{131}_{50}Sn$

4. Contributions to the early history of radioactivity include:

   (a) Henri Becquerel: He discovered radioactivity.

   (b) Marie and Pierre Curie: They discovered the elements polonium and radium.

   (c) Wilhelm Roentgen: He discovered X rays and developed the technique of producing them. While this was not a radioactive phenomenon, it triggered Becquerel's discovery of radioactivity.

   (d) Earnest Rutherford: He discovered alpha and beta particles, established the link between radioactivity and transmutation, and produced the first successful man-made transmutation.

   (e) Otto Hahn and Fritz Strassmann. They were first to produce nuclear fission.

5. Chemical reactions are caused by atoms or ions coming together, so are greatly influenced by temperature and concentration, which affect the number of collisions. Radioactivity is a spontaneous reaction of an individual nucleus, and is independent of such influences.

6. The term isotope is used with reference to atoms of the same element that contain different masses. For example $^{12}_{6}C$ and $^{14}_{6}C$. The term nuclide is used in nuclear chemistry to infer any isotope of any atom.

7. $(5 \times 10^9 \text{years})\left(\frac{1 \text{ half-life}}{7.6 \times 10^7 \text{years}}\right) = 70 \text{ half-lives}$

   Even if plutonium–224 had been present in large quantities five billion years ago, no measureable amount would survive after 70 half-lives.

8.

	charge	mass	nature of particles	penetrating power
Alpha	+2	4 amu	He nucleus	low
Beta	−1	$\frac{1}{1837\text{ amu}}$	electron	moderate
Gamma	0	0	electromagnetic radiation	high

9. Natural radioactivity is the spontaneous disintegration of those radioactive isotopes found in nature. Artificial radioactivity is the spontaneous disintegration of radioactive isotopes produced synthetically by man.

10. A radioactive disintegration series starts with a particular radionuclide and progresses stepwise by alpha and beta emissions to other radionuclides, ending at a stable nuclide. For example:

$$^{238}_{92}\text{U} \xrightarrow{\text{14 steps}} {}^{206}_{82}\text{Pb (stable)}$$

11. Transmutation is the conversion of one element into another by natural or artificial means. The nucleus of an atom is bombarded by various particles (alpha, beta, protons, etc.). The fast moving particles are captured by the nucleus, forming an unstable nucleus, which decays to another kind of atom. For example:

$$^{9}_{4}\text{Be} + {}^{4}_{2}\text{He} \longrightarrow {}^{12}_{6}\text{C} + {}^{1}_{0}\text{n}$$

12. $^{232}_{90}\text{Th} \xrightarrow{-\alpha} {}^{228}_{88}\text{Ra} \xrightarrow{-\beta} {}^{228}_{89}\text{Ac} \xrightarrow{-\beta} {}^{228}_{90}\text{Th} \xrightarrow{-\alpha} {}^{224}_{88}\text{Ra} \xrightarrow{-\alpha} {}^{220}_{86}\text{Rn} \xrightarrow{-\alpha} {}^{216}_{84}\text{Po} \xrightarrow{-\alpha} {}^{212}_{82}\text{Pb}$

$\xrightarrow{-\beta} {}^{212}_{83}\text{Bi} \xrightarrow{-\beta} {}^{212}_{84}\text{Po} \xrightarrow{-\alpha} {}^{208}_{82}\text{Pb}$

13. $^{237}_{93}\text{Np}$ loses seven alpha particles and four beta particles.

Determination of the final product: $^{209}_{83}\text{Bi}$

nuclear charge $= 93 - 7(2) + 4(1) = 83$

mass $= 237 - 7(4) = 209$

14. Decay of bismuth-211

$^{211}_{83}\text{Bi} \longrightarrow {}^{4}_{2}\text{He} + {}^{207}_{81}\text{Tl}$ $\quad$ $^{207}_{81}\text{Tl} \longrightarrow {}^{0}_{-1}\text{e} + {}^{207}_{82}\text{Pb}$

15. The Geiger counter is used to detect the presence of radioactive material. The rays which are emitted enter the Geiger tube and ionize the argon gas present in the tube. The ions cause an electrical discharge in the tube, producing a current pulse, which is amplified. The pulse appears as a signal in the form of audible clicks and/or a flashing light, that can be measured with a calibrated meter.

16. Two Germans, Otto Hahn and Fritz Strassmann, were the first scientists to report nuclear fission. The fission resulted from bombarding uranium nuclei with neutrons.

17. Natural uranium is 99+% U-238. Commercial nuclear reactors use U-235 enriched uranium as a fuel. Slow neutrons will cause the fission of U-235, but not U-238. Fast neutrons are capable of a nuclear reaction with U-238 to produce fissionable Pu-239. A breeder reactor converts nonfissionable U-238 to fissionable Pu-239, and in the process, manufactures more fuel than it consumes.

18. The fission reaction in a nuclear reactor and in an atomic bomb are essentially the same. The difference is that the fissioning is "wild" or uncontrolled in the bomb. In a nuclear reactor, the fissioning rate is controlled by means of moderators, such as graphite, to slow the neutrons and control rods of cadmium or boron to absorb some of the neutrons.

19. A certain amount of fissionable material (a critical mass) must be present before a self-supporting chain reaction can occur. Without a critical mass, too many neutrons from fissions will escape, and the reaction cannot reach a chain reaction status, unless at least one neutron is captured for every fission that occurs.

20. The mass defect is the difference between the mass of an atom and the sum of the masses of the number of protons, neutrons, and electrons in that atom. The energy equivalent of this mass defect is known as the nuclear binding energy.

21. When radioactive rays pass through normal matter, they cause that matter to become ionized (usually by knocking out electrons). Therefore, the radioactive rays are classified as ionizing radiation.

22. Some biological hazards associated with radioactivity are:

(a) High levels of radiation can cause nausea, vomiting, diarrhea, and death. The radiation produces ionization in the cells, particularly in the nucleus of the cells.

(b) Long-term exposure to low levels of radiation can weaken the body and cause malignant tumors.

(c) Radiation can damage DNA molecules in the body causing mutations, which by reproduction, can be passed on to succeeding generations.

23. Strontium-90 has two characteristics that create concern. Its half-life is 28 years, so it remains active for a long period of time (disintegrating by emitting β radiation). The other characteristic is that Sr-90 is chemically similar to calcium, so that it is deposited in bone tissue along with calcium. Red blood cells are produced in the bone marrow. If the marrow is subjected to beta radiation from strontium-90, the red blood cells will be destroyed, increasing the incidence of leukemia and bone cancer.

24. A radioactive "tracer" is a radioactive material, whose presence is traced by a Geiger counter or some other detecting device. Tracers are often injected into the human body, animals, and plants to determine chemical pathways, rates of circulation, etc. For example, use of a tracer could determine the length of time for material to travel from the root sytem to the leaves in a tree.

25. In living species, the ratio of carbon-14 to carbon-12 is constant due to the constant C-14/C-12 ratio in the atmosphere and food sources. When a species dies, life processes stop. The C-14/C-12 ratio decreases with time because C-14 is radioactive and decays according to its half-life, while the amount of C-12 in the species remains constant. Thus, the age of an archaeological artifact containing carbon can be calculated by comparing the C-14/C-12 ratio in the artifact with the C-14/C-12 ratio in the living species.

26. Radioactivity could be used to locate a leak in an underground pipe by using a water soluble tracer element. Dissolve the tracer in water and pass the water through the pipe. Test the ground along the path of the pipe with a Geiger counter until radioactivity from the leak is detected. Dig.

27. The half-life of carbon-14 is 5668 years.

$$(4 \times 10^6 \text{ years})\left(\frac{1 \text{ half-life}}{5668 \text{ years}}\right) = 7 \times 10^2 \text{ half-lives}$$

700 half-lives would pass in 4 million years. Not enough C-14 would remain to allow detection with any degree of reliability. C-14 dating would not prove useful in this case.

28. The correct statements are c, e, f, g, j, k, n, o, q, r, t

(a) Radioactivity was discovered by Henri Becquerel.

(b) An atom of $^{59}_{28}$Ni has 28 protons and 31 neutrons.

(d) The emission of an alpha particle from the nucleus of an atom lowers its atomic number by 2 and lowers its mass number by 4.

(h) The gamma ray has the greatest penetrating power of all the rays emitted from the nucleus of an atom.

(i) Radioactivity is due to an unstable ratio of neutrons to protons in an atom.

(l) The disintegration of $^{226}_{88}$Ra into $^{214}_{83}$Po involves the loss of 3 alpha particles and one beta particle.

(m) If 1.0 g of a radionuclide has a half-life of 7.2 days, the half-life of 0.50 g of that nuclide is 7.2 days.

(p) Radiocarbon dating of archaelogical artifacts is based on a decrease in the C-14/C-12 ratio in the object.

(s) Carbon-14 is produced in the atmosphere.

29.

		Protons	Neutrons	Nucleons
(a)	$^{35}_{17}$Cl	17	18	35
(b)	$^{226}_{88}$Ra	88	138	226

30.

		Protons	Neutrons	Nucleons
(a)	$^{235}_{92}U$	92	143	235
(b)	$^{82}_{35}Br$	35	47	82

31. When a nucleus loses an alpha particle, its atomic number decreases by two, and its mass number decreases by four.

32. When a nucleus loses a beta particle, its atomic number increases by one, and its mass number remains unchanged.

33. Equations for alpha decay:

(a) $^{218}_{85}At \longrightarrow {}^{4}_{2}He + {}^{214}_{83}Bi$

(b) $^{221}_{87}Fr \longrightarrow {}^{4}_{2}He + {}^{217}_{85}At$

34. Equations for alpha decay:

(a) $^{192}_{78}Pt \longrightarrow {}^{4}_{2}He + {}^{188}_{76}Os$

(b) $^{210}_{84}Po \longrightarrow {}^{4}_{2}He + {}^{206}_{82}Pb$

35. Equations for beta decay:

(a) $^{14}_{6}C \longrightarrow {}^{0}_{-1}e + {}^{14}_{7}N$

(b) $^{137}_{55}Cs \longrightarrow {}^{0}_{-1}e + {}^{137}_{56}Ba$

36. Equations for beta decay:

(a) $^{239}_{93}Np \longrightarrow {}^{0}_{-1}e + {}^{239}_{94}Pu$

(b) $^{90}_{38}Sr \longrightarrow {}^{0}_{-1}e + {}^{90}_{39}Y$

37. $^{13}_{6}C + {}^{1}_{0}n \longrightarrow {}^{14}_{6}C$

38. $^{30}_{15}P \longrightarrow {}^{30}_{14}S + {}^{0}_{+1}e$

39. (a) $^{27}_{13}Al + {}^{4}_{2}He \longrightarrow {}^{30}_{15}P + {}^{1}_{0}n$

(b) $^{27}_{14}Si \longrightarrow {}^{0}_{+1}e + {}^{27}_{13}Al$

(c) $^{12}_{6}C + {}^{2}_{1}H \longrightarrow {}^{13}_{7}N + {}^{1}_{0}n$

(d) $^{82}_{35}Br \longrightarrow {}^{82}_{36}Kr + {}^{0}_{-1}e$

40. (a) $^{66}_{29}Cu \longrightarrow {}^{66}_{30}Zn + {}^{0}_{-1}e$

(b) $^{0}_{-1}e + {}^{7}_{4}Be \longrightarrow {}^{7}_{3}Li$

(c) $^{27}_{13}Al + {}^{4}_{2}He \longrightarrow {}^{30}_{14}Si + {}^{1}_{1}H$

(d) $^{85}_{37}Rb + {}^{1}_{0}n \longrightarrow {}^{82}_{35}Br + {}^{4}_{2}He$

41. $(112 \text{ years})\left(\frac{1 \text{ half-life}}{28 \text{ years}}\right) = 4$ half-lives

In 4 half-lives 1/16th $(1/2)^4$ of the starting amount would remain.

$\frac{1.00 \text{ mg Sr-90}}{16} = 0.0625$ mg Sr-90 remains after 112 years.

42. $\frac{240}{2} = 120$; $\frac{120}{2} = 60$; $\frac{60}{2} = 30$; 3 half-lives are required to reduce the count from 240 to 30 counts/min.

$1980 + (3 \times 28) = 2064$. One eighth of the original amount Sr-90 remains. $\left[\left(\frac{1}{2}\right)^3 = \frac{1}{8}\right]$

43. (a) ${}^{235}_{92}\text{U} + {}^{1}_{0}\text{n} \longrightarrow {}^{94}_{38}\text{Sr} + {}^{139}_{54}\text{Xe} + 3\,{}^{1}_{0}\text{n}$ + energy

Mass loss = mass of reactants – mass of products

Mass of reactants = 235.0439 + 1.0087 = 236.0526

Mass of products = 93.9154 + 138.9179 + 3(1.0087) = 235.8594

Mass lost = 236.0526 amu – 235.8594 amu = 0.1932 amu

$$(0.1932 \text{ amu})\left(\frac{1.000 \text{ g}}{6.022 \times 10^{23} \text{ amu}}\right)\left(\frac{9.0 \times 10^{13} \text{ J}}{1.00 \text{ g}}\right) = 2.9 \times 10^{-11} \text{ J/atom U-235}$$

(b) $\left(\frac{2.9 \times 10^{-11} \text{ J}}{\text{atom}}\right)\left(\frac{6.022 \times 10^{23} \text{ atoms}}{\text{mol}}\right) = 1.7 \times 10^{13}$ J/mol

(c) $\left(\frac{0.1932 \text{ amu}}{236.0526 \text{ amu}}\right)(100) = 0.08185\%$ mass loss

44. (a) ${}^{1}_{1}\text{H} + {}^{2}_{1}\text{H} \longrightarrow {}^{3}_{2}\text{He}$ + energy

Mass loss = mass reactants - mass products

Mass reactants = 1.00794 g/mol + 2.01410 g/mol = 3.02204 g/mol

Mass products = 3.01603 g/mol

Mass loss = 3.02204 – 3.01603 = 0.00601 g/mol

$$\left(\frac{0.00601 \text{ g}}{\text{mol}}\right)\left(\frac{9.0 \times 10^{13} \text{ J}}{\text{g}}\right) = 5.4 \times 10^{11} \text{ J/mol}$$

(b) $\left(\frac{0.00601 \text{ g}}{3.02204 \text{ g}}\right)(100) = 0.199\%$ mass loss

45. $(0.0100 \text{ g RaCl}_2)\left(\frac{226.0 \text{ g Ra}}{296.9 \text{ g RaCl}_2}\right)\left(\frac{\$50,000}{1 \text{ g Ra}}\right) = \$381$

46. 100% to 25% requires 2 half-lives. The half-life of C-14 is 5668 years. The specimen will be the age of two half-lives:
(2)(5668 years) = 11340 years old.

47. 16.0 g ⟶ 8.0 g ⟶ 4.0 g ⟶ 2.0 g ⟶ 1.0 g ⟶ 0.50 g

16.0 g to 0.50 g requires five half-lives.

$$\frac{90 \text{ minutes}}{5 \text{ half-lives}} = 18 \text{ minutes/half-life}$$

48. (a) $^{7}_{3}$Li is made up of 3 protons, 4 neutrons, and 3 electrons.

Calculated mass

3 protons	3(1.0073 g)	= 3.0219 g
4 neutrons	4(1.0087 g)	= 4.0348 g
3 electrons	3(0.00055 g)	= 0.0017 g
calculated mass		7.0584 g

Mass defect = calculated mass - actual mass

Mass defect 7.0584 g − 7.0160 g = 0.0424 g/mol

(b) Binding energy

$$\left(\frac{0.0424 \text{ g}}{\text{mol}}\right)\left(\frac{9.0 \text{ x } 10^{13} \text{ J}}{\text{g}}\right) = 3.8 \text{ x } 10^{12} \text{ J/mol}$$

49. $^{235}_{92}\text{U} \longrightarrow {}^{207}_{82}\text{Pb}$

Mass loss: 235 - 207 = 28

Net proton loss (atomic number): 92 p − 82 p = 10 p

The mass loss is equivalent to 7 alpha particles (28/4). A loss of 7 alpha particles gives a loss of 14 protons. A decrease in the atomic number to 78 (14 protons) is due to the loss of 7 alpha particles, (92 − 14 = 78). Therefore, a loss of 4 beta particles is required to increase the atomic number from 78 to 82.

The total loss = 7 alpha particles and 4 beta particles.

50. (a) Geiger counter - radiation passes through a thin glass window into a chamber filled with argon gas and containing two electrodes. Some of the argon ionizes, sending a momentary electrical impulse between the electrodes to the detector. This signal is amplified electronically and read out on a counter or as a series of clicks.

(b) Scintillation counter - radiation strikes a scintillator, which is composed of molecules that emits light in the presence of ionizing radiation. A light sensitive detector counts the flashes and converts them into a digital readout.

(c) Film badge - radiation penetrates a film holder. The silver grains in the film darken when exposed to radiation. The film is developed at regular intervals.

51. (3 days)(24 hours/day) = 72 hours

72 hr + 6 hr = 78 hr

$$\frac{78 \text{ hr}}{13 \frac{\text{hr}}{t_{0.5}}} = 6 \text{ half-lives} \qquad (10 \text{ mg})\left(\frac{1}{2}\right)^6 = 0.16 \text{ mg remaining}$$

52. Fission is the process of splitting a large nucleus into two roughly equal mass pieces. Fission occurs in nuclear reactors, or atomic bombs.

Example: $^{235}_{92}U + ^{1}_{0}n \longrightarrow ^{144}_{54}Xe + ^{90}_{38}Sr + 2\,^{1}_{0}n$

Fusion is the process of combining two relatively small nuclei to form a single larger nucleus. Fusion occurs on the sun, or in a hydrogen bomb.

Example: $^{3}_{1}H + ^{2}_{1}H \longrightarrow ^{4}_{2}He + ^{1}_{0}n + \text{energy}$

53.

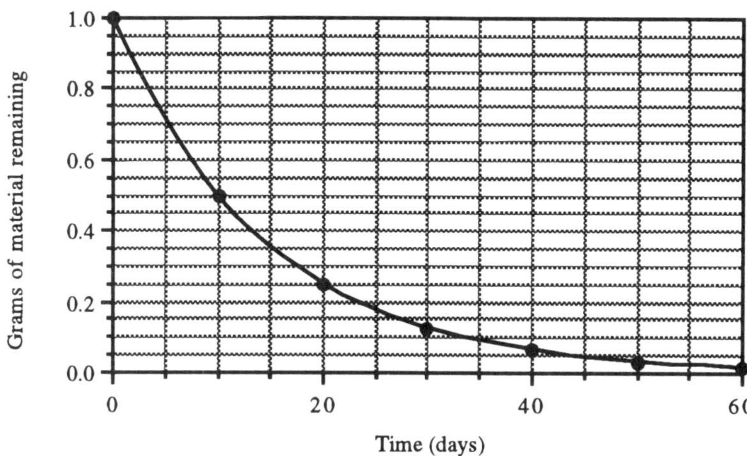

The graph produces a curve for radioactive decay which never actually crosses the x-axis (where mass = 0), it simply approaches that point.

54. (a) $^{235}_{92}U + ^{1}_{0}n \longrightarrow ^{143}_{54}Xe + 3\,^{1}_{0}n + ^{90}_{38}Sr$

(b) $^{235}_{92}U + ^{1}_{0}n \longrightarrow ^{102}_{39}Y + 3\,^{1}_{0}n + ^{131}_{53}I$

(c) $^{14}_{7}N + ^{1}_{0}n \longrightarrow ^{1}_{1}H + ^{14}_{6}C$

55. (a) $H_2O(l) \longrightarrow H_2O(g)$

Energy$_2$: Weakest bond changes requires the least energy.

(b) $H_2(g) + \frac{1}{2}\,O_2(g) \longrightarrow H_2O(g)$

Energy$_1$: medium-sized value involved in interatomic bonds

(c) $^{2}_{1}H + ^{2}_{1}H \longrightarrow ^{3}_{1}H + ^{1}_{1}H$

Energy$_3$: Nuclear process; greatest amount of energy involved

56. $^{236}_{92}U \longrightarrow ^{90}_{38}Sr + 3\ ^{1}_{0}n + ^{143}_{54}Xe$

57. (a) Beta emission: $^{29}_{12}Mg \longrightarrow ^{0}_{-1}e + ^{29}_{13}Al$

(b) alpha emission: $^{150}_{60}Nd \longrightarrow ^{4}_{2}He + ^{146}_{58}Ce$

(c) positron emission: $^{72}_{33}As \longrightarrow ^{0}_{+1}e + ^{72}_{32}Ge$

58. (a) $^{87}_{37}Rb \longrightarrow ^{0}_{-1}e + ^{87}_{38}Sr$

(b) $^{87}_{38}Sr \longrightarrow ^{0}_{+1}e + ^{87}_{37}Rb$

59.

$t_{\frac{1}{2}}$	0	12.5	25.0	37.5	50.0	62.5	75.0	87.5	100. hours
Amount	15.4	7.7	3.85	1.93	0.965	0.483	0.241	0.121	0.0605 mg

Fraction of K-42 remaining $\frac{0.0605\text{ mg}}{15.4\text{ mg}} = 0.00393$ (or 0.393%)

No. After an additional eight half-lives there would be less than one microgram (0.000001 g) remaining.

$(200\text{ hrs})\left(\frac{1\text{ half life}}{12.5\text{ hr}}\right) = 16$ half lives

Amount remaining $= (15.4\text{ mg})\left(\frac{1}{2}\right)^{16}\left(\frac{10^3\ \mu g}{\text{mg}}\right) = 0.235\ \mu g$

60. $(270\text{ years})\left(\frac{1\text{ half life}}{30\text{ years}}\right) = 9$ half lives

$t_{\frac{1}{2}}$	0	30	60	90	120	150	180	210	240	270 years
Amount	7680	3840	1920	960.	480.	240.	120.	60.0	30.0	15.0 g

There would have been 7680 g originally

61. Element 114 would fall under lead on the periodic table. If would be a metal and would most likely form +2 and +4 ions in solution (like lead).

62. 1.00g Co-60

(a) one half-life: $\frac{1.00\text{ g}}{2} = 0.500$ g left

(b) two half-lives: $\frac{0.500\text{ g}}{2} = 0.250$ g left

(c) four half-lives: $2^4 = 16; \frac{1}{16}$ left $\frac{1.00\text{ g}}{16} = 0.0625$ g

(d) ten half-lives: $2^{10} = 1024; \frac{1}{1024}$ left $\frac{1.00\text{ g}}{1024} = 9.77 \times 10^{-4}$ g

63. (a) $^{11}_{5}B \longrightarrow {}^{4}_{2}He + {}^{7}_{3}Li$

(b) $^{88}_{38}Sr \longrightarrow {}^{0}_{-1}e + {}^{88}_{39}Y$

(c) $^{107}_{47}Ag + {}^{1}_{0}n \longrightarrow {}^{108}_{47}Ag$

(d) $^{41}_{19}K \longrightarrow {}^{1}_{1}H + {}^{40}_{18}Ar$

(e) $^{116}_{51}Sb + {}^{0}_{-1}e \longrightarrow {}^{116}_{50}Sn$

64. C-14 content of 1/16 of that in living plants means that four half-lives have passed. ^{14}C half-life is 5668 years.

$\left(\frac{5668 \text{ years}}{\text{half-life}}\right)(4 \text{ half-lives}) = 22{,}672 \text{ years} \quad (2.267 \times 10^4 \text{ years})$

65. 1 Curie = 3.7×10^{10} disintegrations/sec

1 becquerel = 1 disintegration/sec

Therefore there are 3.7×10^{10} becquerels/1 Curie

$\left(\frac{3.7 \times 10^{10} \text{ becquerel}}{1 \text{ Curie}}\right)(1.24 \text{ Curies}) = 4.6 \times 10^{10} \text{ becquerels}$

66. $E = mc^2 \quad {}^{0}_{-1}e + {}^{0}_{+1}e \longrightarrow \text{energy}$

$E = \left(9.1096 \times 10^{-31} \frac{\text{kg}}{\text{electron}}\right)(2 \text{ electrons})(3 \times 10^8 \text{ m/s})^2$

$= 1.6397 \times 10^{-13} \text{ kg} \cdot \text{m}^2/\text{s}^2 = 1.6397 \times 10^{-13} \text{ J}$

CHAPTER 19

CHEMISTRY OF SELECTED ELEMENTS

1. The first element in a family differs significantly from the others because it is smaller and has a higher electronegativity. They also have the smallest number of electron shells shielding the electrons in the valence shell.

2. Carbon has the least metallic character in Group IVA. It is not lustrous or malleable.

3. Reactivity of the alkali metals is based on ionization energy which decreases as the atomic mass increases. Thus, alkali metals with higher mass lose their outer shell electron more easily and are more reactive. For halogens reactivity is based on electronegativity which increases with decreasing atomic mass. Fluorine, the most electronegative halogen has the greatest attraction for electrons and is thus, the most reactive halogen.

4. The process of reduction is used to convert an ore to a free metal.

5. Nonmetals may be prepared by liquefaction of air (N_2), electrolysis of water ($H_2 + O_2$), decomposition of methane (H_2), mined in elemental form(S) and oxidation of halide salt solutions (Cl_2, Br_2, and I_2).

6. In a solution alloy the metals are combined in the molten state and the components mix uniformly and randomly. In an intermetallic compound the alloy is homogeneous and of fixed composition.

   solution alloy examples -- brass, pewter, nitinol

   intermetallic compound examples -- dental amalgam, Co_5Sm, Cr_3Pt

7. Alloys are made to modify the properties of a pure metallic element.

8. Hydrogen is considered to be a family of one because its properties and reactions are unique and do not fit with those of other elements.

9. Potassium reacts more violently than lithium because its ionization energy is considerably less than that of lithium. K has 2 more energy levels than Li allowing the outermost electron to be more easily removed since it is held less tightly to the nucleus.

10. Fr ionization energy would most likely be less than that of Cs. (also see Question 3)

11. The carbon dioxide and water vapor exhaled by the wearer of the breathing apparatus react with the superoxide to form oxygen.

$$4\ KO_2 + 4\ CO_2 + 2\ H_2O \rightarrow 4\ KHCO_3 + 3\ O_2$$

12. The answers will vary:

regulation of pH, intercellular osmotic pressure, regulation of water balance, etc.

13. Group IIA metals have two electrons in their valence shell. Since both must be lost to attain the electron structure of a noble gas, they are less reactive than alkali metals which need only lose one electron to achieve stability.

14. a. Precipitation of Mg^{2+} as $Mg(OH)_2$ from seawater.

$$Mg^{2+} + 2\ OH^- \rightarrow Mg(OH)_2(s)$$

b. Conversion of $Mg(OH)_2$ to $MgCl_2$

$$Mg(OH)_2(s) + 2\ HCl(aq) \rightarrow MgCl_2(aq) + H_2O(l)$$

c. Electrolysis of $MgCl_2$

$$MgCl_2(aq) \xrightarrow{\text{electrolysis}} Mg(s) + Cl_2(g)$$

15. The mortar absorbs CO_2 from the air and the lime reverts back to calcium carbonate.

$$Ca(OH)_2 + CO_2 \rightarrow CaCO_3 + H_2O$$

Sand remains stuck in the material forming a hard substance resistant to wear.

16. Answers will vary.

Mg: present in chlorophyll of green plants; associated with the production and use of adenosine triphosphate (ATP), the central energy molecule of the cell.

Ca: found in inorganic salts in human bones and teeth; involved in muscle contraction and hormone regulation.

17. Barium sulfate is insoluble and is not absorbed so can be seen with X rays while barium chloride is soluble and would not be observable with X rays.

18. When these elements are heated in a flame some ground state electrons are promoted to higher energy levels. When they fall back to ground state the absorbed energy is emitted as the characteristic light.

19. Boron can be used in bulletproof armor, borax (as a cleansing agent), boric acid (as a flame retardant in eye wash), and used in borosilicate glass known as Pyrex.

20. A mordant is a substance which binds to both cloth and dye molecules, adhering the dye to the cloth.

21. When an element exists in two or more forms, each form is an allotrope of the element. Carbon exists as three allotropes, diamond graphite, and buckminsterfullerene.

22. Answers will vary: lipstick, lotion, Silly Putty, car polish.

23. The tendency toward metallic properties increases from top to bottom in Group VA. Nitrogen is a colorless gas while bismuth shows many metal characteristics.

24. Beginning with the atmosphere nitrogen is fixed by bacterial action, combustion or chemical fixation. In the soil nitrogen is converted to nitrates which are absorbed by higher plants and animals and converted to organic compounds. Eventually, the nitrogen is returned to the soil in the form of urea or feces or through bacterial decomposition of dead plants and animals. Some remains in the soil and the rest returns to the atmosphere as free nitrogen.

25. Elements in Group VA which are essential to health and well-being include phosphorus and nitrogen.

26. Free nitrogen is so inert because it is triple bonded and has a very high bond dissociation energy. This property makes nitrogen useful as a nonoxidizing atmosphere for preservation of food, wine, or other articles. It is also used as a liquid coolant to freeze things because of its low boiling point (-196°C).

27. The reactivity of elements in Group VIIA decreases down the column because these elements are nonmetals and therefore their reactivity is governed by electronegativity. The most electronegative element in the group is fluorine and it is also the most reactive.

28. Sulfur compounds are less ionic and more covalent than oxygen compounds.

29. HF is the only hydrogen halide which has the ability to hydrogen bond. These hydrogen bonds require additional energy to break during the phase change from liquid to gas resulting in a higher boiling point for HF.

30. F > Cl > Br > I as oxidizing agents. This is because F is the most electronegative and therefore has the greatest capacity to remove an electron from another substance. As electronegativity decreases down the column so does the capacity to oxidize other substances.

31. Transition metals show similarities across periods as well as in columns. The last electron is entering a *d* or *f* orbital instead of an *s* or *p* orbital in the representative elements.

32. Pig iron is the initial product in steel making and contains several impurities. In steel, these impurities have been removed or lowered to a controlled level.

33. Rusting is the oxidation (corrosion) of iron to hydrated iron(III) oxide. The process is summarized:

$$2\ Fe + O_2 + 2\ H_2O \rightarrow 2\ Fe(OH)_2$$

$$4\ Fe(OH)_2 + O_2 + 2\ H_2O \rightarrow 4\ Fe(OH)_3$$

$$2\ Fe(OH)_3 \rightarrow Fe_2O_3 \bullet x\ H_2O$$

34. Aluminum and magnesium both form oxides which adhere tightly to the surface of the metal protecting it from further corrosion. Iron forms rust which flakes off the surface allowing further corrosion of the metal.

35. In cathodic protection rods of a more reactive metal (such as Zn or Mg) are connected by wires to the object to be protected. This results in an electrochemical cell with the more reactive metal (Zn or Mg) acting as the anode. Oxidation occurs here instead of on the object being protected which is acting as the cathode in the electrochemical cell.

36. Answers will vary. Wires in electrical system; alloys for coins and jewelry; a trace element essential for life; cooking utensils, etc.

37. An object is galvanized to protect its surface from oxidation (corrosion). Zinc is the most common galvanizing metal.

38. The correct statements are b, c, d, e, h, i, j, m, o, p, r, s

    (a) Few of the common metals are found in the free or uncombined state in nature.
    (f) Both calcium and magnesium are essential elements for animals, but only magnesium is essential for plants.
    (g) Very pure iron is too soft for industrial applications.
    (k) The electronegativity of the halogens decreases from top to bottom in the periodic table.
    (l) Calcium plays a central role in bone development in the human body.
    (n) The major use of a fluorine in the United States is for processing uranium.
    (q) The Haber process in used to prepare ammonia on an industrial scale.

39. (a) Sr metal
    (b) Kr nonmetal
    (c) Si metalloid
    (d) W metal

40. (a) Br nonmetal
    (b) As metalloid
    (c) Pd metal
    (d) H nonmetal

41. (a) highest electronegativity O
    (b) smallest atomic radius O
    (c) smallest ionization energy K
    (d) greatest metallic character K

42. (a) highest electronegativity N, Cl
    (b) smallest atomic radius N
    (c) smallest ionization energy Ca
    (d) greatest metallic character Ca

43.

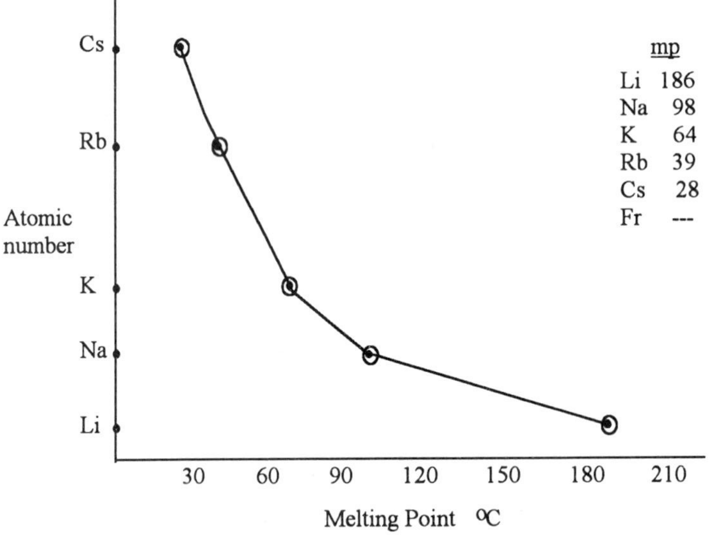

Estimated melting point of Fr is about 0-5 °C.
Francium will be a liquid at room temperature

44.

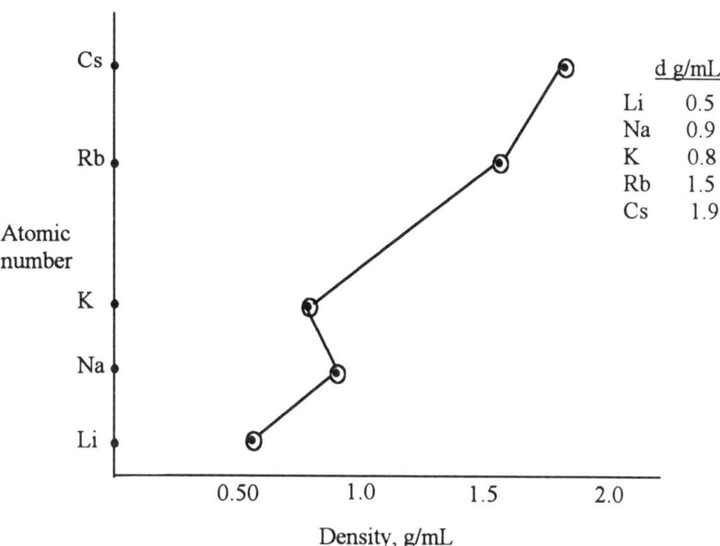

The density of Fr is probably less than 2.0 g/mL. According to density Fr would sink. However, Fr would be extremely reactive with water and would explode upon contact.

45. oxide — Li_2O
hydroxide — $NaOH$
acetate — $KC_2H_3O_2$
hydrogen carbonate — $RbHCO_3$
carbonate — Cs_2CO_3

46. oxide — BeO
hydroxide — $Mg(OH)_2$
acetate — $Ca(C_2H_3O_2)_2$
hydrogen carbonate — $Sr(HCO_3)_2$
carbonate — $BaCO_3$

47. (a) :N:::N:

(b)
```
[    H    ]+
     ..
  H: N: H
     ..
     H
```

(c)
```
     ..
  H: N: H
     ..
     H
```

48. (a) :N:::N:

(b) $[:\ddot{\underset{..}{N}}:]^{3-}$

(c) $\left[\begin{array}{c}:\ddot{O}::N:\ddot{\underset{..}{O}}: \\ :\underset{..}{\ddot{O}}:\end{array}\right]^{-}$

49. (a) A lack of fluoride in the diet could result in greater susceptibility to dental cavities.
 (b) A lack of chloride in the diet could result in difficulties with muscle contraction or nerve transmission.
 (c) A lack of bromide produces no ill effects.
 (d) A lack of iodide in the diet results in enlargement of the thyroid (goiter).

50. (a) A lack of copper in the diet could affect the synthesis of hemoglobin, the development connective tissue, the production of melanin, etc.
 (b) A lack of zinc in the diet could affect certain enzymes in the digestive and respiratory systems.
 (c) A lack of iron in the diet could affect the formation of hemoglobin, and is required for the liver, spleen and bone marrow.
 (d) A lack of calcium in the diet could affect proper formation of teeth and bones, which can lead to osteoporosis. Calcium is also needed for muscle construction, hormone regulation and blood coagulation.

51. (a) $2\ PbO + C \xrightarrow{\Delta} 2\ Pb + CO_2$

 (b) $WO_3 + 3\ H_2 \xrightarrow{\Delta} W + 3\ H_2O$

 (c) $2\ Ag_2O \xrightarrow{\Delta} 4\ Ag + O_2$

52.

$\cdot\ddot{\underset{..}{X}}: + \cdot\ddot{\underset{..}{X}}: \longrightarrow :\ddot{\underset{..}{X}}:\ddot{\underset{..}{X}}:$ covalent (share e^-)

$H\cdot + \cdot\ddot{\underset{..}{X}}: \longrightarrow H:\ddot{\underset{..}{X}}:$ covalent (share e^-)

$Na\cdot + \cdot\ddot{\underset{..}{X}}: \longrightarrow [Na]^+ [:\ddot{\underset{..}{X}}:]^-$ ionic (transfer e^-)

53. $N_2 + 3\ H_2 \rightleftharpoons 2\ NH_3$

Most ammonia made by the Haber process is used to make fertilizer.

54. Hydrogen has an oxidation number of -1 in a hydride. $2\ K + H_2 \rightarrow 2\ KH$

55. oxide $4\ K + O_2 \rightarrow 2\ K_2O$

peroxide $2\ K + O_2 \rightarrow K_2O_2$

superoxide $K + O_2 \rightarrow KO_2$

56. (a) If carbon is burned in a limited supply of O_2, CO is formed.
(b) With an excess of O_2, CO_2 is formed.

57. $S + O_2 \rightarrow SO_2$

$$(150\ \text{g S})\left(\frac{1\ \text{mol}}{32.06\ \text{g}}\right)\left(\frac{1\ \text{mol SO}_2}{1\ \text{mol S}}\right)\left(\frac{22.4\ \text{L}}{1\ \text{mol}}\right) = 1.0 \times 10^2\ \text{L SO}_2$$

58. $$\left(\frac{300{,}000\ \text{L H}_2\text{O}}{1\ \text{hr}}\right)\left(\frac{1\ \text{kg}}{1\ \text{L}}\right)\left(\frac{1000\ \text{g}}{1\ \text{kg}}\right)\left(\frac{0.20\ \text{g Cl}}{1{,}000{,}000\ \text{g H}_2\text{O}}\right) = \frac{60\ \text{g Cl}}{\text{hr}}$$

59. (a) $$\left(\frac{1000\ \text{mL}}{1\ \text{L}}\right)\left(\frac{1.84\ \text{g}}{1\ \text{mL}}\right)\left(\frac{96\ \text{g H}_2\text{SO}_4}{100.\ \text{g}}\right)\left(\frac{1\ \text{mol}}{98.08\ \text{g}}\right) = 18\ M\ \text{H}_2\text{SO}_4$$

(b) $$\left(\frac{1000\ \text{mL}}{1\ \text{L}}\right)\left(\frac{1.42\ \text{g}}{1\ \text{mL}}\right)\left(\frac{70.\ \text{g HNO}_3}{100.\ \text{g}}\right)\left(\frac{1\ \text{mol}}{63.02\ \text{g}}\right) = 16\ M\ \text{HNO}_3$$

(c) $$\left(\frac{1000\ \text{mL}}{1\ \text{L}}\right)\left(\frac{1.19\ \text{g}}{1\ \text{ml}}\right)\left(\frac{37\ \text{g HCl}}{100.\ \text{g}}\right)\left(\frac{1\ \text{mol}}{36.46\ \text{g}}\right) = 12\ M\ \text{HCl}$$

60. Magnesium, strontium, or barium

All these elements are in Group IIA of the periodic table and contain the same outershell electron structure as calcium.

61. (a) $C(s) + O_2(g) \rightarrow CO_2(g)$

(b) Oxygen is consumed and CO_2 concentration increases. Given enough time breathing would become difficult due to a shortage of O_2 and an excess of CO_2.

62. Fe_2O_3 is formed. Fe_2O_3 is insoluble in water and alcohol, but soluble in acids.

63. Chlorine in its elemental form is a gas that is deadly when inhaled because it is capable of destroying both plant and animal tissue. Chlorine as a chloride in salt (NaCl) is in the form of an ion (Cl^-) and incapable of the oxidizing effect it has in its elemental form. Some other harmful elements are lead, mercury, arsenic and bromine.

CHAPTER 20

ORGANIC CHEMISTRY: SATURATED HYDROCARBONS

1. Two of the major reasons for the large number of organic compounds is the ability of carbon to form short or very long chains of atoms covalently bonded together and isomerism.

2. The carbon atom has only two unshared electrons, making two covalent bonds logical, but in CH_4, carbon forms four equivalent bonds. Promoting one 2s electron to the empty 2p orbital would make four bonds possible, but without hybridization, we could not explain the fact that all four bonds in CH_4 are identical, and the bond angles are equal (109.5°).

3. The first ten normal alkanes:

methane	CH_4	hexane	C_6H_{14}
ethane	C_2H_6	heptane	C_7H_{16}
propane	C_3H_8	octane	C_8H_{18}
butane	C_4H_{10}	nonane	C_9H_{20}
pentane	C_5H_{12}	decane	$C_{10}H_{22}$

4. (a) A molecule of ethane, C_2H_6, contains seven sigma bonds.
   (b) A molecule of butane, C_4H_{10}, contains thirteen sigma bonds.
   (c) A molecule of 2-methylpropane also contains thirteen sigma bonds.

5. Advantages: Some freons have low boiling points and therefore are excellent refrigerants. They are stable, nontoxic, nonflammable and noncorrosive.

   Disadvantages: Freons are a major factor in the destruction of the ozone layer once they get into the stratosphere.

6. The following statements are correct: a, b, c, d, g, i, j, l, m, p, and s.

   (e) Hydrocarbons are composed of carbon and hydrogen.

   (f) In the alkane homologous series, the formula of each member differs from its preceding member by CH_2.

   (h) The name for the alkane C_5H_{12} is pentane.

   (k) The name for $CH_3CH_2CH_2CHClCH_3$ is 2-chloropentane.

(n) Chlorocyclohexane and 1-chlorohexane are not isomers.

(o) The products of complete combustion of a hydrocarbon are carbon dioxide and water.

(q) Isobutane and 2-methylpropane are correct names for the same compound.

(r) Two monochlorosubstituted products result from the chlorination of butane.

7. Lewis structures:

(a) CCl_4

```
     ..
    :Cl:
 ..  ..  ..
:Cl: C :Cl:
 ..  ..  ..
    :Cl:
     ..
```

(b) C_2Cl_6

```
      ..   ..
     :Cl::Cl:
 ..  ..    ..  ..
:Cl: C :   C : Cl:
 ..  ..    ..  ..
     :Cl::Cl:
      ..   ..
```

(c) $CH_3CH_2CH_3$

```
    H   H   H
    ..  ..  ..
H : C : C : C : H
    ..  ..  ..
    H   H   H
```

8. Lewis structures:

(a) CH_4

```
    H
    ..
H : C : H
    ..
    H
```

(b) C_3H_8

```
    H   H   H
    ..  ..  ..
H : C : C : C : H
    ..  ..  ..
    H   H   H
```

(c) C_5H_{12}

```
    H   H   H   H   H
    ..  ..  ..  ..  ..
H : C : C : C : C : C : H
    ..  ..  ..  ..  ..
    H   H   H   H   H
```

9. Formulas (a) and (i) are not isomers. They are the same compound.
Formulas (b) and (c) are isomers of C_5H_{12}.
Formulas (f) and (h) are isomers of C_5H_{10}.
Formulas (d), (e), and (g) are isomers of C_6H_{14}.

10. Compounds (b), (e), and (f) are identical. The others are different.

11. The formulas in Exercise 9 contain the following numbers of methyl groups:

(a) 2 (b) 2 (c) 3 (d) 2 (e) 4 (f) 0 (g) 3 (h) 1 (i) 2

12. The formulas in Exercise 10 contain the following numbers of methyl groups:

(a) 4 (b) 4 (c) 4 (d) 4 (e) 4 (f) 4

13. heptane

```
CH3CH2CH2CH2CH2CH2CH3          CH3CH2CH2CH2CHCH3
                                           |
                                           CH3

                                  CH3                  CH3
                                  |                    |
CH3CH2CH2CHCH2CH3        CH3CH2CH2CCH3        CH3CH2CCH2CH3
           |                      |                    |
           CH3                    CH3                  CH3

         CH3                                    CH3  CH3
         |                                      |    |
CH3CH2CHCHCH3            CH3CHCH2CHCH3       CH3CH---CCH3
           |                |     |                  |
           CH3              CH3   CH3                CH3

CH3CH2CHCH2CH3
      |
      CH2CH3
```

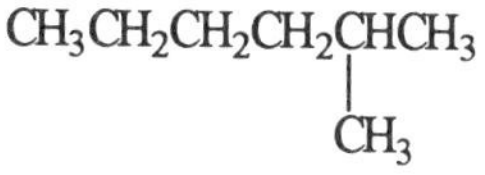

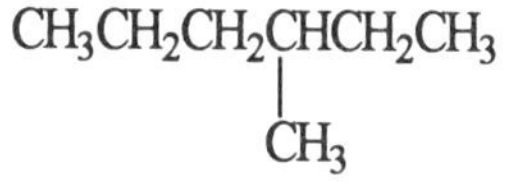

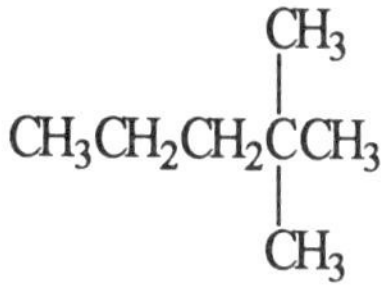

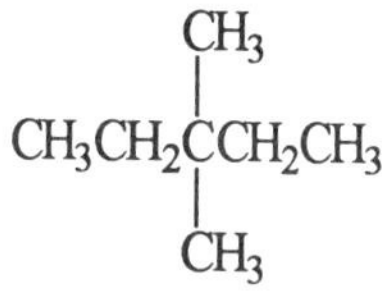

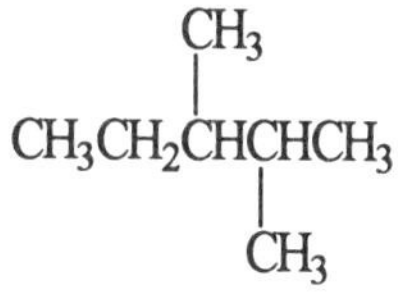

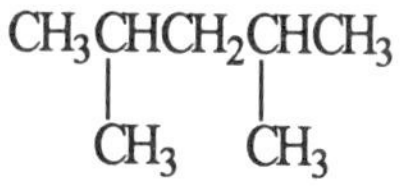

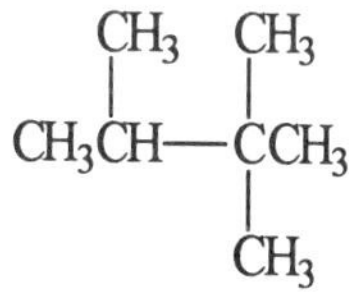

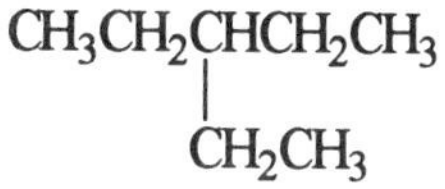

14. hexane

```
CH3CH2CH2CH2CH2CH3             CH3CH2CH2CHCH3
                                         |
                                         CH3

                              CH3                 CH3
                              |                   |
CH3CH2CHCH2CH3           CH3CH2CCH3           CH3CHCHCH3
      |                       |                     |
      CH3                     CH3                   CH3
```

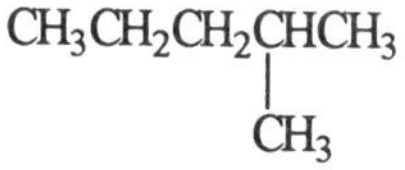

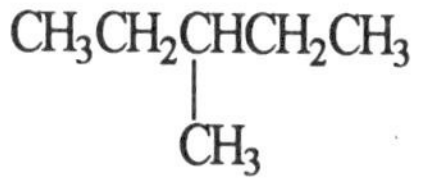

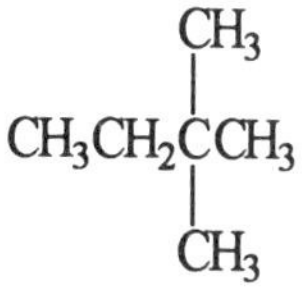

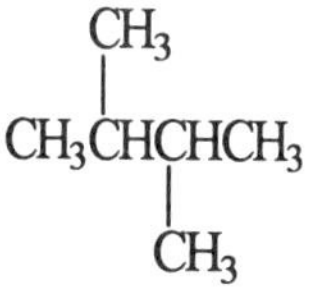

15. Isomers with these formulas:

(a) CH_3Br, one

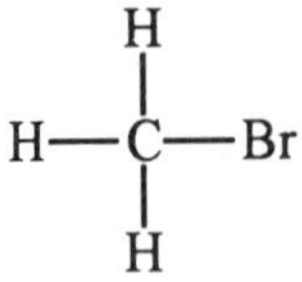

```
     H
     |
H—C—Br
     |
     H
```

(b) C_2H_5Cl, one

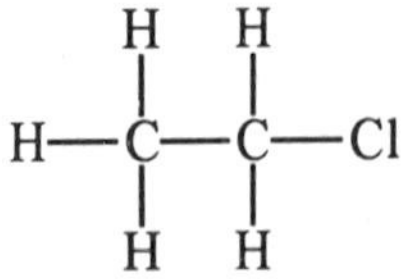

```
     H   H
     |   |
H—C—C—Cl
     |   |
     H   H
```

(c) C_4H_9I, four

```
     H   H   H   H
     |   |   |   |
H—C—C—C—C—I
     |   |   |   |
     H   H   H   H
```

```
     H   H   H   H
     |   |   |   |
H—C—C—C—C—H
     |   |   |   |
     H   H   I   H
```

```
         H
         |
     H—C—H
         |
     H   |   H
     |   |   |
H—C—C—C—I
     |   |   |
     H   H   H
```

```
         H
         |
     H—C—H
         |
     H   |   H
     |   |   |
H—C—C—C—H
     |   |   |
     H   I   H
```

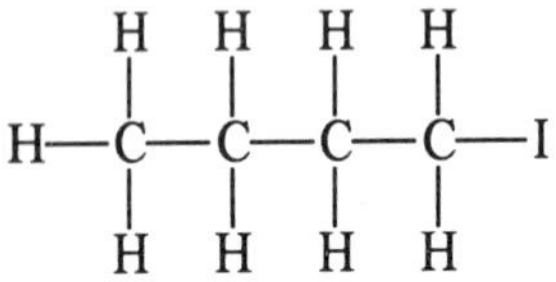

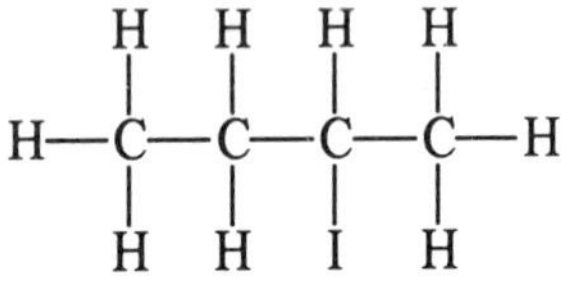

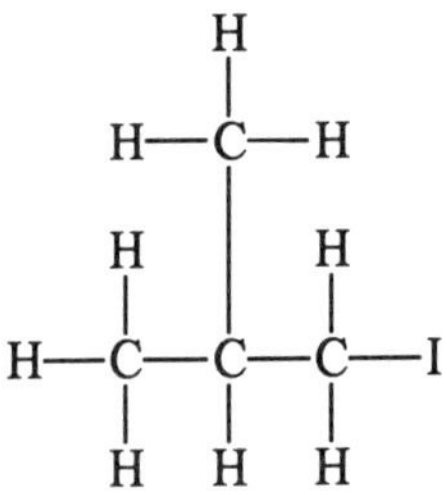

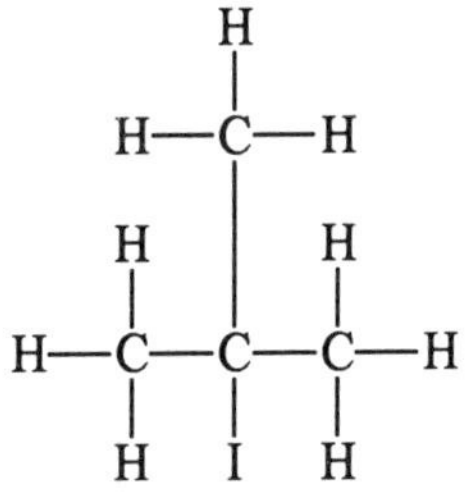

(d) C_3H_6BrCl, five

```
     H   H   H
     |   |   |
H—C—C—C—Br
     |   |   |
     H   H   Cl
```

```
     H   H   H
     |   |   |
H—C—C—C—Br
     |   |   |
     H   Cl  H
```

```
     H   H   H
     |   |   |
H—C—C—C—Cl
     |   |   |
     H   Br  H
```

```
      H   H   H
      |   |   |
Cl—C—C—C—Br
      |   |   |
      H   H   H
```

```
     H   Cl  H
     |   |   |
H—C—C—C—H
     |   |   |
     H   Br  H
```

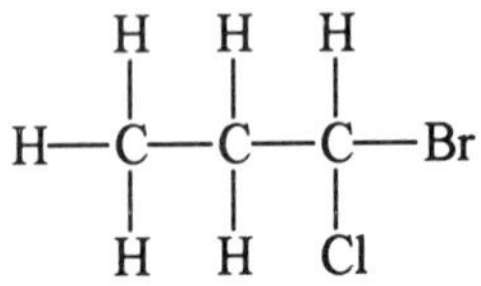

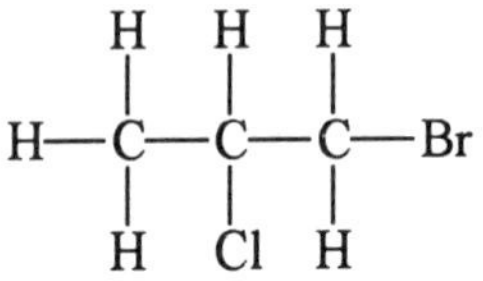

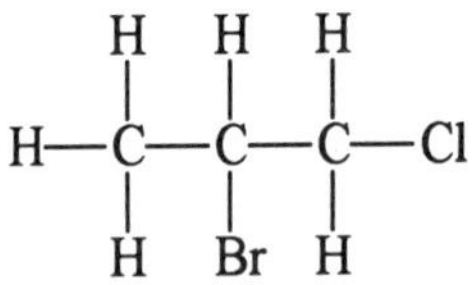

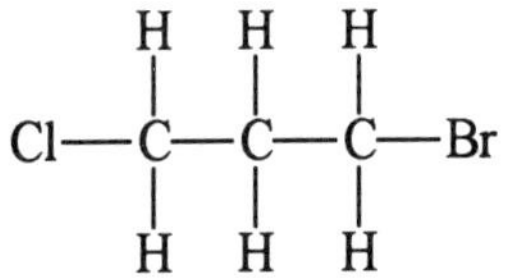

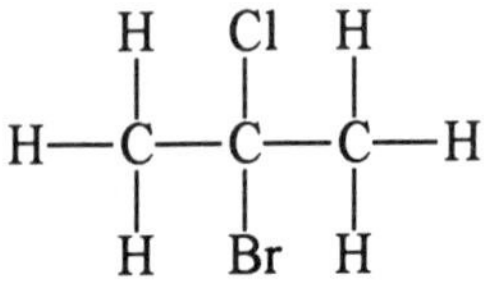

16. (a) CH_2Cl_2, one

```
     H
     |
 H — C — Cl
     |
     Cl
```

(b) C_3H_7Br, two

```
     H   H   H                  H   H   H
     |   |   |                  |   |   |
 H — C — C — C — Br         H — C — C — C — H
     |   |   |                  |   |   |
     H   H   H                  H   Br  H
```

(c) $C_3H_6Cl_2$, four

```
     H   H   H                  H   H   H
     |   |   |                  |   |   |
 H — C — C — C — Cl         H — C — C — C — Cl
     |   |   |                  |   |   |
     H   H   Cl                 H   Cl  H

     H   H   H                  H   Cl  H
     |   |   |                  |   |   |
Cl — C — C — C — Cl         H — C — C — C — H
     |   |   |                  |   |   |
     H   H   H                  H   Cl  H
```

(d) $C_4H_8Cl_2$, nine

```
     H   H   H   H                  H   H   H   H
     |   |   |   |                  |   |   |   |
 H — C — C — C — C — Cl         H — C — C — C — C — Cl
     |   |   |   |                  |   |   |   |
     H   H   H   Cl                 H   H   Cl  H

     H   H   H   H                  H   H   H   H
     |   |   |   |                  |   |   |   |
 H — C — C — C — C — Cl        Cl — C — C — C — C — Cl
     |   |   |   |                  |   |   |   |
     H   Cl  H   H                  H   H   H   H
```

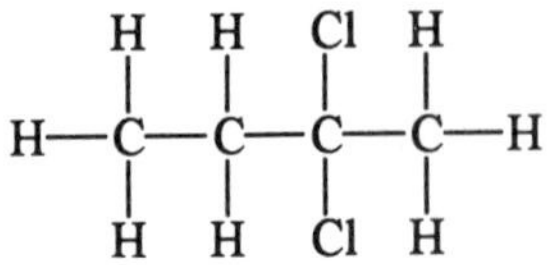

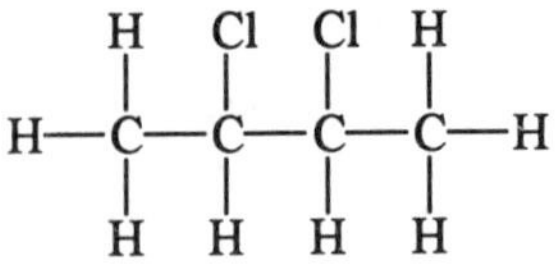

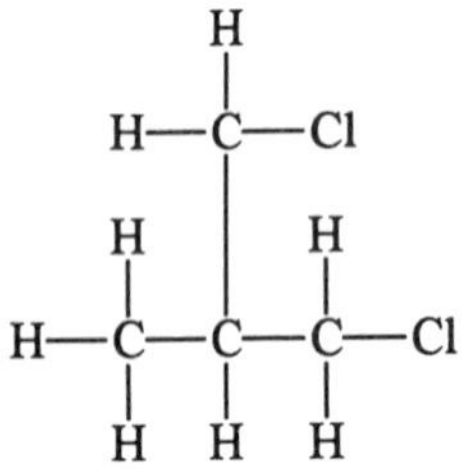

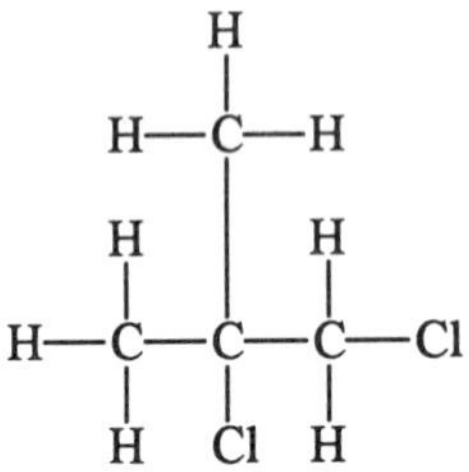

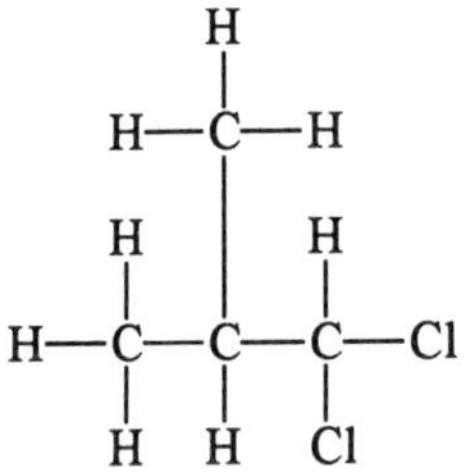

```
     H   H   Cl  H              H   Cl  Cl  H
     |   |   |   |              |   |   |   |
  H—C—C—C—C—H              H—C—C—C—C—H
     |   |   |   |              |   |   |   |
     H   H   Cl  H              H   H   H   H
```

```
         H                          H
         |                          |
     H—C—Cl                    H—C—H
         |                          |
     H   |   H                  H   |   H
     |   |   |                  |   |   |
  H—C—C—C—Cl               H—C—C—C—Cl
     |   |   |                  |   |   |
     H   H   H                  H   Cl  H
```

```
         H
         |
     H—C—H
         |
     H   |   H
     |   |   |
  H—C—C—C—Cl
     |   |   |
     H   H   Cl
```

17. IUPAC names

(a) 1-chloropropane
(b) 2-chloropropane
(c) 2-chloro-2-methylpropane
(d) 2-methylbutane
(e) 2,3-dimethylhexane

18. IUPAC names

(a) chloroethane
(b) 1-chloro-2-methylpropane
(c) 2-chlorobutane
(d) methylcyclopropane
(e) 2,4-dimethylpentane

19. Structural formulas:

(a) 2,4-dimethylpentane

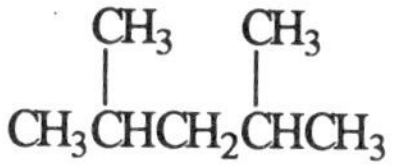

$$\begin{array}{ccccc} & CH_3 & & CH_3 & \\ & | & & | & \\ CH_3 & CH & CH_2 & CH & CH_3 \end{array}$$

(b) 2, 2-dimethylpentane

$(CH_3)_3CCH_2CH_2CH_3$

(c) 3-isopropyloctane

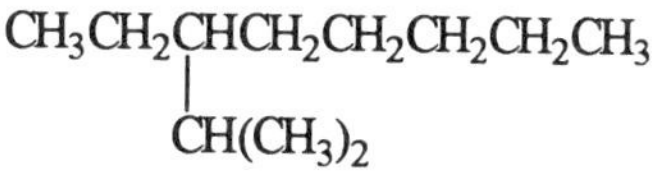

$$\begin{array}{ccc} CH_3CH_2 & CH & CH_2CH_2CH_2CH_2CH_3 \\ & | & \\ & CH(CH_3)_2 & \end{array}$$

(d) 5,6-diethyl-2,7-dimethyl-5-propylnonane

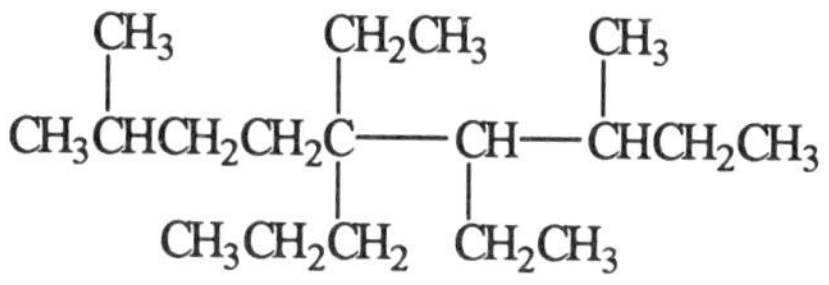

$$\begin{array}{ccccccc} & CH_3 & & CH_2CH_3 & & CH_3 & \\ & | & & | & & | & \\ CH_3 & CH & CH_2CH_2 & C & - CH - & CH & CH_2CH_3 \\ & & & | & | & & \\ & & & CH_3CH_2CH_2 & CH_2CH_3 & & \end{array}$$

20. (a) 4-ethyl-2-methylhexane

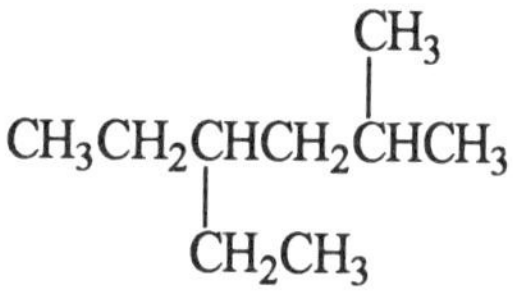

$$\begin{array}{ccccc} & & & CH_3 & \\ & & & | & \\ CH_3CH_2 & CH & CH_2 & CH & CH_3 \\ & | & & & \\ & CH_2CH_3 & & & \end{array}$$

(b) 4-t-butylheptane

$$\begin{array}{ccc} CH_3CH_2CH_2 & CH & CH_2CH_2CH_3 \\ & | & \\ & C(CH_3)_3 & \end{array}$$

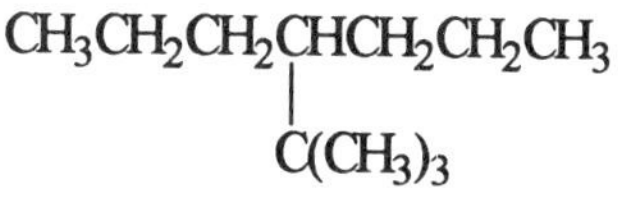

(c) 4-ethyl-7-isopropyl-2,4,8-trimethyldecane

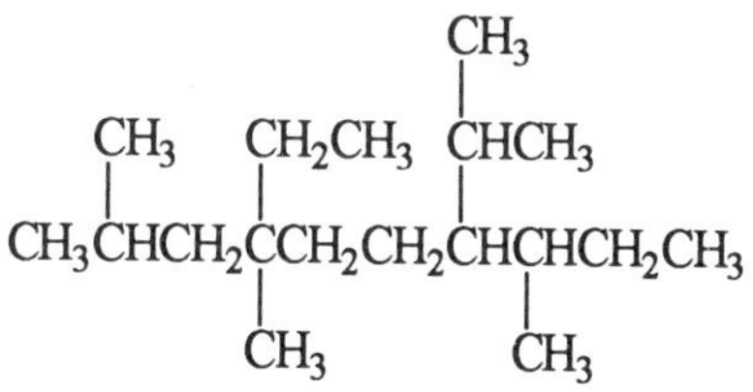

$$\begin{array}{ccccccccc} & & & & & CH_3 & & & \\ & & & & & | & & & \\ & CH_3 & & CH_2CH_3 & & CHCH_3 & & & \\ & | & & | & & | & & & \\ CH_3 & CH & CH_2 & C & CH_2CH_2 & CH & CH & CH_2CH_3 \\ & & & | & & & | & & \\ & & & CH_3 & & & CH_3 & & \end{array}$$

(d) 3-ethyl-2,2-dimethyloctane

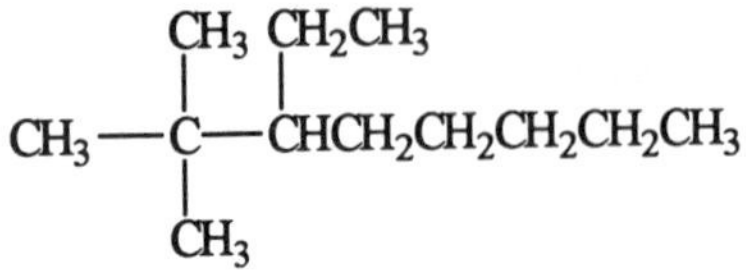

$$CH_3-C(CH_3)_2-CH(CH_2CH_3)CH_2CH_2CH_2CH_2CH_3$$

21. (a) 3-methylbutane $CH_3CH_2CH(CH_3)CH_3$

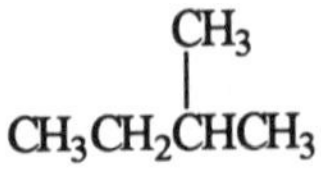

Numbering was done from the wrong end of the molecule. The correct name is 2-methylbutane.

(b) 2-ethylbutane

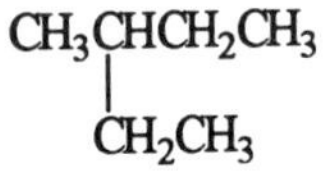

$CH_3CH(CH_2CH_3)CH_2CH_3$

The name is not based on the longest carbon chain (5 carbons). The correct name is 3-methylpentane.

(c) 2-dimethylpentane. Each methyl group needs to be numbered. Depending on the structure, the correct name is 2,2-dimethylpentane; 2,3-dimethylpentane; or 2,4-dimethylpentane.

(d) 1,4-dimethylcyclopentane

CH_3

CH_3

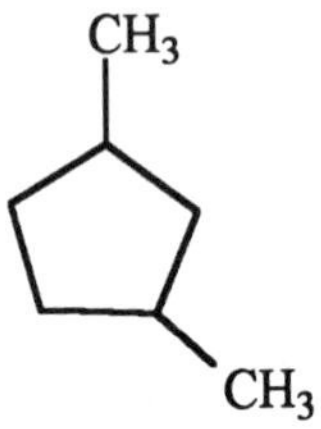

The ring was numbered in the wrong direction. The correct name is 1,3-dimethylcyclopentane.

22. (a) 3-methyl-5-ethyloctane $CH_3CH_2CH(CH_3)CH_2CH(CH_2CH_3)CH_2CH_2CH_3$

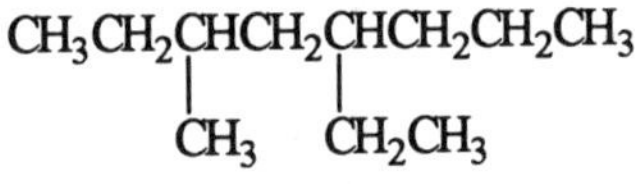

Ethyl should be named before methyl (alphabetical order). The numbering is correct. The correct name is 5-ethyl-3-methyloctane.

(b) 3,5,5-triethylhexane

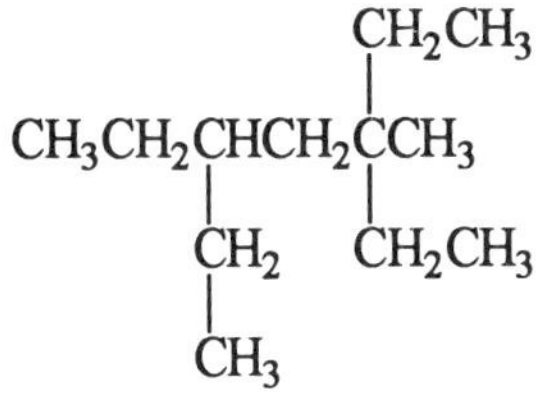

$$\begin{array}{l} \quad\quad\quad\quad\quad\quad\quad CH_2CH_3 \\ \quad\quad\quad\quad\quad\quad\quad | \\ CH_3CH_2CHCH_2CCH_3 \\ \quad\quad\quad | \quad\quad\quad | \\ \quad\quad CH_2 \quad CH_2CH_3 \\ \quad\quad\quad | \\ \quad\quad CH_3 \end{array}$$

The name is not based on the longest carbon chain (7 carbons). The correct name is 3,5-diethyl-3-methylheptane.

(c) 4,4-dimethyl-3-ethylheptane

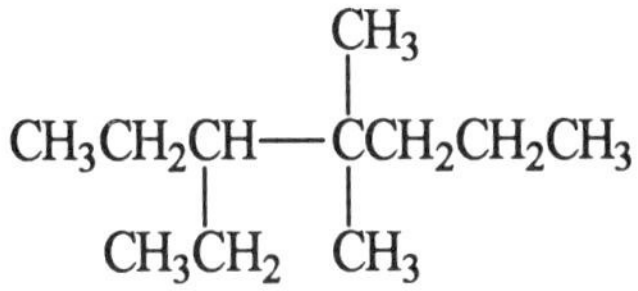

$$\begin{array}{l} \quad\quad\quad\quad\quad\quad\quad CH_3 \\ \quad\quad\quad\quad\quad\quad\quad | \\ CH_3CH_2CH—CCH_2CH_2CH_3 \\ \quad\quad\quad | \quad\quad\quad | \\ \quad CH_3CH_2 \quad CH_3 \end{array}$$

Ethyl should be named before dimethyl (alphabetical order). The correct name is 3-ethyl-4,4-dimethylheptane.

(d) 1,6-dimethylcyclohexane

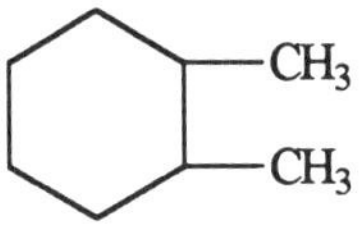

(cyclohexane ring with $-CH_3$ and $-CH_3$ on adjacent carbons)

The ring was numbered in the wrong direction. The correct name is 1,2-dimethylcyclohexane.

23. Structures for the ten dichlorosubstituted compounds of 2-methylbutane:

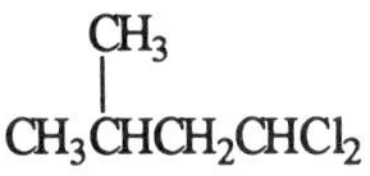

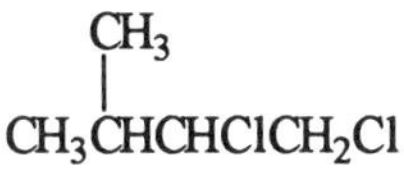

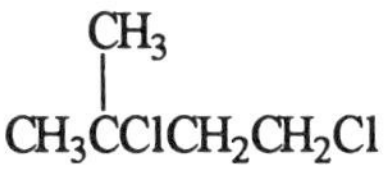

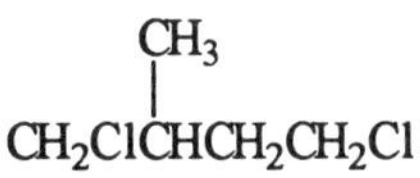

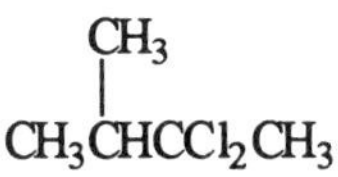

$CH_3CH(CH_3)CH_2CHCl_2$ $\quad$ $CH_3CH(CH_3)CHClCH_2Cl$ $\quad$ $CH_3CCl(CH_3)CH_2CH_2Cl$

$CH_2ClCH(CH_3)CH_2CH_2Cl$ $\quad$ $CH_3CH(CH_3)CCl_2CH_3$ $\quad$ $CH_3CCl(CH_3)CHClCH_3$

$$\underset{CH_2ClCHCHClCH_3}{\overset{CH_3}{|}}\qquad \underset{CHCl_2CHCH_2CH_3}{\overset{CH_3}{|}}\qquad \underset{CH_2ClCClCH_2CH_3}{\overset{CH_3}{|}}$$

$$\underset{CH_2Cl\,CHCH_2CH_3}{\overset{CH_2Cl}{|}}$$

24. $CH_3CH_2CH_2CH_2CH_2CH_2Cl \quad CH_3CH_2CH_2CH_2CHClCH_3 \quad CH_3CH_2CH_2CHClCH_2CH_3$

25. (a) $CH_3CH_2CH_2CH_3 + Cl_2 \xrightarrow{h\nu} CH_3CH_2CHClCH_3 + CH_3CH_2CH_2CH_2Cl + HCl$

    (b) $2\ CH_3CH_2CH_2CH_3 + 13\ O_2 \xrightarrow{\Delta} 8\ CO_2 + 10\ H_2O$

26. (a) $CH_3CH_2CH_3 + Br_2 \xrightarrow{h\nu} CH_3CH_2CH_2Br + CH_3CHBrCH_3 + HBr$

    (b) $CH_3CH_2CH_3 + 5\ O_2 \xrightarrow{\Delta} 3\ CO_2 + 4\ H_2O$

27. The sample had to be a mixture of the two butanes, n-butane and isobutane, to obtain four monobromo products. Each of the butanes yields only two monobromo compounds:

    n-butane yields $CH_3CH_2CH_2CH_2Br$ and $CH_3CH_2CHBrCH_3$

    isobutane yields $\underset{CH_3}{\overset{CH_3CHCH_2Br}{|}}$ and $\overset{CH_3}{|}\!\!\!\underset{Br}{\underset{|}{CH_3CCH_3}}$

28. The formula for dodecane is $C_{12}H_{26}$.

29. $FCH_2CH_2F + Cl_2 \rightarrow FCH_2CHClF + HCl$

30. Data: $\dfrac{1\ \text{gal}}{60\ \text{mi}}$; 60 mi traveled; $\dfrac{19\ \text{mol}\ C_8H_{18}}{\text{gal}}$

    $2\ C_8H_{18} + 25\ O_2 \rightarrow 16\ CO_2 + 18\ H_2O$

    $\dfrac{1\ \text{gal}}{60\ \text{mi}} + 60\ \text{mi} = 1\ \text{gal gasoline used}$

    $19\ \text{mol}\ C_8H_{18} \times \dfrac{16\ \text{mol}\ CO_2}{2\ \text{mol}\ C_8H_{18}} = 1.5 \times 10^2\ \text{mol}\ CO_2$

$PV = nRT \qquad V = \dfrac{nRT}{P}$

$$V = \frac{1.5 \times 10^2 \text{ mol } CO_2 \times 0.0821 \text{ L-atm} \times 293 \text{ K}}{1 \text{ atm} \qquad \text{mol K}} = 3.6 \times 10^3 \text{ L } CO_2$$

31. Cycloalkanes with formulas C_5H_{10}

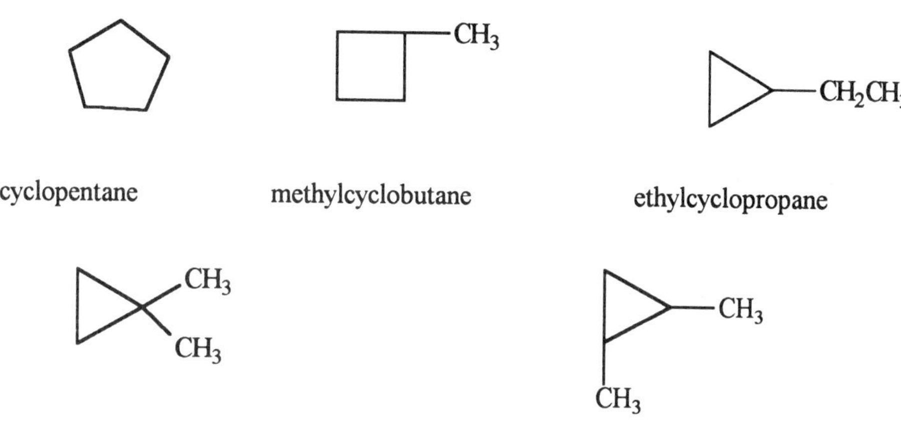

1,1-dimethylcyclopropane 1-2-dimethylcyclopropane

32. (a) elimination

    (b) substitution

    (c) addition

33. It is not possible to distinguish hexane from 3-methylheptane based on solubility in water because both compounds are nonpolar and, thus, insoluble in water.

34. Propane is the most volatile. It has the lowest boiling point of the three alkanes.

35. (a) The compounds are isomers.

    (b) The compounds are not the same and are not isomers.

    (c) The compounds are isomers.

    (d) The compounds are not the same and are not isomers.

36. (a) sp^3 hybrid orbitals in carbon.

    (b) Structural isomers are not possible because only one carbon atom is present.

    (c) dichlorodiflurometha ne

    (d) The closest classification for Freon-12 (CF_2Cl_2) in Table 20.1 is alkyl halide.

37. (a) $CH_3CH_2CH_2CH_2CH_2CH_2CH_2CH_2CH_2CH_2CH_3$ undecane

    $CH_3CH_2CH_2CH_2CH_2CH_2CH_2CH_2CH_2CH_2CH_2CH_2CH_3$ tridecane

    (b) Both compounds are alkanes. They are composed of only carbon and hydrogen and have only single bonds between the carbon atoms. Both formulas agree with the general formula for alkanes, C_nH_{2n+2}.

CHAPTER 21

UNSATURATED HYDROCARBONS

1. The sigma bond in the double bond of ethene is formed by the overlap of two sp^2 electron orbitals and is symmetrical about a line drawn between the nuclei of the two carbon atoms. The pi bond is formed by the sidewise overlap of two p orbitals which are perpendicular to the carbon-carbon sigma bond. The pi bond consists of two electron clouds, one above and one below the plane of the carbon-carbon sigma bond.

2.

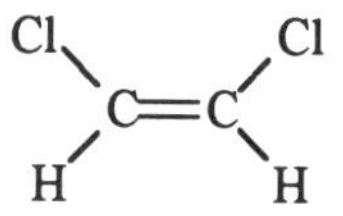

```
Cl        Cl          H         Cl             H   Cl
  \      /             \       /               |   |
   C == C               C == C             H — C — C — H
  /      \             /       \               |   |
H         H          Cl         H              Cl  H
```

cis-1-2-dichloroethene	trans-1, 2-dichloroethene	1, 2-dichloroethane

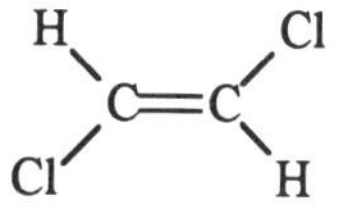

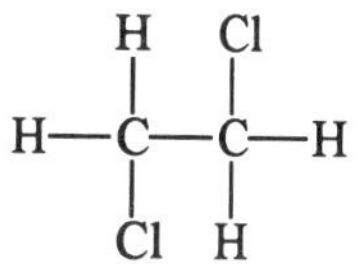

We are able to get cis-trans isomers of ethene because there is no rotation around the carbon-carbon double bond, so the two structures shown are not the same. With 1,2-dichloroethane, there is free rotation of the carbon-carbon single bond. In the structure shown, exchanging the chlorine on the right with either hydrogen atom on that carbon appears to make a different isomer, but rotation makes any of the three positions equivalent.

3. Rubber products deteriorate rapidly in smog-ridden areas because ozone, which is present in smog, causes oxidation of the carbon-carbon double bonds, which changes the properties of the rubber.

4. Acetylene presents two different explosion hazards:

   (a) It can form an explosive mixture with oxygen or air;

   (b) When highly compressed or liquefied, acetylene may decompose violently, either spontaneously or from a slight shock.

5. The small molecule released in an elimination reaction can be reacted with and added to the main elimination product. For example, an alkene can be prepared by release of HX from an alkyl halide. In turn, HX can be added to an alkene to form an alkyl halide.

6. During the 10-year period from 1935 to 1945, the major source of aromatic hydrocarbons shifted from coal tar to petroleum due to the rapid growth of several industries which used aromatic hydrocarbons as raw material. These industries include drugs, dyes, detergents, explosives,

insecticides, plastics, and synthetic rubber. Since the raw material needs far exceeded the aromatics available from coal tar, another source had to be found, and processes were developed to make aromatic compounds from alkanes in petroleum. World War II, which occurred during this period, put high demands on many of these industries, particularly explosives.

7. Benzene does not undergo the typical reactions of an alkene. Benzene does not decolorize bromine rapidly and it does not destroy the purple color of permanganate ions. The reactions of benzene are more like those of an alkane. Reaction of benzene with chlorine requires a catalyst. Benzene does not readily add Cl_2, but rather a hydrogen atom is replaced by a chlorine atom.

$$C_6H_6 + Cl_2 \xrightarrow{Fe} C_6H_5Cl + HCl$$

8. The following statements are correct: a, f, g, h, i, j, k, m, n, o, s, v, w, x, y, z.

   (b) If C_8H_{10} is an open-chain compound, it needs an additional eight hydrogen atoms to become a saturated hydrocarbon.

   (c) Propene and propane are not isomers.

   (d) The pi bond is formed from two *p* orbitals.

   (e) A double bond consists of one sigma and one pi bond.

   (l) The disappearance of the purple color when $KMnO_4$ reacts with an alkene is known as the Baeyer test for unsaturation.

   (p) Alkynes have the general formula C_nH_{2n-2}.

   (q) Cis-trans isomerism occurs in alkenes.

   (r) Geometric isomers are not superimposable on each other.

   (t) The chemical behavior of benzene is very different from that of alkenes.

   (u) Toluene and benzene are not isomers.

9. (a) ethane (b) ethene (c) ethyne

```
    H  H          H   H
    ..  ..        ..  ..
  H:C : C :H      C :: C           H:C:::C:H
    ..  ..        ..  ..
    H  H          H   H
```

10. (a) propane

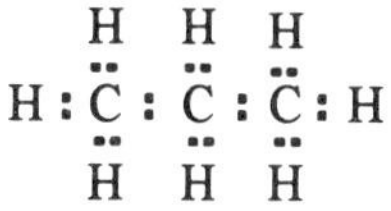

```
   H   H   H
   ..  ..  ..
H: C : C : C :H
   ..  ..  ..
   H   H   H
```

(b) propene

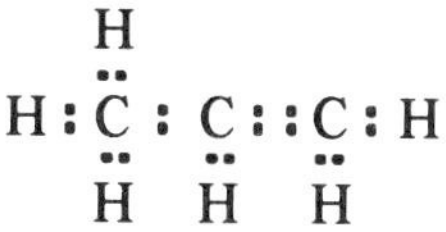

```
   H
   ..
H: C : C::C :H
   ..  ..  ..
   H   H   H
```

(c) propyne

```
   H
   ..
H: C : C:::C :H
   ..
   H
```

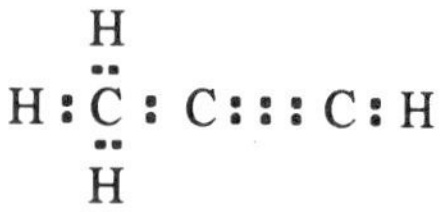

11. Isomeric iodobutenes, C_4H_7I

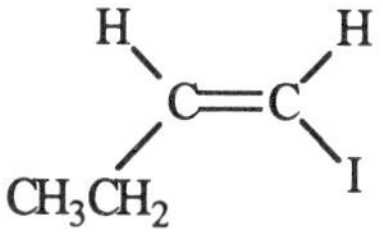

```
      H       H
       \     /
        C=C
       /     \
 CH3CH2       I
```

cis-1-iodo-1-butene

```
      H       I
       \     /
        C=C
       /     \
 CH3CH2       H
```

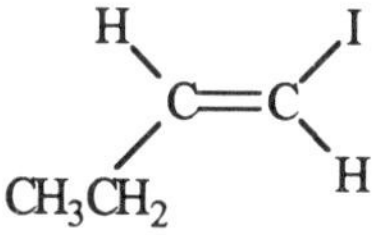

trans-1-iodo-1-butene

$CH_3CH_2CI = CH_2$ 2-iodo-1-butene

$CH_3CHICH = CH_2$ 3-iodo-1-butene

$CH_2ICH_2CH = CH_2$ 4-iodo-1-butene

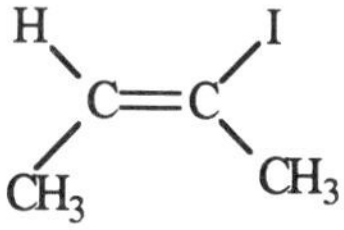

```
   H       I
    \     /
     C=C
    /     \
 CH3       CH3
```

cis-2-iodo-2-butene

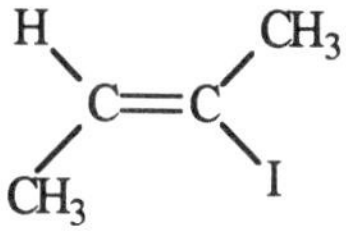

```
   H       CH3
    \     /
     C=C
    /     \
 CH3       I
```

trans-2-iodo-2-butene

```
   H       H
    \     /
     C=C
    /     \
 CH3       CH2I
```

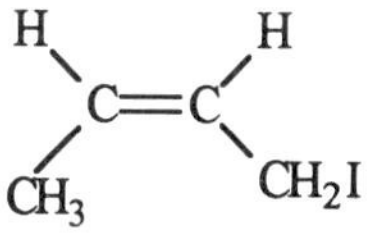

cis-1-iodo-2-butene

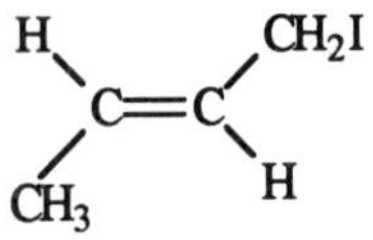

$$\underset{CH_3}{\overset{H}{}}\!\!>C=C<\!\!\underset{H}{\overset{CH_2I}{}}$$ trans-1-iodo-2-butene

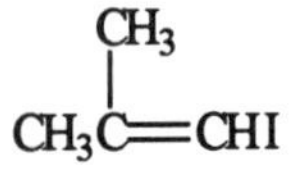

$$CH_3\overset{\overset{CH_3}{|}}{C}=CHI$$ 1-iodo-2-methylpropene

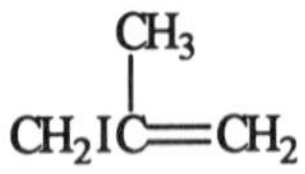

$$CH_2I\overset{\overset{CH_3}{|}}{C}=CH_2$$ 3-iodo-2-methylpropene

12. (a) C_3H_5Cl

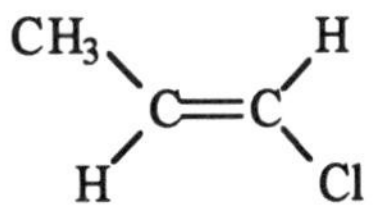

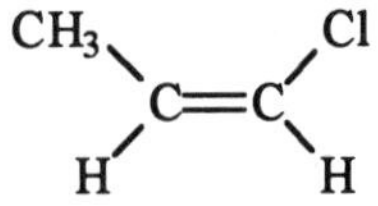

$$\underset{H}{\overset{CH_3}{}}\!\!>C=C<\!\!\underset{Cl}{\overset{H}{}}$$ $$\underset{H}{\overset{CH_3}{}}\!\!>C=C<\!\!\underset{H}{\overset{Cl}{}}$$

$CH_3CCl = CH_2$
$CH_2ClCH = CH_2$

(b) chlorocyclopropane

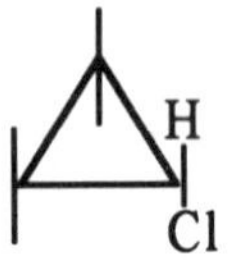

(cyclopropane ring with H and Cl on one carbon)

13. (a)

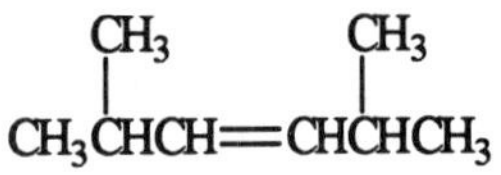

$$CH_3\overset{\overset{CH_3}{|}}{C}HCH=CH\overset{\overset{CH_3}{|}}{C}HCH_3$$ 2,5-dimethyl-3-hexene

(b)

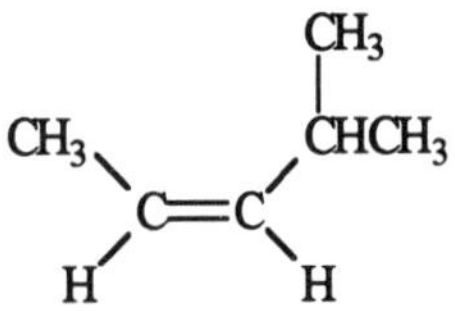

$$\underset{H}{\overset{CH_3}{}}\!\!>C=C<\!\!\underset{H}{\overset{\overset{\overset{CH_3}{|}}{CHCH_3}}{}}$$ cis-4-methyl-2-pentene

(c)  3-penten-1-yne

(d) $CH_3CH_2(H)C{=}C(H)CH_2CH_3$ trans-3-hexene

(e) $CH{\equiv}CCH(CH_3)CH_2CH_3$ 3-methyl-1-pentyne

(f) $CH_3CH(C_6H_5)CH(CH_3)CH_2CH_2CH_3$ 3-methyl-2-phenylhexane

14. (a) $CH_2{=}CHC(CH_2CH_3)(CH_3)CH_2CH_3$ 3-ethyl-3-methyl-1-pentene

(b) $(C_6H_5)HC{=}CH(C_6H_5)$ cis, 1,2-diphenylethene

(c) $CH{\equiv}CCHCH_3$

3-phenyl-1-butyne

(d) cyclopentene

(e) 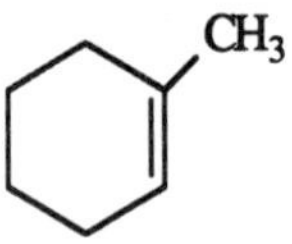 [structure: 1-methylcyclohexene ring with CH_3] 1-methylcyclohexene

(f) 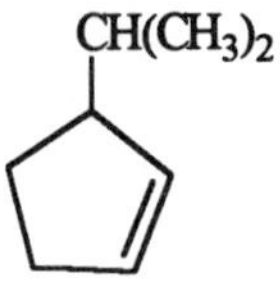 [structure: cyclopentene ring with $CH(CH_3)_2$] 3-isopropylcyclopentene

15. (a) 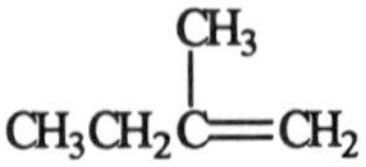 $CH_3CH_2C(CH_3)=CH_2$ — Numbering was started from the wrong end of the structure. Correct name: 2-methyl-1-butene

(b) $CH_3CH_2CH=CHCH_3$ — Numbering was started from the wrong end of the structure. Correct name: 2-pentene

(c) 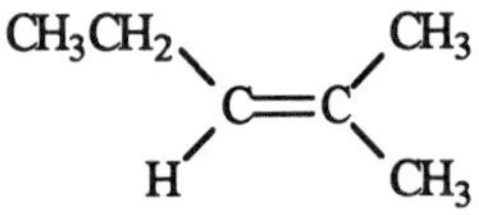 $(CH_3CH_2)(H)C=C(CH_3)_2$ — This compound does not have cis-trans isomers. Correct name: 2-methyl-2-pentene

16. (a) 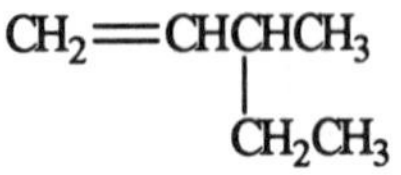 $CH_2=CHCH(CH_2CH_3)CH_3$ — Longest chain contains five carbon atoms. Correct name: 3-methyl-1-pentene

(b) 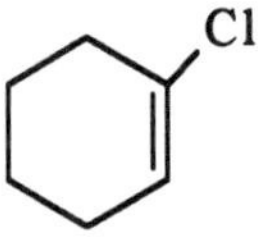 [structure: cyclohexene ring with Cl] The C = C bond in cyclohexene is numbered so that substituted groups have the smallest numbers. Correct name: 1-chlorocyclohexene

(c) $CH_3CH_2CH_2CH=CHCH_3$ — Numbering was started from the wrong end of the structure. Correct name: 2-hexene

17. (a) trans-3-methyl-3-hexene

(b) 3-phenyl-1-propyne

(c) 4, 5-dibromo-2-hexyne

18. (a) cis-4-methyl-2-hexene

(b) 2, 3-dimethyl-2-butene

(c) 3-isopropyl-1-pentene

19. All the hexynes, C_6H_{10}

1-hexyne $CH_3CH_2CH_2CH_2C\equiv CH$

2-hexyne $CH_3CH_2CH_2C\equiv CCH_3$

3-hexyne $CH_3CH_2C\equiv CCH_2CH_3$

3-methyl-1-pentyne $CH_3CH_2CH(CH_3)C\equiv CH$

4-methyl-1-pentyne $CH_3CH(CH_3)CH_2C\equiv CH$

4-methyl-2-pentyne $CH_3CH(CH_3)C\equiv CCH_3$

3, 3-dimethyl-1-butyne $CH_3C(CH_3)_2{-}C\equiv CH$

20. All the pentynes, C_5H_8

1-pentyne $CH_3CH_2CH_2C\equiv CH$

2-pentyne $CH_3CH_2C\equiv CCH_3$

3-methyl-1-butyne $CH_3CH(CH_3)C\equiv CH$

21. Only structure (b) will show cis-trans isomers. $CH_3CH=CHCl$

22. Only structure (c) will show cis-trans isomers. $CH_2ClCH=CHCH_2Cl$

23. (a) $CH_3CH_2CH_2CH=CH_2 + Br_2 \longrightarrow CH_3CH_2CH_2CHBrCH_2Br$

(b) $CH_3CH_2C(CH_3)=CHCH_3 + HI \longrightarrow CH_3CH_2CI(CH_3)CH_2CH_3$

(c) $CH_3CH_2CH{=}CH_2 + H_2O \xrightarrow{H^+} CH_3CH_2CH(OH)CH_3$

(d) $C_6H_5CH{=}CH_2 + H_2 \xrightarrow[\text{1 atm}]{\text{Pt, 25 °C}} C_6H_5CH_2CH_3$

(e) $CH_3CH{=}CHCH_3 + KMnO_4 \xrightarrow[\text{cold}]{H_2O} CH_3CH(OH)CH(OH)CH_3$

24. (a) $CH_3CH_2CH_2CH{=}CH_2 + H_2O \xrightarrow{H^+} CH_3CH_2CH_2CH(OH)CH_3$

(b) $CH_3CH_2CH{=}CHCH_3 + HBr \longrightarrow CH_3CH_2CHBrCH_2CH_3 + CH_3CH_2CH_2CHBrCH_3$

(c) $CH_2{=}CHCl + Br_2 \longrightarrow CH_2BrCHClBr$

(d) $C_6H_5CH{=}CH_2 + HCl \longrightarrow C_6H_5CHClCH_3$

(e) $CH_2{=}CHCH_2CH_3 + KMnO_4 \xrightarrow[\text{cold}]{H_2O} CH_2(OH)CH(OH)CH_2CH_3$

25. (a) $CH_3C{\equiv}CCH_3 + Br_2 \text{ (1 mole)} \longrightarrow CH_3CBr{=}CBrCH_3$

(b) Two-step reaction:

$$CH{\equiv}CH + HCl \longrightarrow CH_2{=}CHCl \xrightarrow{HCl} CH_3CHCl_2$$

(c) $CH_3CH_2CH_2C{\equiv}CH + H_2 \text{ (1 mole)} \longrightarrow CH_3CH_2CH_2CH{=}CH_2$

26. (a) $CH_3C{\equiv}CH + H_2\ (1\ mol) \xrightarrow[1\ atm]{Pt,\ 25^oC} CH_3CH{=}CH_2$

(b) $CH_3C{\equiv}CCH_3 + Br_2\ (2\ mol) \longrightarrow CH_3CBr_2CBr_2CH_3$

(c) Two step reaction

$$CH_3C{\equiv}CH + HCl \longrightarrow CH_3CCl{=}CH_2 \xrightarrow{HCl} CH_3CCl_2CH_3$$

27. When cyclohexene, [structure], reacts with:

(a) Br_2, the product is [structure] 1,2-dibromocyclohexane

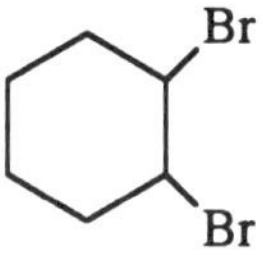

(b) HI, the product is [structure] iodocyclohexane

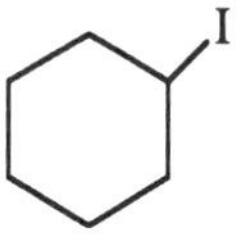

(c) H_2O, H^+, the product is [structure] cyclohexanol

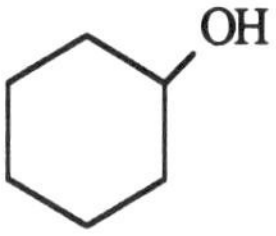

(d) $KMnO_4(aq)$, the product is [structure] cyclohexene glycol or 1,2-dihydroxcyclohexane

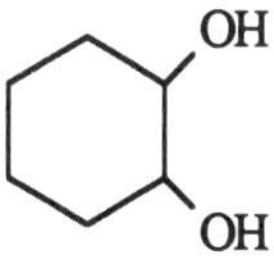

28. When cyclopentene, [structure], reacts with

(a) Cl_2, the product is [structure] 1,2-dichlorcyclopentane

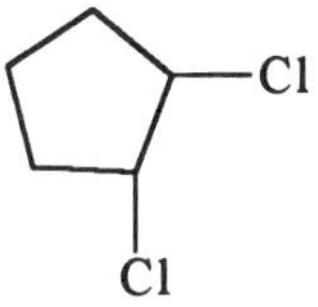

(b) HBr, the product is [cyclopentane ring]–Br bromocyclopentane

(c) H_2, Pt, the product is [cyclopentane ring] cyclopentane

(d) H_2O, H^+, the product is [cyclopentane ring]–OH cyclopentanol

29. Reactions to convert 2-butyne to:

(a) 2,3-dibromobutane

$$CH_3C{\equiv}CCH_3 \ + \ Br_2 \text{ (1 mole)} \longrightarrow CH_3CBr{=}CBrCH_3$$

$$CH_3CBr{=}CBrCH_3 \ + \ H_2 \xrightarrow[\text{1 atm}]{\text{Pt, 25}^{\circ}} CH_3CHBrCHBrCH_3$$

(b) 2,2-dibromobutane

$$CH_3C{\equiv}CCH_3 \ + \ HBr \longrightarrow CH_3CH{=}CBrCH_3 \xrightarrow{HBr} CH_3CH_2CBr_2CH_3$$

(c) 2,2,3,3-tetrabromobutane

$$CH_3C{\equiv}CCH_3 \ + \ Br_2 \text{ (2 moles)} \longrightarrow CH_3CBr_2CBr_2CH_3$$

30. (a) $$CH_3C{\equiv}CCH_2CH_3 \ + \ Cl_2 \text{ (1 mole)} \longrightarrow CH_3CCl{=}CClCH_2CH_3$$

$$CH_3CCl{=}CClCH_2CH_3 \ + \ H_2 \xrightarrow[\text{1 atm}]{25^{\circ}C} CH_3CHClCHClCH_2CH_3$$

(b) $$CH_3C{\equiv}CCH_2CH_3 \ + \ HCl \longrightarrow CH_3CCl{=}CHCH_2CH_3$$

$$CH_3CCl{=}CHCH_2CH_3 \ + \ HCl \longrightarrow CH_3CCl_2CH_2CH_2CH_3$$

Can also yield $CH_3CH_2CCl_2CH_2CH_3$ at the same time.

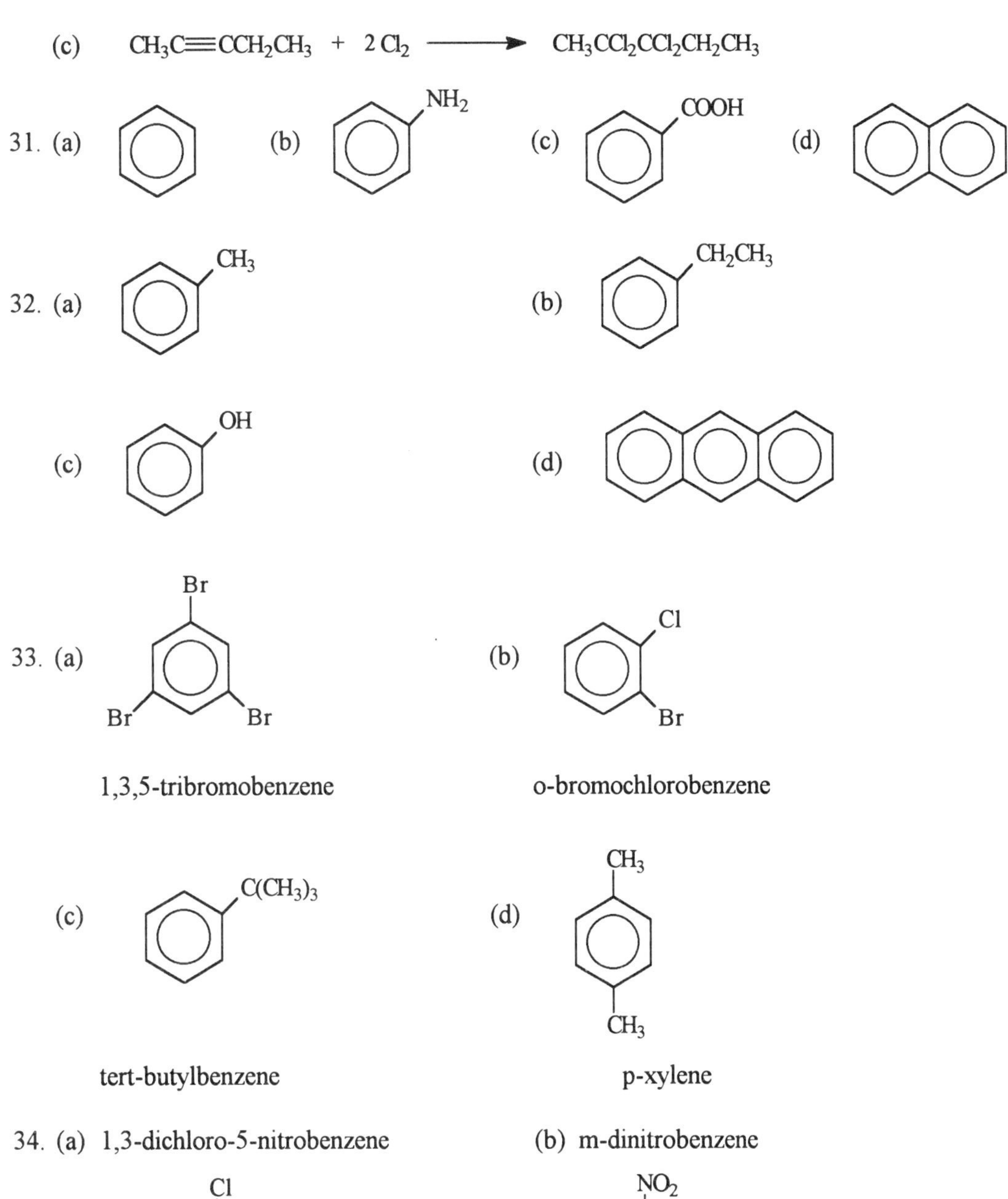

34. (a) 1,3-dichloro-5-nitrobenzene (b) m-dinitrobenzene

Cl

Cl NO_2

NO_2

NO_2

(c) 1,1-diphenylethane

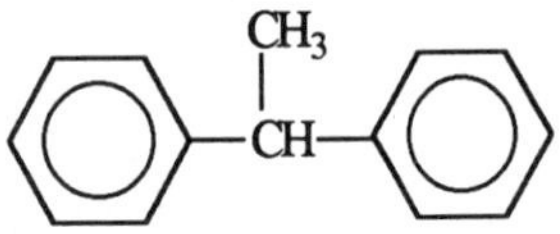

(d) phenylethene (styrene)

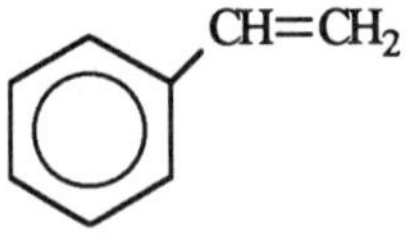

35. (a) dichlorobromobenzenes

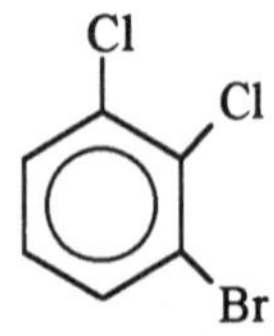

1,2-dichloro-3-bromobenzene

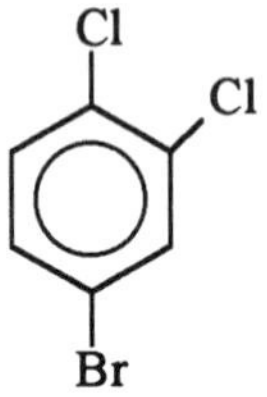

1,2-dichloro-4-bromobenzene

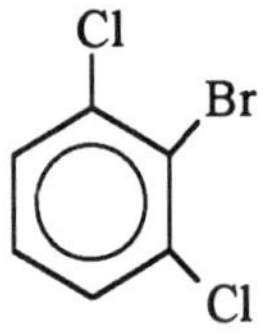

1,3-dichloro-2-bromobenzene

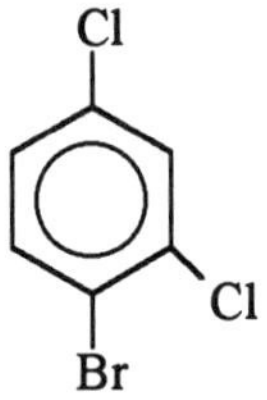

1,3-dichloro-4-bromobenzene

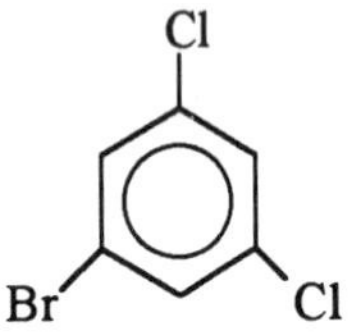

1,3-dichloro-5-bromobenzene

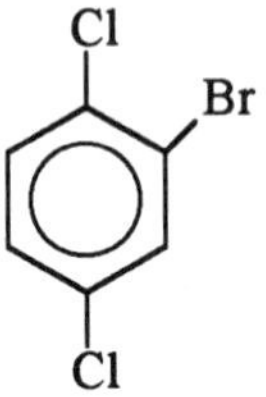

1,4-dichloro-2-bromobenzene

(b) The toluene derivatives of formula C_9H_{12}

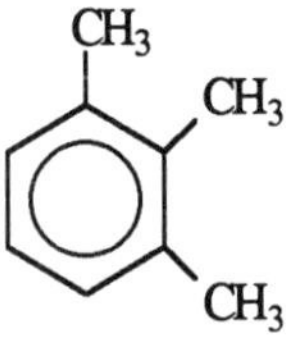

1,2,3-trimethylbenzene

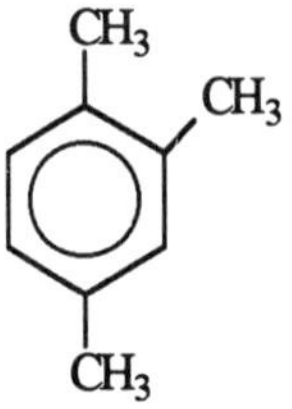

1,2,4-trimethylbenzene

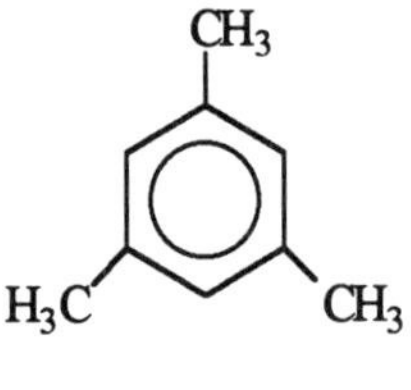

1,3,5-trimethylbenzene

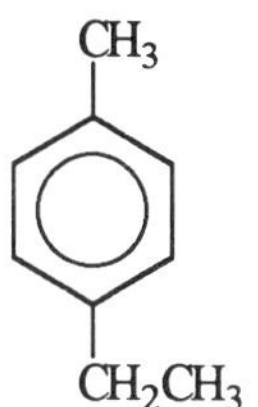

o-ethyltoluene m-ethyltoluene p-ethyltoluene

36. (a) trichlorobenzenes

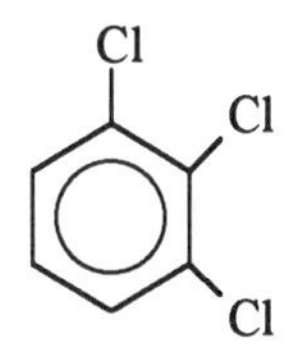

1,2,3-trichlorobenzene

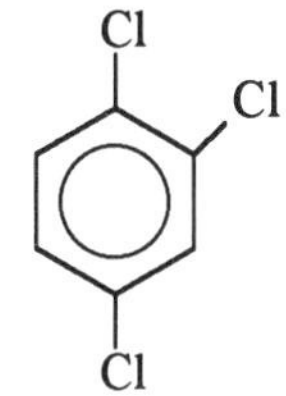

1,2,4-trichlorobenzene

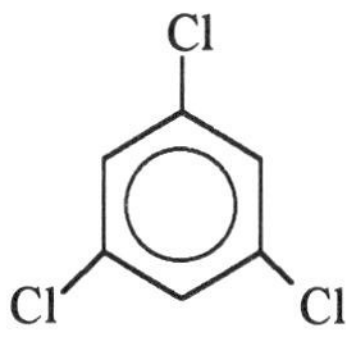

1,3,5-trichlorobenzene

(b) The benzene derivatives of formula C_8H_{10}:

1,2-dimethylbenzene or o-xylene

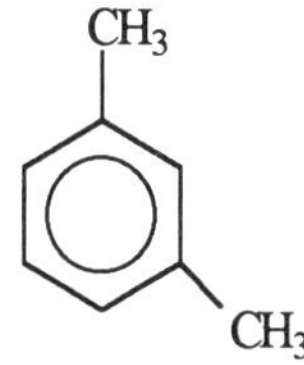

1,3-dimethylbenzene or m-xylene

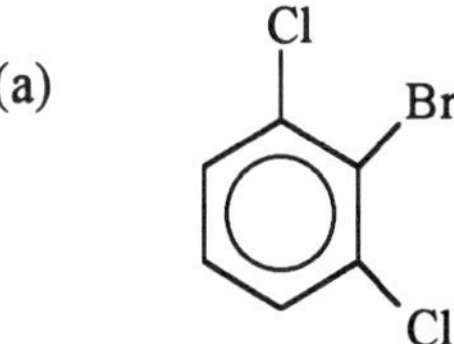

1,4-dimethylbenzene or p-xylene

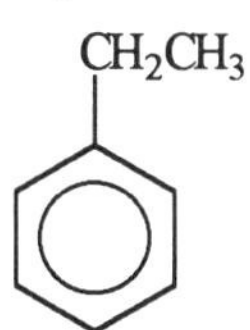

ethylbenzene

37. All isomers that can be written by substituting another chlorine in o-chlorobromobenzene.

(a)

1-bromo-2,6-dichlorobenzene

(b)

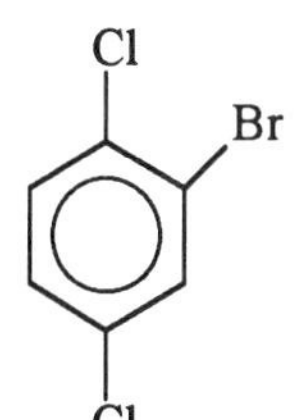

1-bromo-2,5-dichlorobenzene

(c)

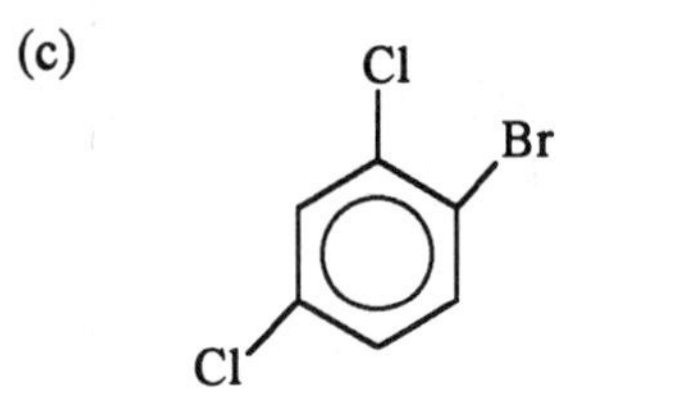

1-bromo-2,5-dichlorobenzene

(d)

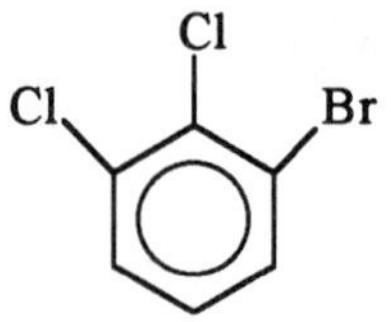

1-bromo-2,3-dichlorobenzene

38. All isomers that can be written substituting a third chlorine atom on o-dichlorobenzene.

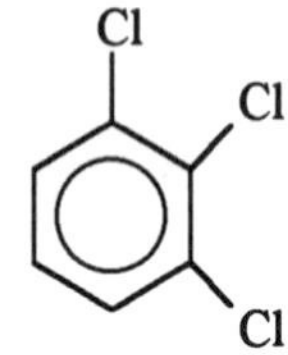

1,2,3-trichlorobenzene

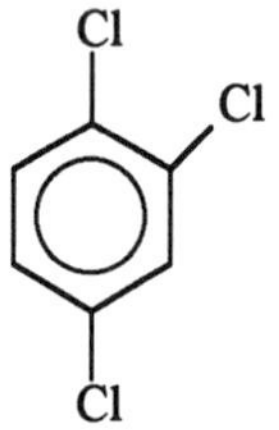

1,2,4-trichlorobenzene

39. (a) p-chloroethylbenzene
    (b) propylbenzene
    (c) m-nitroaniline
    (d) p-bromophenol
    (e) triphenylmethane

40. (a) styrene
    (b) m-nitrotoluene
    (c) 2,4-dibromobenzoic acid
    (d) isopropylbenzene
    (e) 2,4,6-tribromophenol

41. (a)

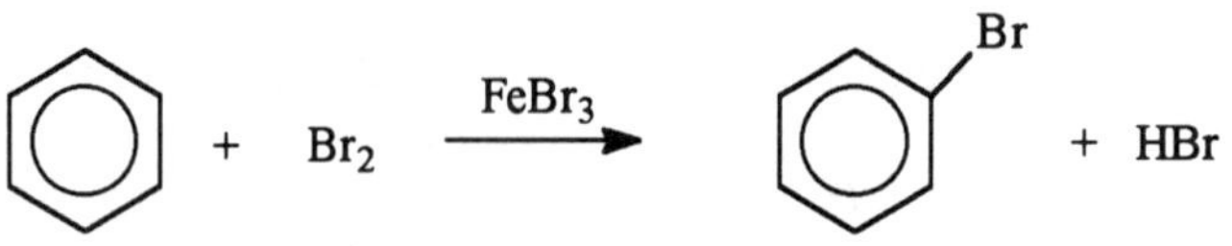

$$+ \quad Br_2 \xrightarrow{FeBr_3} \quad + \quad HBr$$

bromobenzene

(b)

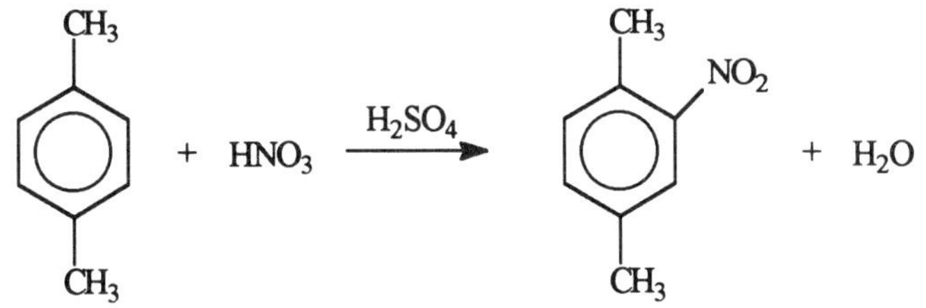

$$\text{1,4-}(CH_3)_2C_6H_4 + HNO_3 \xrightarrow{H_2SO_4} \text{2-}NO_2\text{-1,4-}(CH_3)_2C_6H_3 + H_2O$$

2-nitro-1,4-dimethylbenzene

42. (a)

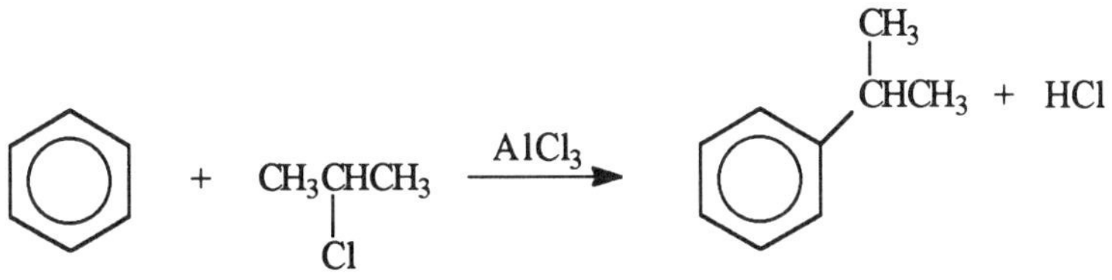

$$C_6H_6 + CH_3CHClCH_3 \xrightarrow{AlCl_3} C_6H_5CH(CH_3)CH_3 + HCl$$

isopropylbenzene

(b)

$$C_6H_5CH_3 + KMnO_4 \xrightarrow{H_2O} C_6H_5COOH$$

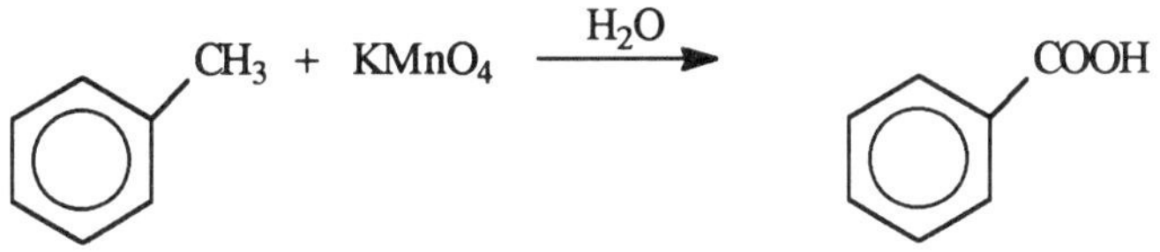

benzoic acid

43. When $CH_2{=}C(CH_3)CH_2CH_2CH_3$ reacts with HBr, two products are possible:

$CH_3CBr(CH_3)CH_2CH_2CH_3$ and $CH_2BrCH(CH_3)CH_2CH_2CH_3$

The first will strongly predominate. This is the product according to Markovnikov's rule and forms because a tertiary carbocation intermediate is more stable than a primary carbocation.

44. Two tests can be used. (1) Baeyer test - hexene will decolorize $KMnO_4$ solution; cyclohexane will not. (2) In the absence of sunlight, hexene will react with and decolorize bromine; cyclohexane will not.

45. Methylpentenes that show geometric isomerism.

$CH_3CH{=}C(CH_3)CH_2CH_3$ — 3-methyl-2-pentene

$CH_3CH{=}CHCH(CH_3)CH_3$ — 4-methyl-2-pentene

46. (a) $\overset{+}{C}H_3$ — methyl carbocation

(b) $CH_3CH_2\overset{+}{C}H_2$ — propyl carbocation

(c) $CH_3\overset{+}{C}(CH_3)CH_3$ — tert-butyl carbocation

(d) $CH_3CH_2CH_2CH_2\overset{+}{C}H_2$ — pentyl carbocation

47. (a) cyclohexene + Br_2 → 1,2-dibromocyclohexane

(b) cyclohexene + HCl → chlorocyclohexane

(c) methylcyclohexene + HCl → 1-chloro-1-methylcyclohexane

(d) 3-methylcyclohexene + HCl ⟶ 2-chloro-1-methylcyclohexane + 3-chloro-1-methylcyclohexane

2-chloro-1-methylcyclohexane   3-chloro-1-methylcyclohexane

48. The reaction mechanism by which benzene is brominated in the presence of $FeBr_3$:

$$FeBr_3 + Br_2 \longrightarrow FeBr_4^- + Br^+$$

benzene + Br^+ ⟶ arenium ion (H, Br on the ring carbon; ring bearing +)

arenium ion (H, Br) + $FeBr_4^-$ ⟶ bromobenzene + $FeBr_3$ + HBr

49. Chemically distinguishing between benzene, 1-hexene, and 1-hexyne

Step 1. Add $KMnO_4$ solution to a sample of each liquid. Benzene is the only one in which the $KMnO_4$ does not lose its purple color.

Step 2. To 0.5 mL samples of 1-hexene and 1-hexyne add bromine solution dropwise until there is no more color change of the bromine (from reddish-brown to colorless). 1-hexyne (with a triple bond) will decolor about twice as many drops of bromine as 1-hexene. Thus the three liquids are identified.

50. (a) $CH_3CHClCH_2CH_3 \xrightarrow{-HCl} CH_2{=}CHCH_2CH_3 + CH_3CH{=}CHCH_3$

(b) $CH_2ClCH_2CH_2CH_2CH_3 \xrightarrow{-HCl} CH_2{=}CHCH_2CH_2CH_3$

(c)

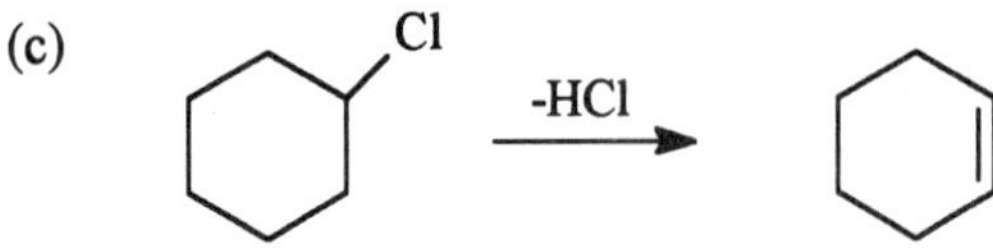

51. Yes, there will be a color change (loss of color of Br_2). The fact that there is no HBr formed indicates that the reaction is not substitution but addition. Therefore, C_4H_8, must contain a carbon-carbon double bond. Three structures are possible.

$CH_3CH_2CH{=}CH_2$ $\qquad$ $CH_3CH{=}CHCH_3$ $\qquad$ $CH_3C(CH_3){=}CH_2$

52. Bayer test: Add $KMnO_4$ solution to each sample. The $KMnO_4$ will lose its purple color with 1-heptene. There will be no reaction (no color change) with 1-heptane.

53. The carbon-carbon bond in the three molecules is different. Ethane has a single bond between carbon atoms formed from the overlap of two sp^3 hybridized orbitals. Ethene has a double bond between carbon atoms. One bond is a sigma bond formed by the overlap of two sp^2 hybridized orbitals, the other is a pi bond formed by the sidewise overlap of two p orbitals. Ethyne has a triple bond between carbon atoms – one sigma bond formed from the overlap of two sp hybridized orbitals and two pi bonds formed from the sidewise overlap of p orbitals.

54.

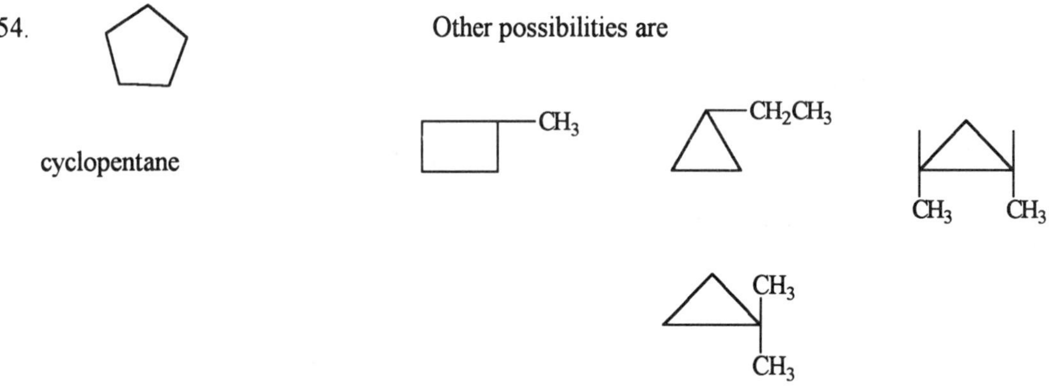

CHAPTER 22

ALCOHOLS, ETHERS, PHENOLS, AND THIOLS

1. The question allows great freedom of choice. These shown here are very simple examples of each type.

   (a) an alkyl halide CH_3CH_2Cl

   (b) a phenol C_6H_5OH

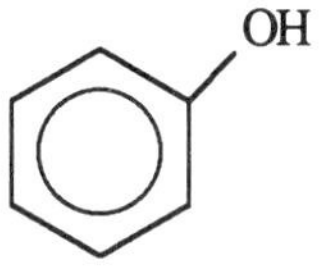

   (c) an ether $CH_3CH_2OCH_2CH_3$

   (d) an aldehyde $CH_3CH{=}O$

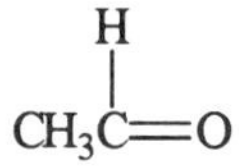

   (e) a ketone $CH_3C({=}O)CH_3$

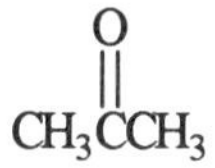

   (f) a carboxylic acid CH_3COOH

   (g) an ester $CH_3C({=}O)\text{—}OCH_2CH_3$

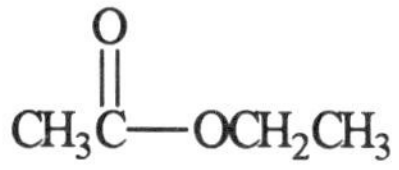

   (h) a thiol CH_3CH_2SH

2. Alkenes are almost never made from alcohols because the alcohols are almost always the higher value material. This is because recovering alkenes from hydrocarbon sources in an oil refinery (primarily catalytic cracking) is a relatively cheap process.

3. Isopropyl alcohol is usually used for rubbing alcohol, rather than n-propyl alcohol, because of price. The easy way to make the alcohol, by adding H_2O to propene, yields isopropyl alcohol rather than n-propyl alcohol. Any process for producing n-propyl alcohol would be much more expensive.

4. Oxidation of primary alcohols yields aldehydes. Further oxidation yields carboxylic acids.

Examples:

$$CH_3CH_2OH \xrightarrow{[O]} CH_3\overset{\overset{O}{\|}}{C}{-}H + H_2O$$

$$CH_3\overset{\overset{O}{\|}}{C}{-}H \xrightarrow{[O]} CH_3\overset{\overset{O}{\|}}{C}{-}OH$$

5. 1,2-Ethanediol is superior to methanol as an antifreeze because of its low volatility. Methanol is much more volatile than water. If the radiator leaks gas under pressure (normally steam), it would primarily leak methanol vapor so you would soon have no antifreeze. Ethylene glycol has a lower volatility than water, so it does not present this problem.

6. Physiological effects of:

   (a) methanol: It is a poisonous liquid capable of causing blindness or death if taken internally. Exposure to methanol vapors is also very dangerous.

   (b) ethanol: It can act as a food, a drug, and a poison. The body can metabolize small amounts of ethanol to produce energy; thus it is a food. It depresses brain functions so that activities requiring skill and judgment are impaired; thus it is a drug. With very large consumption, the depression of brain function can lead to unconsciousness and death; thus it is a poison.

7. The cumene-hydroperoxide synthesis of phenol and acetone:

$$C_6H_6 + CH_3CH{=}CH_2 \xrightarrow{H_2SO_4} C_6H_5CH(CH_3)_2 \xrightarrow{O_2} C_6H_5C(CH_3)_2{-}O{-}O{-}H$$

cumene　　　cumene hydroperoxide

$$\xrightarrow[H_2SO_4]{dilute} C_6H_5OH + CH_3\underset{\underset{O}{\|}}{C}CH_3$$

phenol　　　acetone

8. Low molar mass ethers present two hazards. They are very volatile and their highly flammable vapors form explosive mixtures with air. They also slowly react with oxygen in the air to form unstable explosive peroxides.

9. Ethanol (molar mass = 46.07) is a liquid whereas dimethyl ether (molar mass = 46.07) is a gas because ethanol has a significant amount of hydrogen bonding between molecules in the liquid state, and thus has a much higher boiling point than would be predicted from molar mass alone. Dimethyl ether is not capable of hydrogen bonding, so has low attraction between molecules, making it a gas at room temperature.

10. The correct statements are: a, b, e, g, h, i, j, m, o, q, s.

    (c) Alcohols are less acidic than water.
    (d) Sodium ethoxide cannot be prepared by reacting ethyl alcohol and sodium hydroxide. $CH_3CH_2CH_2OH + NaOH \rightarrow$ No reaction
    (f) Tertiary alcohols are more difficult to oxidize than primary alcohols.
    (k) The product formed when a molecule of water is split out between an alcohol and a carboxylic acid is called an ester.
    (l) When 1-butene is reacted with dilute H_2SO_4, the alcohol formed is 2-butanol.
    (n) The common name for 1,2,3-propanetriol is glycerol.
    (p) Cyclohexanol is a secondary alcohol.
    (r) Aldehydes, but not ketones, may be prepared by the oxidation of primary alcohols.
    (t) Coenzyme A is not an alcohol; it is a coenzyme.

11. (a) methanol CH_3OH

    (b) 3-methyl-1-hexanol

    CH_3 (on C-3)
    $CH_3CH_2CH_2CHCH_2CH_2OH$

    (c) 1,2-propanediol

    CH_3CHCH_2OH
    OH (on C-2)

    (d) 1-phenylethanol

    C_6H_5—$CHCH_3$
    OH

    (e) 2,3-butanediol

    $CH_3CH—CHCH_3$
    OH OH

(f) 2-propanethiol

$$\begin{array}{c}CH_3\underset{|}{C}HCH_3\\ SH\end{array}$$

12. (a) 2-butanol

$$\begin{array}{c}CH_3\underset{|}{C}HCH_2CH_3\\ OH\end{array}$$

(b) 2-methyl-2-butanol

$$\begin{array}{c}CH_3\\ |\\ CH_3\underset{|}{C}CH_2CH_3\\ OH\end{array}$$

(c) 2-propanol

$$\begin{array}{c}CH_3\underset{|}{C}HCH_3\\ OH\end{array}$$

(d) cyclopentanol

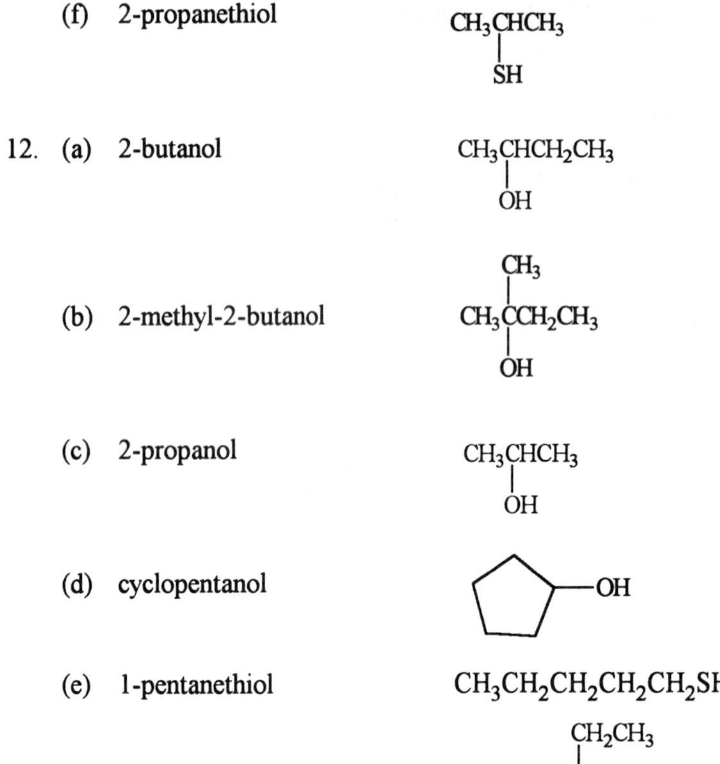

(e) 1-pentanethiol $CH_3CH_2CH_2CH_2CH_2SH$

(f) 4-ethyl-3-hexanol

$$\begin{array}{c}CH_2CH_3\\ |\\ CH_3CH_2\underset{|}{C}HCHCH_2CH_3\\ OH\end{array}$$

13. Eight open chain isomeric alcohols with the formula $C_5H_{11}OH$:

1-pentanol

$CH_3CH_2CH_2CH_2CH_2OH$

2-pentanol

$$\begin{array}{c}OH\\ |\\ CH_3CH_2CH_2CHCH\end{array}$$

3-pentanol

$$\begin{array}{c}OH\\ |\\ CH_3CH_2CHCH_2CH_3\end{array}$$

3-methyl-1-butanol

$$\begin{array}{c}CH_3\\ |\\ CH_3CHCH_2CH_2OH\end{array}$$

2-methyl-1-butanol

$$\begin{array}{c}CH_3\\ |\\ CH_3CH_2CHCH_2OH\end{array}$$

2-methyl-2-butanol

$$\begin{array}{c}CH_3\\ |\\ CH_3CH_2\underset{|}{C}CH_3\\ OH\end{array}$$

3-methyl-2-butanol

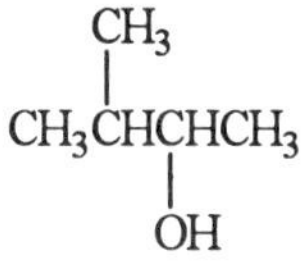

$$\begin{array}{c} CH_3 \\ | \\ CH_3CHCHCH_3 \\ | \\ OH \end{array}$$

2,2-dimethyl-1-propanol

$$\begin{array}{c} CH_3 \\ | \\ CH_3CCH_2OH \\ | \\ CH_3 \end{array}$$

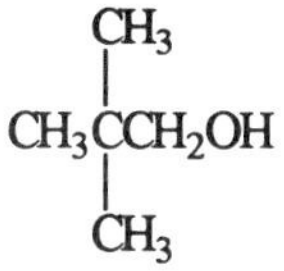

14. (a) C_3H_8O

	Alcohols	Ethers
	$CH_3CH_2CH_2OH$	$CH_3OCH_2CH_3$
	$CH_3CH(OH)CH_3$	

(b) C_4H_8O

	Alcohols	Ethers
	$CH_3CH_2CH_2CH_2OH$	$CH_3OCH_2CH_2CH_3$
	$CH_3CH_2CH(OH)CH_3$	$CH_3OCH(CH_3)_2$
	$(CH_3)_2CHCH_2OH$	$CH_3CH_2OCH_2CH_3$
	$(CH_3)_3C\text{–}OH$	

15. Of the eight compounds in question 13, the primary alcohols are: 1-pentanol; 3-methyl-1-butanol; 2-methyl-1-butanol; and 2,2-dimethyl-1-propanol. The secondary alcohols are: 2-pentanol; 3-pentanol; and 3-methyl-2-butanol. The tertiary alcohol is 2-methyl-2-butanol.

16. The primary alcohols in Question 14 are:

$CH_3CH_2CH_2OH$ $CH_3CH_2CH_2CH_2OH$ $(CH_3)_2CHCH_2OH$

The secondary alcohols are: $CH_3CH(OH)CH_3$ $CH_3CH_2CH(OH)CH_3$

The tertiary alcohol is: $(CH_3)_3C\text{–}OH$

17. The names of the compounds are

(a) 1-butanol (n-butyl alcohol)
(b) 2-propanol (isopropyl alcohol)
(c) 2-methyl-3-phenyl-1-propanol
(d) oxirane (ethylene oxide)
(e) 2-methyl-2-butanol

(f) 2-methylcyclohexanol
(g) 2,3-dimethyl-1,4-butanediol

18. The names of the compounds are

(a) ethanol (ethyl alcohol)
(b) 2-phenylethanol
(c) 3-methyl-3-pentanol
(d) 1-methylcyclopentanol
(e) 3-pentanol
(f) 1,2-propanediol
(g) 4-ethyl-2-hexanol

19. Chief product of dehydration.

(a) 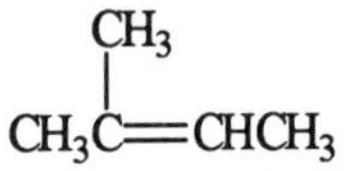 $CH_3C(CH_3)=CHCH_3$ 2-methyl-2-butene

(b) $CH_3CH=CHCH_2CH_3$ 2-pentene

(c)  cyclohexene

20. Chief product of dehydration

(a) 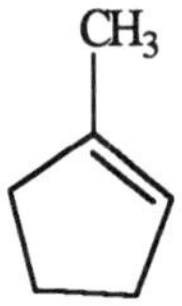 1-methylcyclopentene

(b) $CH_3CH=CHCH_3$ 2-butene

(c) 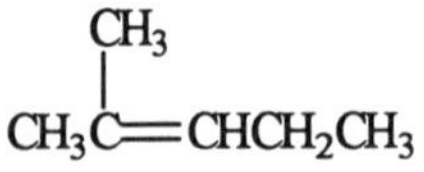 $CH_3C(CH_3)=CHCH_2CH_3$ 2-methyl-2-pentene

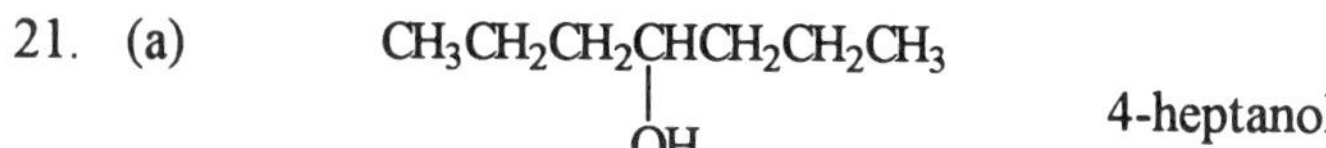

21. (a) $CH_3CH_2CH_2CH(OH)CH_2CH_2CH_3$ 4-heptanol

(b) cyclohexanol

(c) 3-methylcyclobutanol

22. (a) cyclopentanol

(b) 1-methylcyclopentanol

(c) $CH_3CH(CH_3)CH(OH)CH(CH_3)CH_3$ 2,4-dimethyl-3-pentanol

23. Oxidation of a primary alcohol

$CH_3CH_2CH_2CH(CH_2CH_3)CH_2CH_2OH \xrightarrow{[O]} CH_3CH_2CH_2CH(CH_2CH_3)CH_2{-}C({=}O){-}H$ aldehyde

$CH_3CH_2CH_2CH(CH_2CH_3)CH_2CH_2OH \xrightarrow{[O]} CH_3CH_2CH_2CH(CH_2CH_3)CH_2{-}C({=}O){-}OH$ carboxylic acid

24. Oxidation of a secondary alcohol

$$CH_3CH(OH)CH(CH_3)CH_2CH_3 \xrightarrow{[O]} CH_3C(=O)CH(CH_3)CH_2CH_3$$

ketone

25. (a) $CH_3CH{=}CH_2 + H_2O \xrightarrow{H^+} CH_3CH(OH)CH_3$

(b) $CH_3CH_2CH{=}CH_2 + H_2O \xrightarrow{H^+} CH_3CH_2CH(OH)CH_3$

(c) $CH_3CH_2CH{=}CHCH_3 + H_2O \xrightarrow{H^+} CH_3CH_2CH(OH)CH_2CH_3 + CH_3CH_2CH_2CH(OH)CH_3$

26. (a) $CH_3CH{=}CHCH_3 + H_2O \xrightarrow{H^+} CH_3CH_2CH(OH)CH_3$

(b) $CH_3CH_2CH_2CH{=}CH_2 + H_2O \xrightarrow{H^+} CH_3CH_2CH_2CH(OH)CH_3$

(c) $CH_3C(CH_3){=}CHCH_3 + H_2O \xrightarrow{H^+} CH_3C(OH)(CH_3)CH_2CH_3$

27. (a) $CH_3CHBrCH_3$ 2-bromopropane

(b) (cyclohexane ring with Br) bromocyclohexane

(c) $CH_3CH(CH_3)CH_2CH_2Br$ 1-bromo-3-methylbutane

28. (a) $CH_3CH(OH)CH_2CH_3$ 2-butanol

(b) $CH_3CH(OH)CH(CH_2CH_3)CH_2CH_3$ 3-ethyl-2-pentanol

(c) cyclopentyl–OH cyclopentanol

29. (a) $2\ CH_3CH_2OH + H_2SO_4 \xrightarrow{140^o\ C} CH_3CH_2OCH_2CH_3 + H_2O$

diethyl ether

(b) $CH_3CH_2CH_2OH + H_2SO_4 \xrightarrow{180^o\ C} CH_3CH{=}CH_2 + H_2O$

propene

(c) $CH_3CH(OH)CH_2CH_3 \xrightarrow[H_2SO_4]{K_2Cr_2O_7} CH_3C(=O)CH_2CH_3 + H_2O$

2-butanone

(d) $CH_3CH_2C(=O){-}OCH_2CH_3 \xrightarrow[H_2O]{H^+} CH_3CH_2COOH + CH_3CH_2OH$

propanoic acid ethanol

30. (a) $2\ CH_3CH_2OH + 2\ Na \longrightarrow 2\ CH_3CH_2O^- Na^+ + H_2$

(b) $CH_3CH_2CH_2CH_2OH \xrightarrow[H_2SO_4]{K_2Cr_2O_7} CH_3CH_2CH_2C(=O){-}H$

(c) cyclopentyl–$CH{=}CH_2 \xrightarrow[H_2SO_4]{H_2O}$ cyclopentyl–$CH(OH)CH_3$

(d) $CH_3CH_2C(=O){-}OCH_3 + NaOH \longrightarrow CH_3CH_2C(=O){-}O^- Na^+ + CH_3OH$

31. (a) o-methylphenol

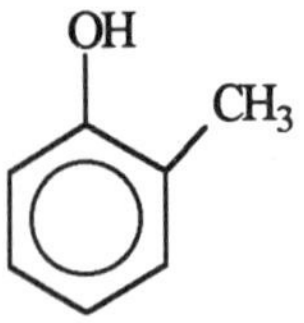

OH, CH_3

(b) 1,3-dihydroxybenzene

OH, OH

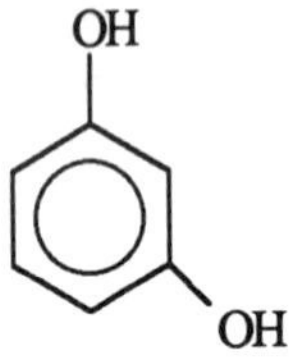

(c) 4-hydroxy-3-methoxybenzaldehyde

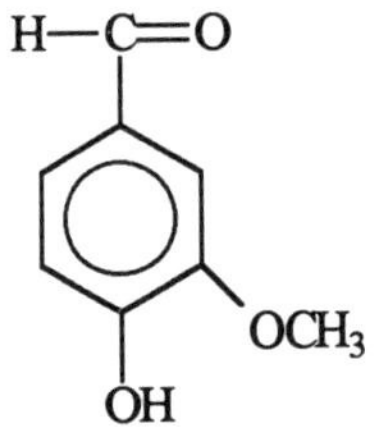

H—C=O, OCH_3, OH

32. (a) p-nitrophenol

OH, NO_2

(b) 2,6-dimethylphenol

OH, H_3C, CH_3

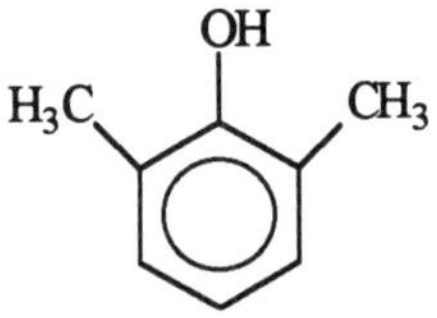

(c) 1,2-dihydroxybenzene

OH, OH

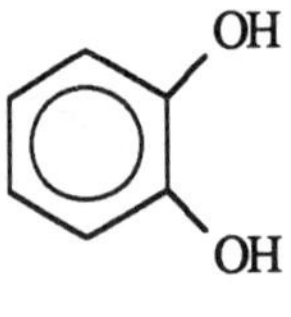

33. (a) phenol
    (b) m-methylphenol
    (c) 2-ethyl-5-nitrophenol
    (d) 4-bromo-2-chlorophenol

34. (a) p-dihydroxybenzene (hydroquinone)
    (b) 2,4-dimethylphenol
    (c) 2,4-dinitrophenol
    (d) m-hexylphenol

35. Order of increasing solubility in water [c (lowest), a, b, d (highest)]

(c) $CH_3CH_2CH_2CH_2CH_3$

(a) $CH_3CH_2OCH_2CH_2CH_3$

(b) $CH_3CH(OH)CH_2CH_2CH_3$

(d) $CH_3CH(OH)CH(OH)CH_2CH_3$

36. Order of decreasing solubility in water: [a (highest), d, c, b (lowest)]

(a) $CH_3CH(OH)CH(OH)CH_2OH$

(d) $CH_3CH(OH)CH_2CH_2OH$

(c) $CH_3CH_2CH_2CH_2OH$

(b) $CH_3CH_2OCH_2CH_3$

37. Fourteen isomeric ethers; $C_6H_{14}O$

$CH_3OCH_2CH_2CH_2CH_2CH_3$

methyl pentyl ether
(1-methoxypentane)

$CH_3OCH(CH_3)CH_2CH_2CH_3$

2-methoxypentane

$CH_3OCH(CH_2CH_3)CH_2CH_3$

3-methoxypentane

$CH_3OCH_2CH(CH_3)CH_2CH_3$

1-methoxy-2-methylbutane

$CH_3OCH_2CH_2CH(CH_3)CH_3$

1-methoxy-3-methylbutane

$CH_3OCH(CH_3)—CH(CH_3)CH_3$

2-methoxy-3-methylbutane

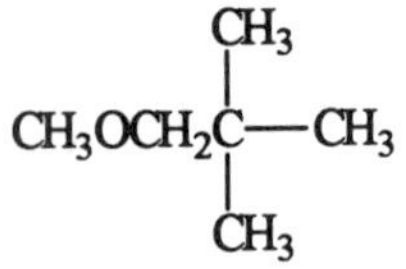

$$CH_3OCH_2\overset{\overset{CH_3}{|}}{\underset{\underset{CH_3}{|}}{C}}—CH_3$$

1-methoxy-2,2-dimethylpropane

$CH_3CH_2OCH_2CH_2CH_2CH_3$

butyl ethyl ether (1-ethoxybutane)

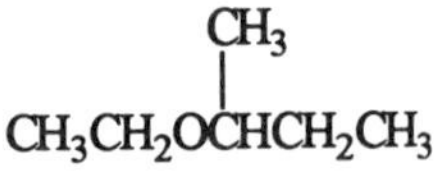

$$CH_3CH_2O\overset{\overset{CH_3}{|}}{C}HCH_2CH_3$$

ethyl sec-butyl ether
(2-ethoxybutane)

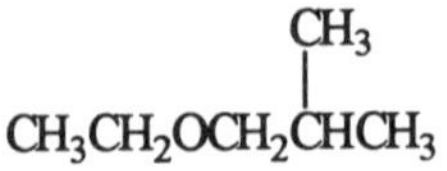

$$CH_3CH_2OCH_2\overset{\overset{CH_3}{|}}{C}HCH_3$$

ethyl isobutyl ether
(1-ethoxy-2-methylpropane)

$$CH_3CH_2O\overset{\overset{CH_3}{|}}{\underset{\underset{CH_3}{|}}{C}}—CH_3$$

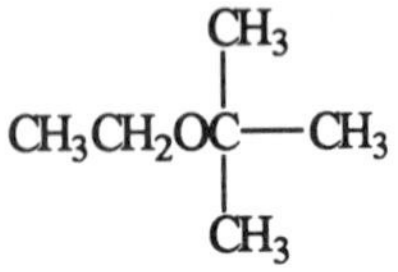

ethyl-t-butyl ether
(2-ethoxy-2-methylpropane)

$CH_3CH_2CH_2OCH_2CH_2CH_3$

di-n-propyl ether
1-propoxypropane

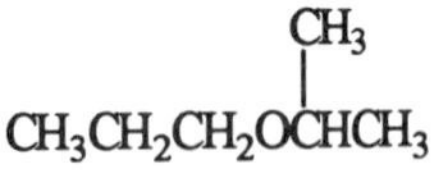

$$CH_3CH_2CH_2O\overset{\overset{CH_3}{|}}{C}HCH_3$$

n-propyl isopropyl ether

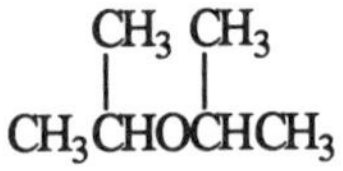

$$CH_3\overset{\overset{CH_3}{|}}{C}HO\overset{\overset{CH_3}{|}}{C}HCH_3$$

diisopropyl ether

38. Six isomeric ethers: $C_5H_{12}O$

$CH_3OCH_2CH_2CH_2CH_3$

methyl n-butyl ether
(1-methoxybutane)

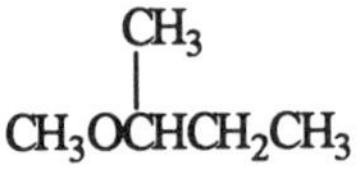

$$CH_3O\overset{\overset{CH_3}{|}}{C}HCH_2CH_3$$

methyl sec-butyl ether
(2-methoxybutane)

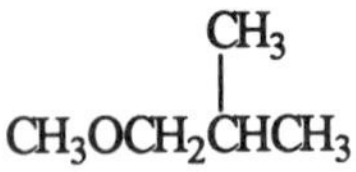

$$CH_3OCH_2\overset{\overset{CH_3}{|}}{C}HCH_3$$

methyl isobutyl ether
(1-methoxy-2-methylpropane)

$$CH_3O\overset{\overset{CH_3}{|}}{\underset{\underset{CH_3}{|}}{C}}CH_3$$

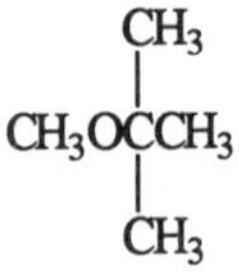

methyl-t-butyl ether
(2-methoxy-2-methylpropane)

$CH_3CH_2OCH_2CH_2CH_3$

ethyl n-propyl ether
(1-ethoxypropane)

$$\begin{array}{c} \quad\quad\quad\quad CH_3 \\ \quad\quad\quad\quad | \\ CH_2CH_2OCHCH_3 \end{array}$$

ethyl isopropyl ether
(2-ethoxypropane)

39. $$\underset{\quad\quad\quad\quad OH}{CH_3CH_2\underset{|}{C}HCH_3} + 6\ O_2 \longrightarrow 4\ CO_2 + 5\ H_2O$$

40. $CH_3CH_2OCH_2CH_3 + 6\ O_2 \longrightarrow 4\ CO_2 + 5\ H_2O$

41. Possible combinations of reactants to make the following ethers by the Williamson synthesis:

(a) $CH_3CH_2OCH_3$ $\quad$ $CH_3ONa + CH_3CH_2Cl$

or $CH_3CH_2ONa + CH_3Cl$

(b) $C_6H_5CH_2OCH_2CH_3$ $\quad$ $C_6H_5CH_2ONa + CH_3CH_2Cl$

or $C_6H_5CH_2Cl + CH_3CH_2ONa$

42. Possible combinations of reactants to make the following ethers by the Williamson synthesis:

(a) $CH_3CH_2CH_2OCH_2CH_2CH_3$ $\quad$ $CH_3CH_2CH_2ONa + CH_3CH_2CH_2Cl$

(b)
$$\begin{array}{l} \ \ CH_3 \\ \ \ | \\ HCOCH_2CH_2CH_3 \\ \ \ | \\ \ \ CH_3 \end{array}$$

$$\begin{array}{l} \ \ CH_3 \\ \ \ | \\ HCONa \\ \ \ | \\ \ \ CH_3 \end{array} + CH_3CH_2CH_2Cl \quad \text{or} \quad CH_3CH_2CH_2ONa + \begin{array}{l} CH_3CHCH_3 \\ \quad\quad | \\ \quad\quad Cl \end{array}$$

43.

OH

OH

$CH_2(CH_2)_4CH_3$

4-hexylresorcinol or 4-hexyl-1,3-dihydroxybenzene

44.

cis-1,2-cyclopentanediol

OH

OH

trans-1,2-cyclopentanediol

45. Increasing acidity

CH_2OH < H_2O < HO ... CH_3

46. The common phenolic structure in the catecholamines is catechol, o-dihydroxybenzene.

OH

OH

47. 1-butanol first adds a proton (H^+) followed by a loss of H_2O to form the butyl carbocation (1° carbocation). Then a hydrogen shift occurs forming the more stable secondary carbocation followed by the loss of a proton to form the product 2-butene.

$$CH_3CH_2CH_2CH_2OH \xrightarrow{H_2SO_4} CH_3CH_2CH_2CH_2CH_2\overset{+}{O}H(-H) \longrightarrow CH_3CH_2CH_2\overset{+}{C}H_2 \text{ (1° carbocation)}$$

$$\longrightarrow CH_3CH_2\overset{+}{C}HCH_3 \text{ (2° carbocation)} \longrightarrow CH_3CH{=}CHCH_3 \text{ (2-butene)}$$

48. Methyl alcohol is converted to formaldehyde by an oxidation reaction:

$$CH_3OH \xrightarrow{[O]} HCH \text{ (with C=O)}$$

49. (a) $CH_3CH(OH)CH_3 \xrightarrow[H^+]{Cr_2O_7^{2-}} CH_3C(=O)CH_3$

(b) $CH_3CH_2CH_2CH{=}CH_2 \xrightarrow[H_2O]{H^+} CH_3CH_2CH_2CH(OH)CH_3$

(c) $CH_3CH_2OH + Na\text{ (metal)} \longrightarrow CH_3CH_2ONa + H_2$

(d) $CH_3CH_2CH{=}CH_2 \xrightarrow[H_2O]{H^+} CH_3CH_2CH(OH)CH_3 \xrightarrow[H^+]{Cr_2O_7^{2-}} CH_3CH_2C(=O)CH_3$

(e) $CH_3CH_2CH_2CH_2OH \xrightarrow[\Delta]{H_2SO_4} CH_3CH{=}CHCH_3 + CH_3CH_2CH{=}CH_2$

$\xrightarrow{HCl} CH_3CH_2CH(Cl)CH_3$

(f) $CH_3CH_2CH_2Cl \xrightarrow{NaOH} CH_3CH_2CH_2OH \xrightarrow[H^+]{Cr_2O_7^{2-}} CH_3CH_2C(=O){-}H$

50. $p\text{-}CH_3C_6H_4OH + NaOH \longrightarrow p\text{-}CH_3C_6H_4O^-\,Na^+ + H_2O$

$$CH_3CH_3 + Cl_2 \xrightarrow{light} CH_3CH_2Cl + HCl$$

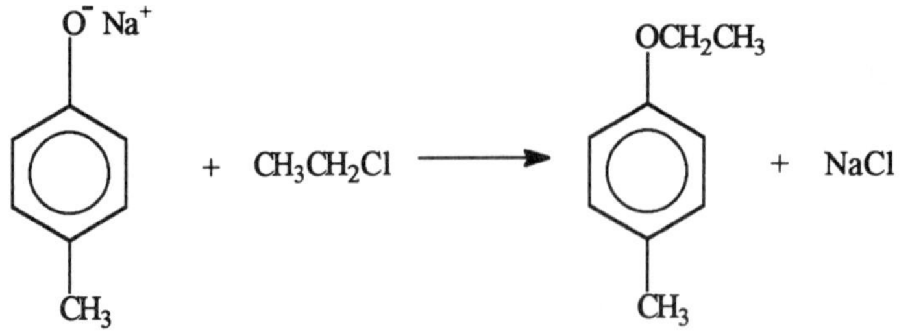

$$p\text{-}CH_3C_6H_4O^-Na^+ + CH_3CH_2Cl \longrightarrow p\text{-}CH_3C_6H_4OCH_2CH_3 + NaCl$$

51. A simple chemical test to distinguish between:

    (a) ethanol and dimethyl ether. Ethanol will react readily with potassium dichromate and sulfuric acid to make acetaldehyde. Visibly, the orange color of the dichromate changes to green. Ethanol reacts with metallic sodium to produce hydrogen gas. Dimethyl ether does not react with either of these reagents.

    (b) 1-pentanol and 1-pentene. 1-pentene will rapidly decolorize bromine as it adds to the double bond. 1-pentanol does not react.

    (c) p-methylphenol has acidic properties so it will react with sodium hydroxide. Methoxybenzene does not have acidic properties and will not react with sodium hydroxide.

52. Differentiation between phenols and alcohols:

    Phenols are acidic compounds; alcohols are not acidic.
    Phenols react with NaOH to form salts; alcohols do not react with NaOH.
    The OH group is bonded to a benzene ring in phenols and to a nonbenzene carbon atom in alcohols.

53. Isomers of $C_8H_{10}O$

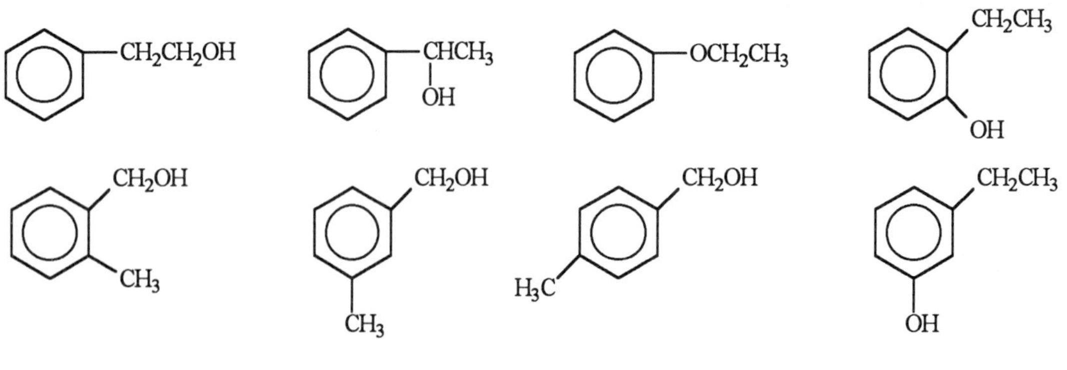

C_6H_5–CH_2CH_2OH $\quad$ C_6H_5–$CHCH_3$ (OH on CH) $\quad$ C_6H_5–OCH_2CH_3 $\quad$ o-CH_2CH_3/OH ring

o-CH_2OH/CH_3 ring $\quad$ m-CH_2OH/CH_3 ring $\quad$ p-CH_2OH/H_3C ring $\quad$ m-CH_2CH_3/OH ring

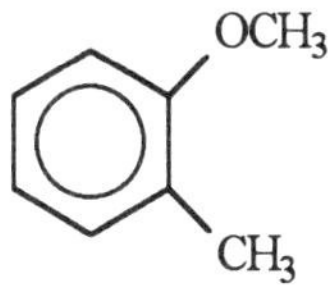

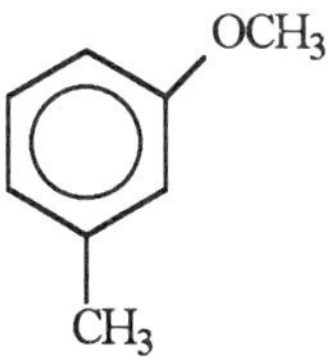

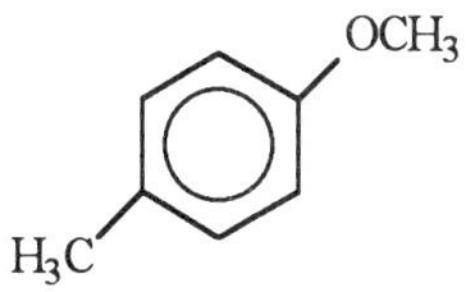

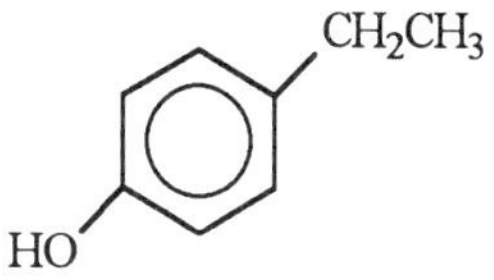

54. Only compound (a) will react with NaOH.

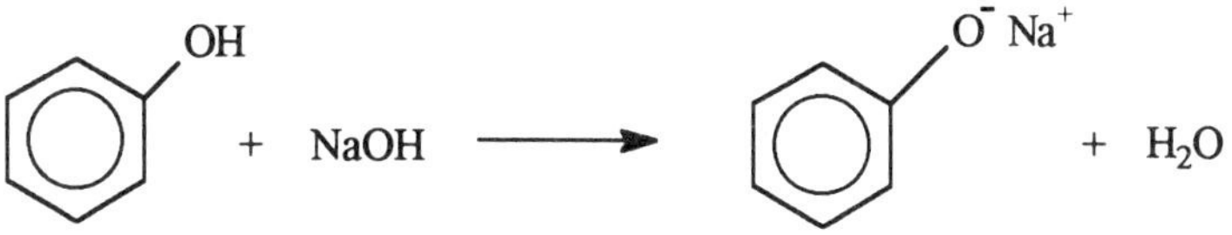

55. Order of increasing boiling points.

1-pentanol	<	1-octanol	<	1,2-pentanediol
138°C		194°C		210°C

All three compounds are alcohols. 1-pentanol has the lowest molar mass and hence the lowest boiling point. 1-octanol has a higher molar mass and therefore a higher boiling point than 1-pentanol. 1,2-pentanediol has two –OH groups and therefore forms more hydrogen bonds than the other two alcohols which causes its higher boiling point.

CHAPTER 23

ALDEHYDES AND KETONES

1. (a) 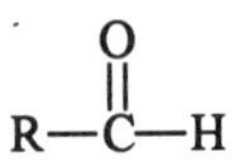 $R-\overset{O}{\overset{\|}{C}}-H$

aldehyde

(b) 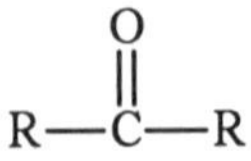 $R-\overset{O}{\overset{\|}{C}}-R$

ketone

(c) 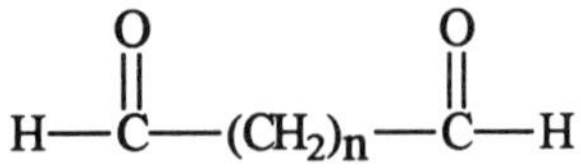 $H-\overset{O}{\overset{\|}{C}}-(CH_2)_n-\overset{O}{\overset{\|}{C}}-H$

a dialdehyde

(d) 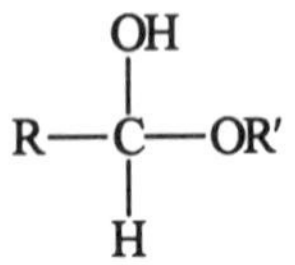 $R-\overset{OH}{\overset{|}{\underset{H}{\underset{|}{C}}}}-OR'$

a hemiacetal

(e) 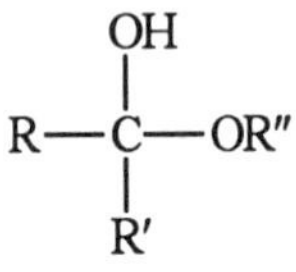 $R-\overset{OH}{\overset{|}{\underset{R'}{\underset{|}{C}}}}-OR''$

a hemiketal

(f) 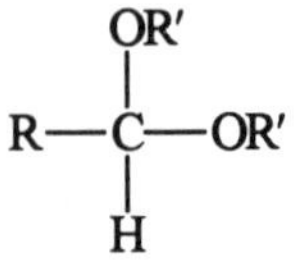 $R-\overset{OR'}{\overset{|}{\underset{H}{\underset{|}{C}}}}-OR'$

an acetal

(g) 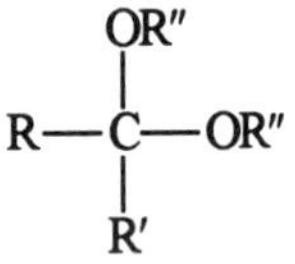 $R-\overset{OR''}{\overset{|}{\underset{R'}{\underset{|}{C}}}}-OR''$

a ketal

(h) 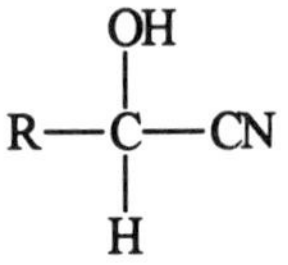 $R-\overset{OH}{\overset{|}{\underset{H}{\underset{|}{C}}}}-CN$

a cyanohydrin

2. Aldehyde and ketones have lower boiling points than alcohols of similar molar mass because they do not form hydrogen bonds, since they contain no –OH groups.

3. Propanal $CH_3CH_2\overset{H}{\overset{|}{C}}=O$ propanone $CH_3-\overset{O}{\overset{\|}{C}}-CH_3$

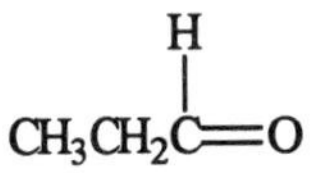

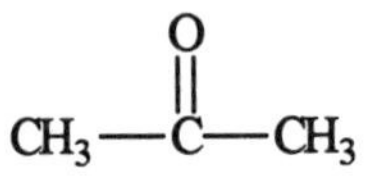

Each has the molecular formula of C_3H_8O, so aldehydes and ketones appear to be isomeric with each other.

Butanal $CH_3CH_2CH_2\overset{H}{\overset{|}{C}}=O$ 2-butanone $CH_3-\overset{O}{\overset{\|}{C}}-CH_2CH_3$

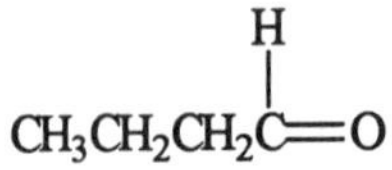

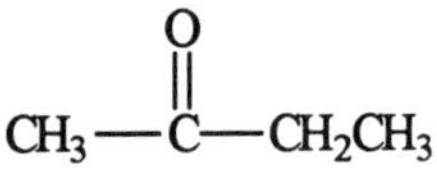

Each has the molecular formula C_4H_8O, so the generalization seems to check out.

4. The correct statements are: b, c, f, h, i, j, l, m, o, q, r, t, v.

(a) The functional group that characterizes aldehydes and ketones is called a carbonyl group.
(d) The higher molar mass aldehydes and ketones are essentially insoluble in water.
(e) Ketones, unlike aldehydes, are not easily oxidized to carboxylic acids.
(g) Diethyl ketone and butyraldehyde do not have the same molecular formula.
(k) Acetals are stable in alkaline solutions but not in acid solutions.
(n) Ethanol may not be distinguished from propanal with Tollens reagent.
(p) The compound C_2H_4O cannot be a ketone.
(s) $C_6H_5CH_2CH(OC_2H_5)_2$ is an acetal.
(u) Of the three compounds ethanal, propanal, and butanal, ethanal has the highest vapor pressure.
(w) Reduction of aldehydes yield primary alcohols.

5. Names of aldehydes.

(a) $H_2C{=}O$ — methanal, formaldehyde

(b) $CH_3CH(CH_3)CH_2C(=O){-}H$ — 3-methylbutanal

(c) $H{-}C(=O)CH_2CH_2C(=O){-}H$ — butanedial

(d) $C_6H_5{-}CH{=}CHC(=O){-}H$ — 3-phenylpropenal

(e) $CH_3(H)C{=}C(H){-}C(=O){-}H$ (CH₃ and CHO on same side) — cis-2-butenal

6. Names of aldehydes

(a) 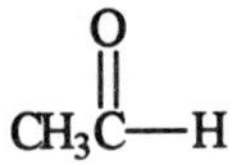 ethanal, acetaldehyde

(b) 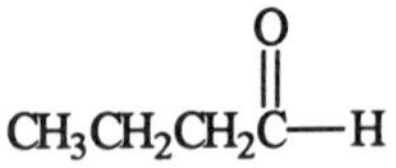 butanal

(c) 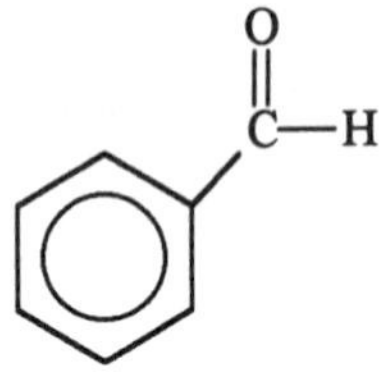 benzaldehyde

(d) 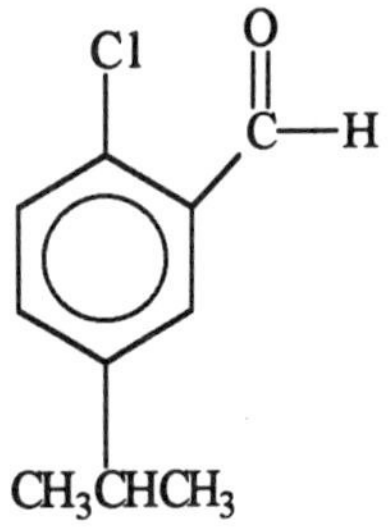 2-chloro-5-isopropylbenzaldehyde

(e) 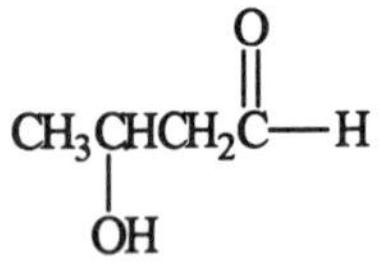 3-hydroxybutanal

7. Names of ketones

(a) propanone, acetone, dimethyl ketone

(b) 1-phenyl-1-propanone, ethyl phenyl ketone

(c) cyclopentanone

(d) 4-hydroxy-4-methyl-2-pentanone

8. Names of ketones

(a) 2-butanone, methyl ethyl ketone (MEK)

(b) 3,3-dimethylbutanone, methyl t-butyl ketone

(c) 2,5-hexanedione

(d) 1-phenyl-2-propanone, methyl benzyl ketone

9. Structural formulas.

(a) 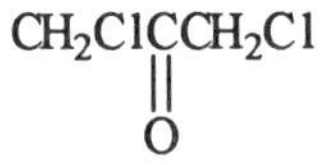 $CH_2Cl\overset{\displaystyle O}{\overset{\|}{C}}CH_2Cl$ (C=O on the central carbon) 1,3-dichloropropanone

(b) 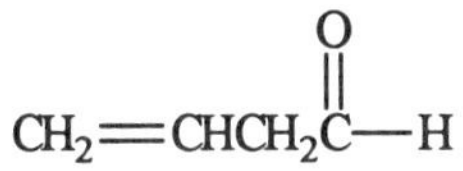 $CH_2{=}CHCH_2\overset{\displaystyle O}{\overset{\|}{C}}{-}H$ 3-butenal

(c) 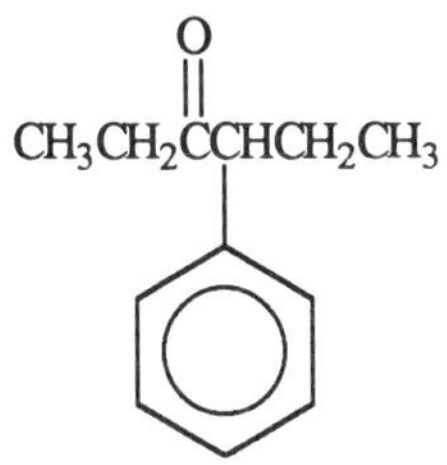 $CH_3CH_2\overset{\displaystyle O}{\overset{\|}{C}}CHCH_2CH_3$ with a phenyl ring on the CH 4-phenyl-3-hexanone

(d) 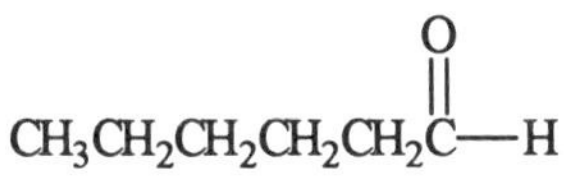 $CH_3CH_2CH_2CH_2CH_2\overset{\displaystyle O}{\overset{\|}{C}}{-}H$ hexanal

(e) 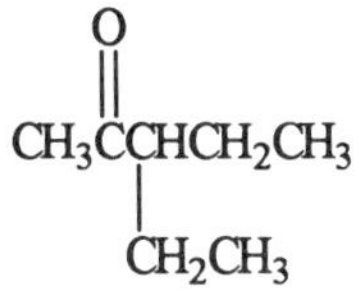 $CH_3\overset{\displaystyle O}{\overset{\|}{C}}\underset{\displaystyle CH_2CH_3}{\underset{|}{C}H}CH_2CH_3$

10. Structural formulas.

(a) 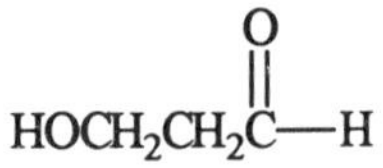 $HOCH_2CH_2C(=O)-H$ 3-hydroxypropanal

(b) 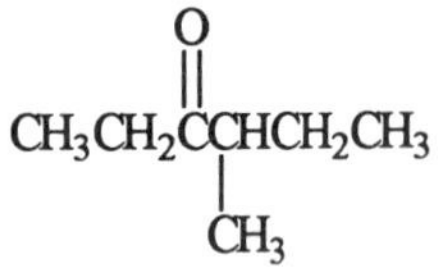 $CH_3CH_2C(=O)CH(CH_3)CH_2CH_3$ 4-methyl-3-hexanone

(c) 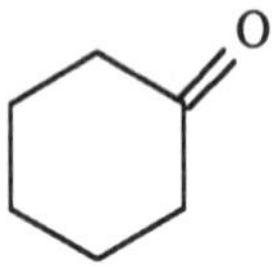 cyclohexanone (six-membered ring with C=O)

(d) 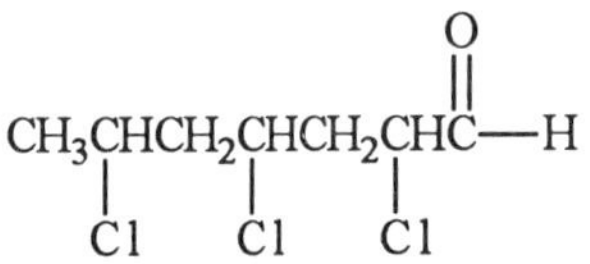 $CH_3CH(Cl)CH_2CH(Cl)CH_2CH(Cl)C(=O)-H$ 2,4,6-trichloroheptanal

(e) 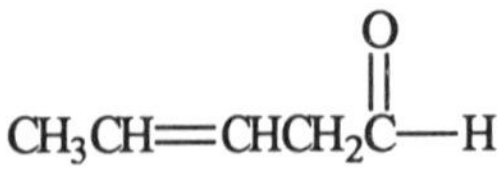 $CH_3CH{=}CHCH_2C(=O)-H$ 3-pentenal

11. Higher boiling point.

(a) 1-pentanol

(b) hexanal

(c) 2-pentanol

(d) butanone

12. Higher boiling point.

(a) pentanal

(b) benzyl alcohol

(c) 2-hexanone

(d) 1-butanol

13. Higher aqueous solubility

(a) 2,5-hexanedione (c) acetaldehyde

(b) propanal

14. Higher aqueous solubility

(a) acetaldehyde (c) propanal

(b) 2,4-pentanedial

15. Equations for the oxidation of:

(a) 3-pentanol

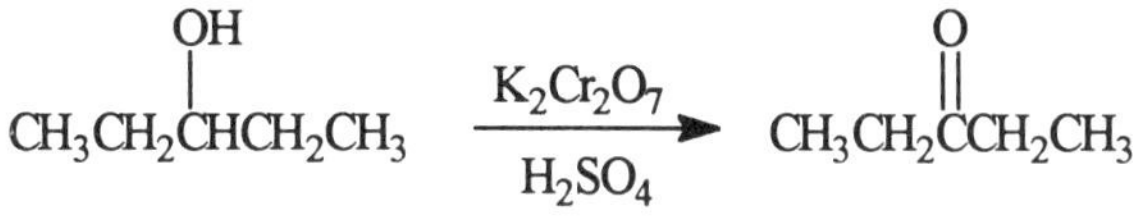

$$CH_3CH_2CH(OH)CH_2CH_3 \xrightarrow[H_2SO_4]{K_2Cr_2O_7} CH_3CH_2C(=O)CH_2CH_3$$

same product for air (O_2) oxidation

(b) 3-methyl-1-hexanol

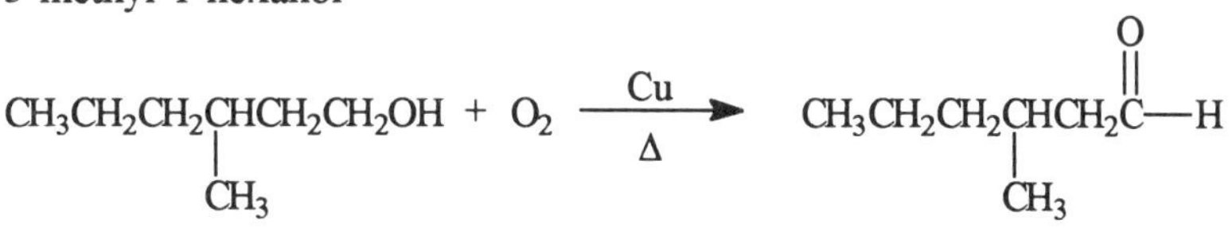

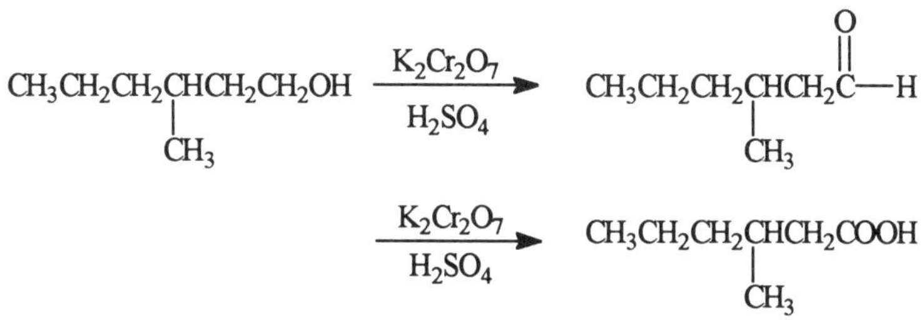

$$CH_3CH_2CH_2CH(CH_3)CH_2CH_2OH + O_2 \xrightarrow[\Delta]{Cu} CH_3CH_2CH_2CH(CH_3)CH_2C(=O)—H$$

$$CH_3CH_2CH_2CH(CH_3)CH_2CH_2OH \xrightarrow[H_2SO_4]{K_2Cr_2O_7} CH_3CH_2CH_2CH(CH_3)CH_2C(=O)—H$$

$$\xrightarrow[H_2SO_4]{K_2Cr_2O_7} CH_3CH_2CH_2CH(CH_3)CH_2COOH$$

16. Equations for the oxidation of:

(a) 1-propanol

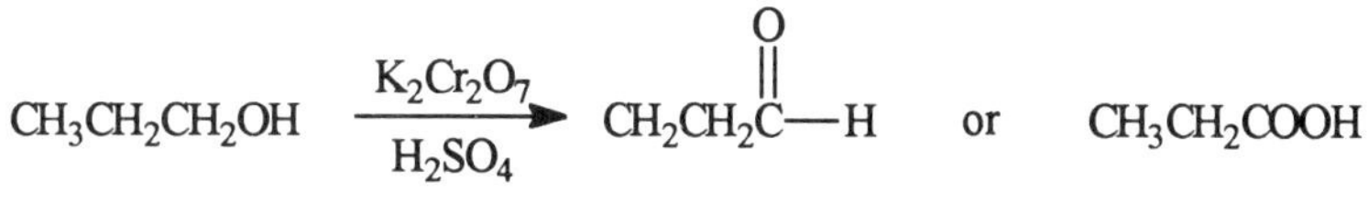

$$CH_3CH_2CH_2OH \xrightarrow[H_2SO_4]{K_2Cr_2O_7} CH_2CH_2C(=O)—H \quad \text{or} \quad CH_3CH_2COOH$$

$$CH_3CH_2CH_2OH + O_2 \xrightarrow[\Delta]{Cu} CH_3CH_2\overset{O}{\overset{\|}{C}}{-}H$$

(b)

$$CH_3\underset{CH_3}{\underset{|}{CH}}{-}\overset{OH}{\overset{|}{\underset{CH_3}{\underset{|}{C}}}}CH_3 + O_2 \xrightarrow[\Delta]{Ag} \text{No reaction (3° alcohol) with either oxidizing agent}$$

17. (a) An aldehyde group, $-\overset{H}{\overset{|}{C}}{=}O$, must be present to give a positive Tollens test.

(b) The visible evidence for a positive Tollens test is the formation of a silver mirror on the inner walls of a test tube.

(c)

$$CH_3\overset{H}{\overset{|}{C}}{=}O + 2\ Ag^+ \xrightarrow[H_2O]{NH_3} CH_3COO^-\ \overset{+}{N}H_4 + 2\ Ag(s) \quad \text{(silver mirror)}$$

18. (a) An aldehyde group, $-\overset{H}{\overset{|}{C}}{=}O$, must be present to give a positive Fehling test.

(b) The visible evidence for a positive Fehling test is the formation of brick red Cu_2O, which precipitates during the reaction.

(c)

$$CH_3\overset{H}{\overset{|}{C}}{=}O + 2\ Cu^{2+} \xrightarrow[H_2O]{NaOH} CH_3COONa + Cu_2O(s) \quad \text{(brick-red)}$$

19. Product(s) when each of the following is reacted with Tollens reagent:

(a) butanal → butanoic acid + Ag(s)

(b) benzaldehyde → benzoic acid + Ag(s)

(c) methyl ethyl ketone → no reaction

20. Product(s) when each of the following is reacted with Fehling reagent:

(a) propanal → propanoic acid + $Cu_2O(s)$

(b) acetone → no reaction

(c) 3-methylpentanal → 3-methylpentanoic acid + $Cu_2O(s)$

21. Aldol condensation

(a) butanal

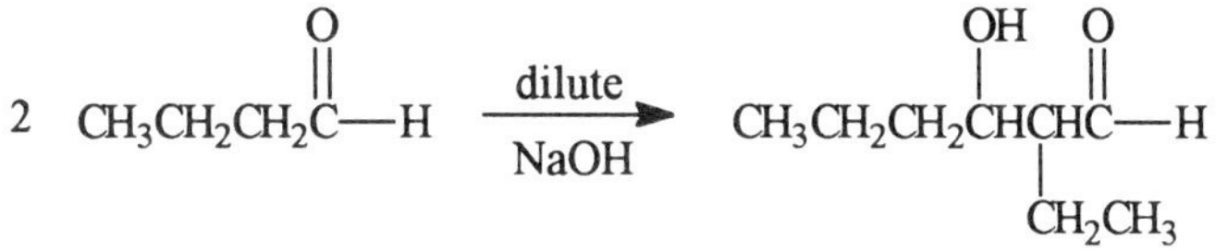

$$2\ CH_3CH_2CH_2C(=O){-}H \xrightarrow[\text{NaOH}]{\text{dilute}} CH_3CH_2CH_2CH(OH)CH(CH_2CH_3)C(=O){-}H$$

(b)

$$2\ C_6H_5{-}CH_2C(=O){-}H \xrightarrow[\text{NaOH}]{\text{dilute}} C_6H_5{-}CH_2CH(OH){-}CH(C_6H_5)C(=O){-}H$$

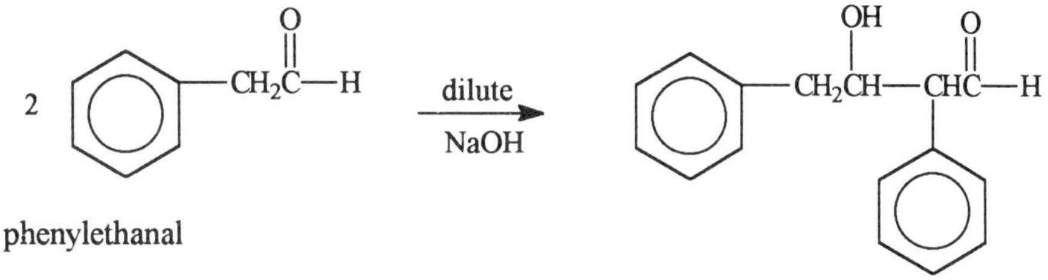

phenylethanal

22. Aldol condensation

(a) 3-pentanone

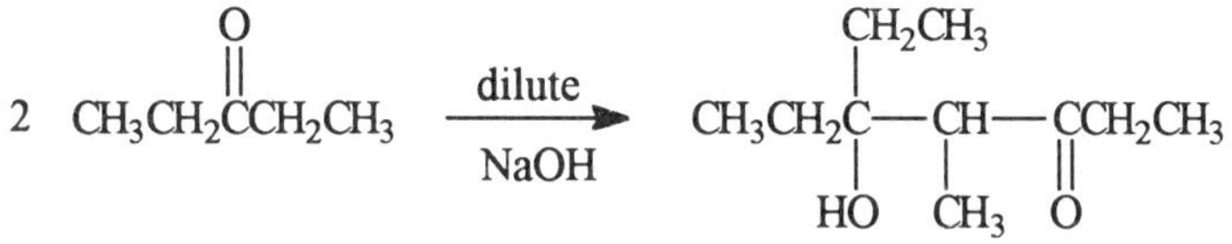

$$2\ CH_3CH_2C(=O)CH_2CH_3 \xrightarrow[\text{NaOH}]{\text{dilute}} CH_3CH_2C(OH)(CH_2CH_3){-}CH(CH_3){-}C(=O)CH_2CH_3$$

(b) propanal

$$2\ CH_3CH_2C(H){=}O \xrightarrow[\text{NaOH}]{\text{dilute}} CH_3CH_2CH(OH)CH(CH_3)C(H){=}O$$

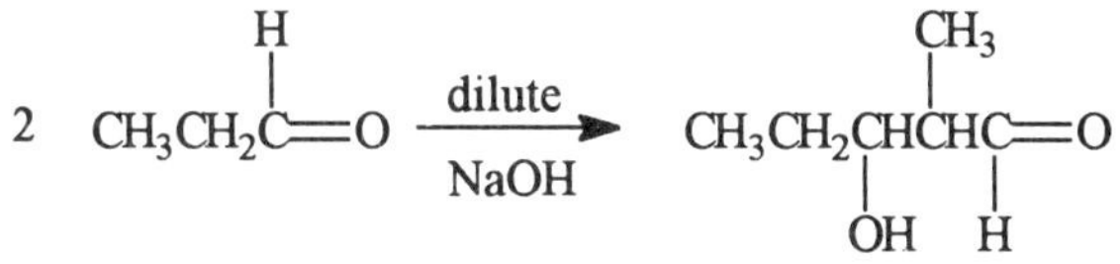

23. The completed equations are:

(a)

$$CH_3\overset{O}{\overset{\|}{C}}CH_3 + \underset{OH\ \ OH}{CH_2CH_2} \xrightleftharpoons{\text{dry HCl}} (CH_3)_2C\begin{matrix} OCH_2 \\ OCH_2 \end{matrix} + H_2O$$

(b)

$$CH_3CH_2\overset{H}{\overset{|}{C}}{=}O + CH_3CH_2OH \xrightleftharpoons{H^+} CH_3CH_2\overset{OH}{\overset{|}{\underset{H}{\underset{|}{C}}}}-OCH_2CH_3$$

(c)

$$CH_3\underset{CH_3}{\underset{|}{C}}HCH_2CH(OCH_3)_2 \xrightarrow[H^+]{H_2O} CH_3\underset{CH_3}{\underset{|}{C}}HCH_2\overset{O}{\overset{\|}{C}}-H + 2\ CH_3OH$$

24. The completed equations are:

(a)

$$CH_3CH_2\overset{H}{\overset{|}{C}}{=}O + 2\ CH_3CH_2CH_2OH \xrightleftharpoons{\text{dry HCl}} CH_3CH_2\overset{OCH_2CH_2CH_3}{\overset{|}{\underset{H}{\underset{|}{C}}}}-OCH_2CH_2CH_3 + H_2O$$

(b)

$$\text{cyclohexanone} + CH_3OH \xrightleftharpoons{H^+} \text{1-methoxycyclohexanol } (C_6H_{10}(OH)(OCH_3))$$

(c)

$$CH_3CH_2CH_2CH(OCH_3)_2 \xrightleftharpoons[H^+]{H_2O} CH_3CH_2CH_2\overset{H}{\overset{|}{C}}{=}O + 2\ CH_3OH$$

25. Sequence of reactions:

(a)

$$CH_3\underset{O}{\underset{\|}{C}}CH_3 + HCN \longrightarrow CH_3\overset{CH_3}{\overset{|}{\underset{OH}{\underset{|}{C}}}}-CN$$

(b)

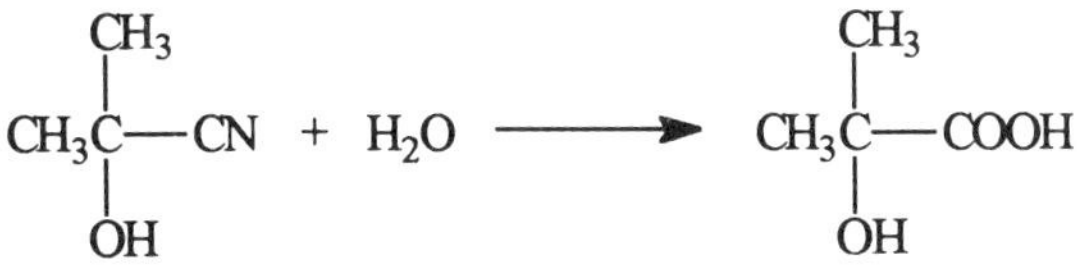

$$CH_3C(CH_3)(OH)\text{—}CN + H_2O \longrightarrow CH_3C(CH_3)(OH)\text{—}COOH$$

(c)

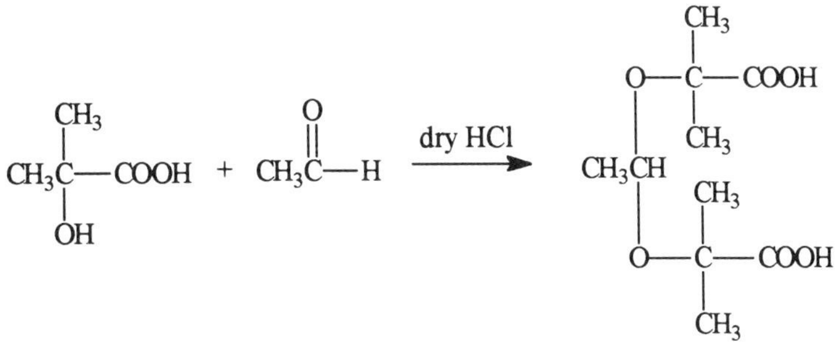

$$CH_3C(CH_3)(OH)\text{—}COOH + CH_3\overset{O}{\overset{\|}{C}}\text{—}H \xrightarrow{\text{dry HCl}} CH_3CH[O\text{—}C(CH_3)_2\text{—}COOH]_2$$

26. Sequence of reactions:

(a)

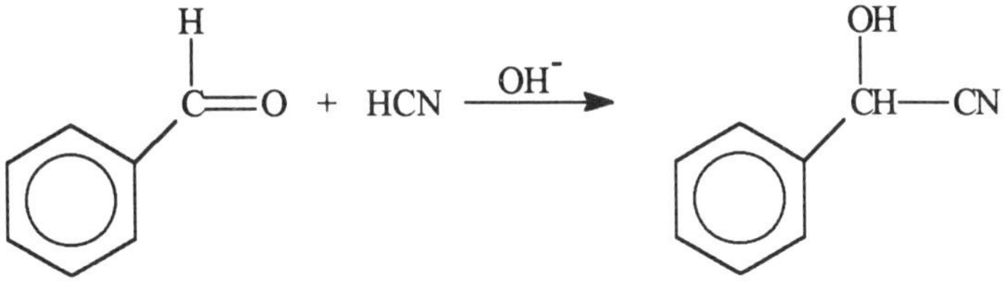

$$C_6H_5\text{—}CH{=}O + HCN \xrightarrow{OH^-} C_6H_5\text{—}CH(OH)\text{—}CN$$

(b)

$$C_6H_5\text{—}CH(OH)\text{—}CN + H_2O \xrightarrow{H^+} C_6H_5\text{—}CH(OH)COOH$$

(c)

$$C_6H_5\text{—}CH(OH)COOH \xrightarrow[H_2SO_4]{K_2Cr_2O_7} C_6H_5\text{—}\overset{O}{\overset{\|}{C}}\text{—}COOH$$

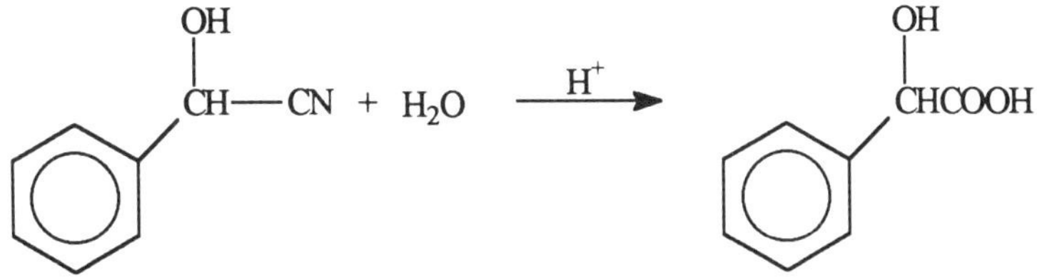

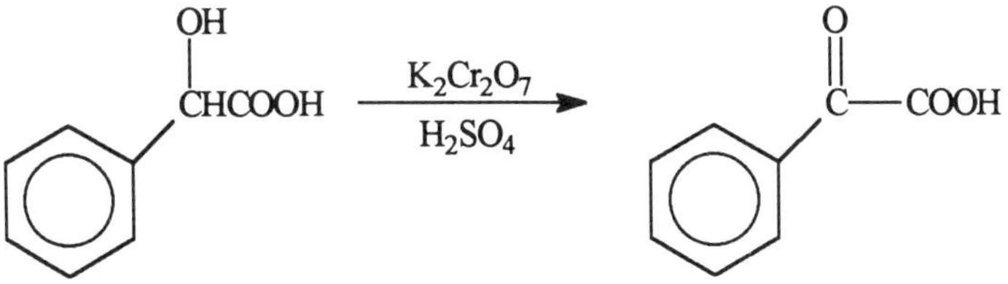

27. 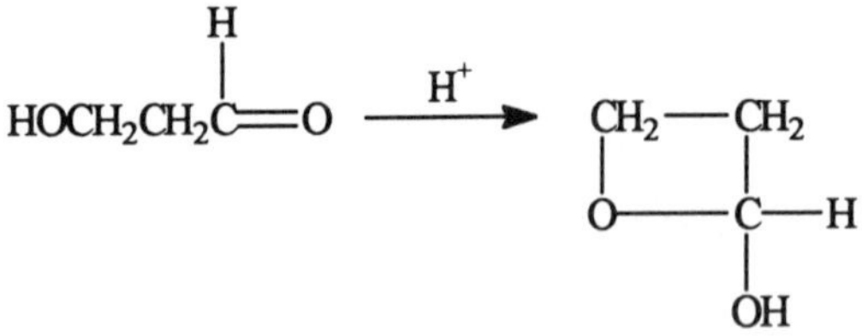 $HOCH_2CH_2C(H){=}O \xrightarrow{H^+}$ cyclic product: ring of CH_2—CH_2—C(H)(OH)—O (four-membered ring: CH_2—CH_2, CH_2—O, O—C—H with OH on C)

28. 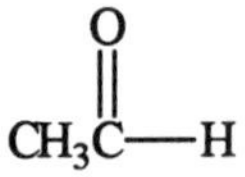 $CH_3C(=O)—H$ ethanal

29. Four aldol condensation products from a mixture of ethanol (E) and propanol (P) are possible.

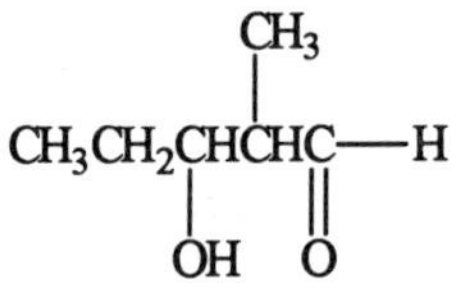

EE P–P EP PE

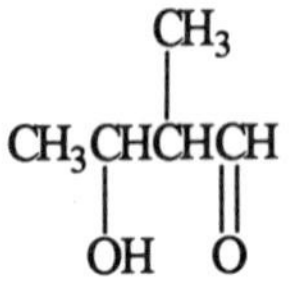

EE is $CH_3CH(OH)CH_2C(=O)—H$

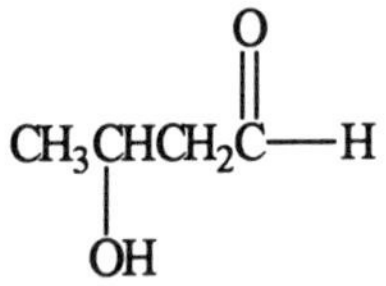

PP is $CH_3CH_2CH(OH)CH(CH_3)C(=O)—H$

EP is $CH_3CH(OH)CH(CH_3)CH(=O)$

PE is $CH_3CH_2CH(OH)CH_2C(=O)—H$

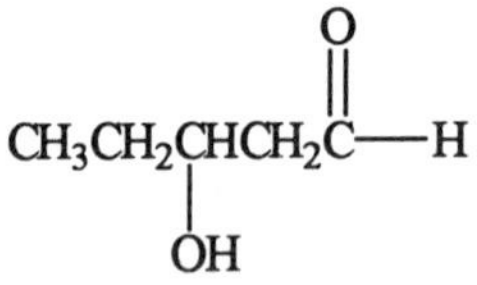

30. The secondary alcohol in compound I is oxidized to the ketone (compound II).

31. (a) 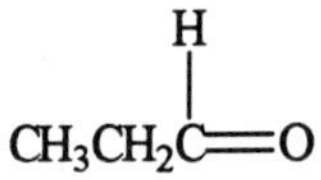 $CH_3CH_2C(H){=}O$ and $CH_3C(=O)CH_3$.

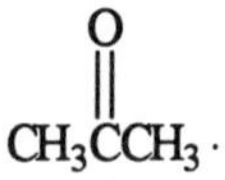

The Tollens test or Fehling test will give positive results (red Cu_2O) with propanal but not with acetone.

(b) $CH_3CH_2\overset{H}{\overset{|}{C}}{=}O$ and $CH_2{=}CH\overset{H}{\overset{|}{C}}{=}O$.

Bromine will decolorize immediately with the second compound (propenal) but not with propanal.

(c) $C_6H_5CH_2CH_2OH$ and $C_6H_5\underset{OH}{\underset{|}{C}}HCH_3$

Oxidize both compounds: 2-phenylethanol will give 2-phenylethanal and 1-phenylethanol will give methyl phenyl ketone. 1-phenylethanal will give a positive Tollens or Benedict test and methyl phenyl ketone will not give a positive test.

32. $$CH_3\overset{H}{\overset{|}{C}}{=}O + HCN \longrightarrow CH_3\underset{OH}{\underset{|}{\overset{H}{\overset{|}{C}}}}{-}CN \xrightarrow[H^+]{H_2O} CH_3\underset{OH}{\underset{|}{C}}H{-}COOH$$

33. $$C_6H_5\overset{OH}{\overset{|}{C}}HC{\equiv}N \longrightarrow C_6H_5\overset{O}{\overset{\|}{C}}{-}H + HCN$$

benzaldehyde

34. Pyruvic acid is changed to lactic acid by a reduction reaction.

35. The structure of the product is $$\underset{}{\overset{OH}{\overset{|}{C}}H_2}\underset{OH}{\underset{|}{C}}H\overset{OH}{\overset{|}{C}}H\underset{OH}{\underset{|}{C}}H\overset{O}{\overset{\|}{C}}\underset{OH}{\underset{|}{C}}H_2$$

36. The alcohols which should be oxidized to give these ketones.

(a) 3-pentanone: 3-pentanol $CH_3CH_2CH(OH)CH_2CH_3$

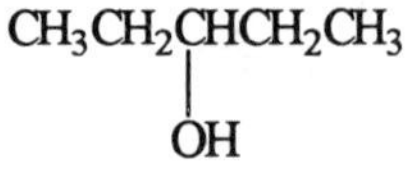

(b) methyl ethyl ketone: 2-butanol $CH_3CH(OH)CH_2CH_3$

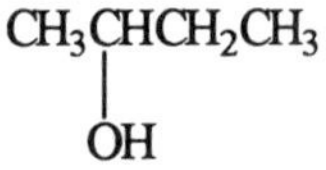

(c) 4-phenyl-2-butanone: 4-phenyl-2-butanol $C_6H_5-CH_2CH_2CH(OH)CH_3$

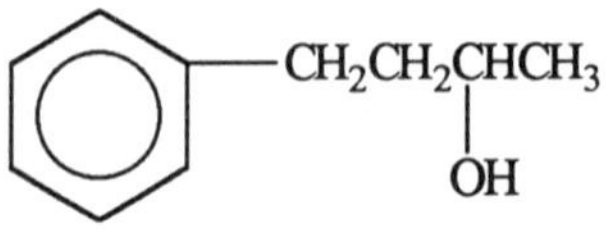

37. Isomeric aldehydes and ketones of formula $C_6H_{12}O$

<u>Aldehydes</u>

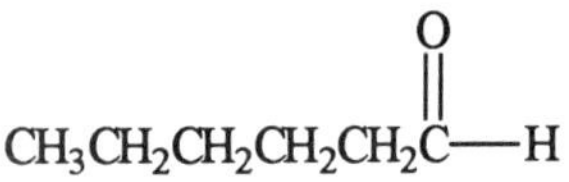

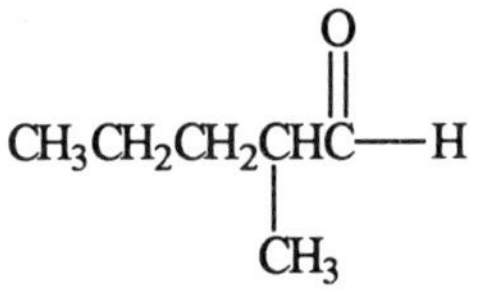

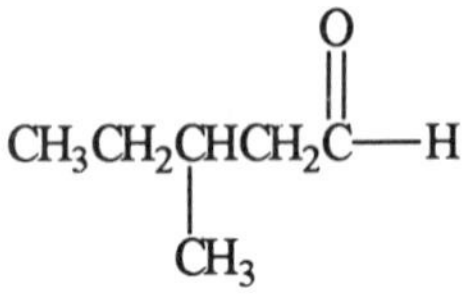

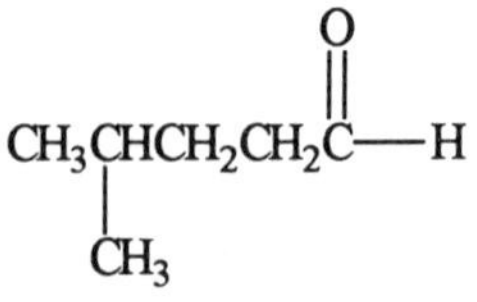

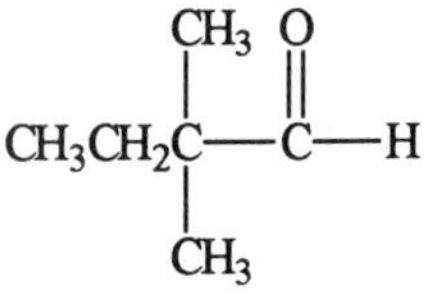

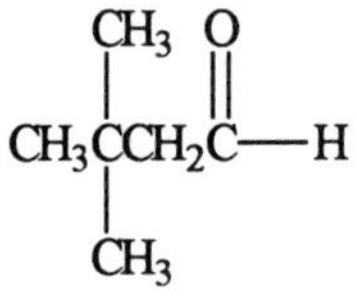

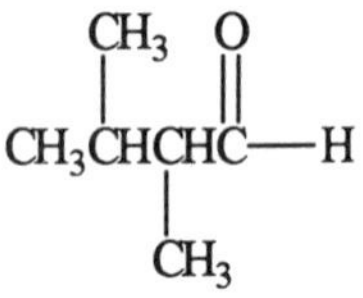

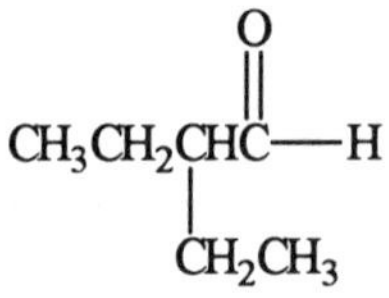

$CH_3CH_2CH_2CH_2CH_2C(=O)-H$

$CH_3CH_2CH_2CH(CH_3)C(=O)-H$

$CH_3CH_2CH(CH_3)CH_2C(=O)-H$

$CH_3CH(CH_3)CH_2CH_2C(=O)-H$

$CH_3CH_2C(CH_3)_2-C(=O)-H$

$CH_3C(CH_3)_2CH_2C(=O)-H$

$CH_3CH(CH_3)CH(CH_3)C(=O)-H$

$CH_3CH_2CH(CH_2CH_3)C(=O)-H$

Ketones

$CH_3CH_2CH_2CH_2C(=O)CH_3$ $\qquad$ $CH_3CH_2CH_2C(=O)CH_2CH_3$

$CH_3CH_2CH(CH_3)C(=O)CH_3$ $\qquad$ $CH_3CH(CH_3)CH_2C(=O)CH_3$

$CH_3CH_2C(=O)CH(CH_3)CH_3$ $\qquad$ $CH_3C(CH_3)_2C(=O)CH_3$

38. Benzaldehyde isomers of formula $C_9H_{10}O$.

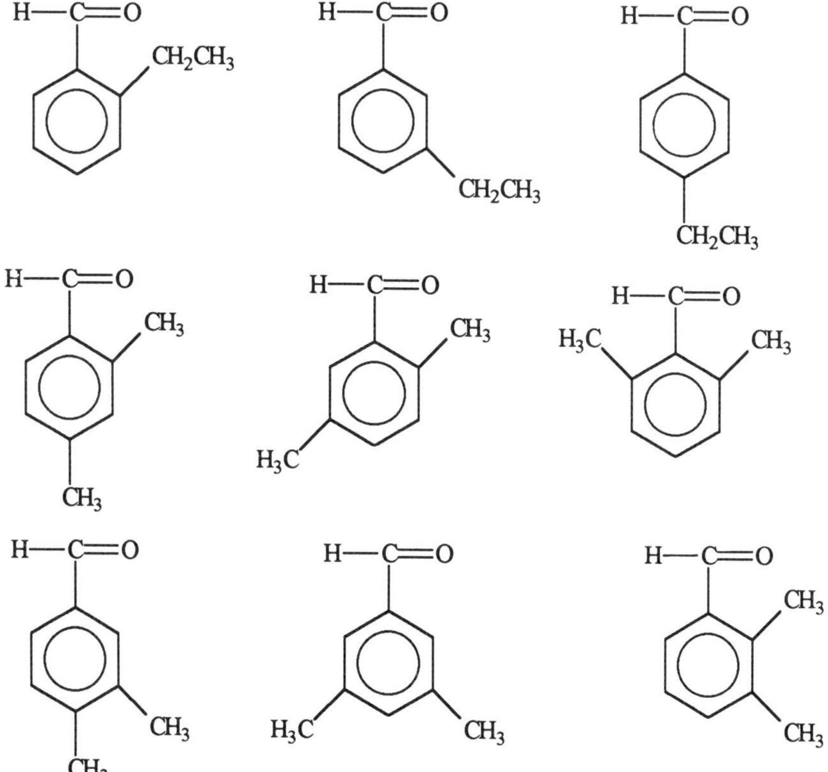

CHAPTER 24

CARBOXYLIC ACIDS AND ESTERS

1. (a) CH_3COOH

(b) 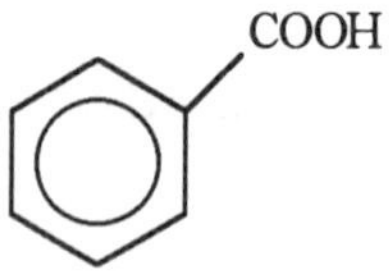 C_6H_5COOH (benzene ring with COOH)

(c) $CH_3CH(OH)COOH$

(d) $CH_3CH(NH_2)COOH$

(e) $CH_2(Cl)CH_2COOH$

(f) $HOOC{-}COOH$

(g) $CH_2{=}CHCOOH$

(h) $CH_3C(=O){-}OCH_3$

(i) CH_3CN

(j) CH_3COONa

(k) $CH_3C(=O){-}Cl$

(l) 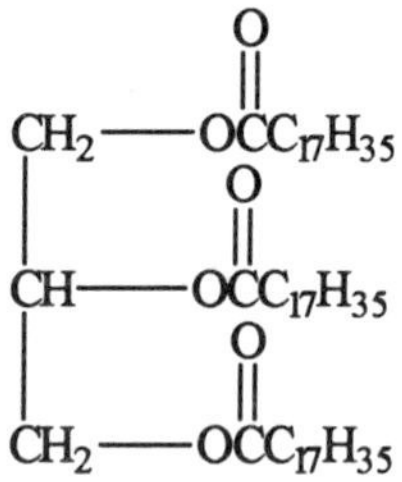 $CH_2{-}OC(=O)C_{17}H_{35}$ / $CH{-}OC(=O)C_{17}H_{35}$ / $CH_2{-}OC(=O)C_{17}H_{35}$

(m) $CH_3(CH_2)_{16}COONa$

2. The butyric acid solution would be expected to have the more objectionable odor because salts normally exhibit little or no odor. Salts are ionic and therefore have low volatility. For example, dilute solutions of acetic acid (vinegar) have considerable odor, but sodium acetate does not.

3. Which has the greater solubility in water?

   (a) Propanoic acid is more soluble than methyl propanoate.
   (b) Sodium palmitate is more soluble than palmitic acid.
   (c) Sodium stearate is more soluble than barium stearate.
   (d) Sodium phenoxide is more soluble than phenol.

4. (a) The major difference between fats and oils is that fats are solids at room temperature, oils are liquids. The fatty acids in the molecules are mostly saturated in fats; more unsaturated in oils.

(b) Soaps are the sodium salts of high molar mass fatty acids. Syndets are synthetic detergents and occur in several different forms. They have cleansing action similar to soaps, but have different structural and solubility characteristics, such as being soluble in hard water. Some syndets also contain long hydrocarbon chains.

(c) Hydrolysis is the breaking apart of an ester in the presence of water to form an alcohol and a carboxylic acid. Mineral acids or digestive enzymes are used to speed the process. Saponification breaks the ester apart using sodium hydroxide to form alcohols and salts of carboxylic acids.

5. Like a soap, a detergent contains a grease soluble component and a water soluble component. The grease soluble component of many molecules dissolves in the grease film. The water soluble components attract water, causing small droplets of grease-bearing dirt to break loose and float away.

6. The principal advantage of synthetic detergents over soap is that the syndets do not form insoluble precipitates with the ions in hard water (Ca^{2+}, Mg^{2+}, Fe^{3+}).

7. Medicinal effects of aspirin. Aspirin acts in the body as an antipyretic, an analgesic, and as an antiinflammatory agent.

   Risks of using aspirin. Irritation of the stomach lining; inhibit blood clotting; can prolong labor in giving birth; can bring on the development of Reyes syndrome.

8. Methyl salicylate (pain relieving liniments)
   Phenyl salicylate (protective coating for pills)

9. The most common substitutes for aspirin are acetaminophen and Ibuprofen. Acetaminophen acts as an analgesic and as an antipyretic. Ibuprofen acts as a prostaglandin inhibitor.

10. The correct statements are: a, b, f, g, h, i, j, k, l, m, n, p. q.

   (c) The name for $CH_3CH_2CHBrCH_2COOH$ is β-bromovaleric acid.
   (d) Acetic acid is a weaker acid than hydrochloric acid.
   (e) Sodium benzoate is more soluble in water than benzoic acid.
   (o) The presence of unsaturation in the acid component of a fat tends to lower its melting point compared with the corresponding saturated compound.
   (r) Phosphoric anhydride serves as a temporary store for metabolic energy.
   (s) The hydrophobic end of a detergent is not water soluble.
   (t) Saponification is the hydrolysis of a fat or an ester in a basic medium.
   (u) Methyl salicylate is an ester.

11. Names

(a) hexanoic acid

(b) fumaric acid

(c) o-phthalic acid

(d) stearic acid

(e) sodium propanoate

12. Names

(a) o-chlorobenzoic acid

(b) oleic acid

(c) m-toluic acid

(d) 2-hydroxybutanoic acid

(e) ammonium propanoate

13. Structures

(a) $CH_3(CH_2)_4COOH$ hexanoic acid

(b) 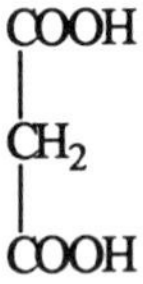 COOH–CH_2–COOH malonic acid

(c) 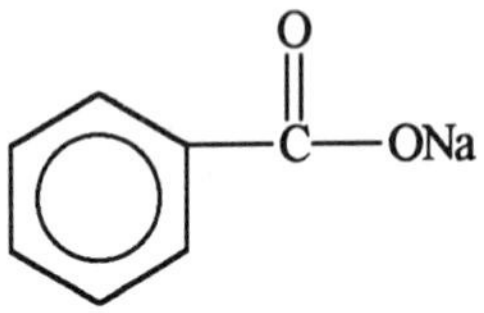 C_6H_5–C(=O)–ONa sodium benzoate

(d) 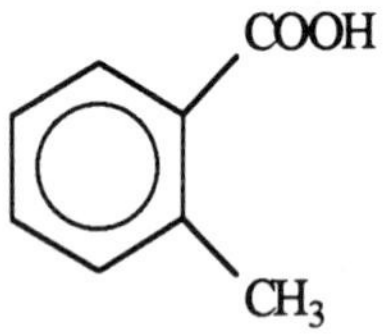 o-toluic acid (benzene ring with COOH and CH_3 on adjacent carbons)

(e) $CH_3(CH_2)_{16}COOH$ stearic acid

(f) $CH_3CH(Cl)COOH$ 2-chloropropanoic acid

(g) $CH_3CH_2CH_2COOK$ potassium butyrate

14. Structures

(a) $HOOC{-}COOH$ oxalic acid

(b) $CH_3CH_2CH_2CH_2COOH$ pentanoic acid

(c) $C_6H_4(COOH)_2$ (1,2-) o-phthalic acid

(d) $CH_3CH_2CH{=}CHCH_2CH{=}CHCH_2CH{=}CH(CH_2)_7COOH$ linolenic acid

(e) COONa

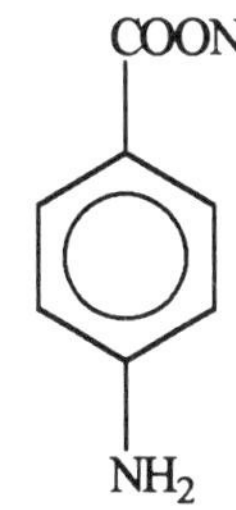

NH_2

p-amino-sodium benzoate

(f) $CH_3CH_2COONH_4$ ammonium propanoate

(g) $CH_3CH(OH)CH_2COOH$ β-hydroxybutyric acid

15. Increasing pH means increasing basicity

HCl < CH_3COOH < phenol (C_6H_5OH) < NaCl < NH_3 < NaOH

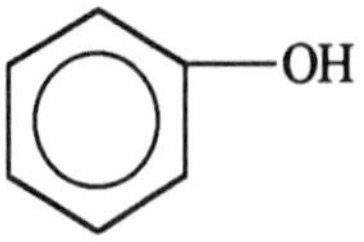

Increasing pH ⟶

16. Decreasing pH means increasing acidity

KOH > NH_3 > KBr > phenol (C_6H_5OH) > HCOOH > HBr

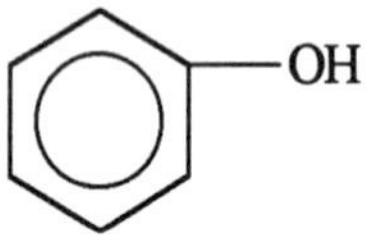

Decreasing pH ⟶

17. IUPAC and common names

(a) methyl propenoate — methyl acrylate

(b) ethyl butanoate — ethyl butyrate

(c) phenyl-2-hydroxybenzoate — phenyl salicylate

18. IUPAC and common names

(a) methyl methanoate — methyl formate

(b) propyl benzoate — propyl benzoate

(c) ethyl propanoate — ethyl propionate

19. Structural formulas

(a) 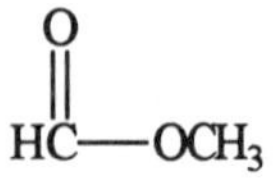 $HC(=O)—OCH_3$

(b) 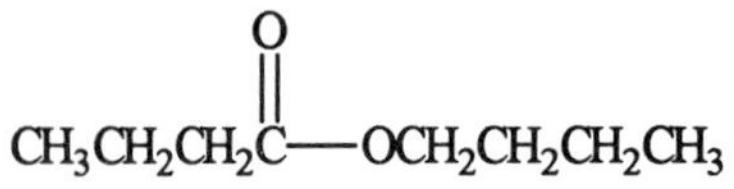 $CH_3CH_2CH_2C(=O)—OCH_2CH_2CH_2CH_3$

(c) $C_6H_5C(=O){-}OCH_2CH_3$

20. Structural formulas

(a) $CH_3C(=O){-}OCH_2CH_2CH_3$

(b) $C_6H_5C(=O){-}OCH_3$

(c) $CH_3CH_2CH_2CH_2CH_2C(=O){-}OCH_2CH_3$

21. (a) $CH_3(CH_2)_7CH{=}CH(CH_2)_7COOH + H_2 \xrightarrow{Ni} CH_3(CH_2)_{16}COOH$

stearic acid

(b) $C_6H_5CHO \xrightarrow[H_2SO_4]{Na_2Cr_2O_7} C_6H_5COOH$

benzoic acid

(c) $CH_3CH_2COOH + NaOH \longrightarrow CH_3CH_2COONa + H_2O$

sodium propanoate

(d) $CH_3(CH_2)_3CH_2C(=O){-}OCH_2CH_2CH_3 + NaOH \xrightarrow{\Delta}$

$CH_3(CH_2)_3CH_2COONa + CH_3CH_2CH_2OH$

sodium hexanoate 1-propanol

(e) $C_6H_5CH_2CH_2CH_3 \xrightarrow[\text{NaOH, }\Delta]{NaMnO_4} C_6H_5COONa + CH_3COONa$

sodium benzoate   sodium acetate

22. (a) $m\text{-}CH_3C_6H_4CH_2CH_3 \xrightarrow[\text{NaOH, }\Delta]{NaMnO_4} m\text{-}C_6H_4(COONa)_2$

sodium-m-phthalate

(b) $HOOCCH{=}CHCOOH + H_2 \xrightarrow{Ni} HOOCCH_2CH_2COOH$

succinic acid

(c) $CH_3(CH_2)_7CH{=}CH(CH_2)_7COOH + NaOH \longrightarrow$

$CH_3(CH_2)_7CH{=}CH(CH_2)_7COONa + H_2O$

sodium oleate

(d) $CH_3CH_2CH_2\overset{O}{\overset{\|}{C}}{-}H \xrightarrow[H_2SO_4]{Na_2Cr_2O_7} CH_2CH_2CH_2COOH$

butanoic acid

(e) $C_6H_5\overset{O}{\overset{\|}{C}}{-}O{-}C_6H_5 + NaOH \xrightarrow{\Delta} C_6H_5COONa + C_6H_5ONa$

sodium benzoate   sodium phenolate

23. Structural formula for the ester that when hydrolyzed would yield:

(a) methanol and acetic acid

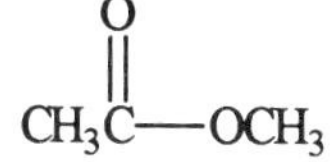

(b) ethanol and formic acid $HC(=O)—OCH_2CH_3$

(c) 2-propanol and benzoic acid $C_6H_5—C(=O)—O—CH(CH_3)—CH_3$

24. Structural formula for the ester that when hydrolyzed would yield:

(a) methanol and propanoic acid $CH_3CH_2C(=O)—OCH_3$

(b) 1-octanol and acetic acid $CH_3C(=O)—OCH_2(CH_2)_6CH_3$

(c) ethanol and butanoic acid $CH_3CH_2CH_2C(=O)—OCH_2CH_3$

25. Structural formulas for the reactants that will yield the following esters:

(a) methyl palmitate $CH_3OH + CH_3(CH_2)_{14}COOH$

(b) phenyl propionate $C_6H_5OH + CH_3CH_2COOH$

(c) dimethyl succinate $CH_3OH + HOOCCH_2CH_2COOH$

26. Structural formulas for the reactants that will yield the following esters:

(a) isopropyl formate

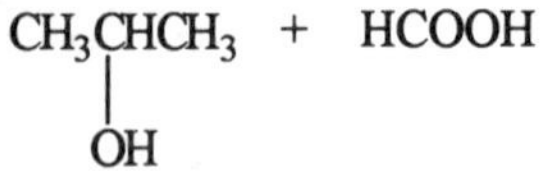

$$\underset{\displaystyle OH}{CH_3\overset{|}{C}HCH_3} + HCOOH$$

(b) diethyl adipate

$$CH_3CH_2OH + HOOC(CH_2)_4COOH$$

(c) benzyl benzoate

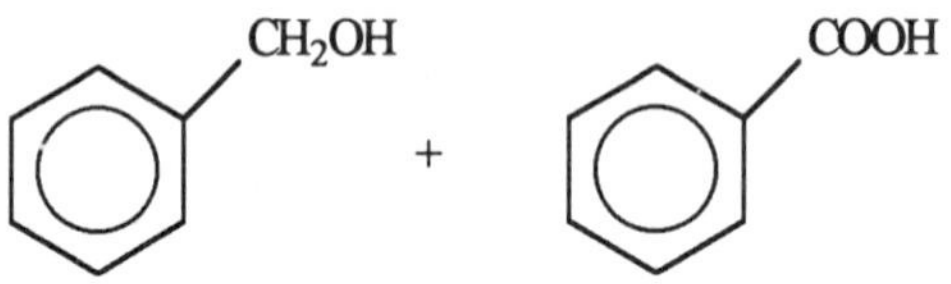

$$C_6H_5CH_2OH + C_6H_5COOH$$

27. Structural formulas for the organic products formed in the following reactions:

(a)

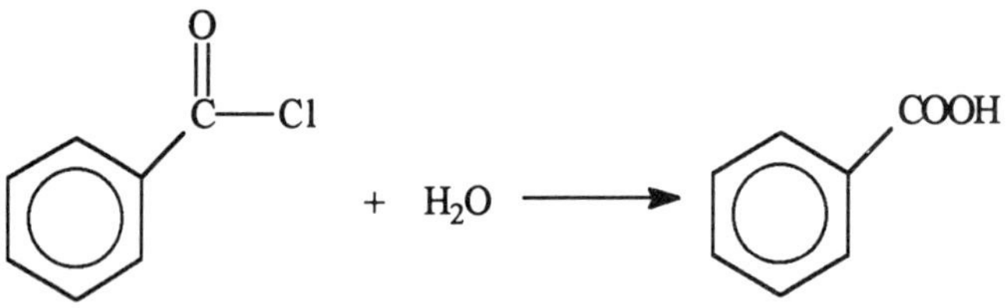

$$C_6H_5\overset{\displaystyle O}{\overset{\|}{C}}—Cl + H_2O \longrightarrow C_6H_5COOH$$

(b) $CH_2{=}CHCOOH + Br_2 \longrightarrow CH_2BrCHBrCOOH$

(c)

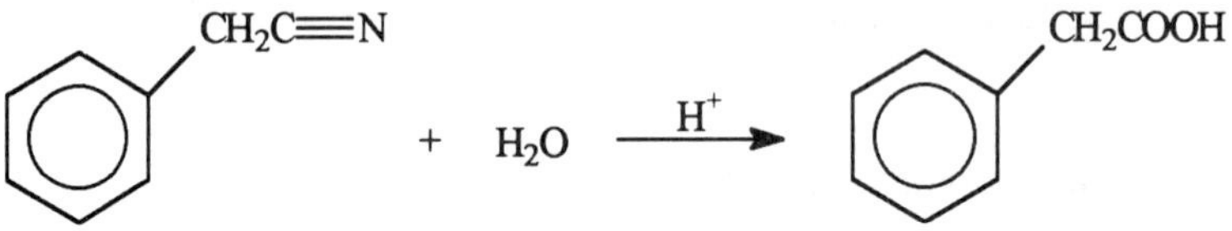

$$C_6H_5CH_2C{\equiv}N + H_2O \xrightarrow{H^+} C_6H_5CH_2COOH$$

(d)

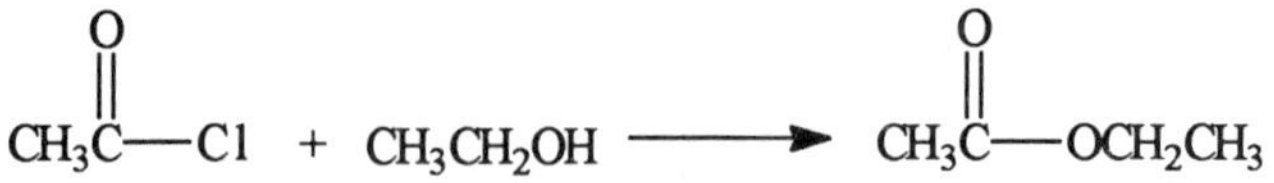

$$CH_3\overset{\displaystyle O}{\overset{\|}{C}}—Cl + CH_3CH_2OH \longrightarrow CH_3\overset{\displaystyle O}{\overset{\|}{C}}—OCH_2CH_3$$

(e)

$$CH_3\overset{\displaystyle O}{\overset{\|}{C}}—Cl + NH_3 \longrightarrow CH_3\overset{\displaystyle O}{\overset{\|}{C}}—NH_2$$

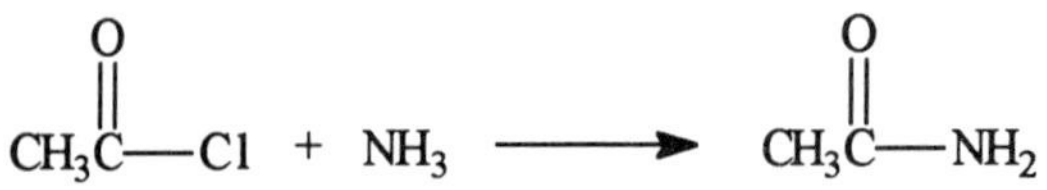

28. Structural formulas for the organic products formed in the following reactions:

(a) $C_6H_5COCl + NH_3 \longrightarrow C_6H_5CONH_2$

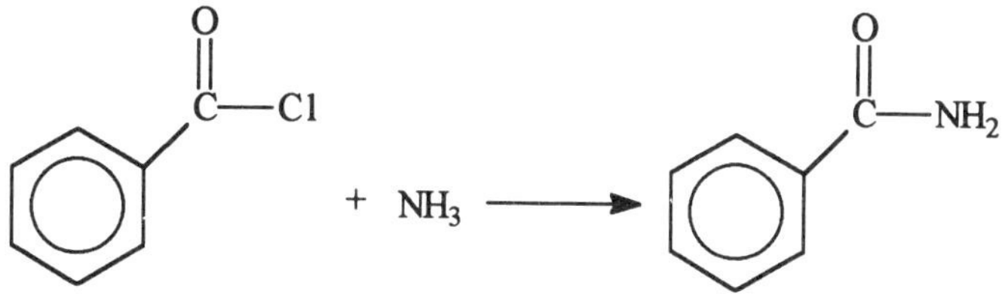

(b) $CH_3CH_2COOH + SOCl_2 \longrightarrow CH_3CH_2C(=O){-}Cl$

(c) $CH_3CH_2C(=O){-}Cl + CH_3OH \longrightarrow CH_3CH_2C(=O){-}OCH_3$

(d) $CH_3CH_2CH_2C{\equiv}N + H_2O \xrightarrow{H^+} CH_3CH_2CH_2COOH$

(e) $C_6H_5CH_2CH_2COOH + SOCl_2 \longrightarrow C_6H_5CH_2CH_2C(=O){-}Cl$

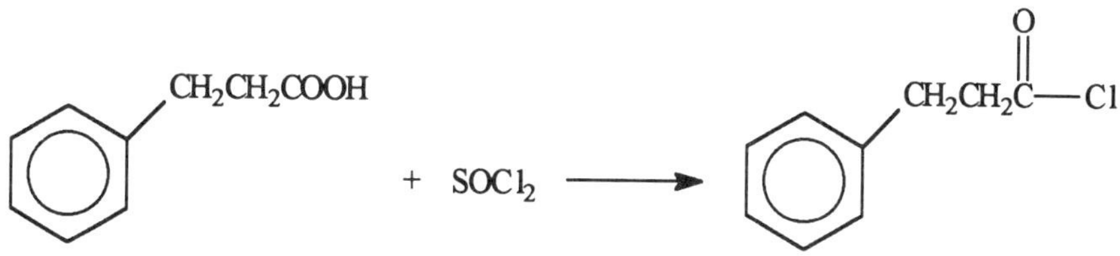

29. Simple test to distinguish between:

(a) Sodium benzoate and benzoic acid: Sodium benzoate is water soluble; benzoic acid is not.

(b) Maleic acid and malonic acid. Maleic acid has a carbon-carbon double bond, it will readily add and decolorize bromine; malonic acid will not decolorize bromine.

30. Simple tests to distinguish between:

(a) benzoic acid and ethyl benzoate; benzoic acid is an odorless solid; ethyl benzoate is a fragrant liquid.

(b) succinic acid and fumaric acid: fumaric acid has a carbon-carbon double bond and will readily add and decolorize bromine; succinic acid will not decolorize bromine.

31. Structural formulas for the products of the following reactions:

(a) $CH_3CH(COOH)_2 \xrightarrow{150^\circ C} CH_3CH_2COOH + CO_2$

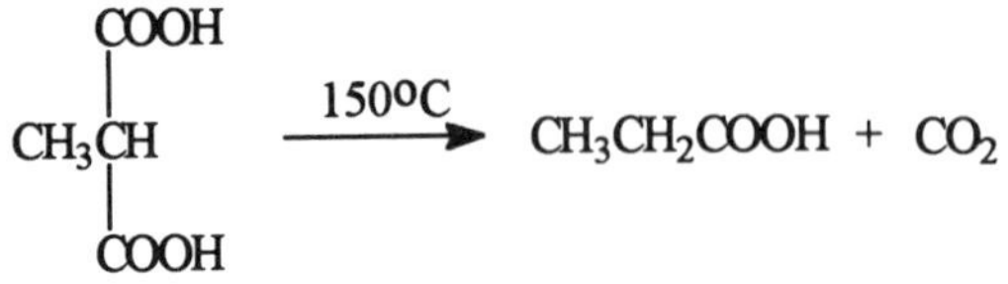

(b) $C_6H_5-\overset{O}{\overset{\|}{C}}-OCH_2CH_3 + H_2O \xrightarrow{H^+} C_6H_5COOH + CH_3CH_2OH$

(c) $CH_3CH_2COOH + C_6H_5OH \xrightarrow{H^+} CH_3CH_2\overset{O}{\overset{\|}{C}}-O-C_6H_5$

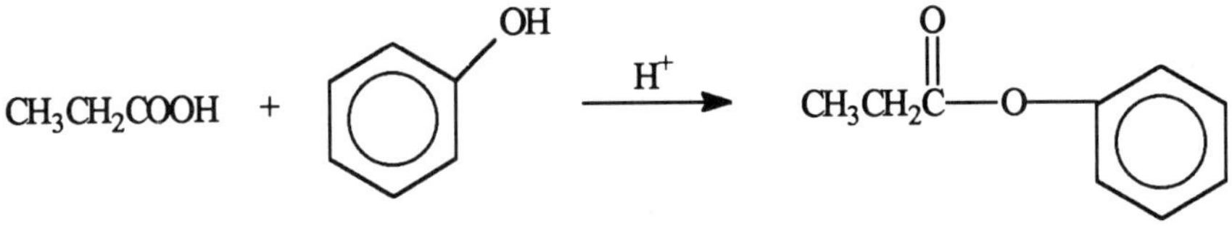

32. (a) $HOOC-CH_2-CH_2-CH_2-COOH \xrightarrow{\Delta}$ cyclic anhydride (CH_2–CH_2–CH_2 with two $C=O$ joined through O)

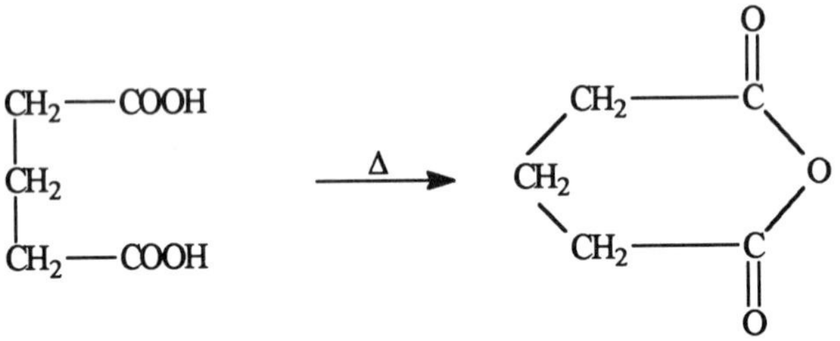

glutaric anhydride

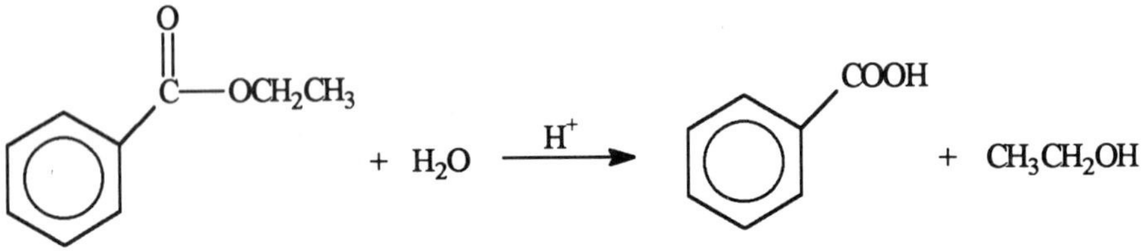

(b) $HOOC-CH_2-COOH + 2\ CH_3CH_2OH \xrightarrow{H^+} CH_3CH_2OOC-CH_2-COOCH_2CH_3$

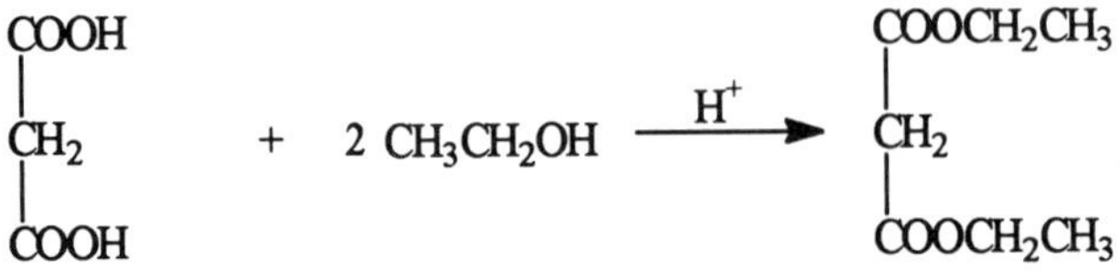

diethyl malonate

(c) $C_6H_5O\text{—}\overset{\displaystyle O}{\overset{\|}{C}}CH_3 + H_2O \xrightarrow[\Delta]{H^+} C_6H_5OH + CH_3COOH$

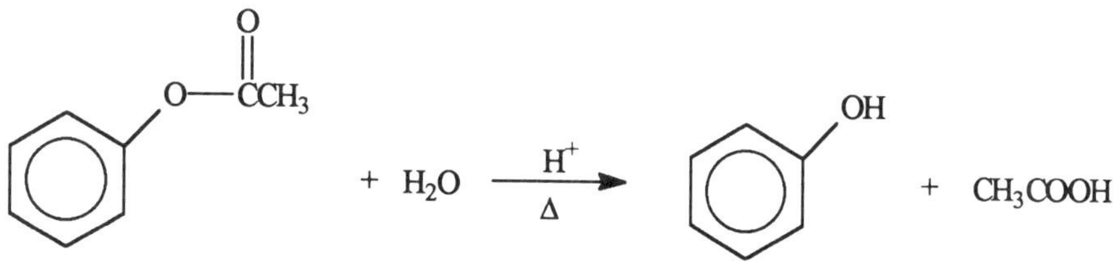

phenol acetic acid

33. Structure of *cis*, *cis*, *cis*-linolenic acid:

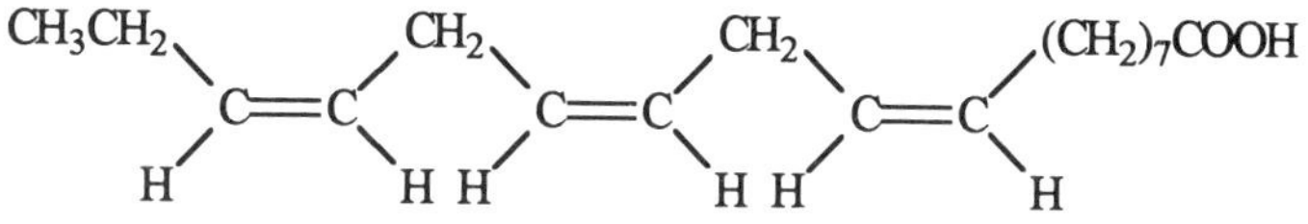

$CH_3CH_2(H)C{=}C(H)CH_2(H)C{=}C(H)CH_2(H)C{=}C(H)(CH_2)_7COOH$

Structure of *trans, trans, trans*-linolenic acid:

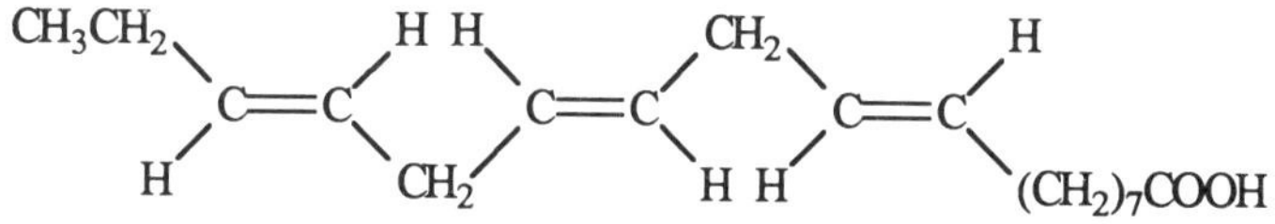

$CH_3CH_2(H)C{=}C(H)CH_2(H)C{=}C(H)CH_2(H)C{=}C(H)(CH_2)_7COOH$

34. (a) *cis*-oleic acid

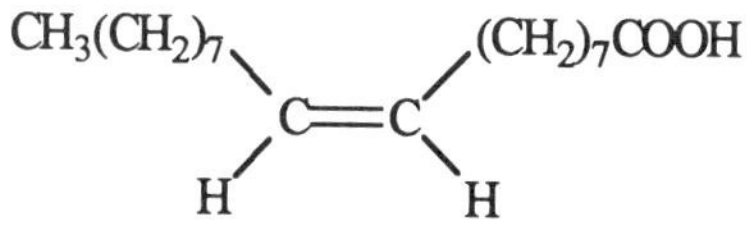

$CH_3(CH_2)_7(H)C{=}C(H)(CH_2)_7COOH$

(b) *cis, cis*-linoleic acid

$CH_3(CH_2)_4(H)C{=}C(H)CH_2(H)C{=}C(H)(CH_2)_7COOH$

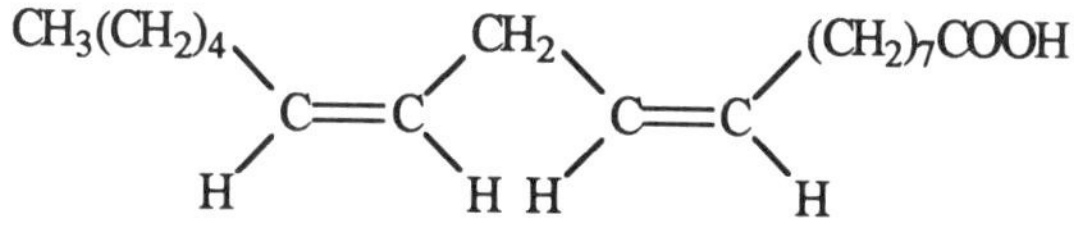

35. $CH_3(CH_2)_{12}COONa$ would be more useful than $CH_3(CH_2)_{12}COOH$ as a cleansing agent in soft water. Both have a long hydrocarbon molecule which would dissolve in the fat, but the acid is not water soluble, whereas the sodium salt is soluble in water.

36. Sodium lauryl sulfate would be more effective as a detergent in hard water than sodium propyl sulfate because the hydrocarbon chain is only three carbons long in the latter, not long enough to dissolve grease well.

37. Only (a), hexadecyltrimethyl ammonium chloride would be a good detergent in water. It is cationic.

38. Only (c) $CH_3(CH_2)_{10}CH_2O(CH_2CH_2O)_7CH_2CH_2OH$, would be a good detergent in water. It is monionic.

39. (a)

$$\mathrm{HO{-}\overset{\overset{\displaystyle O}{\|}}{\underset{\underset{\displaystyle OH}{|}}{P}}{-}OH} + HOCH_2CH_3 \xrightarrow{H^+} \mathrm{HO{-}\overset{\overset{\displaystyle O}{\|}}{\underset{\underset{\displaystyle OH}{|}}{P}}{-}OCH_2CH_3} + H_2O$$

(b)

$$\mathrm{HO{-}\overset{\overset{\displaystyle O}{\|}}{\underset{\underset{\displaystyle OH}{|}}{P}}{-}OCH_2CH_3} + CH_3CH_2CH_2OH \xrightarrow{H^+} \mathrm{HO{-}\overset{\overset{\displaystyle O}{\|}}{\underset{\underset{\displaystyle OCH_2CH_2CH_3}{|}}{P}}{-}OCH_2CH_3} + H_2O$$

[from 39(a).]

40. (a)

$$\mathrm{HO{-}\overset{\overset{\displaystyle O}{\|}}{\underset{\underset{\displaystyle OH}{|}}{P}}{-}OH} + 3\ CH_3OH \xrightarrow{H^+} \mathrm{CH_3O{-}\overset{\overset{\displaystyle O}{\|}}{\underset{\underset{\displaystyle OCH_3}{|}}{P}}{-}OCH_3} + 3\ H_2O$$

(b)

$$\mathrm{HO{-}\overset{\overset{\displaystyle O}{\|}}{\underset{\underset{\displaystyle OH}{|}}{P}}{-}OH} + CH_3OH \xrightarrow{H^+} \mathrm{CH_3O{-}\overset{\overset{\displaystyle O}{\|}}{\underset{\underset{\displaystyle OH}{|}}{P}}{-}OH} + H_2O$$

$$\mathrm{CH_3O{-}\overset{\overset{\displaystyle O}{\|}}{\underset{\underset{\displaystyle OH}{|}}{P}}{-}OH} + \mathrm{CH_3CH_2\underset{\underset{\displaystyle OH}{|}}{C}HCH_3} \xrightarrow{H^+} \mathrm{CH_3O{-}\overset{\overset{\displaystyle O}{\|}}{\underset{\underset{\displaystyle OH}{|}}{P}}{-}O\overset{\overset{\displaystyle CH_3}{|}}{C}HCH_2CH_3}$$

41. Grams and moles of sodium benzoate:

$$0.001 \times 1\ \text{lb} \times \frac{454\ \text{g}}{1\ \text{lb}} = 0.5\ \text{g sodium benzoate}$$

$$0.5\ \text{g}\ NaC_7H_5O_2 \times \frac{1\ \text{mol}}{145.1\ \text{g}} = 3 \times 10^{-3}\ \text{mol}\ NaC_7H_5O_2$$

42.

$$\text{salicylic acid (o-}HOC_6H_4COOH) + CH_3\overset{O}{\overset{\|}{C}}-O-\overset{O}{\overset{\|}{C}}CH_3 \longrightarrow \text{o-}CH_3\overset{O}{\overset{\|}{C}}OC_6H_4COOH + CH_3COOH$$

salicylic acid | acetic anhydride | aspirin (acetylsalicylic acid) | acetic acid

43. The ester would be propyl propanoate $CH_3CH_2\overset{O}{\overset{\|}{C}}OCH_2CH_2CH_3$

Compound A is an acid; B is an alcohol. If B is oxidized to an acid which is the same as A, then A and B both have the same carbon structure, three carbon atoms each. The acid A must be propanoic acid, CH_3CH_2COOH. The alcohol can be 1-propanol or 2-propanol. Only 1-propanol can be oxidized to propanoic acid which is the same as compound A.

44. $Cl-\overset{O}{\overset{\|}{C}}-Cl$

The functional groups are acid chlorides.

The acid from which phosgene is derived is carbonic acid, $HO-\overset{O}{\overset{\|}{C}}-OH$

45.

$$CH_2-O-\overset{O}{\overset{\|}{C}}(CH_2)_7CH{=}CH(CH_2)_7CH_3$$
$$CH-O-\overset{O}{\overset{\|}{C}}(CH_2)_7CH{=}CHCH_2CH{=}CH(CH_2)_4CH_3$$
$$CH_2-O-\overset{O}{\overset{\|}{C}}(CH_2)_7CH{=}CHCH_2CH{=}CH(CH_2)_4CH_3$$

soybean oil

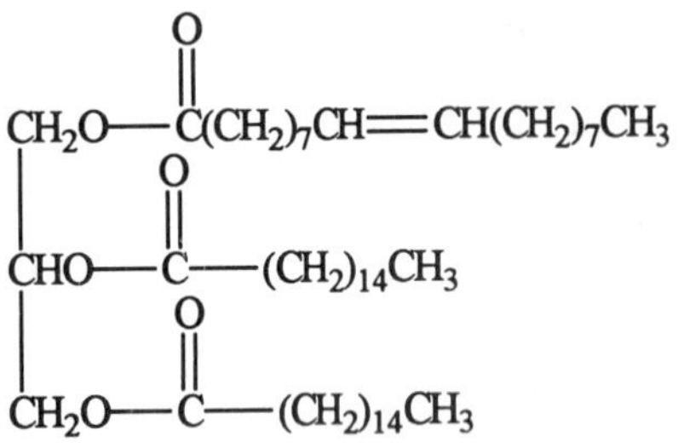

$$\begin{array}{l} \quad\quad\quad\;\; O \\ \quad\quad\quad\;\; \| \\ CH_2O-C(CH_2)_7CH{=}CH(CH_2)_7CH_3 \\ \;|\quad\quad\quad\; O \\ \;|\quad\quad\quad\; \| \\ CHO-C-(CH_2)_{14}CH_3 \\ \;|\quad\quad\quad\; O \\ \;|\quad\quad\quad\; \| \\ CH_2O-C-(CH_2)_{14}CH_3 \end{array}$$

palm oil

The bonded order of the fatty acids to glycerol may vary.

46.

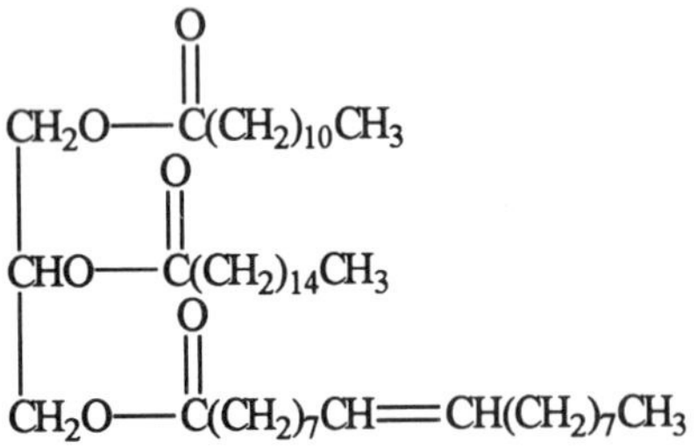

$$\begin{array}{l} \quad\quad\quad\;\; O \\ \quad\quad\quad\;\; \| \\ CH_2O-C(CH_2)_{10}CH_3 \\ \;|\quad\quad\quad\; O \\ \;|\quad\quad\quad\; \| \\ CHO-C(CH_2)_{14}CH_3 \\ \;|\quad\quad\quad\; O \\ \;|\quad\quad\quad\; \| \\ CH_2O-C(CH_2)_7CH{=}CH(CH_2)_7CH_3 \end{array}$$

There would be two other triacylglcerols containing all three of these acids. Each of the three acids can be attached to the middle carbon of the glycerol.

47. Names and formulas of products are

(a)

CH_2OH	and	$CH_3(CH_2)_{10}COOH$	lauric acid
$\vert$ $CHOH$		$CH_3(CH_2)_{14}COOH$	palmitic acid
$\vert$ CH_2OH		$CH_3(CH_2)_7CH{=}CH(CH_2)_7COOH$	oleic acid
glycerol			

(b)

CH_2OH	and	$CH_3(CH_2)_{10}CH_2OH$	1-dodecanol (lauryl alcohol)
$\vert$ $CHOH$		$CH_3(CH_2)_{14}CH_2OH$	1-hexadecanol (cetyl alcohol)
$\vert$ CH_2OH		$CH_3(CH_2)_{16}CH_2OH$	1-octadecanol (stearyl alcohol)
glycerol			

(c)  glycerol and potassium salts:

$$\begin{array}{l} CH_2OH \\ | \\ CHOH \\ | \\ CH_2OH \end{array}$$

glycerol

and

$CH_3(CH_2)_{10}COOK$	potassium laurate
$CH_3(CH_2)_{14}COOK$	potassium palmitate
$CH_3(CH_2)_7CH{=}CH(CH_2)_7COOK$	potassium oleate

(d)

$$\begin{array}{l} CH_2O-\overset{O}{\overset{\|}{C}}(CH_2)_{10}CH_3 \\ | \\ CHO-\overset{O}{\overset{\|}{C}}(CH_2)_{14}CH_3 \\ | \\ CH_2O-\overset{O}{\overset{\|}{C}}(CH_2)_{16}CH_3 \end{array}$$

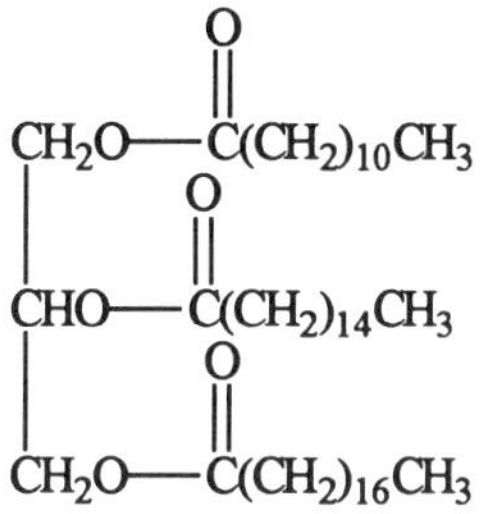

lauroylpalmitoylsteroylglycerol

48.

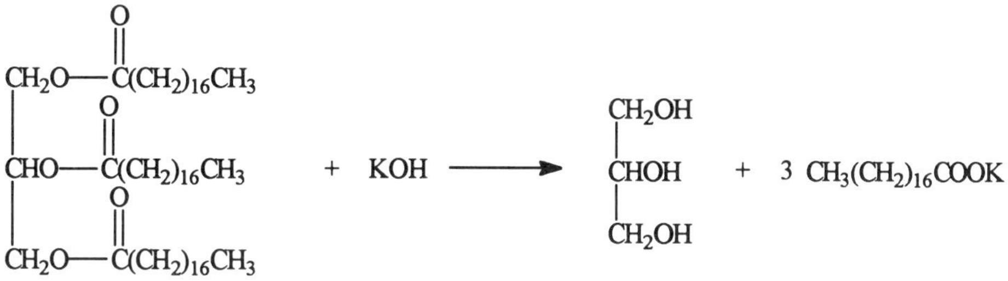

$$\begin{array}{l} CH_2O-\overset{O}{\overset{\|}{C}}(CH_2)_{16}CH_3 \\ | \\ CHO-\overset{O}{\overset{\|}{C}}(CH_2)_{16}CH_3 \\ | \\ CH_2O-\overset{O}{\overset{\|}{C}}(CH_2)_{16}CH_3 \end{array} + KOH \longrightarrow \begin{array}{l} CH_2OH \\ | \\ CHOH \\ | \\ CH_2OH \end{array} + 3\ CH_3(CH_2)_{16}COOK$$

glycerol tristearate → glycerol + potassium stearate

The solubility in water will change considerably. Glycerol tristearate is insoluble in water but the products glycerol and potassium stearate (a salt) are soluble in water.

49. Synthesis of:

(a) acetic acid

$$CH_3CH_2OH \xrightarrow[H_2SO_4]{K_2Cr_2O_7} CH_3COOH$$

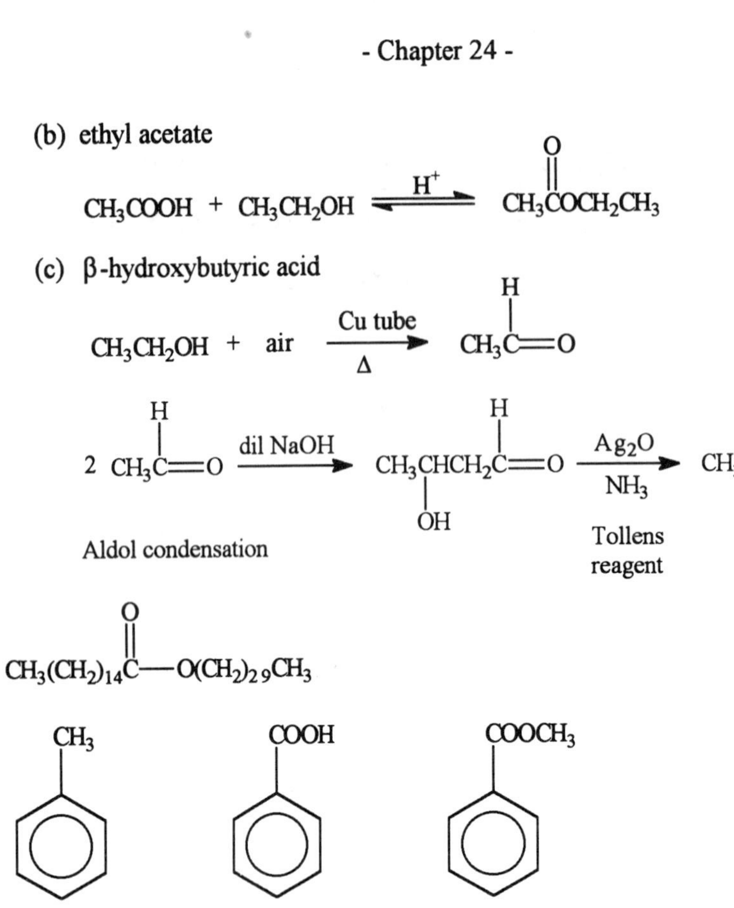

(b) ethyl acetate

$$CH_3COOH + CH_3CH_2OH \underset{}{\overset{H^+}{\rightleftharpoons}} CH_3\overset{O}{\overset{\|}{C}}OCH_2CH_3$$

(c) β-hydroxybutyric acid

$$CH_3CH_2OH + \text{air} \xrightarrow[\Delta]{\text{Cu tube}} CH_3\overset{H}{\overset{|}{C}}{=}O$$

$$2\ CH_3\overset{H}{\overset{|}{C}}{=}O \xrightarrow{\text{dil NaOH}} CH_3\underset{OH}{\underset{|}{C}}HCH_2\overset{H}{\overset{|}{C}}{=}O \xrightarrow[NH_3]{Ag_2O} CH_3\underset{OH}{\underset{|}{C}}HCH_2COOH$$

Aldol condensation

Tollens reagent

50. $CH_3(CH_2)_{14}\overset{O}{\overset{\|}{C}}{-}O(CH_2)_{29}CH_3$

51. A: toluene ($C_6H_5CH_3$) B: benzoic acid (C_6H_5COOH) C: methyl benzoate ($C_6H_5COOCH_3$)

A B C

52.

$$CH_3(CH_2)_{12}\overset{O}{\overset{\|}{C}}{-}O\underset{CH_3}{\underset{|}{C}}HCH_3$$

isopropyl myristate

53. $C_{10}H_{12}O_2 \longrightarrow C_7H_8O + C_3H_6O_2$

A (has a benzene ring) B (alcohol) C (acid)

$$CH_3CH_2\overset{O}{\overset{\|}{C}}{-}OCH_2{-}C_6H_5 \qquad C_6H_5CH_2OH \qquad CH_3CH_2COOH$$

A B C

54. Smell both samples. Butanoic acid has an unpleasant rancid odor; ethyl butanoate has a pleasant odor of pineapple.

55. Each molecule of triolein requires three molecules of H_2 (one for each double bond)

(a) $1.00 \text{ kg triolein} \times \frac{10^3 \text{ g}}{\text{kg}} \times \frac{1 \text{ mol triolein}}{885.4 \text{ g triolein}} \times \frac{3 \text{ mol } H_2}{1 \text{ mol triolein}} \times \frac{22.4 \text{ L}}{1 \text{ mol } H_2} = 75.9 \text{ L } H_2$

(b) $1.00 \text{ kg triolein} \times \frac{891.5 \text{ g tristearin}}{885.4 \text{ g triolein}} = 1.01 \text{ kg tristearin}$

56. The statement is false. When methyl propanoate is hydrolyzed, propanoic acid and methanol are formed.

57. Esters of formula $C_5H_{10}O_2$

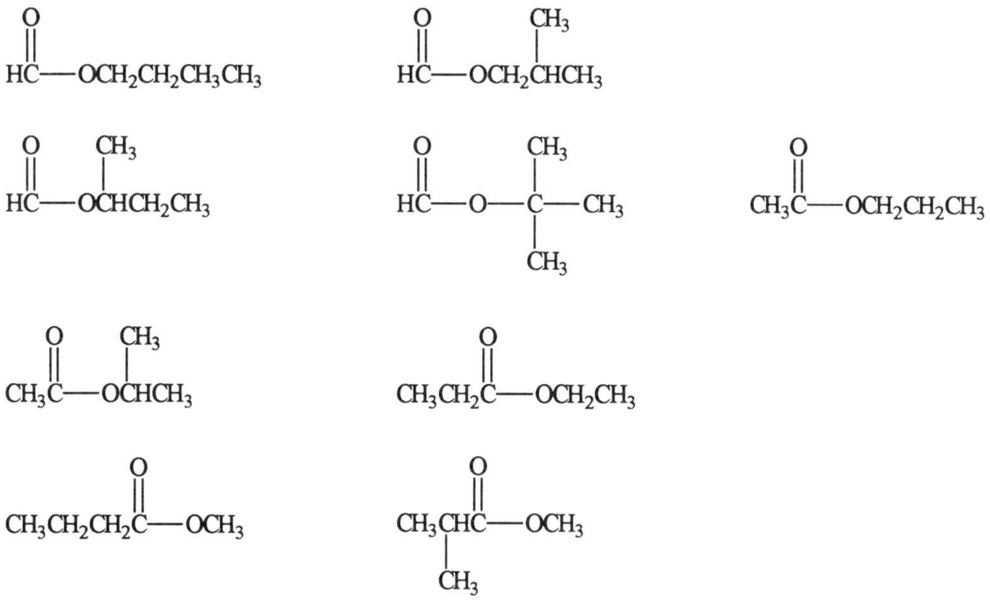

$$H\overset{\overset{\displaystyle O}{\|}}{C}\text{—}OCH_2CH_2CH_3CH_3$$

$$H\overset{\overset{\displaystyle O}{\|}}{C}\text{—}OCH_2\overset{\overset{\displaystyle CH_3}{|}}{C}HCH_3$$

$$H\overset{\overset{\displaystyle O}{\|}}{C}\text{—}O\overset{\overset{\displaystyle CH_3}{|}}{C}HCH_2CH_3$$

$$H\overset{\overset{\displaystyle O}{\|}}{C}\text{—}O\text{—}\underset{\underset{\displaystyle CH_3}{|}}{\overset{\overset{\displaystyle CH_3}{|}}{C}}\text{—}CH_3$$

$$CH_3\overset{\overset{\displaystyle O}{\|}}{C}\text{—}OCH_2CH_2CH_3$$

$$CH_3\overset{\overset{\displaystyle O}{\|}}{C}\text{—}O\overset{\overset{\displaystyle CH_3}{|}}{C}HCH_3$$

$$CH_3CH_2\overset{\overset{\displaystyle O}{\|}}{C}\text{—}OCH_2CH_3$$

$$CH_3CH_2CH_2\overset{\overset{\displaystyle O}{\|}}{C}\text{—}OCH_3$$

$$CH_3\underset{\underset{\displaystyle CH_3}{|}}{C}H\overset{\overset{\displaystyle O}{\|}}{C}\text{—}OCH_3$$

58. The common functional group that all three compounds (aspirin, acetaminophen, and ibuprofen) have in common is an aromatic ring.

CHAPTER 25

AMIDES AND AMINES: ORGANIC NITROGEN COMPOUNDS

1.

$$CH_3\overset{\overset{\displaystyle O}{\|}}{C}—NH_2$$

ethanamide

$$CH_3CH_2NH_2$$

ethyl amine

In an amide the NH_2 group is bonded to a carbonyl group. In an amine the NH_2 group is bonded to a saturated carbon atom.

2. Amides: Unsubstituted amides (except formamide) are solids at room temperature. Many are odorless and colorless. Low molar-mass amides are water soluble. Solubility in water decreases as the molar mass increases. Amides are neutral compounds. The NH_2 group is capable of hydrogen bonding.

   Amines: Low molar-mass amines are flammable gases with an ammonia-like odor. Aliphatic amines up to six carbon atoms are water soluble. Many amines have a "fishy" odor and many have very foul odors. Aromatic amines occur as liquids and solids. Soluble aliphatic amines give basic solutions. Aromatic amines are less soluble in water and less basic than aliphatic amines. The NH_2 group is capable of hydrogen bonding.

3. Amines and alcohols of similar molar masses have approximately the same water solubility due to the fact that both of these classes of compounds can hydrogen bond with water.

4. a) Heterocyclic compounds are those in which all the atoms in the ring are not alike.

   b) The number of heterocyclic rings in each of the compounds is:

   (i) purine, 2 (ii) ampicillin, 2 (iii) methadone, 0 (iv) nicotine, 2

5. Functional groups in:

   (a) procaine hydrochloride: aromatic amine; ester; quaternary ammonium ion

   (b) cocaine: 3° amine; two ester groups

   (c) nicotinamide: amide; heterocyclic amine

   (d) methamphetamine: 2° amine

6. The correct statements are: h, j, m, o, p.

   (a) The common name for CH_3CONH_2 is acetamide.

   (b) The general formula for a primary amine is RNH_2.

   (c) Urea is the primary way the body excretes nitrogen.

   (d) Heterocyclic compounds are ring compounds in which all the atoms in the ring are not the same.

   (e) Alkaloids are nitrogen containing alkaline compounds that are derived from plants and show physiological activity.

   (f) Barbiturates are synthetic drugs classified as sedatives.

   (g) Amphetamines are stimulants to the central nervous system.

   (i) The name for $[(CH_3)_2CH]_2NH$ is diisopropyl amine.

   (k) Most amines have unpleasant odors.

   (l) Lactic acid is an α-hydroxy acid.

   (n) Isopropylamine is a primary amine.

7. (a) NH_2 — benzene ring — CH_3

   (b) $CH_3CH(NH_2)CH_2CH_3$

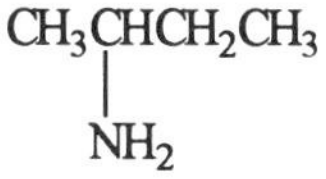

   $CH_3CHCH_2CH_3$
   NH_2

   (c) $O{=}C{-}NHCH_3$ — benzene ring — Br

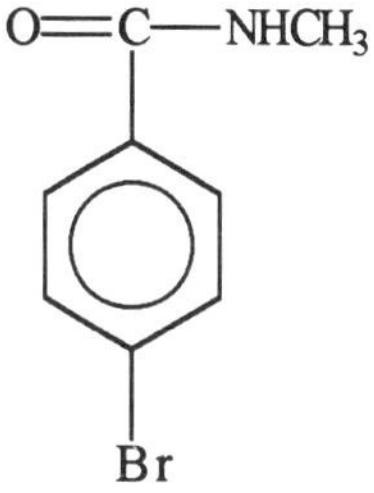

8. (a) $CH_3CH(NH_2)CH_2CHClCH_3$

   $CH_3CHCH_2CHClCH_3$
   NH_2

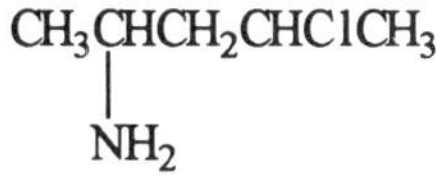

   (b) benzene ring — $C(=O){-}N(CH_2CH_3)_2$

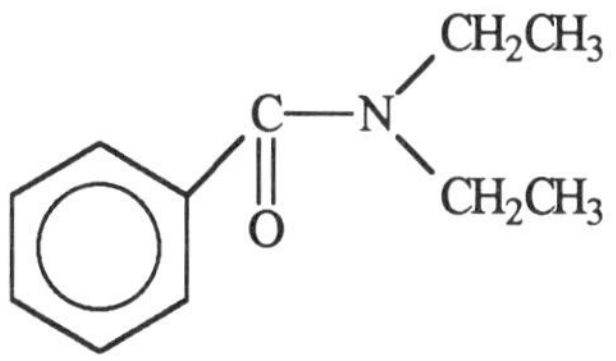

(c)

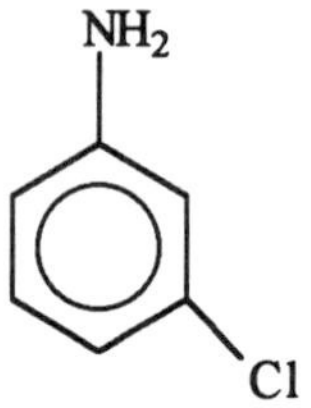

NH_2 on a benzene ring with Cl (meta)

9. Names

(a) acetamide
(b) 4-amino-2-pentanol
(c) m-methyl-N-methylbenzamide

10. Names

(a) 1-amino-3-methylbutane
(b) formamide
(c) N-phenylpropanamide

11. Increasing solubility in water.

$CH_3CH_2C(=O)—N(CH_3)_2 < CH_3CH_2C(=O)—NHCH_3 < CH_3CH_2C(=O)—NH_2$

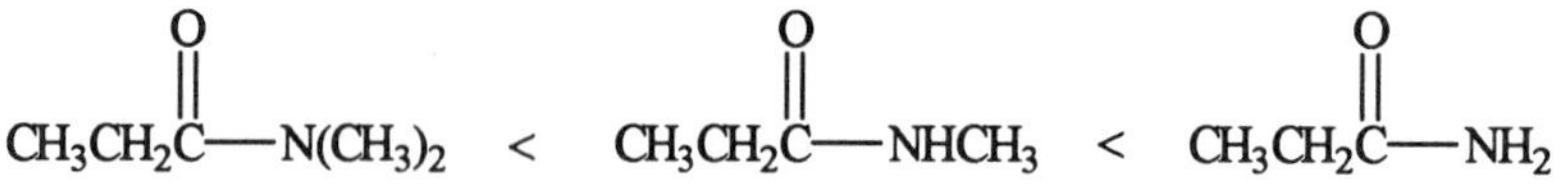

12. Increasing solubility in water.

$C_6H_5C(=O)—NH_2 < CH_3(CH_2)_4C(=O)—NH_2 < CH_3C(=O)—NH_2$

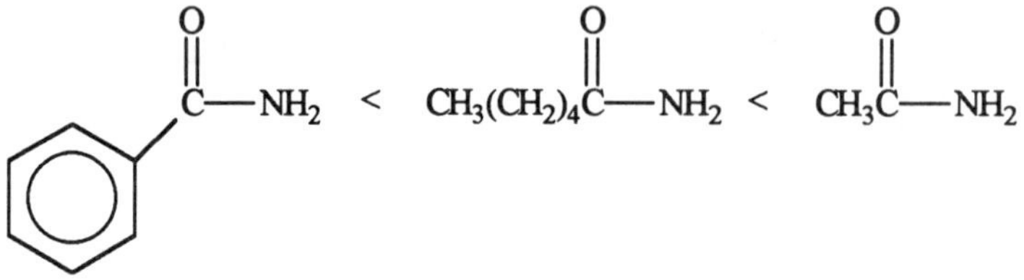

13. Hydrogen bonding in water.

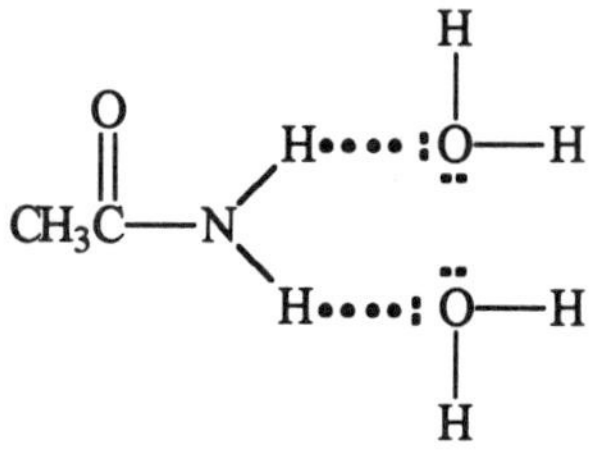

$CH_3C(=O)—N$ with H••••:O(H)—H and H••••:O(H)—H and $CH_3C(=O)—N(H)—H$ with :N••••H—Ö:(H)

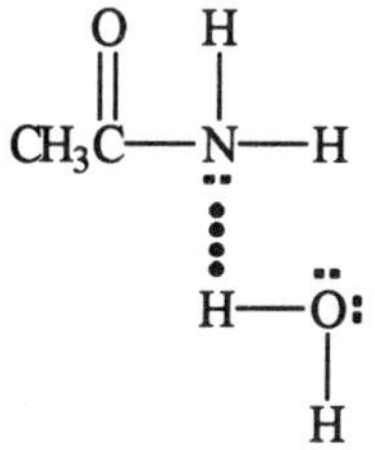

14. (Several possibilities) Hydrogen bonding in water.

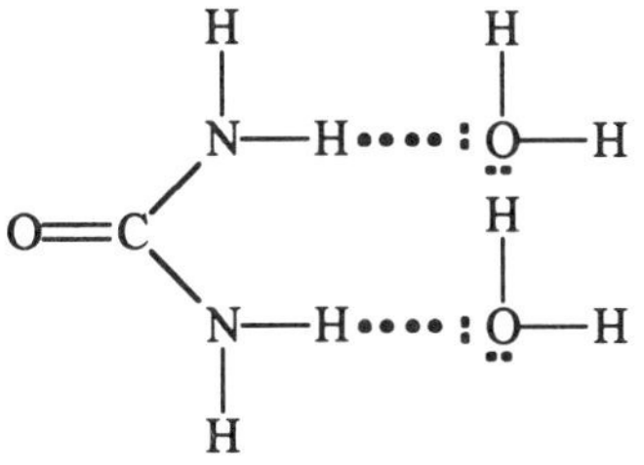

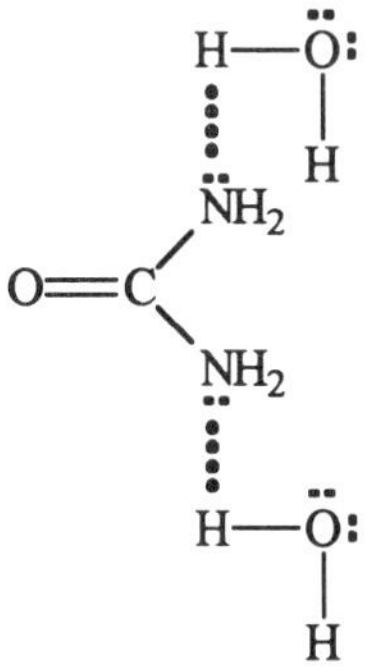

$$\text{O}=\text{C}\begin{cases}\text{N(H)}-\text{H}\cdots\cdots:\ddot{\text{O}}(\text{H})-\text{H} \\ \text{N(H)}-\text{H}\cdots\cdots:\ddot{\text{O}}(\text{H})-\text{H}\end{cases}$$

$$\text{O}=\text{C}\begin{cases}\ddot{\text{N}}\text{H}_2\cdots\cdots\text{H}-\ddot{\text{O}}:-\text{H} \\ \ddot{\text{N}}\text{H}_2\cdots\cdots\text{H}-\ddot{\text{O}}:-\text{H}\end{cases}$$

15. (a) neither acid or base
    (b) base
    (c) base
    (d) acid
    (e) neither acid or base
    (f) acid

16. (a) neither acid or base
    (b) base
    (c) base
    (d) neither acid or base
    (e) neither acid or base
    (f) acid

17. Organic products

(a) $CH_3C(=O)-OH + CH_3NH_3^+$

(b) $CH_3C(=O)-NHCH(CH_3)_2$

(c) $H_2NCH_2CH_2CH_2CH_2COONa$

18. Organic products

(a) $CH_3C(=O)-N(CH_2CH_3)_2$

(b) 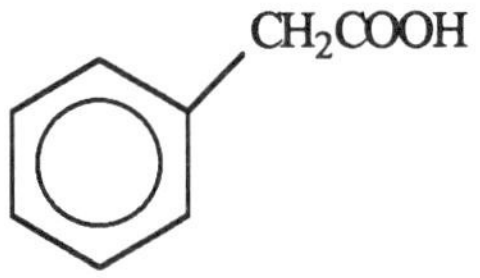 $C_6H_5CH_2COOH + CH_3\overset{+}{N}H_2CH_3$

(c) 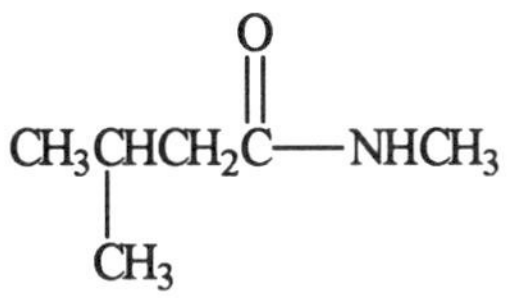 $CH_3CH(CH_3)CH_2C(=O)-NHCH_3$

19. Structures of amines with formula $C_4H_{11}N$.

$CH_3CH_2CH_2CH_2NH_2$ — 1°

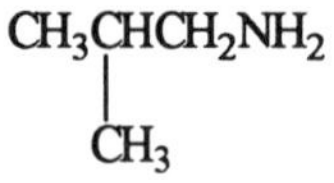

$CH_3CHCH_2NH_2$
|
CH_3
— 1°

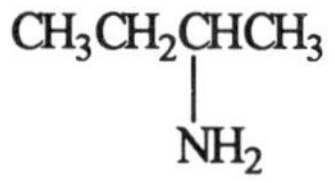

$CH_3CH_2CHCH_3$
|
NH_2
— 1°

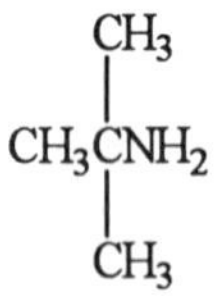

CH_3
|
CH_3CNH_2
|
CH_3
— 1°

$CH_3CH_2CH_2NHCH_3$ — 2°

$(CH_3)_2CHNHCH_3$ — 2°

$CH_3CH_2NHCH_2CH_3$ — 2°

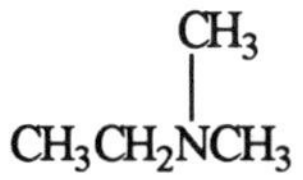

CH_3
|
$CH_3CH_2NCH_3$
— 3°

20. Structures of amines with formula C_3H_9N.

$CH_3CH_2CH_2NH_2$ — 1°

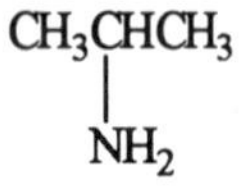

CH_3CHCH_3
|
NH_2
— 1°

$CH_3CH_2NHCH_3$ — 2°

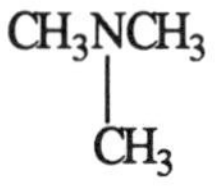

CH_3NCH_3
|
CH_3
— 3°

21. Classification of amines

(a) primary

(b) tertiary

(c) primary

22. Classification of amines

(a) secondary

(b) tertiary

(c) secondary

23. The triethylamine solution in 1.0 M NaOH would have the more objectionable odor because it would be in the form of the free amine, while in the acid solution the amine would form a salt that will have little or no odor.

24. The 2-aminopropane solution in 1.0 M KOH would have the more objectionable odor because it would be in the form of the free amine, while in the acid solution the amine would form a salt that will have little or no odor.

25. Names

(a) CH_3NHCH_3 dimethyl amine

(b) 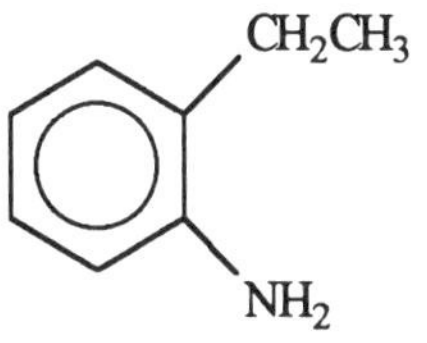 CH_2CH_3 / NH_2 (benzene ring) o-ethylaniline

(c) 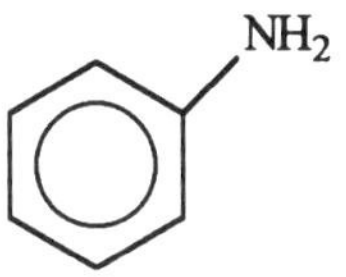 NH_2 (benzene ring) aniline

(d) $(C_2H_5)_4N^+I^-$ tetraethylammonium iodide

(e) 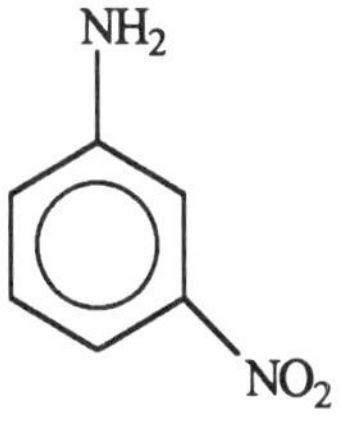 NH_2 / NO_2 (benzene ring) m-nitroaniline

(f) 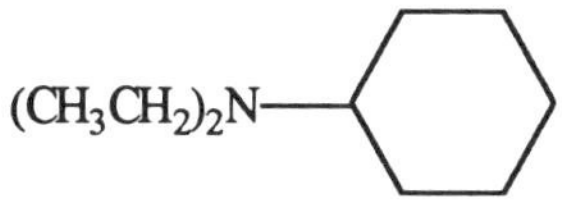 $(CH_3CH_2)_2N$— (cyclohexane ring) cyclohexyldiethylamine

26. Names

(a) 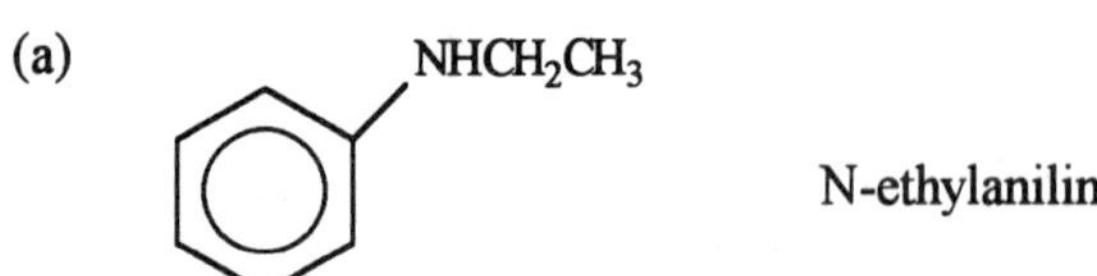

N-ethylaniline

(b) $CH_3CH_2CHCH_3$ with NH_2 on C-2 — 2-aminobutane

(c) C_6H_5–NH–C_6H_5 — diphenylamine

(d) $CH_3CH_2NH_3^+\ Br^-$ — ethylammonium bromide

(e) N (pyridine ring) — pyridine

(f) $CH_3C(=O)$—$NHCH_2CH_3$ — N-ethylacetamide

27. Structural formulas

(a) $CH_3CH_2NHCH_3$

(b) C_6H_5–NH_2

(c) $H_2NCH_2CH_2CH_2CH_2NH_2$

(d) $CH_3CH_2NCH(CH_3)_2$ (with CH_3 on N)

(e) 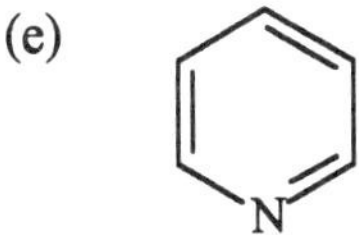

28. Structural formulas

(a) $CH_3CH_2CH_2CH_2NCH_2CH_2CH_2CH_3$ (with $CH_2CH_2CH_2CH_3$ on N)

(b) $C_6H_5\overset{+}{N}H_2\ Cl^-$ (with CH_3 on N)

(c) $CH_3CH_2NH_3^+Cl^-$

(d) $CH_3CH_2CH_2CHCH_2OH$ (with NH_2 on CH)

(e) $CH_3CH_2CH_2C(CH_3)_2—CH(CH_3)—NH—C_6H_5$

29. (a) $CH_3CH_2CH_2Br + NH_3 \text{ (excess)} \longrightarrow CH_3CH_2CH_2NH_2$

(b) $CH_3CH_2CH_2Br + KCN \longrightarrow CH_3CH_2CH_2CN \xrightarrow[Ni]{H_2} CH_3CH_2CH_2CH_2NH_2$

(c) 4-ethylnitrobenzene (NO_2 para to CH_2CH_3) $\xrightarrow[HCl]{Sn}$ 4-ethylaniline (NH_2 para to CH_2CH_3)

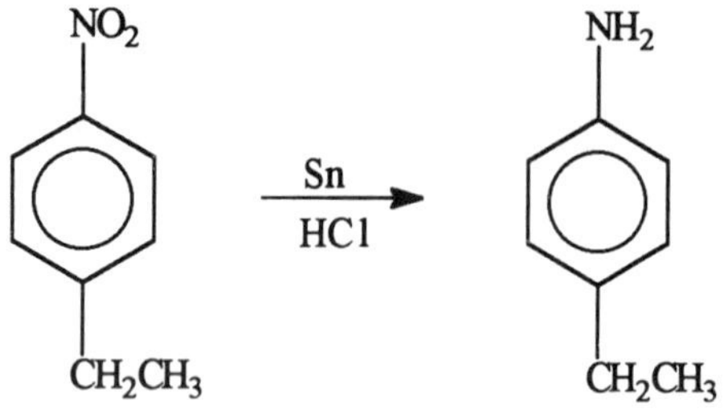

(d) $CH_3CH_2CH_2NH_2$ + $CH_3C(=O)—Cl$ ⟶ $CH_3C(=O)—NHCH_2CH_2CH_3$

30. Organic products

(a) $CH_3CH_2C(=O)—NH_2$

(b) $CH_3CH_2CH_2CH_2NH_2$

(c) H_3C–C_6H_4–NH_2 (para)

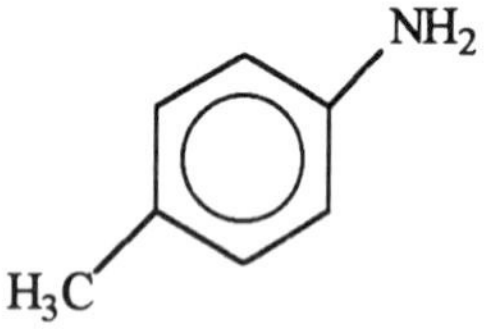

(d) piperidine N–$C(=O)$–CH_3

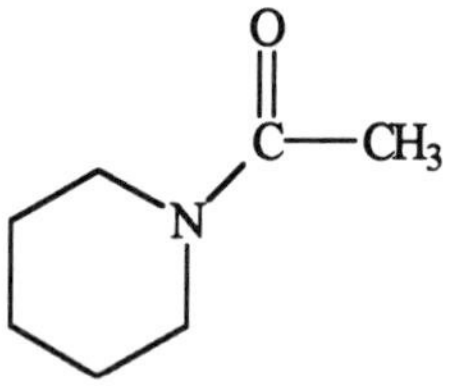

31. The main reason why trimethylamine has a lower boiling point than 1-aminopropane and ethylmethylamine is that trimethylamine cannot hydrogen bond because it has no hydrogen atoms bonded to the nitrogen atom. The other two amines do have hydrogen atom(s) bonded to the nitrogen atom and their molecules can hydrogen bond, which results in higher boiling points.

32. Amphetamines: stimulate the central nervous system; used to treat depression, narcolepsy, and obesity.

Tranquilizers: used to modify psychotic behavior, relieve pressure and anxiety.

Antibacterial agents: antibiotics

33. $\left(1.0 \times 10^{-7}\ \frac{\text{mol}}{\text{L}}\right)(5.0\ \text{L}) = 5.0 \times 10^{-7}$ mol epinephrine

$(5.0 \times 10^{-7}$ mol epinephrine$)(183.2$ g/mol$) = 9.2 \times 10^{-5}$ g epinephrine

34. $-\overset{\overset{\displaystyle O}{\|}}{C}-NH_2$ amide

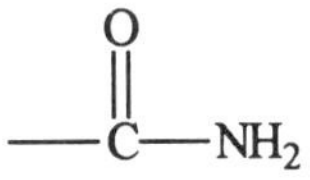

Some classes of biochemicals that contain an amide are:

- Antibacterial agents such as ampicillin
- B-vitamins such as niacin (nicotinamide)
- Barbiturates such as nembutal
- Tranquilizers such as valium (diazepam)

35. Lemon juice, being acidic, will react with the basic amines forming salts, which are soluble and can be washed away with water.

36.

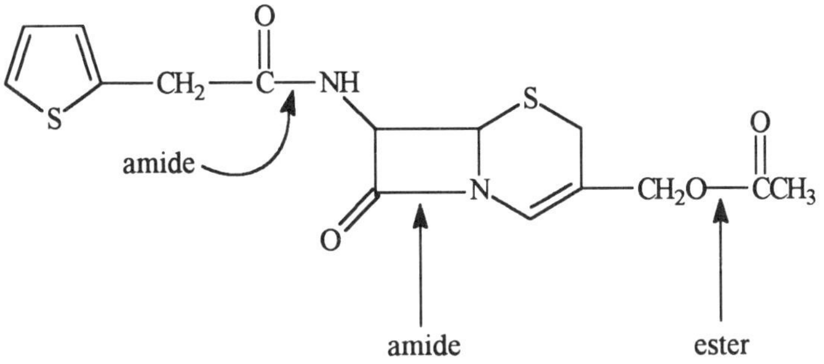

amide

amide

ester

37.

$C_6H_5NH_2$ + HCl ⟶ $C_6H_5NH_3^+Cl^-$

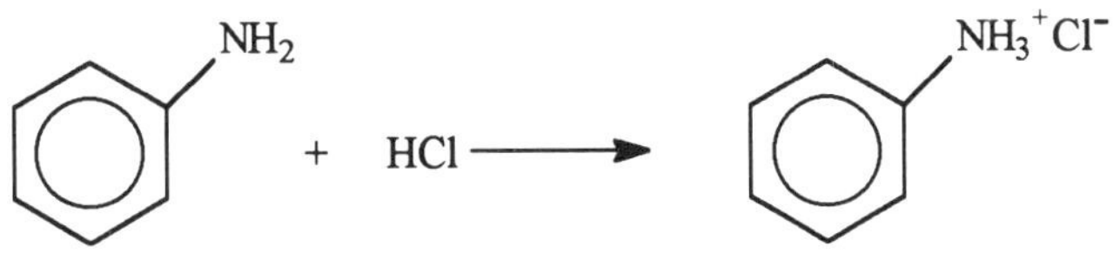

38. $H_2N–CH_2CH_2CH_2CH_2–NH_2$

Putrescine has two primary amine bonds.

39. Drugs are given as ammonium salts because the salts are soluble in water.

40.

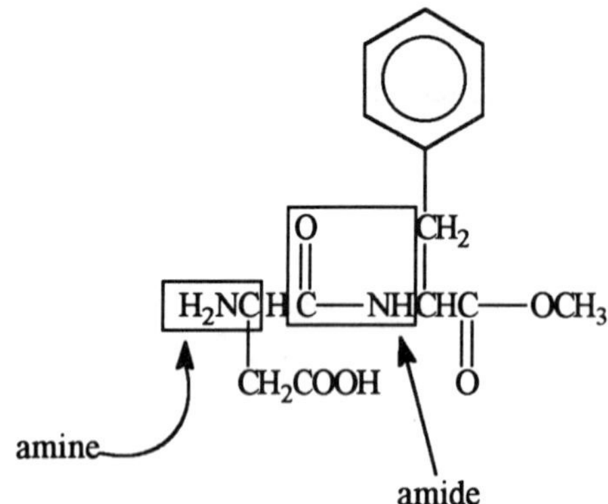

CHAPTER 26

POLYMERS: MACROMOLECULES

1. An addition polymer is one that is produced by the successive addition of repeating monomer molecules. A condensation polymer is one that is formed from monomer molecules in a reaction which splits out water or some other simple molecule. Condensation polymerization usually involves two different monomers.

2. Those polymers which soften on reheating are thermoplastic polymers; those which set to an infusible solid and do not soften on reheating are thermosetting polymers.

3. In order to form a thermosetting polymer, there must be cross linking of polymer chains. This requires that some monomer molecules be trifunctional. If all molecules are only bifunctional, then only long chains can result, and they will not be thermosetting.

4. Vulcanization is the process of heating raw rubber with sulfur. Sulfur atoms are introduced as cross links between the polymeric chains.

5. Silicone polymers are more resistant to high temperatures than the usual organic polymers because they contain silicon-oxygen chains which, like quartz, are stable at high temperature.

6. The correct statements are a, b, d, f, i, k.

   (c) The monomers of Nylon-66 are a dicarboxylic acid and a diamine.

   (e) 1,6-diaminohexane is a primary amine.

   (g) Vulcanization was invented by Charles Goodyear.

   (h) The monomer for polystyrene is $C_6H_5-CH{=}CH_2$

   (j) Bakelite is a polymer of phenol and formaldehyde.

7. How many ethylene units are in a polyethylene that has a molar mass of approximately 25,000?

   Polyethylene $\left(CH_2-CH_2\right)_n$

Molar mass of one unit = 2(12.01 g/mol) + 4(1.008 g/mol) = 28.05 g/mol

$$25{,}000 \text{ g/mol} \times \frac{\text{ethylene unit}}{28.05 \text{ g/mol}} = 8.9 \times 10^2 \text{ ethylene units}$$

8. Polystyrene $\left[\!-CH_2-CH(C_6H_5)-\right]_n$ C_8H_8

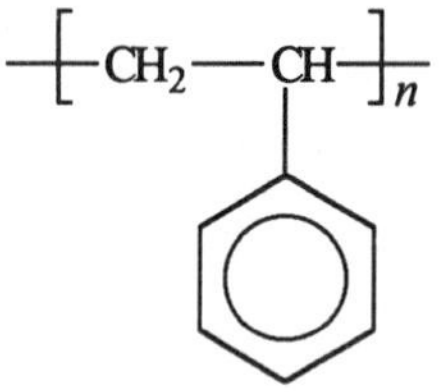

molar mass of 1 unit: 8(12.01 g/mol) + 8(1.008 g/mol) = 104.1 g/mol

$$(3000 \text{ units})\left(\frac{104.1 \text{ g/mol}}{\text{unit}}\right) = 3 \times 10^5 \text{ g/mol} = \text{molar mass}$$

9. Structural formulas for:

(a) Dacron

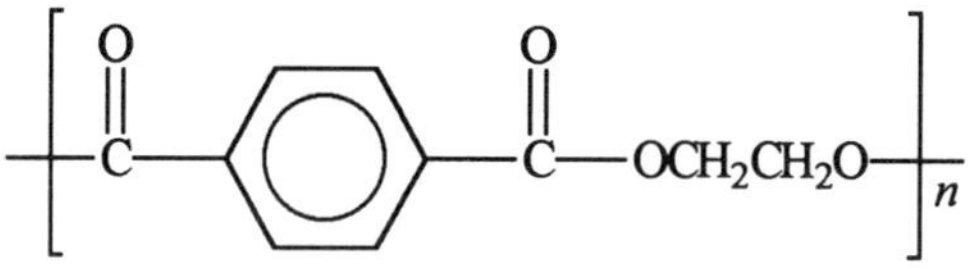

$$\left[-\overset{O}{\overset{\|}{C}}-C_6H_4-\overset{O}{\overset{\|}{C}}-OCH_2CH_2O-\right]_n$$

(b) Nylon-66

$$\left[-\overset{O}{\overset{\|}{C}}(CH_2)_4\overset{O}{\overset{\|}{C}}-NH(CH_2)_6NH-\right]_n$$

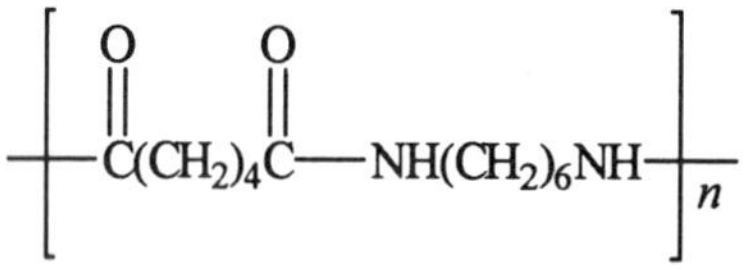

(c) Bakelite

OH OH
CH2 CH2 CH2
CH2 CH2
OH
CH2 CH2 CH2

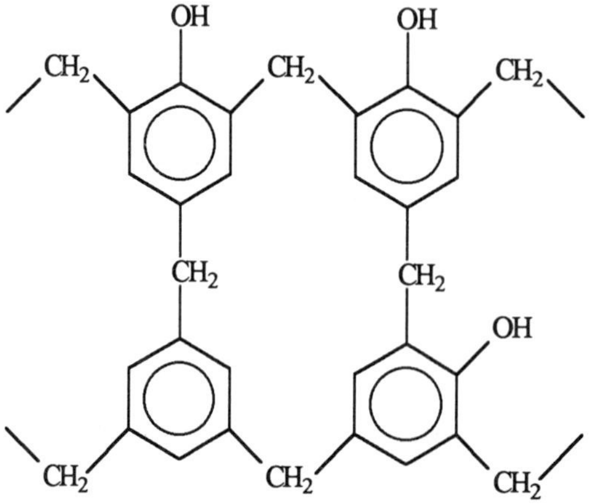

(d) Polyurethane

$$\left[-\underset{\underset{O}{\|}}{C}-NH-C_6H_4-NH-\underset{\underset{O}{\|}}{C}-OCH_2CH_2O- \right]_n$$

10. Structural formulas for:

(a) Saran $\left(-CH_2CCl_2- \right)_n$

(b) Orlon $\left(-CH_2\underset{\underset{CN}{|}}{CH}- \right)_n$

(c) Teflon $\left(-CF_2CF_2- \right)_n$

(d) Polystyrene $\left(-CH_2\underset{\underset{C_6H_5}{|}}{CH}- \right)_n$

(e) Lucite $\left(-CH_2-\overset{\overset{CH_3}{|}}{\underset{\underset{\underset{\underset{O}{\|}}{COCH_3}}{|}}{C}}- \right)_n$

11. polypropylene free radical (2 units)

Starting with RO• free radical $RO CH_2\underset{\underset{CH_3}{|}}{CH}-CH_2\underset{\underset{CH_3}{|}}{CH}\bullet$

12. polyethylene free radical (2 units)

Starting with RO• free radical $ROCH_2CH_2CH_2CH_2\bullet$

13. Formulas of the polymers that can be formed from:

(a) propylene $\left(-CH_2-\underset{\underset{CH_3}{|}}{CH}- \right)_n$

(b) 2-methylpropene $\left(-CH_2-\overset{\overset{CH_3}{|}}{\underset{\underset{CH_3}{|}}{C}}- \right)_n$

(c) 2-butene $\left(-\underset{\underset{CH_3}{|}}{CH}-\underset{\underset{CH_3}{|}}{CH}- \right)_n$

14. Formulas of the polymers that can be formed from:

(a) ethylene $\left(CH_2-CH_2 \right)_n$ (b) chloroethene $\left(CH_2-CHCl \right)$

(c) 1-butene $\left(CH_2-\underset{\displaystyle CH_2CH_3}{\underset{|}{CH}} \right)_n$

15. Two possible ways in which vinyl chloride can polymerize to form polyvinyl chloride:

$$\left(CH_2\underset{Cl}{\underset{|}{C}}HCH_2\underset{Cl}{\underset{|}{C}}HCH_2\underset{Cl}{\underset{|}{C}}HCH_2\underset{Cl}{\underset{|}{C}}H \right)_n$$

$$\left(CH_2\underset{Cl}{\underset{|}{C}}H-\underset{Cl}{\underset{|}{C}}HCH_2CH_2\underset{Cl}{\underset{|}{C}}H-\underset{Cl}{\underset{|}{C}}HCH_2 \right)_n$$

(other possibilities also)

16. Two possible ways in which acrylonitrile can polymerize to form Orlon.

$$\left(CH_2\underset{CN}{\underset{|}{C}}HCH_2\underset{CN}{\underset{|}{C}}HCH_2\underset{CN}{\underset{|}{C}}HCH_2\underset{CN}{\underset{|}{C}}H \right)_n$$

$$\left(CH_2\underset{CN}{\underset{|}{C}}H-\underset{CN}{\underset{|}{C}}HCH_2CH_2\underset{CN}{\underset{|}{C}}H-\underset{CN}{\underset{|}{C}}HCH_2 \right)_n$$

17. Natural rubber (all cis)

$$\left(\overset{CH_2}{\underset{CH_3}{>}}C=C\overset{CH_2-CH_2}{\underset{H}{<}} \overset{}{\underset{CH_3}{>}}C=C\overset{CH_2-CH_2}{\underset{H}{<}} \overset{}{\underset{CH_3}{>}}C=C\overset{CH_2}{\underset{H}{<}} \right)_n$$

18. Gutta percha (all trans)

$$\left(\overset{CH_2}{\underset{CH_3}{>}}C=C\overset{H}{\underset{CH_2-CH_2}{<}} \overset{CH_3}{\underset{}{>}}C=C\overset{CH_2-CH_2}{\underset{H}{<}} \overset{}{\underset{CH_3}{>}}C=C\overset{H}{\underset{CH_2}{<}} \right)_n$$

19. Yes, smog in the atmosphere contains ozone and ozone attacks natural rubber at the site of the double bond causing "age hardening" and cracking.

20. Yes, styrene-butadiene rubber contains carbon-carbon double bonds and is attacked by the ozone in smog causing "age hardening" and cracking.

21. Chemical structures for the monomers of:

(a) natural rubber $CH_2{=}C(CH_3)CH{=}CH_2$

(b) synthetic rubber (SBR) 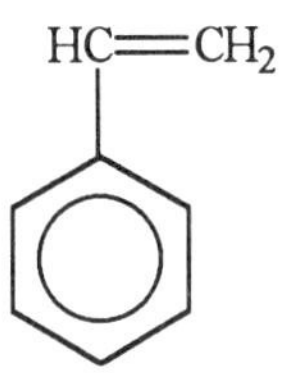 $HC{=}CH_2$ (with phenyl ring attached to HC) and $CH_2{=}CHCH{=}CH_2$

22. Chemical structures for the monomers of:

(a) synthetic natural rubber $CH_2{=}C(CH_3)CH{=}CH_2$

(b) neoprene rubber $CH_2{=}CCl{-}CH{=}CH_2$

23. (a) Mass percent of nitrogen in Nylon-66.

$$\left(\underset{\substack{\| \\ O}}{C} - (CH_2)_4 - \underset{\substack{\| \\ O}}{C} - NH(CH_2)_6NH \right)_n$$ molar mass = 226.3 g/mol

$$\frac{28.02 \text{ g/mol N}}{226.3 \text{ g/mol}} \times 100 = 12.38\% \text{ N}$$

(b) Mass percent of chlorine in polyvinyl chloride $\left(CH_2CHCl \right)_n$ molar mass = 62.49 g/mol

$$\frac{35.45 \text{ g/mol Cl}}{62.49 \text{ g/mol}} \times 100 = 56.73\% \text{ Cl}$$

24. (a) Mass percent of oxygen in Dacron

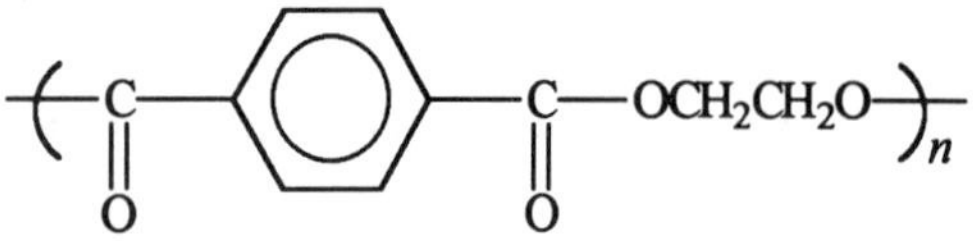

$$\left(\underset{\underset{O}{\|}}{C} - C_6H_4 - \underset{\underset{O}{\|}}{C} - OCH_2CH_2O \right)_n$$

molar mass = 192.2 g/mol

$$\frac{64.00 \text{ g/mol O}}{192.2 \text{ g/mol}} \times 100 = 33.30\% \text{ O}$$

(b) Mass percent of fluorine in Teflon.

$$\left(CF_2CF_2 \right)_n$$

molar mass = 100.0 g/mol

$$\frac{76.00 \text{ g/mol F}}{100.0 \text{ g/mol}} \times 100 = 76.00\% \text{ F}$$

25. Nitrile rubber (Buna N)

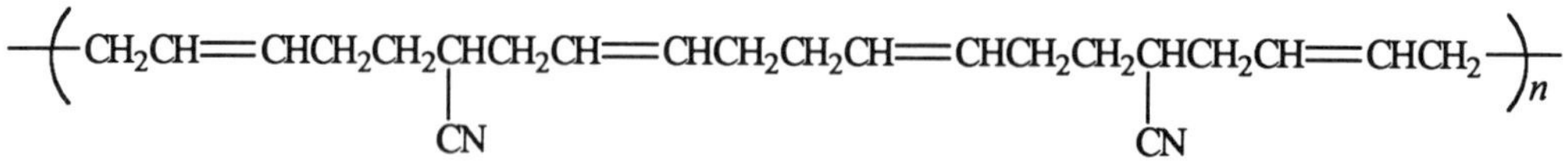

$$\left(CH_2CH{=}CHCH_2CH_2\underset{\underset{CN}{|}}{C}HCH_2CH{=}CHCH_2CH_2CH{=}CHCH_2CH_2\underset{\underset{CN}{|}}{C}HCH_2CH{=}CHCH_2 \right)_n$$

(other structures are possible)

26. (a) Isotactic polypropylene

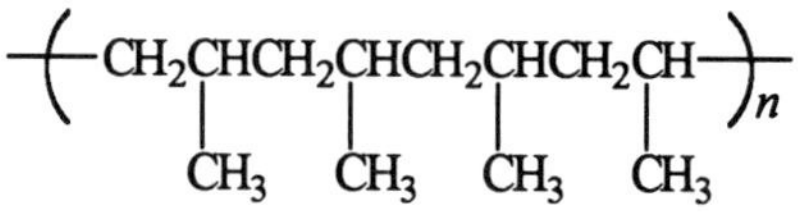

$$\left(CH_2\underset{\underset{CH_3}{|}}{C}HCH_2\underset{\underset{CH_3}{|}}{C}HCH_2\underset{\underset{CH_3}{|}}{C}HCH_2\underset{\underset{CH_3}{|}}{C}H \right)_n$$

(b) Another form of polypropylene

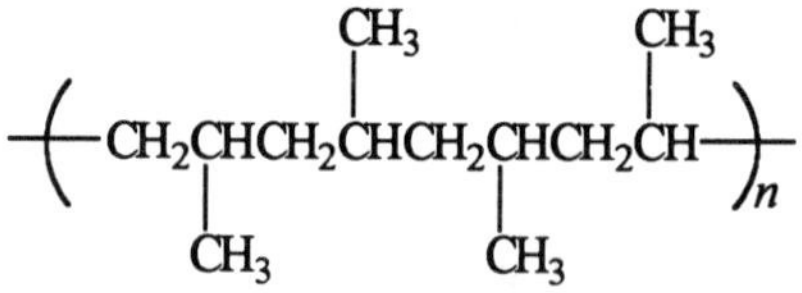

$$\left(CH_2\underset{\underset{CH_3}{|}}{C}HCH_2\overset{\overset{CH_3}{|}}{C}HCH_2\underset{\underset{CH_3}{|}}{C}HCH_2\overset{\overset{CH_3}{|}}{C}H \right)_n$$

(other structures are possible)

27. Polymer

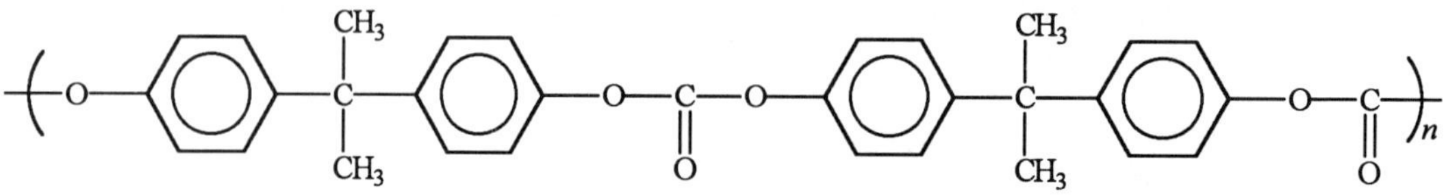

$$\left(O - C_6H_4 - \underset{\underset{CH_3}{|}}{\overset{\overset{CH_3}{|}}{C}} - C_6H_4 - O - \underset{\underset{O}{\|}}{C} - O - C_6H_4 - \underset{\underset{CH_3}{|}}{\overset{\overset{CH_3}{|}}{C}} - C_6H_4 - O - \underset{\underset{O}{\|}}{C} \right)_n$$

28. Polymer

$$\left(\underset{\underset{O}{\|}}{C} - C_6H_4 - \underset{\underset{O}{\|}}{C} - NH - C_6H_4 - NH - \underset{\underset{O}{\|}}{C} - C_6H_4 - \underset{\underset{O}{\|}}{C} - NH - C_6H_4 - NH \right)_n$$

29. In phenol, three positions, ortho, ortho, and para to the OH group are used in the reaction to form a thermosetting polymer. However, in p-cresol the para position is occupied by a methyl group and cannot react with formaldehyde. This leaves the p-cresol molecule as a bifunctional monomer, which results in a more linear, thermoplastic polymer.

30. Foam or sponge rubber materials achieve the foam by incorporating chemicals that release a gas within the material during the polymerization or the molding process. For example, in spongy polyurethane, water added during the polymerization reacts with the isocyanate to produce carbon dioxide, which causes the polymer to foam.

31. Polystyrene has the highest mass percent carbon. $\left(CH_2 - \underset{\underset{C_6H_5}{|}}{CH} \right)$ C_8H_8
molar mass = 104.1 g/mol

$$\frac{96.08 \text{ g C}}{104.1 \text{ g}} \times 100 = 92.30\% \text{ C}$$

32. A glyptol polyester would more likely be thermosetting, because the glycerol is trifunctional, and would thus allow cross linking between chains in forming ester linkages.

33. Nylon-6 polymers

$$\left(NHCH_2CH_2CH_2CH_2\overset{\overset{O}{\|}}{C} - NHCH_2CH_2CH_2CH_2\overset{\overset{O}{\|}}{C} \right)_n$$

34. The latex polymer belong to the vinyl class of polymers.

$$\left(CH_2 - \underset{\underset{\underset{\underset{CH_2CH_3}{|}}{O}}{|}}{\underset{|}{C=O}}{CH} - CH_2 - \underset{\underset{\underset{\underset{CH_2CH_3}{|}}{O}}{|}}{\underset{|}{C=O}}{CH} - CH_2 - \underset{\underset{\underset{\underset{CH_2CH_3}{|}}{O}}{|}}{\underset{\|}{C=O}}{CH} \right)_n$$

35. Cyanoacrylate ester polymer

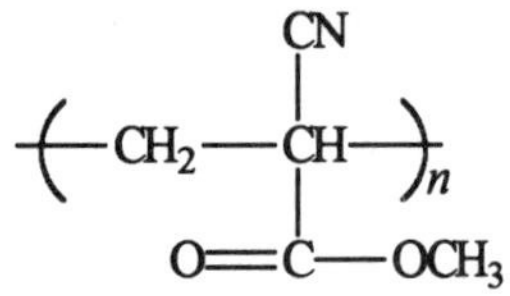

$$\left(-CH_2-\underset{\underset{O=C-OCH_3}{|}}{\overset{\overset{CN}{|}}{CH}}-\right)_n$$

36. (a) From 2,3-dimethyl-1,3-butadiene, $CH_2=\underset{H_3C}{\underset{|}{C}}-\underset{CH_3}{\underset{|}{C}}=CH_2$

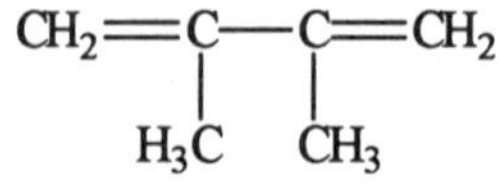

this polymer can be made $\left(-CH_2\underset{H_3C}{\underset{|}{C}}=\underset{CH_3}{\underset{|}{C}}CH_2-\right)_n$

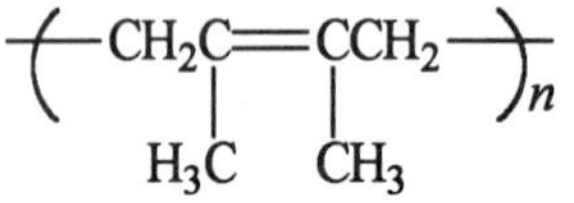

(b) If produced by the free-radical mechanism, a random mixture of cis and trans connections are made. It is possible, using catalysts, for the reaction to proceed by an ionic mechanism which will give a stereochemically controlled polymer.

37. It is easier to recycle a thermoplastic polymer because they are linear and can easily be reformed on heating. On the other hand, the cross linkages in thermosetting polymers are not easily broken or reformed.

38. $CH_2=CH_2$ $CF_2=CF_2$

ethylene tetrofluoroethylene

$$\left(-CH_2-CH_2-CF_2-CF_2-CH_2-CH_2-CF_2-CF_2-\right)_n$$

(many other possible structures)

39. Monomers of the given polymers:

(a) 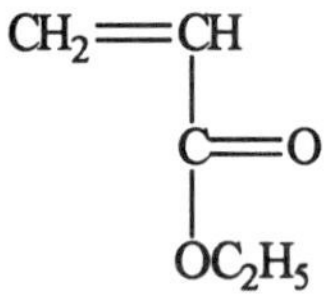 $CH_2=\underset{\underset{\underset{OC_2H_5}{|}}{C=O}}{\underset{|}{CH}}$ a polyvinyl

(b) $HO-\overset{\overset{O}{\|}}{C}-C_6H_4-\overset{\overset{O}{\|}}{C}-OH$ and $HOCH_2-C_6H_4-CH_2OH$

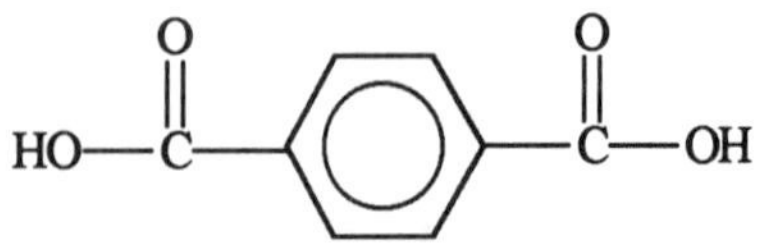

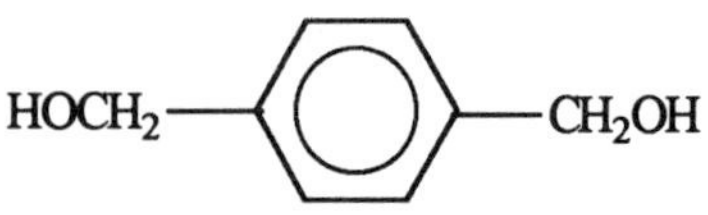

a polyester

(c) 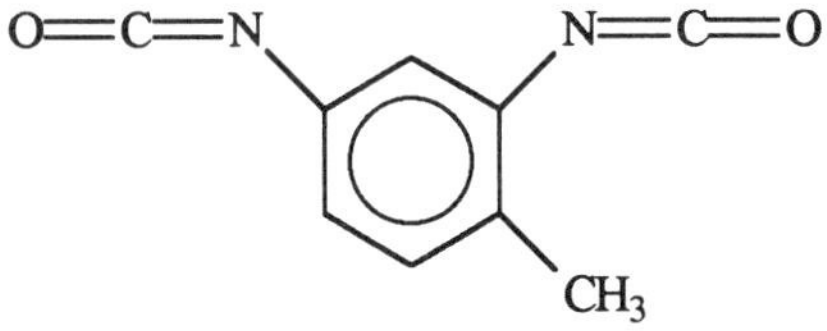 O=C=N-(2,4-toluene diisocyanate: benzene ring bearing two N=C=O groups and a CH_3) and HO—CH_2CH_2—OH

a polyurethane

(d) $HO—\overset{O}{\overset{||}{C}}(CH_2)_8\overset{O}{\overset{||}{C}}—OH$ and $H_2N(CH_2)_6NH_2$

a polyamide

40. (a) Quiana is a polyamide

(b) monomers

$HOOC(CH_2)_6COOH$ and H_2N—(cyclohexane)—CH_2—(cyclohexane)—NH_2

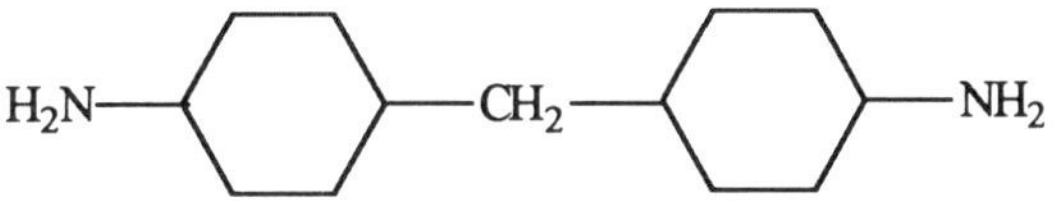

41. The methylmethacrylate polymers result from an addition reaction because no small molecule is split out and the polymer forms in a manner similar to that of polyethylene.

CHAPTER 27

STEREOISOMERISM

1. An asymmetric carbon atom is one to which four different atoms or groups are attached and is a center of dissymmetry in a molecule. In the following three compounds, the asymmetric carbon atoms are marked with an asterisk. (These are merely three examples; there are an infinite number of compounds which contain one asymmetric carbon atom.)

$$\begin{array}{ccccc} & & H & & H & & \\ & & | & & | & & \\ Cl & — & \overset{*}{C} & — & C & — & H \\ & & | & & | & & \\ & & Br & & Br & & \end{array} \qquad \begin{array}{ccccccc} & & H & & H & & Cl & & \\ & & | & & | & & | & & \\ H & — & C & — & \overset{*}{C} & — & C & — & Cl \\ & & | & & | & & | & & \\ & & Cl & & Cl & & Cl & & \end{array} \qquad CH_3\overset{*}{C}H(OH)CH_2OH)$$

2. When the axes of two pieces of polaroid film are parallel, you have maximum brightness of the light passing through both. When one piece has been rotated by 90° the polaroid appears black, indicating very little light passing through.

3. A necessary and sufficient condition for a compound to show enantiomerism is that the compound not be superimposable on its mirror image.

4. Enantiomers are nonsuperimposable mirror image isomers. Diastereomers are stereoisomers that are not enantiomers (not mirror image isomers).

5. Physical properties of a pair of enantiomers

	(+) 2-methyl-1-butanol	(-) 2-methyl-1-butanol
specific rotation	+5.76°	-5.76°
boiling point	129°C	129°C
density	0.819 g/mL	0.819 g/mL

6. The correct statements are: a, e, g, i, j

   (b) Many natural products are optically active.
   (c) Cis-trans isomers are one class of stereoisomers.
   (d) J. H. van't Hoff received the first Nobel prize in chemistry in 1901.
   (f) The compound $CH_3CHBrCHBrCH_2OH$ has four optical isomers.
   (h) Diastereomers do not have identical melting points.

7. Enantiomers have identical physical properties except their effect on polarized light, so they cannot be separated by ordinary chemical and physical means.

8. Diastereomers do not have identical physical properties, so the differences form a basis for chemical or physical separation. Differences of boiling point, freezing point, and solubilities would be most commonly used.

9. The objects that are chiral: (a) your ear; (b) a pair of pliers; (c) a coiled spring; (d) the letter b.

10. The objects that are chiral: (a) a wood screw; (c) the letter g; (d) this textbook.

11. Number of asymmetric carbon atoms.

    (a) 1 (b) 1 (c) 2 (d) 1

12. Number of asymmetric carbon atoms.

    (a) 0 (b) 0 (c) 3 (d) 2

13. Which compounds will show optical activity?

    (a) $CH_3CH_2CH_2CH_2CH_2Cl$ — no optical activity

    (b) $CH_3CH_2CHClCH_2CH_3$ — no optical activity

    (c) $CH_3CH_2CH_2C(CH_3)(Cl)CH_3$ (the C bearing CH_3 above and Cl below) — no optical activity

    (d) $CH_3\overset{*}{C}HClCH_2CH(CH_3)CH_3$ — will show optical activity (* asymmetric carbon)

14. Which compounds will show optical activity?

    (a) $CH_3CH_2CH_2\overset{*}{C}HClCH_3$ — will show optical activity (* asymmetric carbon)

    (b) $CH_3CH_2CH_2\overset{*}{C}H(CH_3)CH_2Cl$ — will show optical activity (* asymmetric carbon)

(c) $CH_3CH_2\overset{*}{C}HClCHCH_3$ with CH_3 on the CH — will show optical activity (* asymmetric carbon)

(d) $CH_3CH_2CClCH_2CH_3$ with CH_3 on the CCl — no optical activity

15. Glucose, which has four asymmetric carbon atoms will have 16 possible stereoisomers. This can be determined from 2^n. $2^4 = 16$.

16. Fructose, which has three asymmetric carbon atoms, will have 8 possible stereoisomers. This can be determined from 2^n. $2^3 = 8$.

17. The two projection formulas (A) and (B) are the same compound, for it takes two changes to make (B) identical to (A).

(A)	(B)	1st change in (B) (H and CH_3)	2nd change in (B) (F and CH_3)
top: H; left: Br; right: F; bottom: CH_3	top: CH_3; left: Br; right: H; bottom: F	top: H; left: Br; right: CH_3; bottom: F	top: H; left: Br; right: F; bottom: CH_3

18. The two projection formulas (A) and (B) are the same compound, for it takes two charges to make (B) identical to (A)

(A)	(B)	1st change in (B) (H and CH_3)	2nd change in (B) (F and CH_3)
top: H; left: Br; right: F; bottom: CH_3	top: CH_3; left: Br; right: H; bottom: F	top: H; left: Br; right: CH_3; bottom: F	top: H; left: Br; right: F; bottom: CH_3

19. (–) - lactic acid is

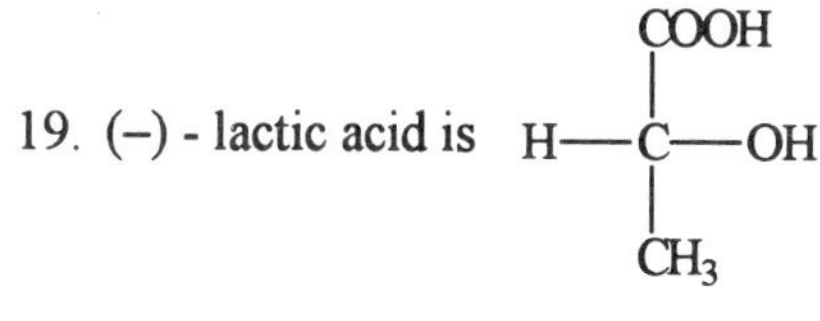

(+) - lactic acid is

$$\begin{array}{c} COOH \\ | \\ HO—C—H \\ | \\ CH_3 \end{array}$$

(a)
$$\begin{array}{c} CH_3 \\ HO—+—H \\ COOH \end{array}$$

(b)
$$\begin{array}{c} OH \\ H—+—COOH \\ CH_3 \end{array}$$

(e)
$$\begin{array}{c} COOH \\ CH_3—+—H \\ OH \end{array}$$

(c)
$$\begin{array}{c} COOH \\ H—+—CH_3 \\ OH \end{array}$$

(f)
$$\begin{array}{c} H \\ CH_3—+—OH \\ COOH \end{array}$$

(d)
$$\begin{array}{c} CH_3 \\ H—+—OH \\ COOH \end{array}$$

20. (+) - alanine
$$\begin{array}{c} COOH \\ H_2N—+—H \\ CH_3 \end{array}$$

(–) - alanine
$$\begin{array}{c} COOH \\ H—+—NH_2 \\ CH_3 \end{array}$$

(b)
$$\begin{array}{c} NH_2 \\ H—+—COOH \\ CH_3 \end{array}$$

(a)
$$\begin{array}{c} CH_3 \\ H_2N—+—H \\ COOH \end{array}$$

(c)
$$\begin{array}{c} COOH \\ H—+—CH_3 \\ NH_2 \end{array}$$

(e)
$$\begin{array}{c} COOH \\ CH_3—+—H \\ NH_2 \end{array}$$

(d)

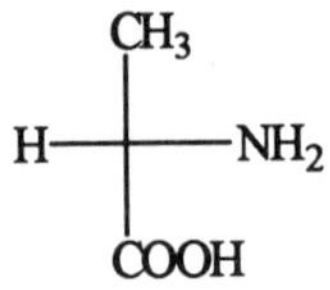

$$\begin{array}{c} CH_3 \\ H-\!\!\!\!+\!\!\!\!-NH_2 \\ COOH \end{array}$$

(f)

$$\begin{array}{c} H \\ CH_3-\!\!\!\!+\!\!\!\!-NH_2 \\ COOH \end{array}$$

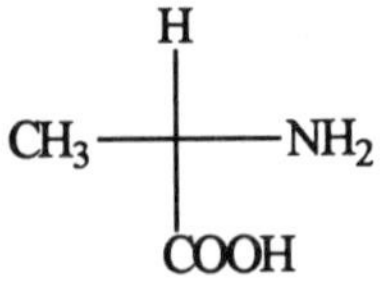

21. All possible stereoisomers of the following compounds, with enantiomers and meso compounds labeled:

(a) 1,2-dibromopropane

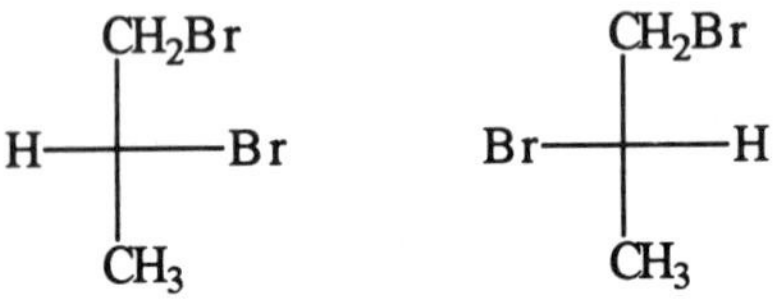

$$\begin{array}{c} CH_2Br \\ H-\!\!\!\!+\!\!\!\!-Br \\ CH_3 \end{array} \qquad \begin{array}{c} CH_2Br \\ Br-\!\!\!\!+\!\!\!\!-H \\ CH_3 \end{array}$$

enantiomers

(b) 2-butanol

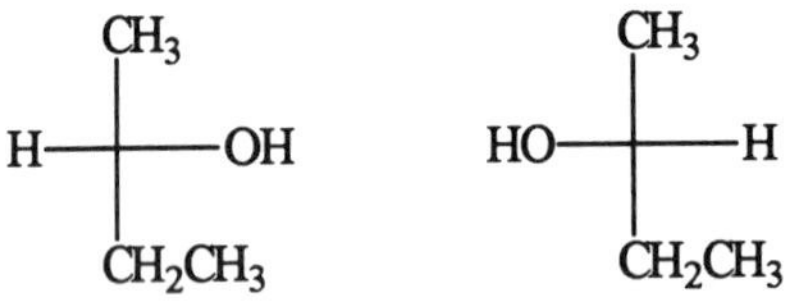

$$\begin{array}{c} CH_3 \\ H-\!\!\!\!+\!\!\!\!-OH \\ CH_2CH_3 \end{array} \qquad \begin{array}{c} CH_3 \\ HO-\!\!\!\!+\!\!\!\!-H \\ CH_2CH_3 \end{array}$$

enantiomers

(c) 3-chlorohexane

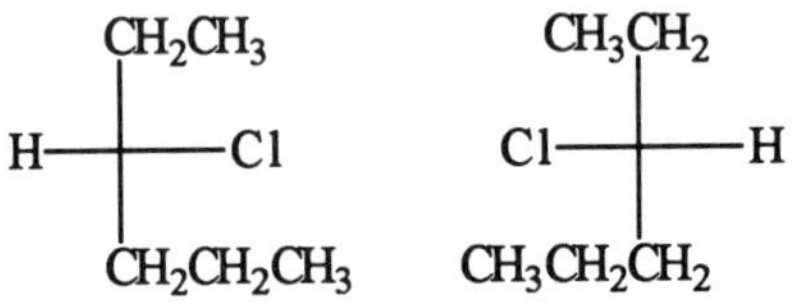

$$\begin{array}{c} CH_2CH_3 \\ H-\!\!\!\!+\!\!\!\!-Cl \\ CH_2CH_2CH_3 \end{array} \qquad \begin{array}{c} CH_3CH_2 \\ Cl-\!\!\!\!+\!\!\!\!-H \\ CH_3CH_2CH_2 \end{array}$$

enantiomers

There are no meso compounds.

22. All possible stereoisomers of the following compounds with enantiomers and meso compounds labeled.

(a) 2,3-dichlorobutane

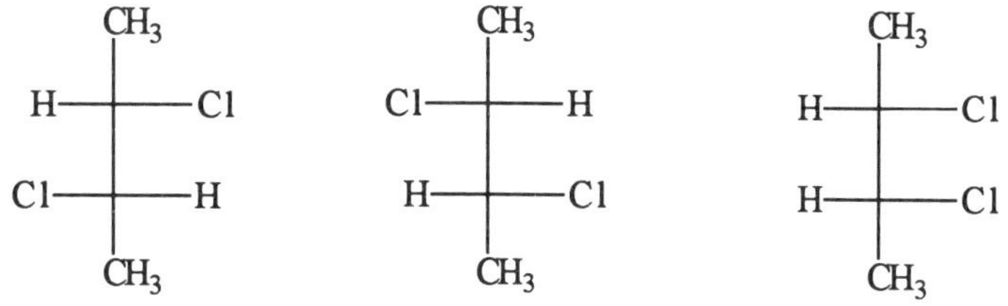

CH_3	CH_3	CH_3
H—+—Cl	Cl—+—H	H—+—Cl
Cl—+—H	H—+—Cl	H—+—Cl
CH_3	CH_3	CH_3

enantiomers (first two); meso (third)

(b) 2,4-dibromopentane

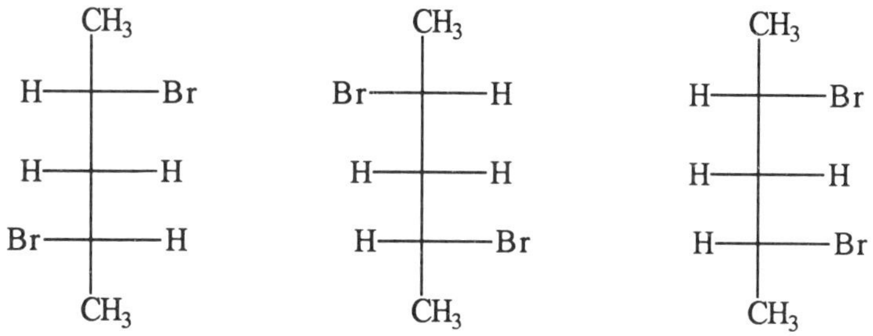

CH_3	CH_3	CH_3
H—+—Br	Br—+—H	H—+—Br
H—+—H	H—+—H	H—+—H
Br—+—H	H—+—Br	H—+—Br
CH_3	CH_3	CH_3

enantiomers (first two); meso (third)

(c) 3-hexanol

CH_2CH_3	CH_3CH_2
H—+—OH	HO—+—H
$CH_2CH_2CH_3$	$CH_3CH_2CH_2$

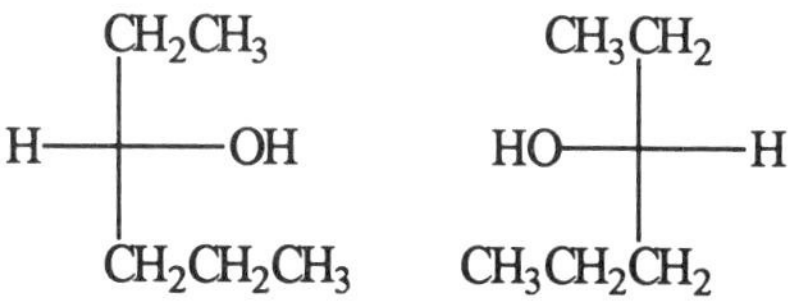

enantiomers

23. All the stereoisomers of 1,2,3-trihydroxybutane:

A: CH_2OH; H—|—OH; H—|—OH; CH_3

B: CH_2OH; HO—|—H; HO—|—H; CH_3

C: CH_2OH; H—|—OH; HO—|—H; CH_3

D: CH_2OH; HO—|—H; H—|—OH; CH_3

Compounds A and B, and C and D are pairs of enantiomers. There are no meso compounds. Pairs of diastereomers are A and C, A and D, B and C, and B and D.

24. All the stereoisomers of 3,4-dichloro-2-methylpentane

A: $CH(CH_3)_2$; H—|—Cl; H—|—Cl; CH_3

B: $CH(CH_3)_2$; Cl—|—H; Cl—|—H; CH_3

C: $CH(CH_3)_2$; H—|—Cl; Cl—|—H; CH_3

D: $CH(CH_3)_2$; Cl—|—H; H—|—Cl; CH_3

Compounds A and B, and C and D are pairs of enantiomers. There are no meso compounds. Pairs of diastereomers are A and C, A and D, B and C, and B and D.

25. The four stereoisomers of 2-hydroxy-3-pentene.

cis: CH_3; H—C—OH; H—C; ‖; H—C; CH_3

cis: CH_3; HO—C—H; H—C; ‖; H—C; CH_3

trans: CH_3; H—C—OH; H—C; ‖; C—H; CH_3

trans: CH_3; HO—C—H; H—C; ‖; C—H; CH_3

The two cis compounds are enantiomers and the two trans compounds are enantiomers.

26. The four stereoisomers of 2-chloro-3-hexene.

```
    CH3             CH3             CH3             CH3
    |               |               |               |
H—C—Cl        Cl—C—H         H—C—Cl        Cl—C—H
    |               |               |               |
H—C             H—C             H—C             H—C
    ||              ||              ||              ||
H—C             H—C               C—H             C—H
    |               |               |               |
    CH2CH3          CH2CH3          CH2CH3          CH2CH3
    cis             cis             trans           trans
```

The two cis compounds are enantiomers and the two trans compounds are enantiomers.

27. (a)

$CH_3CH_2CH_2CHCl_2$	$CH_3CH_2CHCl\,CH_2Cl$	$CH_3CHClCH_2CH_2Cl$
(a)	(b)	(c)
$CH_2ClCH_2CH_2CH_2Cl$	$CH_3CH_2CCl_2CH_3$	$CH_3CHClCHClCH_3$
(d)	(e)	(f)
$CH_3CH(CH_3)CHCl_2$	$CH_3CCl(CH_3)CH_2Cl$	$CH_2ClCH(CH_3)CH_2Cl$
(g)	(h)	(i)

(b) (b) is chiral

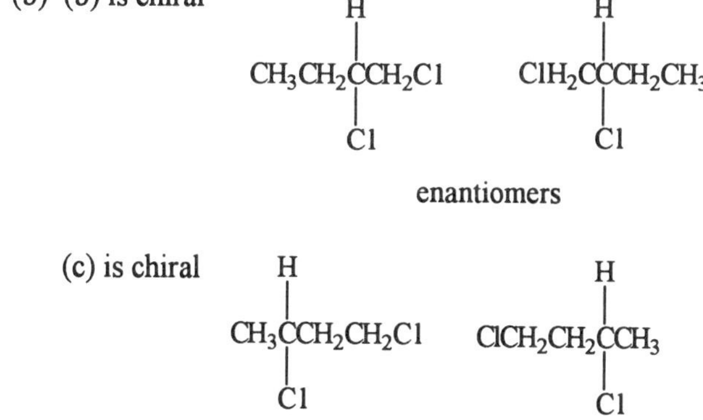

(c) is chiral

enantiomers

(f) can be both chiral and meso

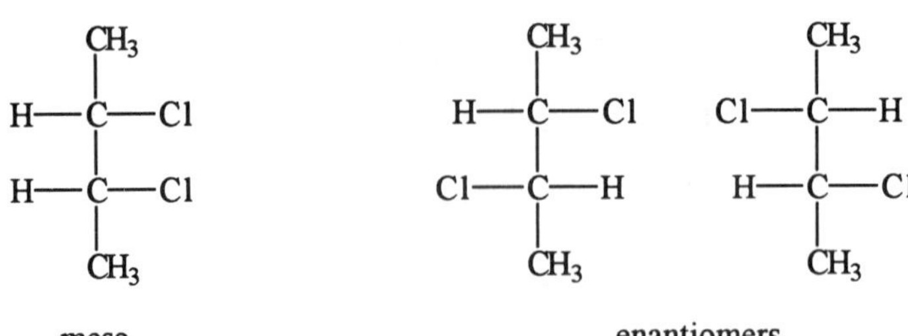

meso enantiomers

(a), (d), (e), (g), (h), and (i) are achiral.

28. (a) $CH_3CH_2CHBr_2$ (a) $CH_3CHBrCH_2Br$ (b)

$CH_3CBr_2CH_3$ (c) $CH_2BrCH_2CH_2Br$ (d)

(b) (b) is chiral

CH_2Br / H—+—Br / CH_3 CH_2Br / Br—+—H / CH_3

enantiomers

(a), (c), and (d) are achiral; there are no meso compounds.

29. Assume (+)-2-bromopentane is

CH_3 / H—C—Br / CH_2 / CH_2 / CH_3

All possible isomers formed when (+)-2-bromopentane is further brominated to dibromopentanes:

```
     CH2Br            CH3              CH3              CH3
      |                |                |                |
  H---C---Br     Br---C---Br      H---C---Br       H---C---Br
      |                |                |                |
     CH2              CH2          H---C---Br      Br---C---H
      |                |                |                |
     CH2              CH2              CH2              CH2
      |                |                |                |
     CH3              CH3              CH3              CH3

      A                B                C                D
```

```
     CH3              CH3              CH3
      |                |                |
  H---C---Br      H---C---Br       H---C---Br
      |                |                |
     CH2              CH2              CH2
      |                |                |
  H---C---Br     Br---C---H            CH2
      |                |                |
     CH3              CH3             CH2Br

      E                F                G
```

Compounds A, C, D, F, G would be optically active; B has no asymmetric carbon atom; E is a meso compound.

30. Assume (+)-2-chlorobutane is

```
      CH3
       |
      CH2
       |
   H---C---Cl
       |
      CH3
```

All possible isomers formed when (+)-2-chlorobutane is further chlorinated to dichlorobutane:

```
     CH3            CH2Cl           CH3            CH3             CH3
     |              |               |              |               |
     CH2            CH2             CH2         H—C—Cl         Cl—C—H
     |              |               |              |               |
  H—C—Cl         H—C—Cl         Cl—C—Cl        H—C—Cl          H—C—Cl
     |              |               |              |               |
     CH2Cl          CH3             CH3            CH3             CH3
     A              B               C              D               E
```

Compounds A, B, and E would be optically active; C does not have an asymmetric carbon atom; D is a meso compound.

31. Neither of the products, 1-chloropropane or 2-chloropropane, have an asymmetric carbon atom so neither product would rotate polarized light.

32. If 1-chlorobutane and 2-chlorobutane were obtained by chlorinating butane, and then distilled, they would be separated into the two fractions, because their boiling points are different. 1-chlorobutane has no asymmetric carbon, so would not be optically active. 2-chlorobutane would exist as a racemic mixture (equal quantities of enantiomers) because substitution of Cl for H on carbon-2 gives equal amounts of the two enantiomers. Distillation would not separate the enantiomers because their boiling points are identical. The optical rotation of the two enantiomers of the 2-chlorobutane fraction would exactly cancel, and thus would not show optical activity.

33. Compounds (a) and (d) are meso.

```
        COOH                 CH3
         |                    |
   H ----+---- OH        H ---+--- Cl
         |                    |
   HO ---+---- COOH      H ---+--- Br
         |                    |
         H               H ---+--- Cl
                              |
                             CH3
```

34. Compound (d) is meso.

```
          H
          |
   Br ----+---- CH3
          |
   Cl ----+---- CH3
          |
   Br ----+---- CH3
          |
          H
```

35. (b) is chiral

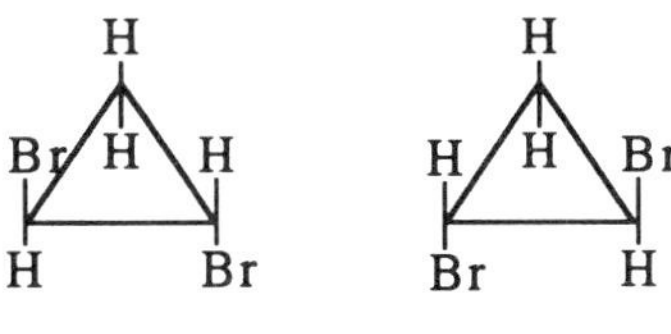

enantiomers

36. (c) is chiral

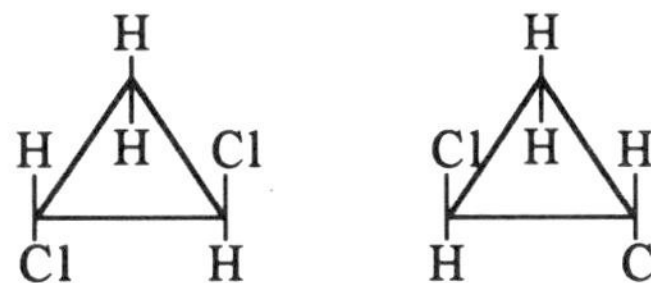

enantiomers

37. If four different groups were attached to a central carbon atom in a planar arrangement, it would not rotate polarized light because there would be a plane of symmetry in the molecule. No such plane of symmetry is possible when the four different groups are arranged in a tetrahedral structure.

38.

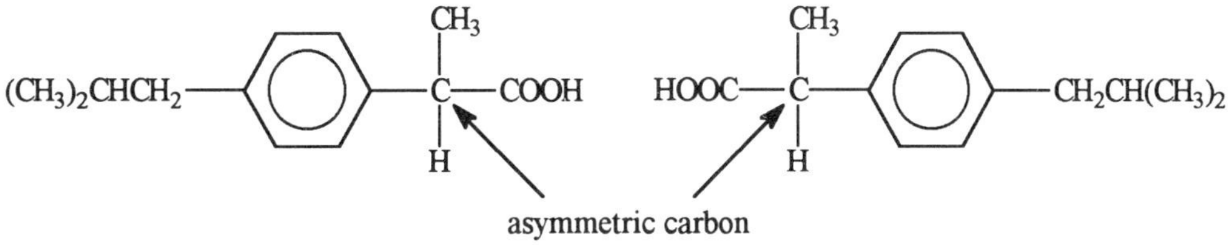

$(CH_3)_2CHCH_2-C_6H_4-C(CH_3)(H)-COOH$ and $HOOC-C(CH_3)(H)-C_6H_4-CH_2CH(CH_3)_2$

asymmetric carbon

39. (a) A chiral primary alcohol of formula $C_5H_{12}O$.

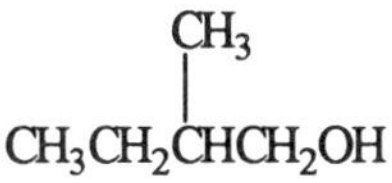

$CH_3CH_2CH(CH_3)CH_2OH$

(b) A compound with three primary alcohol groups is chiral, and has the formula $C_6H_{14}O_3$.

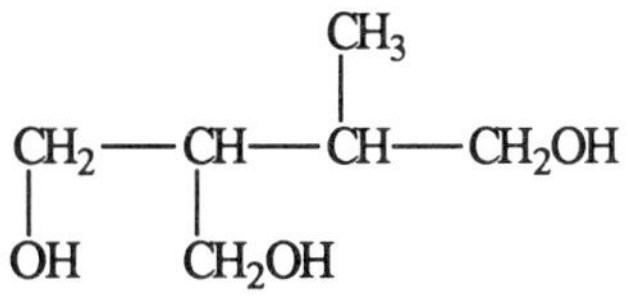

$HOCH_2-CH(CH_2OH)-CH(CH_3)-CH_2OH$

40. (a)

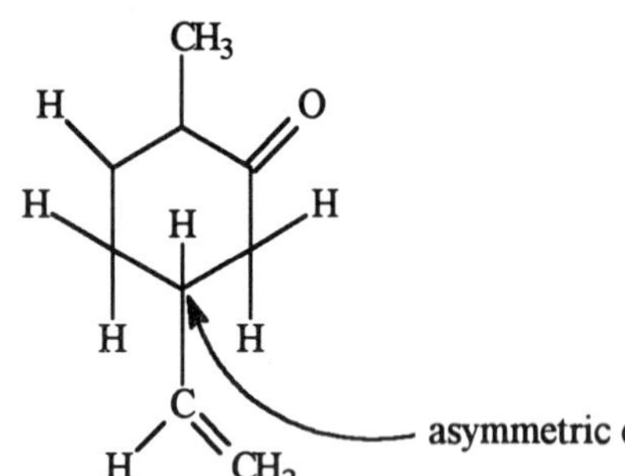

caraway

(b) The spearmint molecule is the optical isomer of the caraway molecule and differs from it in structure at the asymmetric carbon atom.

41. A compound of formula $C_3H_8O_2$:

(a) is chiral; contains two OH groups $\quad CH_3CH(OH)CH_2OH$

(b) is chiral; contains one OH group $\quad CH_3—O—CH(OH)CH_3$

(c) is achiral; contains two OH groups $\quad CH_2(OH)CH_2CH_2(OH)$

42. Ephedrine has two asymmetric carbons and can have four stereoisomers. This number is calculated using 2^n. $2^2 = 4$.

43.

$H_2N—C(H)(CH_3)—C(=O)—OH$ (asymmetric carbon)

$HO—C(=O)—C(H)(NH_2)—CH_3$ (mirror image)

asymmetric carbon $\qquad$ mirror image

44. (–) placed in front of a name is used to indicate the rotation of plane-polarized light to the left, and (+) for rotation to the right. Therefore, we can write (–)-methorphan for levomethorphan and (+)-methorphan for dextromethorphan. There is no obvious correlation between the structures of enantiomers and the direction in which they rotate plane polarized light.

45. Stereoisomer structures

(a) 2-bromo-3-chlorobutane

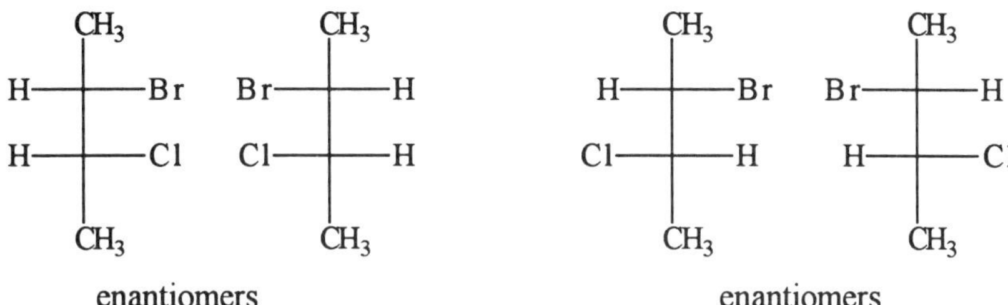

(b) 2,3,4-trichloro-1-pentanol

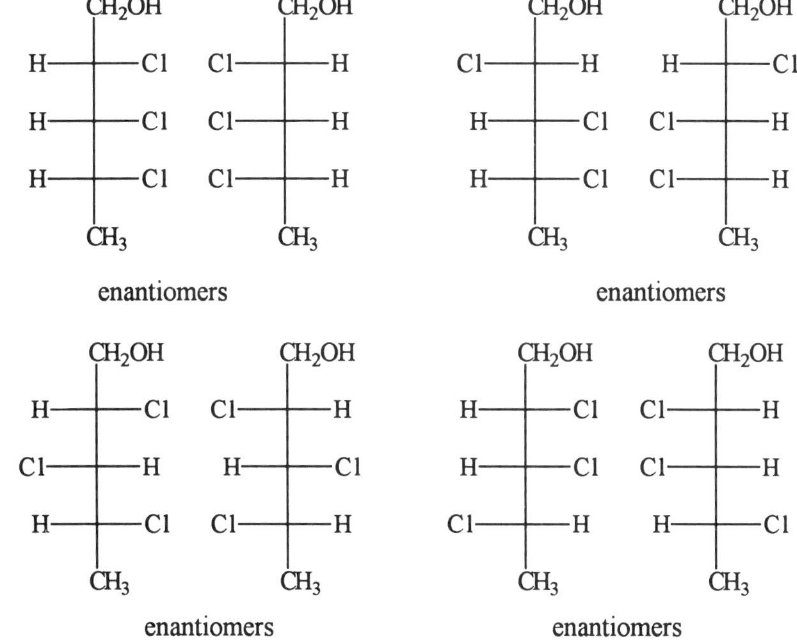

There are no meso compounds.

CHAPTER 28

CARBOHYDRATES

1. In general, the carbohydrate carbon oxidation state determines the carbon's metabolic energy content. The more oxidized a carbon is, the less energy it can provide in biological systems.

2. The notations D and L in the name of a carbohydrate specify the configuration on the last asymmetric carbon. If that configuration is the same as D-glyceraldehyde, then the carbohydrate is designated as D. If it is the same as L-glyceraldehyde, it is designated as L.

3. The notations (+) and (–) in the name of a carbohydrate specify whether the compound rotates the plane of polarized light to the right (+) or to the left (–).

4. Galactosemia is the inability of infants to metabolize galactose. The galactose concentration increases markedly in the blood and also appears in the urine. Galactosemia causes vomiting, diarrhea, enlargement of the liver, and often mental retardation. If not recognized a few days after birth it can lead to death.

5. There are four pairs of epimers among the D-aldohexoxes in Figure 28.1. They are: allose and altrose; glucose and mannose; gulose and idose; and galactose and talose.

6. A carbohydrate forms a five member or six member heterocyclic ring (one oxygen atom, the rest carbon atoms). If it forms a five-member ring, it is termed a furanose, after the compound furan. If it forms a six-membered ring it is termed a pyranose, after the compound pyran.

Furan   Pyran

7. α-D-glucopyranose and β-D-glucopyranose differ in the configuration at the number 1 carbon in the cyclic structure. In the open-chain structure, carbon 1 is the aldehyde group, and is not asymmetric. In the cyclic structure that carbon contains a hemiacetal structure, which is asymmetric. When the ring forms, carbon 1 can have two configurations leading to the two structures called α and β.

8. The cyclic forms of monosaccharides are hemiacetals, because the number one carbon has an ether and an alcohol group; whereas a glycoside is an acetal which contains two ether linkages.

9. Mutarotation is the phenomenon by which the α or β form of a sugar, when in solution, will undergo change to reach an equilibrium mixture, not necessarily 50%-50%, of the two forms. To achieve this equilibrium, the chain must open up and then reclose. On closing, it has the possibility of closing in either the α or β form as the equilibrium mixture is achieved.

10. Major sources:

(a) sucrose:	sugar beets and sugar cane
(b) lactose:	milk
(c) maltose:	sprouting grain and partially hydrolyzed starch

11. The following parts are related to the eight D-aldohexoses shown in the text (Figure 28.1):

(a) Four pairs will each produce the same osazone. They are allose and altrose; glucose and mannose; gulose and idose; and galactose and talose.

(b) If each of the aldohexoses is oxidized by nitric acid to dicarboxylic acids, allose and galactose would become meso forms.

allose		$\xrightarrow{HNO_3}$			galactose		$\xrightarrow{HNO_3}$		
	CHO			COOH		CHO			COOH
H—	—OH		H—	—OH	H—	—OH		H—	—OH
H—	—OH		H—	—OH	HO—	—H		HO—	—H
H—	—OH		H—	—OH	HO—	—H		HO—	—H
H—	—OH		H—	—OH	H—	—OH		H—	—OH
	CH_2OH			COOH		CH_2OH			COOH

allose galactose

(c) Names and structures of enantiomers of D-altrose and D-idose:

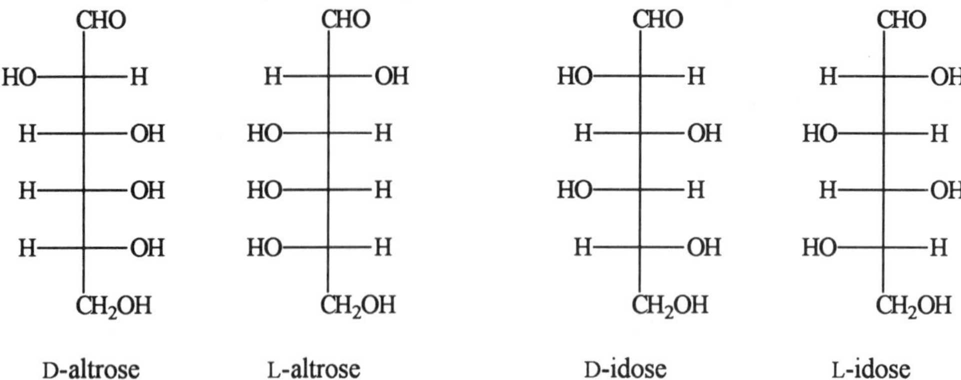

D-altrose L-altrose D-idose L-idose

12. Invert sugar is sweeter than sucrose because it is a 50-50 mixture of fructose and glucose. Glucose is somewhat less sweet than sucrose, but fructose is much sweeter, so the mixture is sweeter.

13. Amylose is a linear polymer of D-glucopyranose units linked by α-1,4-glycosidic bonds while cellulose is a linear polymer of D-glucopyranose units linked by β-1,4-glycosidic bonds. Amylose molecules take the shape of a "coil" while cellulose molecules form fibers.

14. In the Benedict test, both concentrated and dilute glucose solutions will give a precipitate. However, the concentrated glucose solution will produce a redder precipitate while the dilute glucose solution will appear more greenish-yellow.

15. The two main components of starch are amylose and amylopectin. They are both composed of glucose units joined by α-1,4-glycosidic linkages. The difference is that amylopectin also has branching which occurs through α-1,6-glycosidic linkages about every 25 glucose units.

16. The correct statements are: c, e, g, h, i, l, n, p, q, r, s, v, w, x. The others are not correct for these reasons:

    (a) α-D-glucopyranose and β-D-glucopyranose differ only at the anomeric carbon. Enantiomers (mirror-images) must differ at all chiral centers.

    (b) The carbon of a secondary alcohol has an oxidation number of 0, while the carbon of a primary alcohol has an oxidation number of -1. Thus, the primary alcohol carbon is more reduced.

    (d) To be epimers, D-threose and L-threose would have to be identical at carbon numbers three and four, but they are opposites, not identical, thus they are not epimers.

    (f) Osazones eliminate any differences at carbons one and two, but D-glucose and L-glucose would be opposites at carbons three, four and five after forming the osazones.

(j) D-mannitol is a sugar alcohol and is obtained from D-mannose by reduction, not oxidation.

(k) Methyl glucosides are acetals, are stable in alkaline solutions, and thus the ring does not open to form an aldehyde with reducing ability.

(m) Fructose is a hexose, a monosaccharide, a ketose, but not an aldose.

(o) The disaccharide found in mammalian milk is lactose. Galactose is one of the two monosaccharides in lactose.

(t) Invert sugar is composed of both glucose and fructose and is not as sweet as pure fructose.

(u) Sucrose is not a reducing sugar because it lacks the functional groups which will react in the Benedict test (for example, aldehydes, ketones adjacent to hydroxyl groups, hemiacetals, hemiketals).

(y) Aspartame can provide metabolic energy but much less than would be derived from an amount of sucrose needed for an equivalent sweetness.

(z) The function of mucopolysaccharides depends on their ability to act as acids, not bases.

17. (a)

$$\begin{array}{l} CH_2OH \\ | \\ C{=}O \\ | \\ CH_2OH \end{array}$$

There are no asymmetric carbon atoms in dihydroxyacetone.

(b) If dihydroxyacetone is reacted with hydrogen in the presence of a platinum catalyst, the product will be:

$$\begin{array}{ccc} & CH_2OH & \\ & | & \\ H- & C & -OH \\ & | & \\ & CH_2OH & \end{array}$$

18. (a) D-glyceraldehyde

H—C=O
|
H—C—OH
|
CH_2OH

L-glyceraldehyde

H—C=O
|
HO—C—H
|
CH_2OH

(b) If D-glyceraldehyde is reacted with hydrogen in the presence of a platinum catalyst, the product will be:

CH_2OH
|
H—C—OH
|
CH_2OH

19. The structure of an epimer of D-mannose will differ from D-mannose at one asymmetric carbon atom. The epimer could differ at carbon 2 <u>or</u> carbon 3 <u>or</u> carbon 4 <u>or</u> carbon 5. One possible answer is

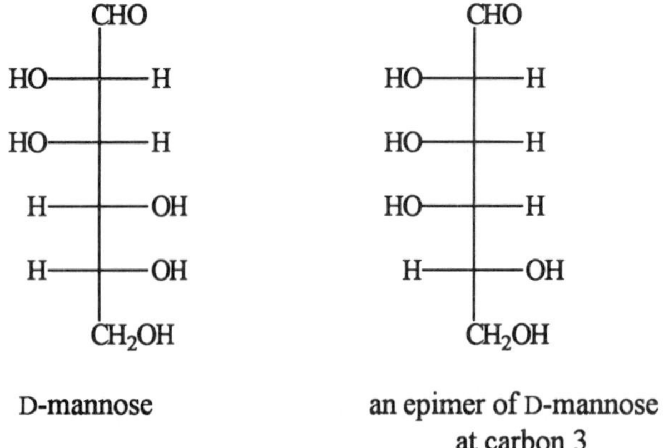

D-mannose

an epimer of D-mannose at carbon 3

20. The structure of an epimer of D-galactose will differ from D-galactose at one asymmetric carbon atom. The epimer could differ at carbon 2 <u>or</u> carbon 3 <u>or</u> carbon 4 <u>or</u> carbon 5. One possible answer is

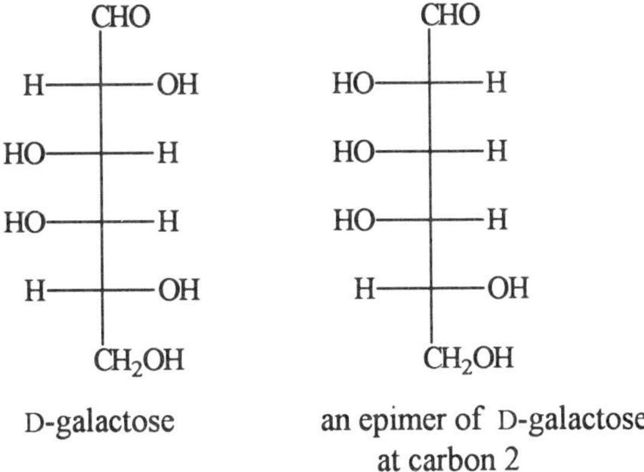

D-galactose

an epimer of D-galactose at carbon 2

21. The enantiomers are L-galactose, L-mannose and L-ribose which are mirror images of D-galactose, D-mannose and D-ribose.

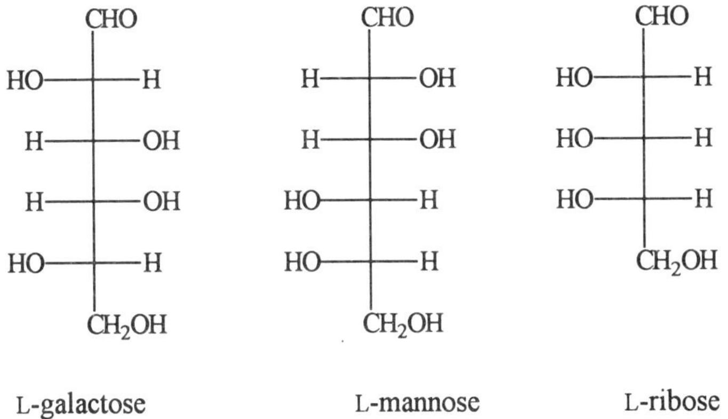

L-galactose L-mannose L-ribose

22. The enantiomers are L-glucose, L-fructose and L-2-deoxyribose which are mirror images of D-glucose, D-fructose and D-2-deoxyribose

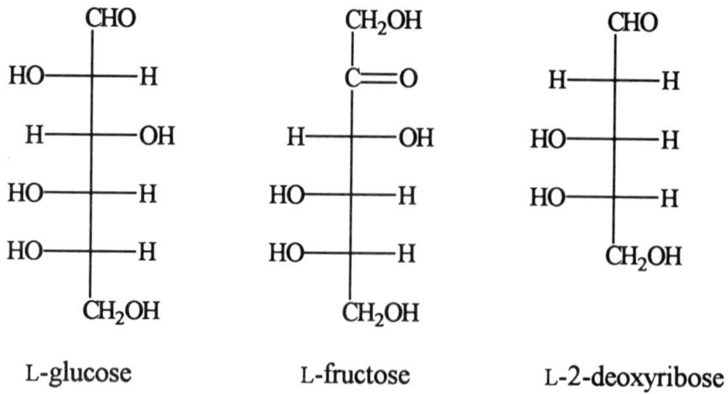

23. Either the Fischer projection formula or Haworth formulas are satisfactory.

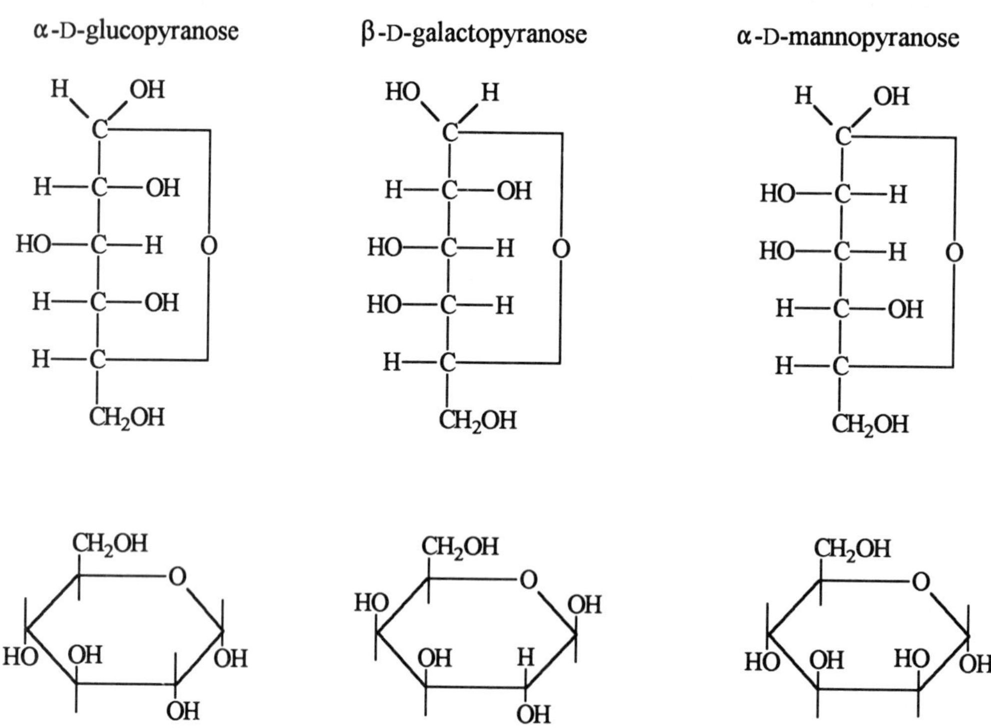

24. Either the Fischer projection formulas or Haworth formulas are satisfactory.

β-D-glucopyranose α-D-galactopyranose β-D-mannopyranose

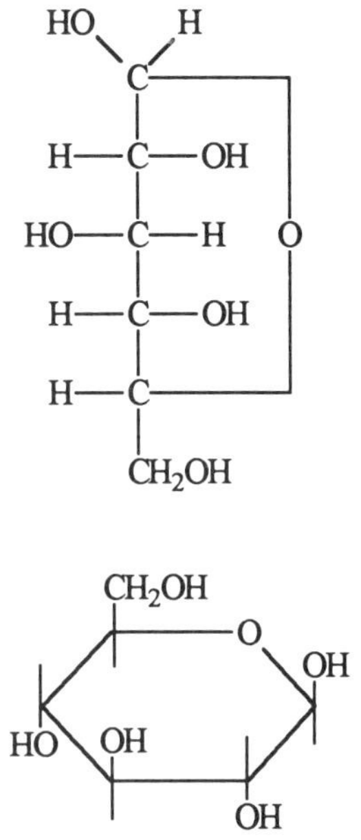

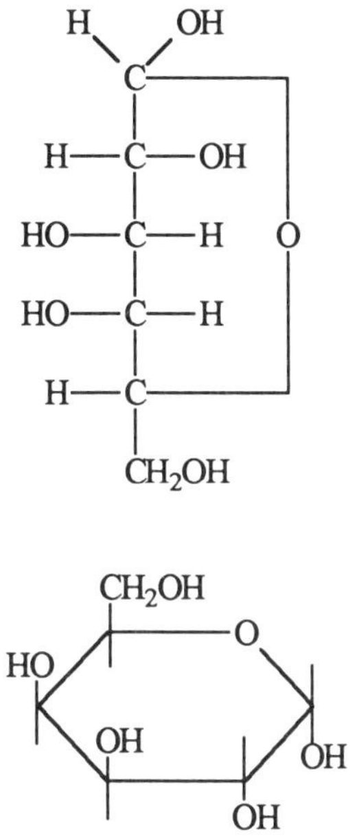

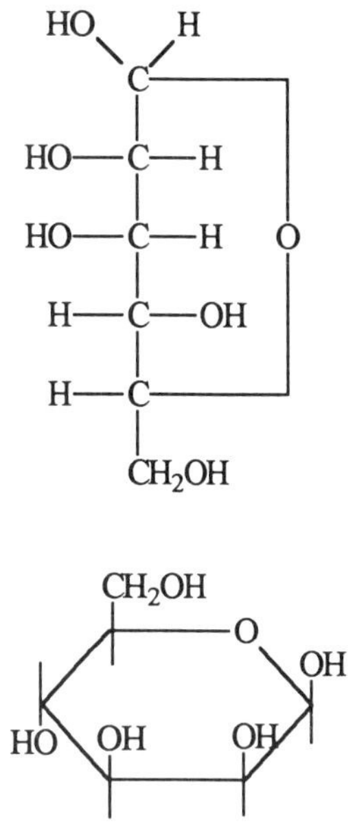

25. The structure of dihydroxyacetone is

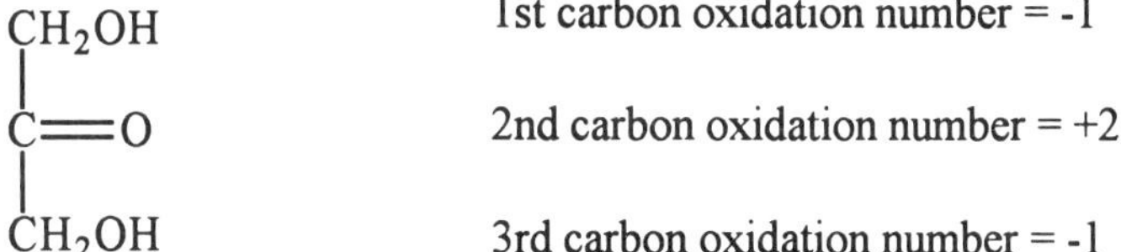

1st carbon oxidation number = -1

2nd carbon oxidation number = +2

3rd carbon oxidation number = -1

Carbon 2 has the most positive oxidation number and is the most oxidized.

26. The structure of L-glyceraldehyde is

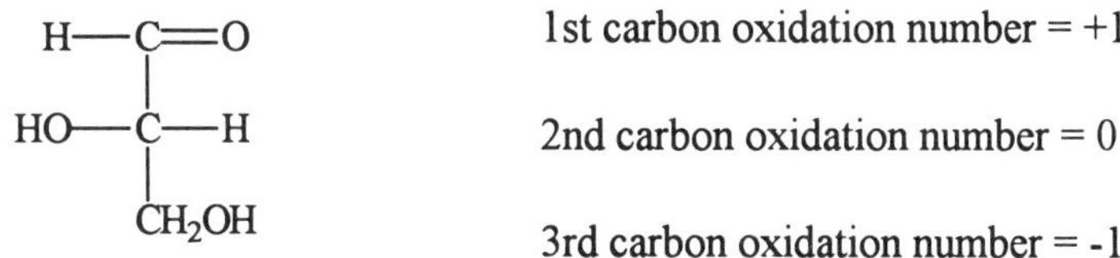

1st carbon oxidation number = +1

2nd carbon oxidation number = 0

3rd carbon oxidation number = -1

Carbon number 3 has the most negative oxidation number and is the most oxidized.

27. Kiliani-Fischer synthesis of D-glucose from the proper D-tetrose (D-erythrose)

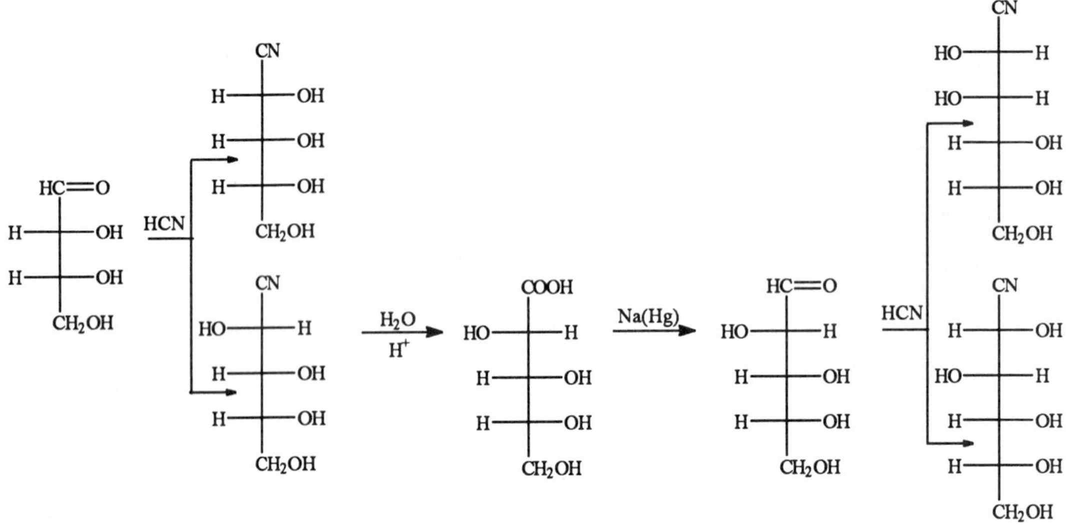

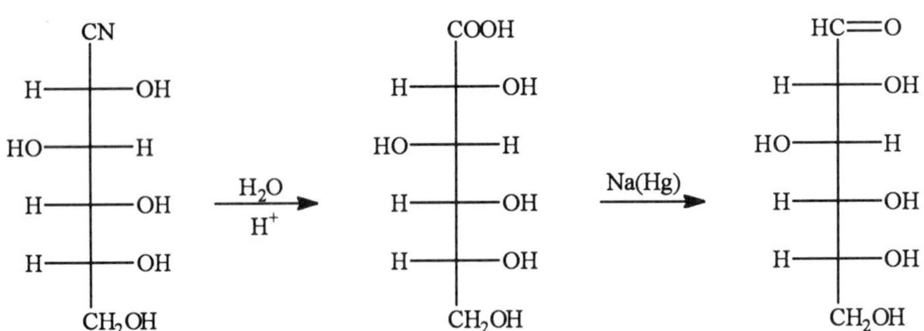

D-glucose

28. Kiliani-Fischer synthesis of D-ribose starts with the proper D-triose, D-glyceraldehyde.

HC=O
H—C—OH
CH_2OH

HCN →

CN
HO—C—H
H—C—OH
CH_2OH

CN
H—C—OH
H—C—OH
CH_2OH

$\xrightarrow[H^+]{H_2O}$

COOH
H—C—OH
H—C—OH
CH_2OH

$\xrightarrow{Na(Hg)}$

HC=O
H—C—OH
H—C—OH
CH_2OH

HCN →

CN
HO—C—H
H—C—OH
H—C—OH
CH_2OH

CN
H—C—OH
H—C—OH
H—C—OH
CH_2OH

CN
H—C—OH
H—C—OH
H—C—OH
CH_2OH

$\xrightarrow[H^+]{H_2O}$

COOH
H—C—OH
H—C—OH
H—C—OH
CH_2OH

$\xrightarrow{Na(Hg)}$

HC=O
H—C—OH
H—C—OH
H—C—OH
CH_2OH

D-ribose

29. Yes, D-2-deoxymannose is the same as D-2-deoxyglucose. It was at carbon 2 that the structures of D-mannose and D-glucose were different, so if the carbon 2 OH group is changed to H, the difference has been reduced.

30. No, D-2-deoxygalactose is not the same as D-2-deoxyglucose. D-galactose differs from D-glucose at carbon 4, so replacement of the carbon 2 OH with an H does not makes these two sugars identical.

31. The monosaccharide composition of:

(a) sucrose: one glucose and one fructose unit
(b) glycogen: many glucose units
(c) amylose: many glucose units
(d) maltose: two glucose units

32. The monosaccharide composition of:

(a) lactose: one glucose and one galactose unit
(b) amylopectin: many glucose units
(c) cellulose: many glucose units
(d) sucrose: one glucose and one fructose unit

33. Both cellobiose and isomaltose are disaccharides composed of two glucose units. However, in cellobiose monosaccharides are linked by a β-1,4-acetal bond while for isomaltose the linkage is α-1,6.

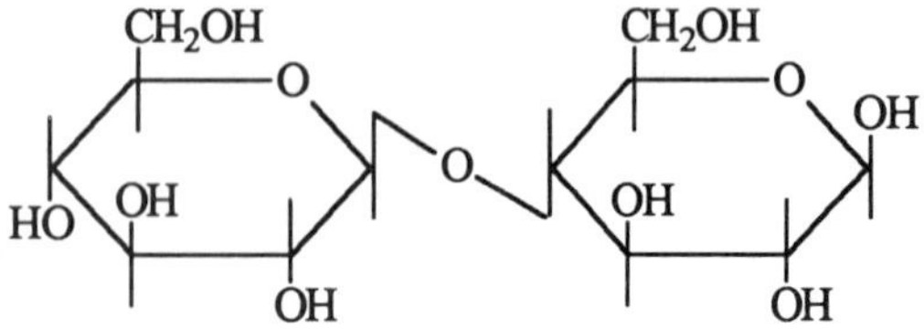

cellobiose

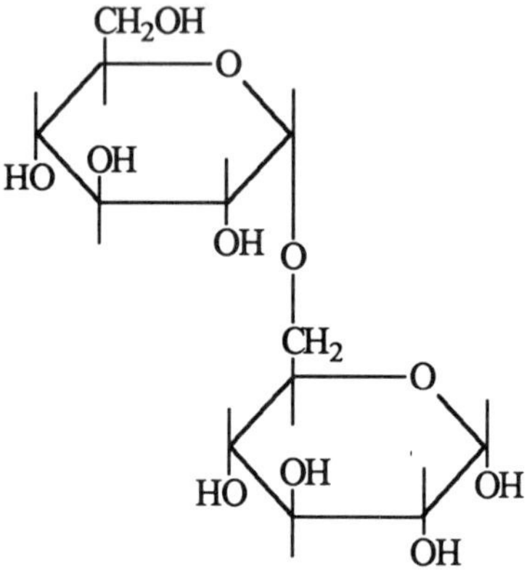

isomaltose

34. Both maltose and isomaltose are disaccharides composed of two glucose units. The glucose units in maltose are linked by an α-1,4-glycosidic bond while the glucose units at isomaltose are linked by an α-1,6-glycosidic bond.

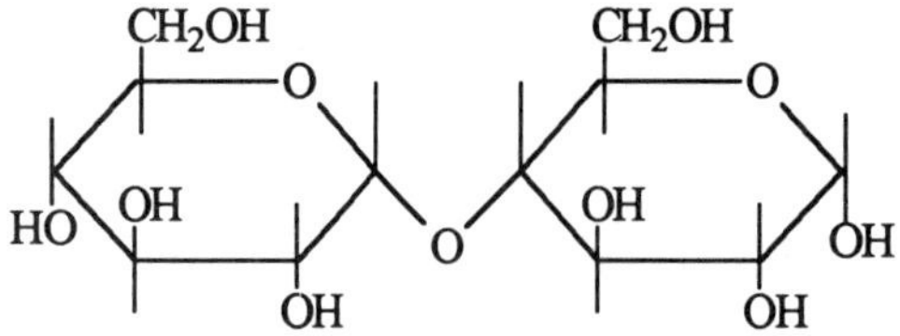

maltose

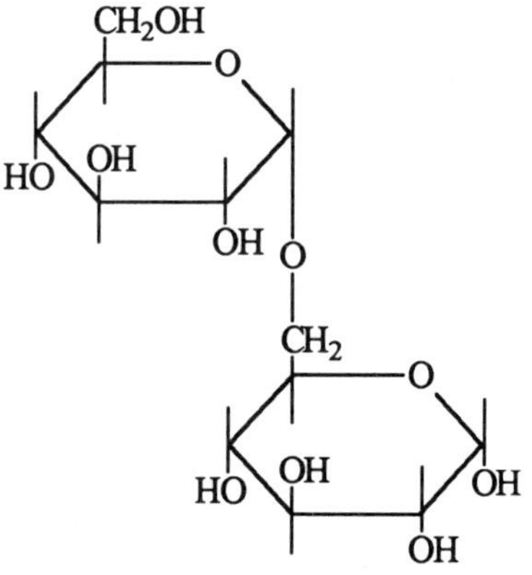

isomaltose

35. Lactose will show mutarotation; sucrose will not. The hemiacetal structure in lactose will open allowing mutarotation. Since sucrose has an acetal structure and no hemiacetal, it will not undergo mutarotation.

36. Both maltose and isomaltose will show mutarotation. Both disaccharides contain a hemiacetal structure which will open allowing mutarotation.

37.

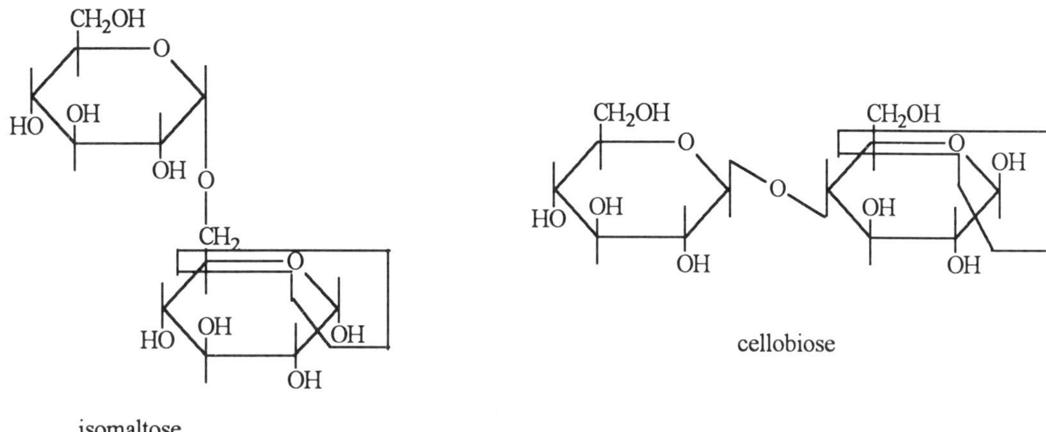

isomaltose

cellobiose

The circled hemiacetal structures allow these two disaccharides to be reducing sugars.

38.

maltose

lactose

The circled hemiacetal structures allow these two disaccharides to be reducing sugars.

39. The systematic name for isomaltose is α-D-glucopyranosyl-(1,6)-α-D-glucopyranose. The systematic name for cellobiose is β-D-glucopyranosyl-(1,4)-β-D-glucopyranose.

40. The systematic name for maltose is α-D-glucopyranosyl-(1,4)-α-D-glucopyranose. The systematic name for lactose is β-D-galactopyranosyl-(1,4)-α-D-glucopyranose.

41. The principal differences and similarities between the members of the following pairs:

 (a) D-glucose and D-fructose. Glucose is an aldose; fructose is a ketose. Glucose often forms a pyranose ring structure; fructose commonly is found in a furanose structure. Both are hexoses and both are reducing sugars.

 (b) Maltose and sucrose. Maltose is composed of two glucose units; sucrose is composed of one glucose and one fructose unit. Maltose is a reducing sugar; sucrose is not. Both are common disaccharides, with formulas $C_{12}H_{22}O_{11}$.

 (c) Cellulose and glycogen. Cellulose is composed of glucose units linked by β-1,4-glycosidic linkages; glycogen is composed of glucose units linked by α-1,4-glycosidic linkages. Glycogen is much more readily hydrolyzed or digested than cellulose. Both are large polymers of glucose. Glycogen is of animal origin; cellulose is from plants.

42. The principal differences and similarities between members of the following pairs:

 (a) D-ribose and D-2-deoxyribose. The D-2-deoxyribose has no OH group on the number 2 carbon, only 2 hydrogen atoms. Both are five carbon sugars.

 (b) Amylose and amylopectin. Amylose is a straight chain polysaccharide; amylopectin has branched chains, and more monomer units per molecule. Both are large polysaccharides composed of α-D-glucose units.

 (c) Lactose and isomaltose. These sugars are disaccharides. Lactose is composed of one galactose unit and one glucose unit while isomaltose is composed of two glucose units. The monosaccharide units in lactose are linked by a β-1,4-glycosidic bond while the units in isomaltose are linked by an α-1,6-glycosidic bond. Both disaccharides also contain hemiacetal structures.

43.

```
H—C=O                         COOH
H——OH                      H——OH
HO——H                      HO——H
HO——H  + HNO3  ———>   HO——H
H——OH                      H——OH
  CH2OH                        COOH
```

D-galactose $+ HNO_3 \longrightarrow$ mucic acid (galactaric acid)

44.

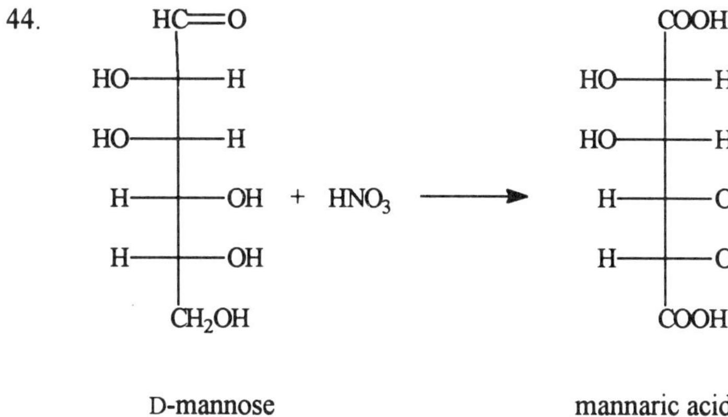

45. The formulas for the four L-ketohexoses are

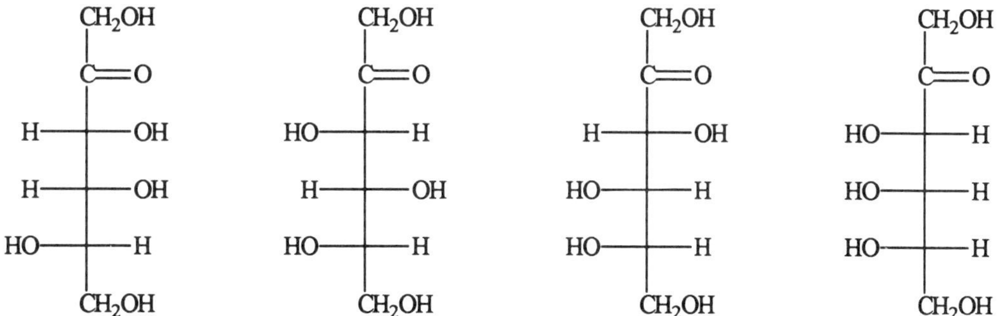

Phenylhydrazine reacts at carbon 1 and carbon 2 of a monosaccharide to form an osazone. Since the four L-ketohexoses differ only at carbons 3 and 4, these differences are unaffected by the phenylhydrazine reaction. All four osazones will be different.

46. The formulas for the four L-aldopentoses are

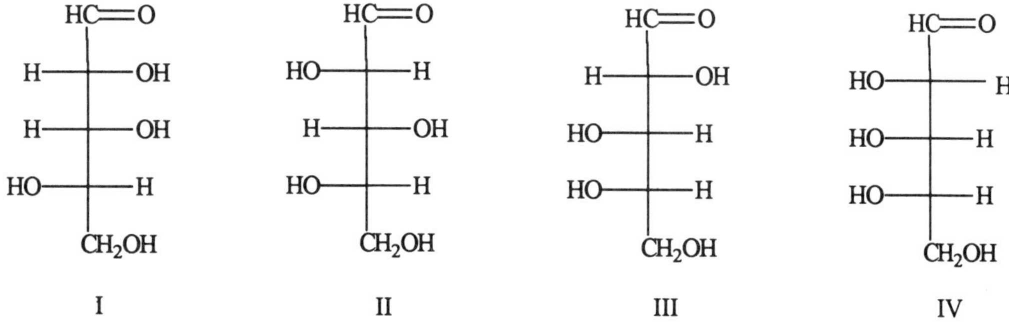

Pairs of monosaccharides which differ only at carbon 1 and/or carbon 2 will form the same osazone upon reaction with phenylhydrazine. Thus, structures I and II will form the same osazone. Structures III and IV will form the same osazone.

47. (a) β-D-mannopyranosyl-(1,4)-β-D-galactopyranose

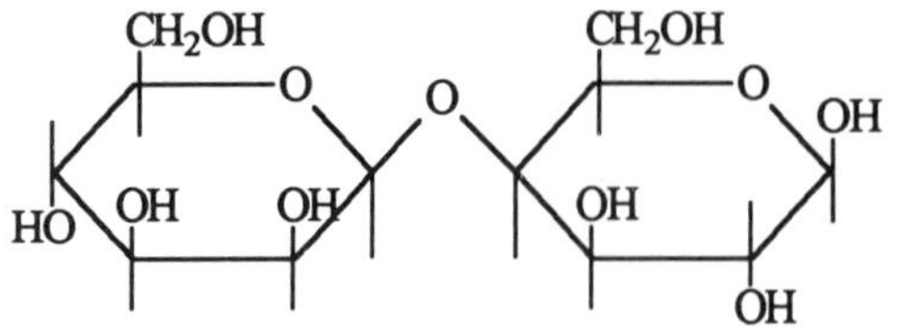

(b) β-D-galactopyranosyl-(1,6)-α-D-glucopyranose

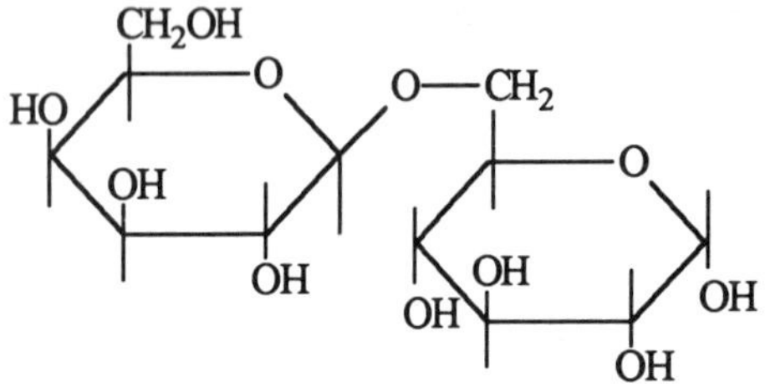

48. The Haworth formula for β-D-glucopyranosyl-(1,4)-α-D-galactopyranose is

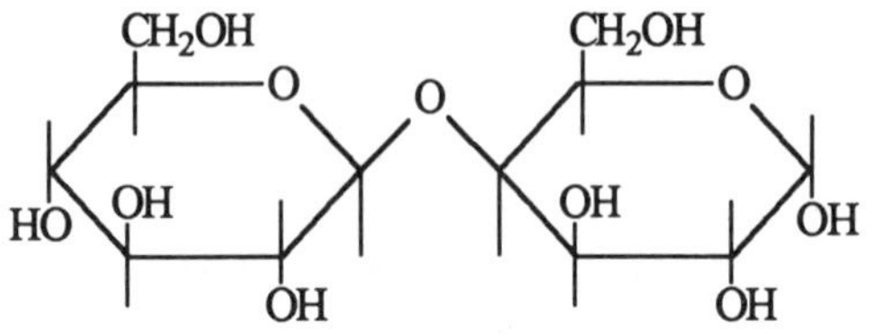

The Haworth formula for β-D-galactopyranosyl-(1,6)-β-D-mannopyranose is

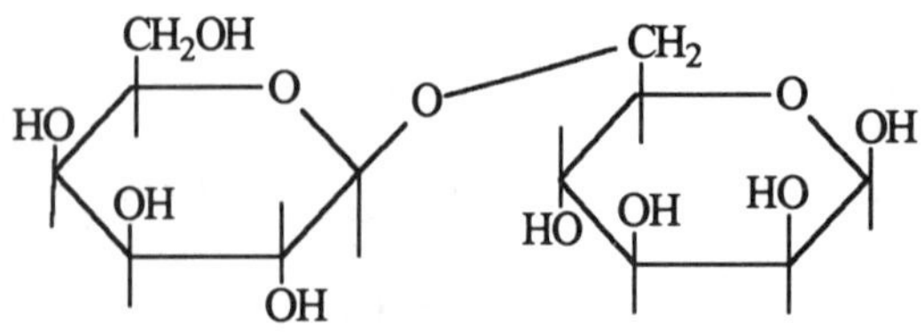

49. D-mannose and D-galactose should provide about equal amounts of metabolic energy. The carbons in each carbohydrate have equivalent oxidation numbers and, thus, will release equivalent energy when they are oxidized.

50. Aspartame supplies many fewer calories than sucrose. In addition, oral bacteria cannot use aspartame as efficiently as sucrose and will form fewer dental carries.

51. The two α-D-glucopyranose units of trehalose have the following structure:

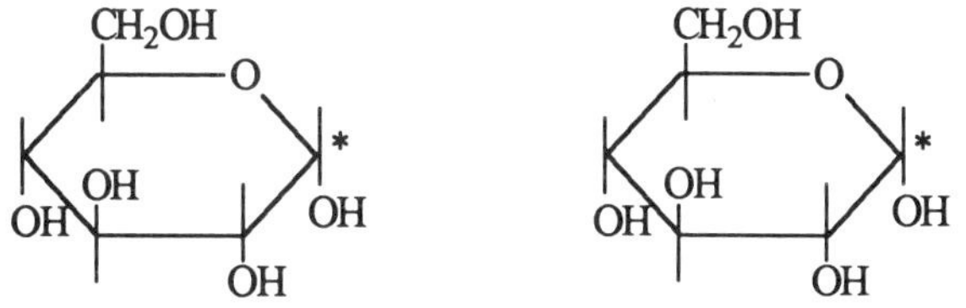

Since trehalose is a nonreducing disaccharide, each α-D-glucopyranose unit must use its hemiacetal structure to make the glycosidic linkage. The two starred carbons are involved in the glycosidic linkage.

52. Mucopolysaccharides serve a protective, lubricant function by absorbing water and taking on a slimy, spongy consistency. The acidic nature of mucopolysaccharides means these polymers have many negative charges at physiological pH. Since negative charges repel each other, the polymer chains will move far apart making large holes to be filled by water, much like a sponge.

53. The hydrolysis of glycosidic linkages is acid catalyzed. Because lemon juice contains acid it will cause sucrose to be hydrolyzed to glucose and fructose. The candy will become sweeter because fructose tastes sweeter than sucrose.

54. High-fructose corn syrup is produced by breaking down some corn starch polymers to D-glucose monomers which are then converted to D-fructose, a very sweet monosaccharide.

55. (a) The sugar acid could be most easily derived from β-D-mannopyranose.

(b)

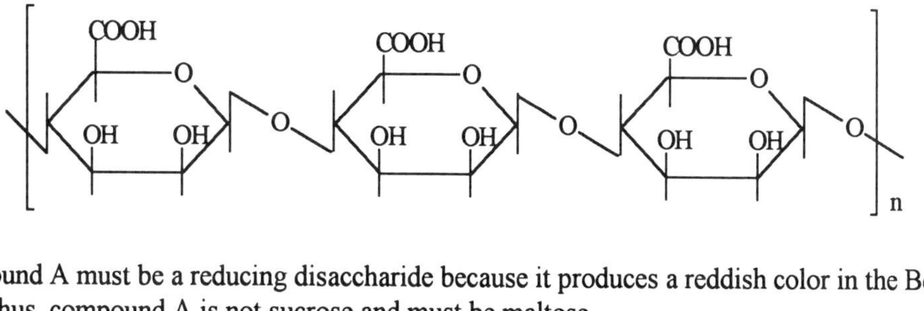

56. Compound A must be a reducing disaccharide because it produces a reddish color in the Benedict test. Thus, compound A is not sucrose and must be maltose.

57. (a) All starred carbons (4) in the following compound are chiral:

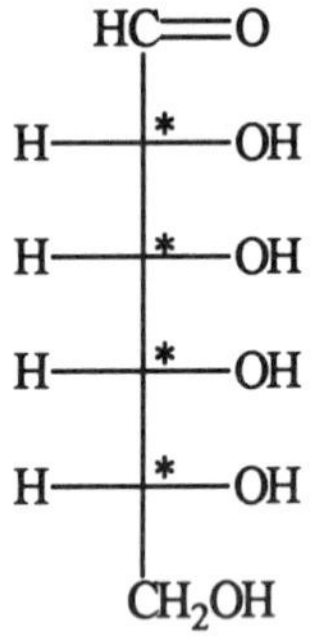

HC=O
H—*—OH
H—*—OH
H—*—OH
H—*—OH
CH_2OH

(b)

CH_2OH * O * * OH OH * * OH OH OH

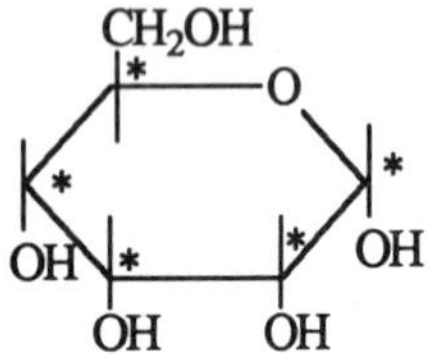

(c) All starred carbons (5) in the structure drawn in part (b) are chiral carbons.

58. If the compound (I) shown below rotates light 25° to the right, its enantiomer will rotate light 25° to the left.

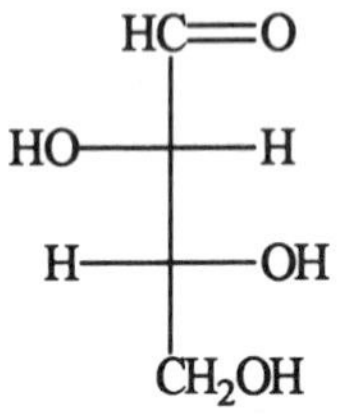

HC=O
HO—H
H—OH
CH_2OH
I

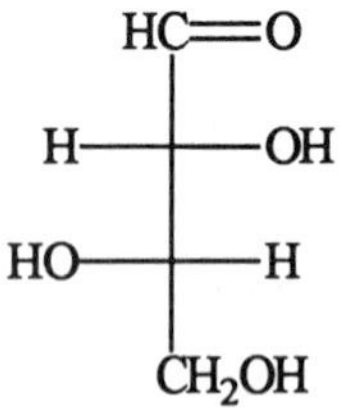

HC=O
H—OH
HO—H
CH_2OH
enantiomer of I

These compounds are not epimers because they differ at more than one chiral carbon.

59. A nonreducing disaccharide composed of two molecules of α-D-galactopyranose can have no hemiacetal structures. Thus, the hemiacetal structure of one α-D-galactopyranose must be used to form the glycosidic link to the hemiacetal structure of the other α-D-galactopyranose unit.

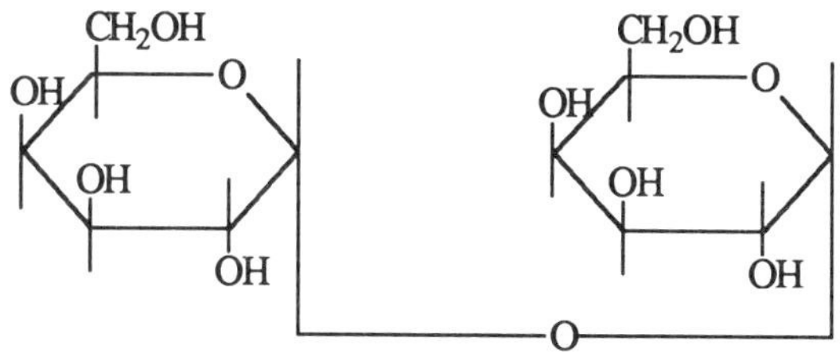

α-D-galactopyranosyl-(1,1)-α-D-galactopyranose

60. The β-1,4-glycosidic linkage in cellulose allows this polysaccharide to form fibers which are not digestible by humans. Starch (amylose and amylopectin) is composed of polysaccharides which have α-1,4-glycosidic linkages and α-1,6-glycosidic linkages; these polysaccharides can be digested.

61. (a) D-galactose and D-glucose differ only at carbon 4. Thus, D-galactose must be changed at carbon 4 to be converted to D-glucose.

    (b) D-galactose is an epimer of D-glucose.

62. No, the classmate should not be believed. Although D-glucose and D-mannose are related as epimers, it is pairs of enantiomers which yield equal and opposite optical rotation.

CHAPTER 29

LIPIDS

1. The lipids, which are dissimilar substances, are arbitrarily classified as a group on the basis of their solubility in fat solvents and their insolubility in water.

2. Although caproic acid, $CH_3(CH_2)_4COOH$, has the same number of polar bonds as stearic acid, $CH_3(CH_2)_{16}COOH$, caproic acid has a shorter nonpolar hydrocarbon chain and, therefore is more water soluble.

3. Arachidonic acid is used to synthesize the eicosanoids, hormone-like substances, which include the prostaglandins, the leukotrienes, the prostacyclins and the thromboxanes.

4. The three essential fatty acids are linoleic, linolenic, and arachidonic acids. Diets lacking these fatty acids lead to impaired growth and reproduction, and skin disorders such as eczema and dermatitis.

5. Fats contain more biochemical energy than carbohydrates because (a) the carbons of fat are more reduced than those found in carbohydrates and (b) there are more reduced carbons per gram of fat than from the same mass of carbohydrate.

6. Aspirin relieves inflammation by blocking the conversion of arachidonic acid to prostaglandins.

7. Waxes serve as a protective coating because they are very hydrophobic. They do not dissolve in water and many compounds cannot pass through a wax coating.

8. A membrane lipid must be (a) partially hydrophobic to act as a barrier to water and (b) partially hydrophilic, so that the membrane can interact with water along its surface.

9. Phospholipids are mainly produced in the liver.

10. The four classes of eicosanoids are prostaglandins, prostacyclins, thromboxanes and leukotrienes.

11. In general, a membrane lipid will have both hydrophilic and hydrophobic parts. Sphingomyelin can serve as a membrane lipid because its phosphate and choline groups are hydrophilic and its two long carbon chains are hydrophobic.

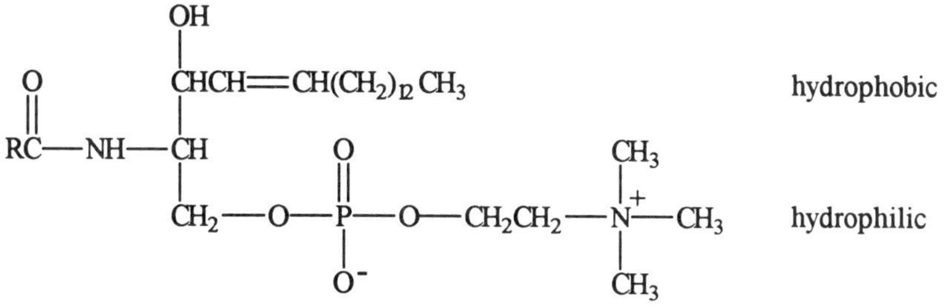

```
                 OH
                 |
 O               CHCH=CH(CH2)12CH3                              hydrophobic
 ||              |
RC—NH—CH       O                   CH3
                 |         ||                   |+
                 CH2—O—P—O—CH2CH2—N—CH3          hydrophilic
                           |                    |
                           O-                   CH3
```

sphingomyelin

12. Atherosclerosis is the deposition of cholesterol and other lipids on the inner walls of the large arteries. These deposits, called plaque, accumulate, making the arterial passages narrower and narrower. Blood pressure increases as the heart works to pump sufficient blood through the restricted passages. This may lead to a heart attack, or the rough surface can lead to coronary thrombosis.

13. Dietary cholesterol is transported first to the liver where it is bound to other lipids and proteins to form the very low density lipoprotein (VLDL). This aggregate moves through the blood stream delivering lipids to various tissues. As lipids are removed, the VLDL is converted to a low density lipoprotein (LDL). Cells needing cholesterol can absorb LDL. Cholesterol is transported back to the liver by the high density lipoprotein (HDL).

14. Dietary fish oils provide fatty acids which inhibit formation of thromboxanes, compounds which participate in blood clotting.

15. HDL is a cholesterol scavenger, picking up this steroid in the serum and returning it to the liver.

16. Because the interior of a lipid bilayer is very hydrophobic, molecules with hydrophilic character only cross the lipid bilayer with difficulty. A lipid bilayer acts as a barrier to water-soluble compounds.

17. Liposomes can package drugs so that only the target organ/tissue is exposed to the drug's effects.

18. Both facilitated diffusion and active transport catalyze movement of compounds through membranes. Active transport requires an input of energy while facilitated diffusion does not.

19. All steroids possess a 17-carbon unit structure containing four fused rings known as the steroid nucleus. Many steroids could be shown; these are two common ones:

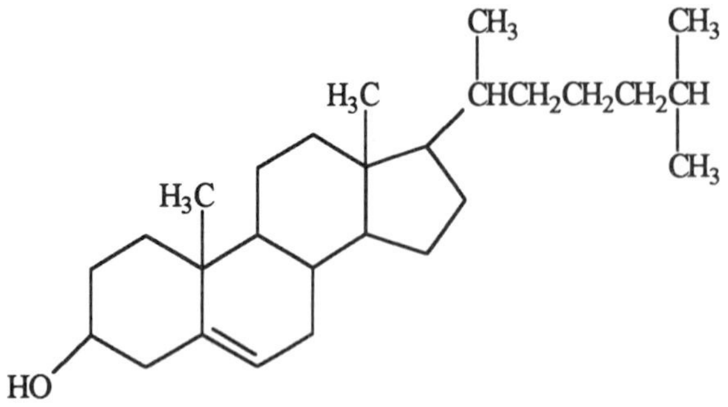

cholesterol

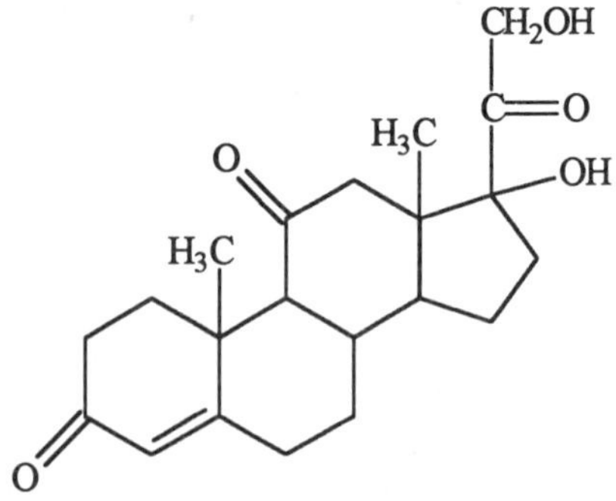

cortisone

20. The correct statements are: d, e, g, j, l, m and n. The others are not correct for these reasons:

(a) In general, a lipid is hydrophobic and does not interact well with water.
(b) As a class, lipids are relatively large molecules (containing more than about 10 carbons per molecule).
(c) A triacylglycerol is high in biochemical energy because this molecules contains many reduced carbons.
(f) A sphingolipid may contain carbohydrate but does not always have this component.
(h) Facilitated diffusion will only allow a net movement of molecules from areas of high concentration to areas of lower concentration. This process requires no net energy input.
(i) A lipid bilayer has a hydrophobic interior.
(k) Testosterone is actually the precursor of estradiol.

21. A triacylgylcerol containing one unit each of palmitic, stearic, and oleic acids:

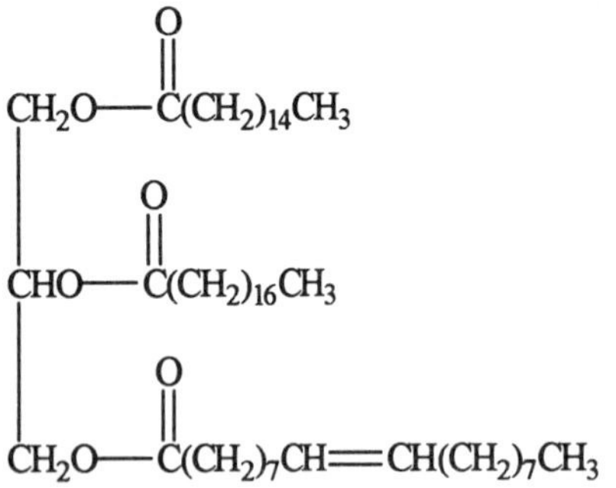

$CH_2O-C(=O)(CH_2)_{14}CH_3$ palmitic acid

$CHO-C(=O)(CH_2)_{16}CH_3$ stearic acid

$CH_2O-C(=O)(CH_2)_7CH=CH(CH_2)_7CH_3$ oleic acid

There would be two other triacylglycerols possible from these same components. Since the top and bottom attachments are equivalent, it only matters which acid is attached to the middle carbon of glycerol.

22. A triacylgylcerol containing two units of palmitic acid and one unit of oleic acid:

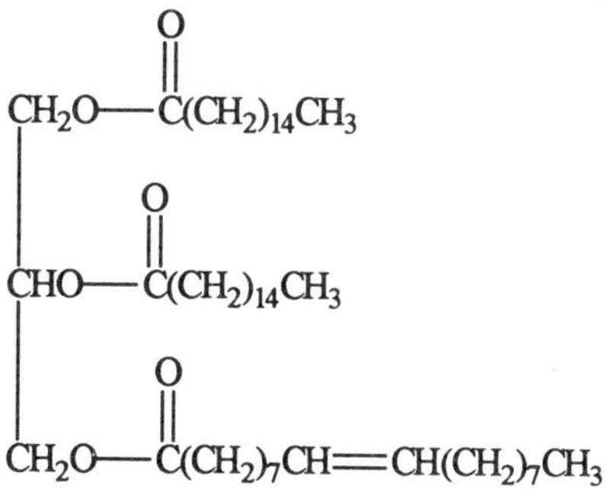

$$\begin{array}{ll} CH_2O\!-\!\overset{O}{\overset{\|}{C}}(CH_2)_{14}CH_3 & \text{palmitic acid} \\ | & \\ CHO\!-\!\overset{O}{\overset{\|}{C}}(CH_2)_{14}CH_3 & \text{palmitic acid} \\ | & \\ CH_2O\!-\!\overset{O}{\overset{\|}{C}}(CH_2)_7CH{=}CH(CH_2)_7CH_3 & \text{oleic acid} \end{array}$$

There is one other possible triacylglycerol with the same components; this triacylglycerol would have palmitic acid units at both ends with the oleic acid unit bound to the middle carbon of glycerol.

23. No, a triacylglycerol that contains three units of lauric acid would be <u>less</u> hydrophobic than a triacylglycerol that contains three units of stearic acid. Smaller molecules tend to be less hydrophobic than larger molecules, and, a triacylglycerol with three lauric acid units is smaller than a triacylglycerol with three stearic acid units.

24. Yes, a triacylglycerol that contains three units of stearic acid would be more hydrophobic than a triacylglycerol that contains three units of myristic acid. Larger molecules tend to be more hydrophobic than smaller molecules and, a triacylglycerol with three stearic acid units is larger than a triacylglycerol with three myristic acid units.

25. Hydrolysis of a triacylglycerol will yield three fatty acids and glycerol. The products will be:

$$\begin{array}{ll} CH_2OH & \\ | & \\ CHOH & \text{glycerol} \\ | & \\ CH_2OH & \end{array}$$

$CH_3(CH_2)_{14}COOH$ palmitic acid

$CH_3(CH_2)_7CH{=}CH(CH_2)_7COOH$ oleic acid

$CH(CH_2)_4CH{=}CHCH_2CH{=}CH(CH_2)_7COOH$ linoleic acid

26. Hydrolysis of a triacylglycerol will yield three fatty acids and glycerol. The products will be:

$$\begin{array}{l} CH_2OH \\ | \\ CHOH \\ | \\ CH_2OH \end{array} \qquad \text{glycerol}$$

$CH_3(CH_2)_7CH{=}CH(CH_2)_7COOH$ oleic acid

$CH_3(CH_2)_{14}COOH$ palmitic acid

$CH_3(CH_2)_{16}COOH$ stearic acid

27. The phospholipid structure is:

$$\begin{array}{l} CH_2O{-}\overset{O}{\overset{\|}{C}}(CH_2)_{14}CH_3 \\ | \\ CHO{-}\overset{O}{\overset{\|}{C}}(CH_2)_{14}CH_3 \\ | \\ CH_2O{-}\overset{O}{\overset{\|}{\underset{O^-}{\underset{|}{P}}}}{-}OCH_2CH_2NH_3^+ \end{array}$$

The phosphoric acid and ethanolamine must be linked to the bottom glycerol carbon. Since a typical phospholipid contains two fatty acid units, a palmitic acid unit must be linked to the top glycerol carbon and another to the middle glycerol carbon.

28. The phospholipid structure is:

$$\begin{array}{l} CH_2O{-}\overset{O}{\overset{\|}{C}}(CH_2)_{16}CH_3 \\ | \\ CHO{-}\overset{O}{\overset{\|}{C}}(CH_2)_{16}CH_3 \\ | \\ CH_2O{-}\overset{O}{\overset{\|}{\underset{O^-}{\underset{|}{P}}}}{-}OCH_2CH_2\overset{CH_3}{\overset{|}{\underset{CH_3}{\underset{|}{N^+}}}}{-}CH_3 \end{array}$$

The phosphoric acid and choline units must be linked to the bottom-most carbon. Since a typical phospholipid contains two fatty acid units, one stearic acid unit must be linked to the top glycerol carbon and a second stearic acid unit must be linked to the middle glycerol carbon.

29. The sphingolipid structure is:

OH

|

$CHCH{=}CH(CH_2)_{12}CH_3$

| O
| ||
$CHNH{-}C(CH_2)_7CH{=}CH(CH_2)_7CH_3$

| O CH_3
| || |+
$CH_2O{-}P{-}OCH_2CH_2N{-}CH_3$
         | |
         O^- CH_3

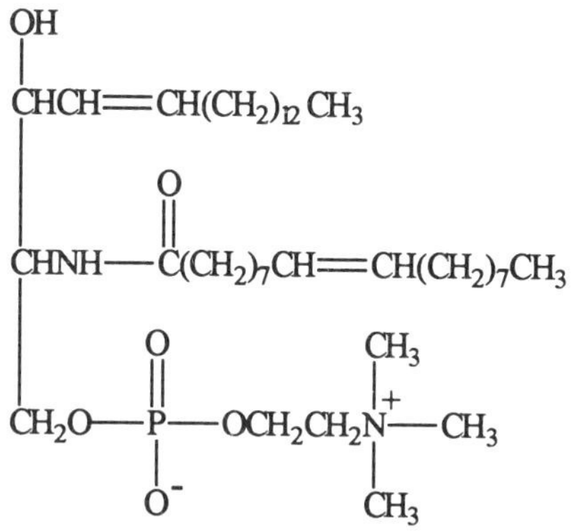

The phosphate and choline units are linked to the bottom-most carbon of sphingosine (when the molecules is written as above). The oleic acid unit is then linked to the nitrogen to form the sphingolipid.

30. The sphingolipid structure is:

OH

|

$CHCH{=}CH(CH_2)_{12}CH_3$

| O
| ||
$CHNH{-}C(CH_2)_{16}CH_3$

| O
| ||
$CH_2O{-}P{-}OCH_2CH_2\overset{+}{N}H_3$
         |
         O^-

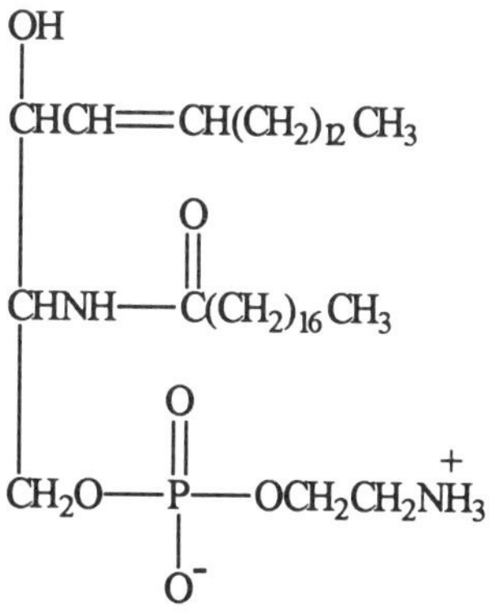

The phosphate and ethanolamine units are linked to the bottom-most sphingosine carbon (when the molecule is written as above). The stearic acid unit is then linked to the nitrogen to form the sphingolipid.

31. The glycolipid structure is:

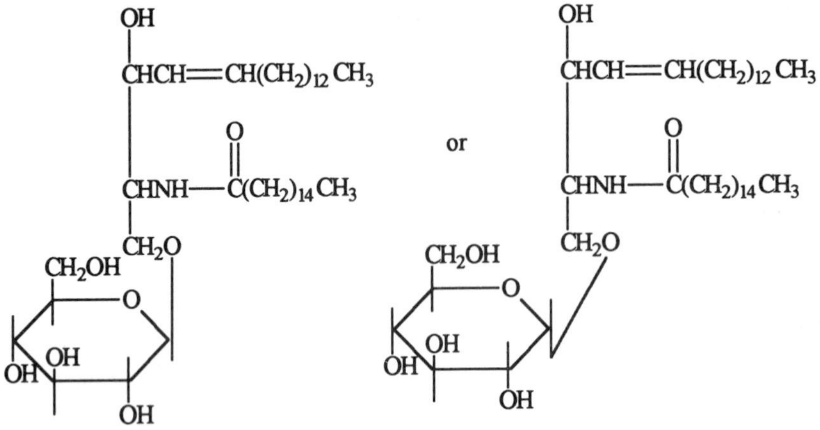

The D-glucose unit is linked to the bottom-most sphingosine carbon (when the molecule is written as above). The palmitic acid unit is then linked to the nitrogen to form the glycolipid.

32. The glycolipid structure is:

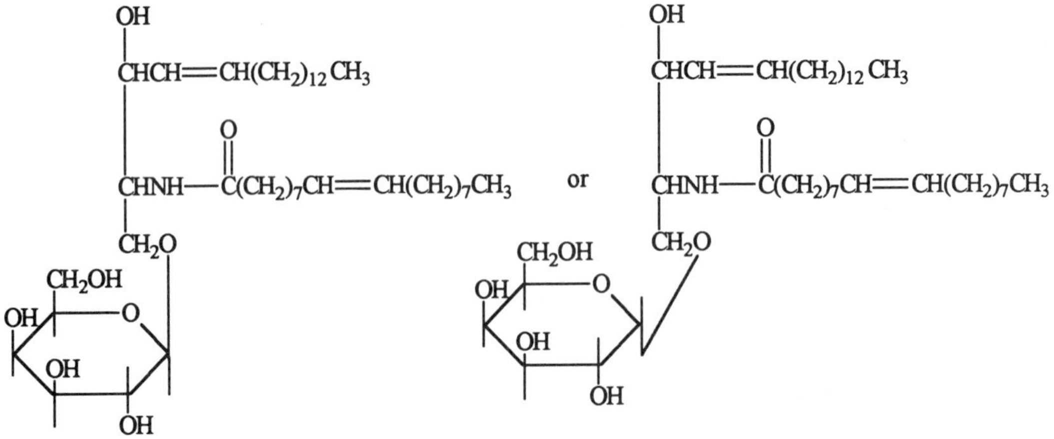

The D-galactose is linked to the bottom-most sphingosine carbon (when the molecule is written as above). The oleic acid unit is then linked to the nitrogen to form the glycolipid.

33.

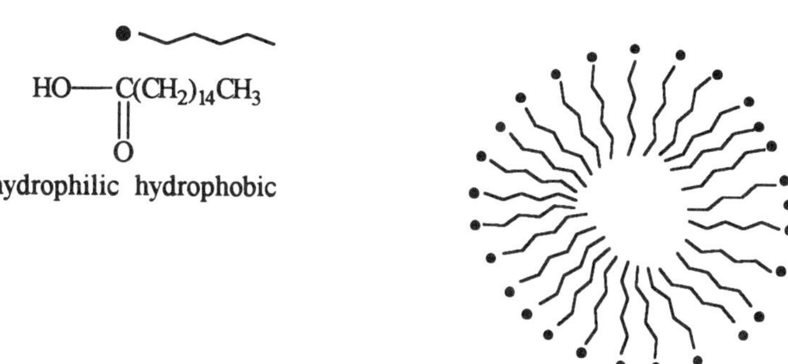

The micelle is a sphere with the fatty acid carboxyl groups on the hydrophilic exterior and the fatty acid alkyl chains in the hydrophobic interior. A condensed structural formula has been used to show the palmitic acid.

34.

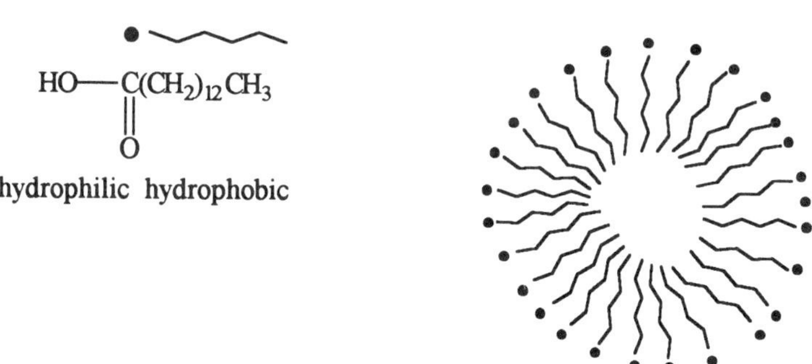

The micelle is a sphere with the fatty acid carboxyl groups on the hydrophilic exterior and the fatty acid alkyl groups in the hydrophobic interior. A condensed structural formula has been used to show the myristic acid.

35. LDL (low density lipoprotein) differs from VLDL (very low density lipoprotein) in that

(a) VLDL has a lower density than LDL.
(b) VLDL is formed in the liver while LDL is formed from VLDL as the lipoproteins circulate in the blood.
(c) VLDL is larger than LDL.

36. HDL (high density lipoprotein) differs from LDL (low density lipoprotein) in that

(a) LDL has a lower density than HDL.

(b) LDL delivers cholesterol to peripheral tissues while HDL scavenges cholesterol and returns it to the liver.

37.

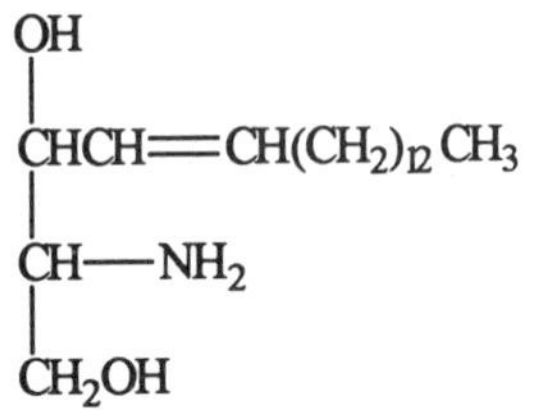

$$\begin{array}{l} OH \\ | \\ CHCH{=}CH(CH_2)_{12}CH_3 \\ | \\ CH{-}NH_2 \\ | \\ CH_2OH \end{array} \qquad \begin{array}{l} O \\ \| \\ CH_2{-}O{-}CR \\ | \\ CH{-}OH \\ | \\ CH_2OH \end{array}$$

sphingosine 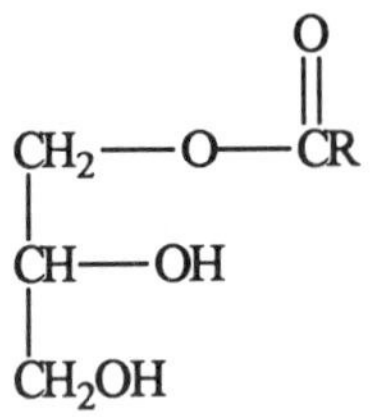 monacylglycerol

Sphingosine is similar to the monoacyglycerol in that (a) both molecules contain a long hydrophobic chain, (b) the sphingosine amino group reacts with a fatty acid as does the secondary alcohol of the monoacylglycerol and, (c) for both compounds, the primary alcohol can react further with either acids (to form esters) or sugars (to form acetals).

38.

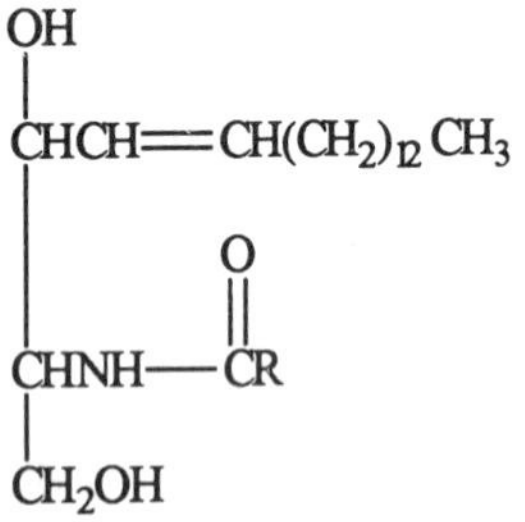

$$\begin{array}{l} OH \\ | \\ CHCH{=}CH(CH_2)_{12}CH_3 \\ | \qquad\quad O \\ | \qquad\quad \| \\ CHNH{-}CR \\ | \\ CH_2OH \end{array} \qquad \begin{array}{l} O \\ \| \\ CH_2O{-}CR \\ | \qquad\ O \\ | \qquad\ \| \\ CHO{-}CR \\ | \\ CH_2OH \end{array}$$

sphingosine and fatty acid unit 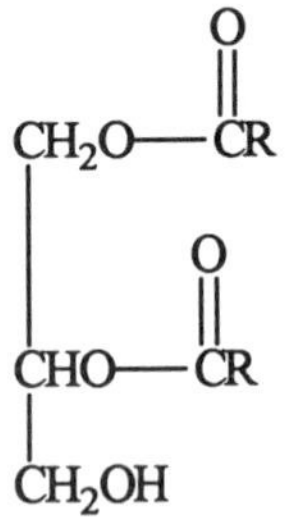 diacylglycerol

(a) Both compounds have two long hydrophobic chains.

(b) For both compounds, the primary alcohol can react further with either acids (to form esters) or sugars (to form acetals).

39. Sodium ion will move from a region of high concentration to a region of low concentration as it moves from a 0.1 M solution across a membrane to a 0.001 M solution. This process does not require energy and can be accomplished by facilitated diffusion.

40. Phosphate ion will move from a region of low concentration to a region of high concentration as it moves from a 0.1 M solution across a membrane to 0.5 M solution. This process requires energy and can be accomplished by active transport.

41. Thromboxanes acts as vasoconstrictors and stimulate platelet aggregation while prostaglandins cause the redness, swelling and pain associated with tissue inflammation.

42. Thromboxanes act as vasoconstrictors and stimulate platelet aggregation while leukotrienes have been associated with many of the symptoms of an allergy attack (e.g., an asthma attack).

43. Ibuprofen (an NSAID) blocks the oxidation of arachidonic acid to form prostaglandins which, in turn, can cause inflammation, redness, and swelling.

44. Both olestra and natural fats contain fatty acid units. However, in natural fats the fatty acid units are linked to glycerol while in olestra the fatty acid units are linked to sucrose.

45. (a) The three essential fatty acids are:

linoleic acid	$CH_3(CH_2)_4CH{=}CHCH_2CH{=}CH(CH_2)_7COOH$
linolenic acid	$CH_3CH_2CH{=}CHCH_2CH{=}CHCH_2CH{=}CH(CH_2)_7COOH$
arachidonic acid	$CH_3(CH_2)_4(CH{=}CHCH_2)_4CH_2CH_2COOH$

(b) A diet which is missing the essential fatty acids will lead to impaired growth and reproduction as well as skin disorders such as eczema and dermatitis.

46. Olestra is not shaped like a natural fat and is not attacked by digestive enzymes. Thus, olestra passes through stomach and intenstines undigested.

47. This meal is changed by increasing the fat content from 10 grams to 15 grams, an increase of 5 grams. Since each gram of fat yields an average of 9.5 Cal, an increase of 5 grams equates to an increase of 47.5 Cal (5 grams × 9.5 Cal/g). Thus, this meal will now contain 234 Cal + 47.5 Cal = 282 Cal.

CHAPTER 30

AMINO ACIDS, POLYPEPTIDES, AND PROTEINS

1. The amino acids of proteins are called alpha amino acids because the amine group is always attached to the alpha carbon atom, that is, the carbon atom next to the carboxyl group, COOH.

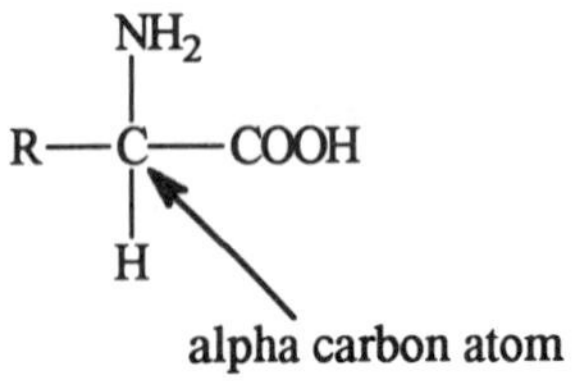

$$R{-}\overset{\displaystyle NH_2}{\underset{\displaystyle H}{\overset{|}{\underset{|}{C}}}}{-}COOH$$

alpha carbon atom

2. All amino acids and proteins contain carbon, hydrogen, oxygen, and nitrogen. Sulfur is contained in some of the amino acids, and thus in most proteins.

3. Proteins from some foods are of greater nutritional value than others because they are "complete", which means they contain all eight essential amino acids, those which the human body cannot synthesize.

4. The amino acids which are essential to humans are isoleucine, leucine, lysine, methionine, phenylalanine, threonine, tryptophan, and valine.

5. Amino acids are amphoteric because the carboxyl group can react with a base to form a salt, or the amine group can react with an acid to form a salt. They are optically active because the alpha carbon is asymmetric, except for glycine. They commonly have the L configuration at carbon two, as in L-serine.

6. At its isoelectric point, a protein molecule must have an equal number of positive and negative charges.

7. (a) Primary structure. The number, kind, and sequence of amino acid units comprising the polypeptide chain making up a molecule.

   (b) Secondary structure. The helix and pleated sheet structures of Pauling and Corey due to the hydrogen bonding between the O of C=O groups and the H of the N–H groups in the polypeptide chains.

   (c) Tertiary structure. The distinctive and characteristic conformation or shape of a protein molecule.

(d) Quaternary structure. The three dimensional shape formed by an aggregate of protein subunits found in some complex proteins.

8. The sulfur-containing amino acid, cysteine, has the special role in protein structure of creating disulfide bonding between polypeptide chains which helps control the shape of the molecule.

9. The major structural difference between hemoglobin and myoglobin is that hemoglobin is composed of four subunits while myoglobin only contains one. Hemoglobin's quaternary structure allows a more effective control of oxygen transport than is possible with myoglobin.

10. Both the α-helix and β-pleated sheet are examples of secondary protein structures. The α-helix forms a tube composed of a spiraling polypeptide chain while the β-pleated sheet forms a plane composed of polypeptide chains aligned roughly parallel to each other.

11. Hydrolysis breaks the peptide bonds, thus disrupting the primary structure of the protein. Denaturation involves alteration or disruption of the secondary, tertiary, or quaternary but not of the primary structure of proteins.

12. Amino acids containing a benzene ring give a positive xanthoproteic test (formation of yellow-colored reaction products). Among the common amino acids, these would include phenylalanine, tryptophan, and tyrosine.

13. The visible evidence observed in the:

    (a) Xanthoproteic reaction is a yellow-colored reaction product.

    (b) Biuret test is a violet color.

    (c) Ninhydrin test is a blue solution with all amino acids except proline and hydroxyproline, both of which produce a yellow solution.

14. Protein column chromatography uses a column packed with polymer beads (solid phase) through which a protein solution (liquid phase) is passed. Proteins separate based on differences in how they react with the solid phase. The proteins move through the column at different rates and can be collected separately.

15. (a) Thin layer chromatography is a way of separating substances based on a differential distribution between two phases, the liquid phase and the solid phase.

    (b) A strip (or sheet) is prepared with a thin coating (layer) of dried alumina or other adsorbent. A tiny spot of solution containing a mixture of amino acids is placed near the bottom of the strip. After the spot dries, the bottom edge of the strip is placed in a suitable solvent. The solvent rises in the strip, carrying the different amino acids upwards at different rates. When the solvent front nears the top, the strip is removed from the solvent and dried.

(c) Ninhydrin is the reagent used to locate the different amino acids on the strip.

16. The correct statements are: b, c, d, f, g, i, m, p, r, s, t and u. The others are incorrect for these reasons:

(a) Proteins are needed for growth and/or maintenance of body tissue.

(e) All amino acids except glycine have an asymmetric carbon atom and are therefore optically active.

(h) A compound with three amino acids is known as a tripeptide. The name is not based on the number of peptide linkages, but on the number of amino acids linked.

(j) The primary structure of a protein is the number, kind, and sequence of amino acids making up the protein.

(k) A protein's primary structure is the protein's sequence of amino acids and helps to determine the protein's tertiary structure (the overall three dimensional arrangement of the polypeptide chain).

(l) The amino acid residues in a peptide chain are numbered beginning with the N-terminal amino acid.

(n) Collagen is an example of a structural protein.

(o) β-pleated sheets often form a "twisted sheet" conformation.

(q) Denaturation involves the destruction of a protein's native three dimensional structure but does not cause the peptide bonds to break.

(v) The presence of the β-amyloid protein may be one of the causes of Alzheimer's disease not the absence of this protein.

17. D-alanine

```
      COOH
       |
H —— C —— NH2
       |
      CH3
```

L-alanine
(form commonly found in proteins)

```
         COOH
          |
H2N —— C —— H
          |
         CH3
```

18. D-serine

L-serine
(form commonly found in proteins)

COOH
|
H—C—NH_2
|
CH_2OH

COOH
|
H_2N—C—H
|
CH_2OH

19. $6.0 \text{ g nitrogen} \times \dfrac{100.\text{ g protein}}{16 \text{ g nitrogen}} \times \dfrac{1}{100.0 \text{ g food product}} \times 100 = 38\%$ protein.

20. $5.2 \text{ g nitrogen} \times \dfrac{100.\text{ g protein}}{16 \text{ g nitrogen}} \times \dfrac{1}{250.0 \text{ g hamburger}} \times 100 = 13\%$ protein

21. The structural formula for threonine at its isoelectric point is: CH_3—CH(OH)—CH($\overset{+}{N}H_3$)—COO^-

22. The structural formula for asparagine at its isoelectric point is: $NH_2C(=O)$—CH_2—CH($\overset{+}{N}H_3$)—COO^-

23. For phenylalanine:

(a) zwitterion formula

C_6H_5—$CH_2CH(\overset{+}{N}H_3)COO^-$

(b) formula in 0.1 M H_2SO_4

C_6H_5—$CH_2CH(\overset{+}{N}H_3)COOH$

(c) formula in 0.1 M NaOH

C_6H_5—$CH_2CH(NH_2)COO^-$

24. For tryptophan:

(a) zwitterion formula

(indol-3-yl)—$CH_2CH(\overset{+}{N}H_3)COO^-$ (ring N—H)

(b) formula in 0.1 M H_2SO_4

(indol-3-yl)—$CH_2CH(\overset{+}{N}H_3)COOH$ (ring N—H)

(c) formula in 0.1 M NaOH

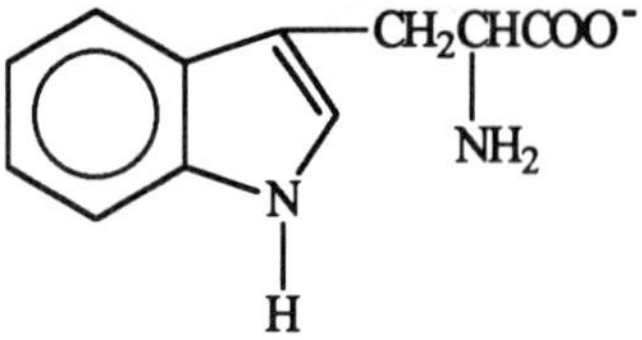

$$\text{(indol-3-yl)}-CH_2\underset{\displaystyle NH_2}{\underset{|}{C}}HCOO^-$$

25. Ionic equations showing how alanine acts as a buffer towards:

(a) H^+

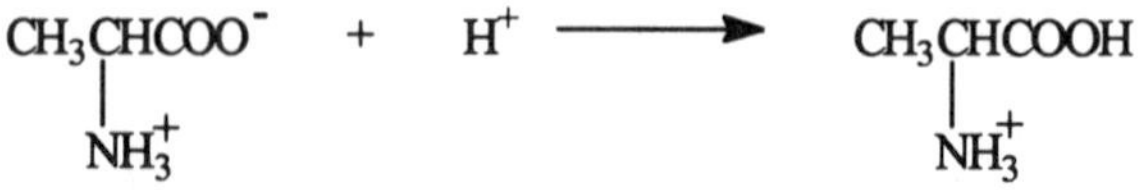

$$CH_3\underset{\displaystyle NH_3^+}{\underset{|}{C}}HCOO^- + H^+ \longrightarrow CH_3\underset{\displaystyle NH_3^+}{\underset{|}{C}}HCOOH$$

(b) OH^-

$$CH_3\underset{\displaystyle NH_3^+}{\underset{|}{C}}HCOO^- + OH^- \longrightarrow CH_3\underset{\displaystyle NH_2}{\underset{|}{C}}HCOO^- + H_2O$$

26. Ionic equations showing how leucine acts as a buffer towards:

(a) H^+

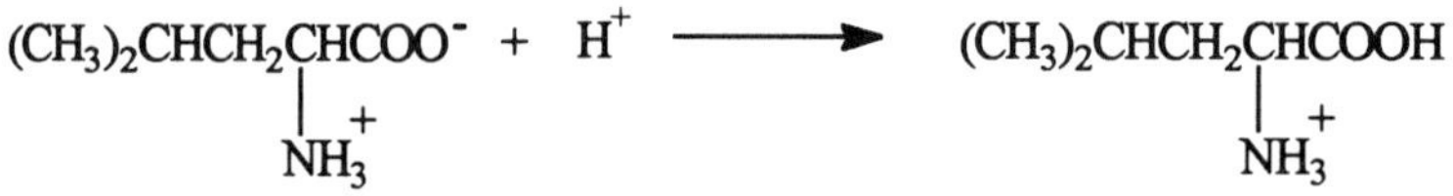

$$(CH_3)_2CHCH_2\underset{\displaystyle NH_3^+}{\underset{|}{C}}HCOO^- + H^+ \longrightarrow (CH_3)_2CHCH_2\underset{\displaystyle NH_3^+}{\underset{|}{C}}HCOOH$$

(b) OH^-

$$(CH_3)_2CHCH_2\underset{\displaystyle NH_3^+}{\underset{|}{C}}HCOO^- + OH^- \longrightarrow (CH_3)_2CHCH_2\underset{\displaystyle NH_2}{\underset{|}{C}}HCOO^- + H_2O$$

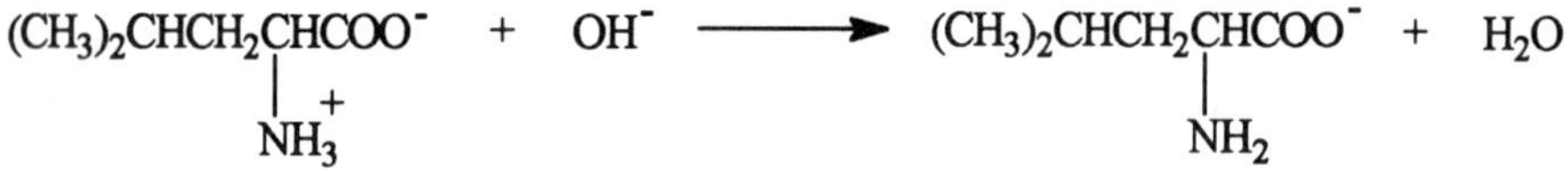

27. Methionine will have the following structure at its isoelectric point:

$$CH_3SCH_2CH_2\underset{\displaystyle NH_3^+}{\underset{|}{C}}HCOO^-$$

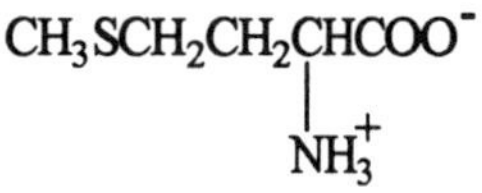

28. Valine will have the following structure at its isoelectric point:

$$(CH_3)_2CH\underset{\displaystyle NH_3^+}{\underset{|}{C}}HCOO^-$$

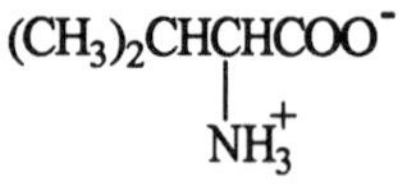

29. The two dipeptides containing serine and alanine:

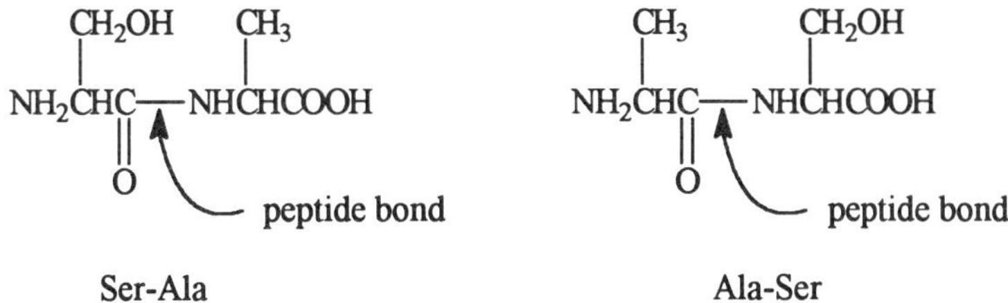

Ser-Ala Ala-Ser

30. The two dipeptides containing glycine and threonine:

$NH_2CH_2\overset{O}{\overset{\|}{C}}\text{—}NH\overset{CH_3CHOH}{CH}COOH$ (arrow: peptide bond)

Gly-Thr

$NH_2\overset{CH_3CHOH}{CH}\overset{O}{\overset{\|}{C}}\text{—}NHCH_2COOH$ (arrow: peptide bond)

Thr-Gly

31. (a) glycylglycine $NH_2CH\overset{O}{\overset{\|}{C}}\text{—}NHCH_2COOH$

(b) alanylglycylserine $NH_2\overset{CH_3}{CH}\overset{O}{\overset{\|}{C}}\text{—}NHCH_2\overset{O}{\overset{\|}{C}}\text{—}NH\overset{CH_2OH}{CH}COOH$

(c) glycylserylglycine $NH_2CH_2\overset{O}{\overset{\|}{C}}\text{—}NH\overset{CH_2OH}{CH}\overset{O}{\overset{\|}{C}}\text{—}NHCH_2COOH$

32. (a) alanylalanine $NH_2\overset{CH_3}{CH}\overset{O}{\overset{\|}{C}}\text{—}NH\overset{CH_3}{CH}COOH$

(b) serylglycylglycine

$$\underset{\substack{|\\ CH_2OH}}{}\ NH_2\overset{CH_2OH}{\overset{|}{C}}H\underset{\underset{O}{\|}}{C}\text{—}NHCH_2\underset{\underset{O}{\|}}{C}\text{—}NHCH_2COOH$$

(c) serylglycylalanine

$$NH_2\overset{CH_2OH}{\overset{|}{C}}H\underset{\underset{O}{\|}}{C}\text{—}NHCH_2\underset{\underset{O}{\|}}{C}\text{—}NH\overset{CH_3}{\overset{|}{C}}HCOOH$$

33. All the possible tripeptides containing one unit each of glycine, phenylalanine, and leucine:

Gly-Phe-Leu Gly-Leu-Phe Phe-Gly-Leu

Phe-Leu-Gly Leu-Gly-Phe Leu-Phe-Gly

34. All the possible tripeptides containing one unit each of tyrosine, aspartic acid, and alanine:

Tyr-Asp-Ala Tyr-Ala-Asp Asp-Tyr-Ala

Asp-Ala-Tyr Ala-Tyr-Asp Ala-Asp-Tyr

35. The oxygen atom in the peptide bond and the hydrogen atom in the peptide bond participate in hydrogen bonding to hold together a β-pleated sheet:

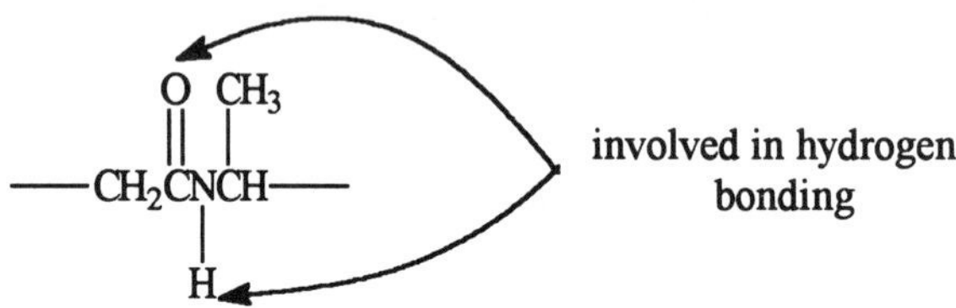

36. The oxygen atom in the peptide bond and the hydrogen bond in the peptide bond participate in hydrogen bonding to hold together an α-helix:

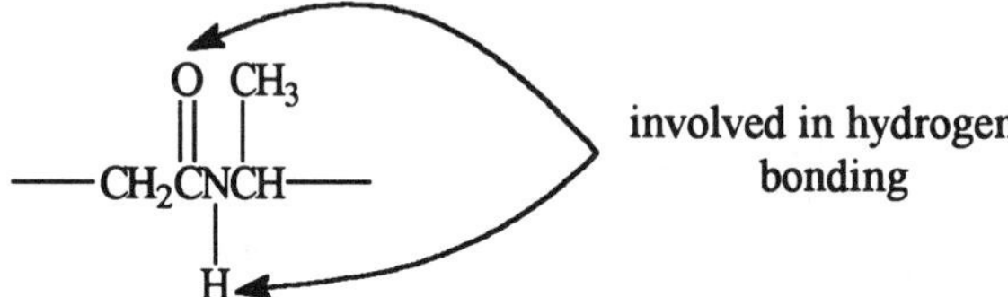

37. Tertiary protein structure is usually held together by bonds between amino acid side chains. Serine side chains will hydrogen bond to each other:

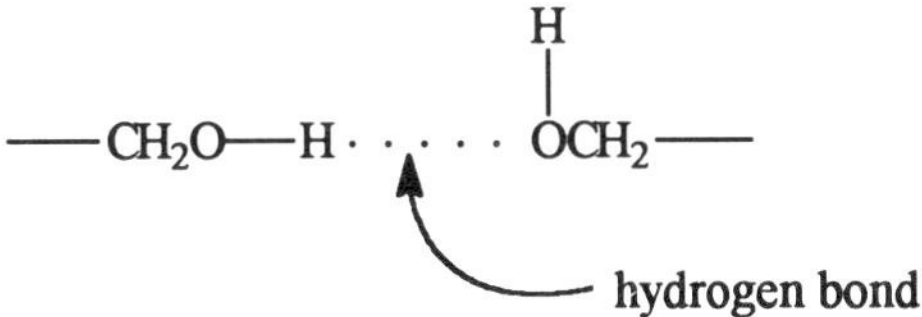

38. Tertiary protein structure is usually held together by bonds between amino acid side chains. At pH = 7, the lysine side chain will contain a positive charge, $\overset{+}{N}H_3CH_2CH_2CH_2CH_2^-$, and the aspartic acid side chain will contain a negative charge, $^-OOCCH_2$–. These two side chains will be held together by an ionic bond:

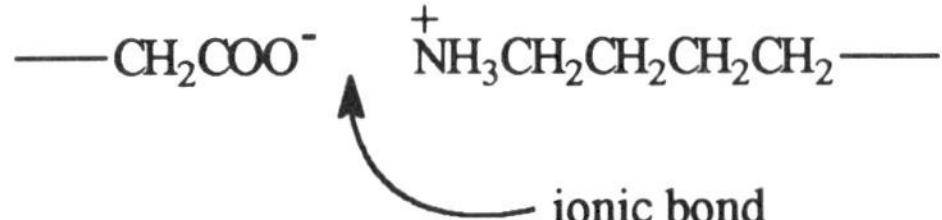

39. The tripeptide, Gly-Ala-Thr, will

(a) react with Sanger's reagent; the N-terminal amino acid, glycine, will be identified.
(b) not react to give a positive xanthoproteic test because there are no aromatic amino acids in this tripeptide.
(c) react with ninhydrin to give a blue solution.

40. The tripeptide, Gly-Ser-Asp, will

(a) react with Sanger's reagent; the N-terminal amino acid, glycine, will be identified.
(b) not react to give a positive xanthoproteic test because there are no aromatic amino acids in this tripeptide.
(c) react with ninhydrin to give a blue solution.

41. Hydrolysis breaks the peptide bonds. One water molecule will react with each peptide bond, a hydrogen atom attaches to the nitrogen to complete the amino group and an –OH group attaches to the carboxyl carbon. The tripeptide, Ala-Phe-Asp, will hydrolyze to yield the following:

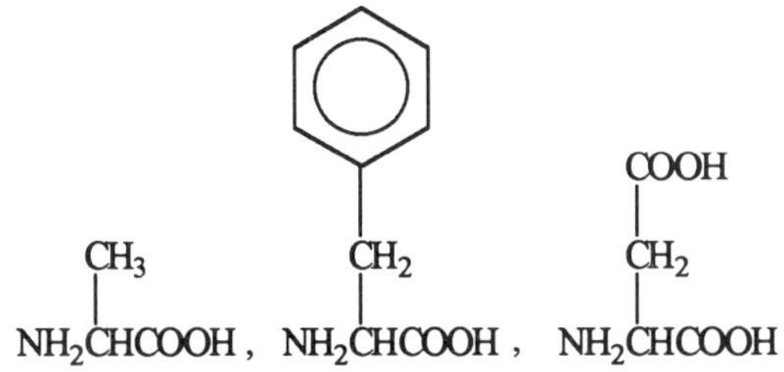

42. Hydrolysis breaks the peptide bonds. One water molecule will react with each peptide bond, a hydrogen atom attaches to the nitrogen to complete the amino group and an –OH group attaches to the carboxyl carbon. The tripeptide, Ala-Glu-Tyr, will hydrolyze to yield the following:

$NH_2CH(CH_3)COOH$, $NH_2CH(CH_2CH_2COOH)COOH$, $NH_2CH(CH_2C_6H_4OH)COOH$

43. $$\frac{1 \text{ mol Fe}}{1 \text{ mol cytochrome c}} \times \frac{55.85 \text{ g}}{\text{mol Fe}} \times \frac{100. \text{ g cytochrome c}}{0.43 \text{ g Fe}} = 1.3 \times 10^4 \frac{\text{g}}{\text{mol}}$$

The molar mass of cytochrome c is 1.3×10^4 g/mole.

44. $$\frac{4 \text{ mol Fe}}{1 \text{ mol hemoglobin}} \times \frac{55.85 \text{ g}}{\text{mol Fe}} \times \frac{100. \text{ g hemoglobin}}{0.33 \text{ g Fe}} = 6.8 \times 10^4 \frac{\text{g}}{\text{mol}}$$

The molar mass of hemoglobin is 6.8×10^4 g/mole.

45. The sequence of amino acids in the heptapeptide is:

Gly - Phe - Leu

Phe - Ala - Gly Leu - Ala - Thr

Phe - Ala - Gly - Phe - Leu - Ala - Thr

46. The amino acid sequence of the heptapeptide is:

Phe - Gly - Tyr

Ala - Leu - Phe

Phe - Ala - Ala

Phe - Ala - Ala - Leu - Phe - Gly - Tyr

47. The amino acid sequence of the nonapeptide is:

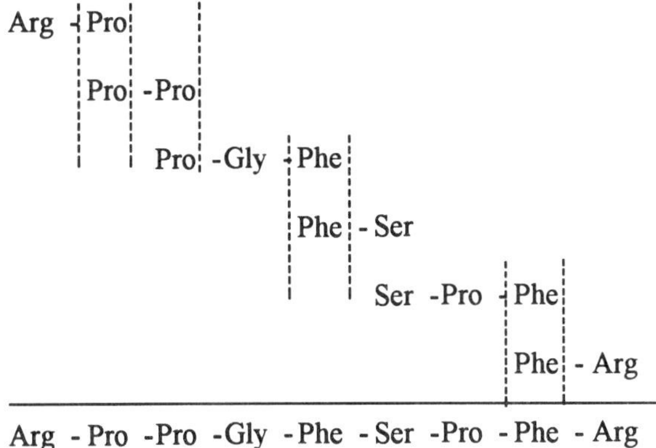

48. (a) Arginine will not migrate to either electrode in an electrolytic cell at a pH of 10.8.

    (b) Arginine will migrate towards the positive electrode at pH greater than 10.8, that is, more basic than its isoelectric point.

49. Yes, all proteins have a primary structure (a sequence of amino acids linked by peptide bonds) but need not form regular, three dimensional secondary structures such as the α-helix or β-pleated sheet.

    No, all proteins must have a primary structure because, by definition, these molecules are polymers composed of a sequence of amino acids linked by peptide bonds (a primary structure).

50. A small protein like ribonuclease (or myoglobin) would have a small number of protein domains, probably only one. Small proteins (e.g., myoglobin, ribonuclease) commonly fold into one globular unit (one domain) while larger proteins will fold into more than one globular unit (more than one domain).

51. The steroisomers of threonine:

COOH	COOH	COOH	COOH
H—C—NH_2	H_2N—C—H	H—C—NH_2	H_2N—C—H
H—C—OH	HO—C—H	HO—C—H	H—C—OH
CH_3	CH_3	CH_3	CH_3

52. The immunoglobulin hypervariable regions allow the body to produce millions of different immunoglobulins, each with its distinct amino acid sequence and unique antigen binding site.

53. (a) The structure of alanine at pH = 9.0 would be

$$\begin{array}{c} CH_3 \\ | \\ NH_2CHCOO^- \end{array}$$

(b) The structure of lysine at pH = 9.0 would be $\overset{+}{N}H_3CH_2CH_2CH_2CH_2\underset{\underset{NH_3^+}{|}}{C}HCOO^-$

(c) The net charge on lysine at pH = 9.0 would be positive (see the structure of lysine in part (b)).

54. Nineteen dipeptides can be written with glycine on the N-terminal side. Another nineteen are possible with glycine on the C-terminal end. Finally, one dipeptide can be written with two glycines giving a total of thirty-nine.

55. Vasopressin will have a higher isoelectric point than oxytocin. Vasopressin has two different amino acids as compared with oxytocin, a phenylalanine instead of an isoleucine and an arginine instead of a leucine. Thus, vasopressin has one additional basic amino acid (arginine) which will cause the vasopressin isoelectric point to be higher than the oxytocin isoelectric point.

56. Leucine

$$(CH_3)_2CHCH_2\underset{\underset{NH_2}{|}}{C}HCOOH$$

Alanine

$$CH_3\underset{\underset{NH_2}{|}}{C}HCOOH$$

Glutamic Acid

$$HOOCCH_2CH_2\underset{\underset{NH_2}{|}}{C}HCOOH$$

Glutamic acid is the only one of these three amino acids with polar bonds in its side chain. Thus, glutamic acid will be the most polar.

CHAPTER 31

ENZYMES

1. Activation energy is the energy barrier to chemical reaction and is measured as the difference between the reactant(s) energy level and the transition state energy level. Enzymes lower the activation energy barrier and, thus, increase biochemical reaction rates.

2. Enzymes are proteins. As noted in the answer to question 1, enzymes serve as catalysts for chemical reactions in the body.

3. A coenzyme is the nonprotein part of a conjugated enzyme. An apoenzyme is the protein part of a conjugated enzyme.

4. Sucrase catalyzes the hydrolysis of sucrose; lactase catalyzes the hydrolysis of lactose; maltase catalyzes the hydrolysis of maltose.

5. The six general classes of enzymes are: (a) oxidoreductases, (b) transferases, (c) hydrolases, (d), lyases, (e) isomerases, and (f) ligases.

6. The reaction rate often can be increased (1) by increasing the reactant (substrate) concentration and (2) by raising the temperature.

7. The lock-and-key hypothesis states that an enzyme active site is rigidly shaped to fit the substrate molecule(s) (as a lock fits a key). The induced-fit model envisions an active site which flexes to make a best fit with the substrate(s).

8. When substrates bind to an active site, they may be converted more easily to products because (1) the substrates are close together (proximity catalysis); (2) the substrates are oriented to best react (productive binding hypothesis); (3) when the substrates bind to the active site, they change shape to be more like products (strain hypothesis).

9. Both an enzyme substrate and an enzyme inhibitor may bind to the active site. However, a substrate will react to form product while an inhibitor will only block the active site from further reaction.

10. The correct statements are: a, c, f, g, h, j, k, l, m, and n. The others are incorrect for these reasons.

    (b) Enzymes decrease the activation energy for a reaction.
    (d) The reactant of an enzyme-catalyzed reaction is called the substrate.

(e) Doubling the reactant concentration does not always double the rate of an enzyme-catalyzed reaction. As the reaction rate approaches V_{max}, a change in reactant concentration has a smaller and smaller impact on the rate.
(i) The induced-fit model requires a flexible enzyme.

11. reaction rate = $\left(\frac{0.005\text{ M}}{3.5\text{ min}}\right)\left(\frac{1\text{ min}}{60\text{ sec}}\right) = 2 \times 10^{-5}$ M/s

12. reaction rate = (0.02 M)/(8 min) $\left(\frac{0.02\text{ M}}{8\text{ min}}\right)\left(\frac{1\text{ min}}{60\text{ s}}\right) = 4 \times 10^{-5}$ M/s

13. The turnover number for lysozyme shows that this enzyme converts 0.5 reactant to products every second. Thus, in 1 min,

(0.5 reactant converted to products/s) $\left(\frac{60\text{ s}}{\text{min}}\right)$ (1 min) = 30 reactants converted to products

14. The turnover number for pepsin shows that this enzyme converts 1.2 reactants to products every second. Thus, in 5 min,

(1.2 reactants converted to products/s) $\left(\frac{60\text{ s}}{\text{min}}\right)$ (5 min) (3 pepsin molecules) = 1×10^3 reactants converted to products by three pepsin molecules

15. The lock-and-key hypothesis predicts that an enzyme active site should complement the shape of the substrate. Thus, the first enzyme would have a narrow, extended active site to fit n-butyl alcohol ($CH_3CH_2CH_2CH_2OH$). For the substrate, 2-methyl-2-propanol

```
         CH3
          |
CH3 —— C —— OH
          |
         CH3
```

the enzyme active site would need to be almost as wide as it is long.

16. The lock-and-key hypothesis predicts that an enzyme active site should complement the shape of the substrate. The first enzyme would have a narrow active site which is long enough to fit the four carbon carboxylic acid, butanoic acid ($CH_3CH_2CH_2COOH$). The second enzyme would have a narrow active site which would be shorter because it need only fit the two carbon carboxylic acid, acetic acid (CH_3COOH).

17. At low temperatures, fewer substrate molecules (reactants) have the energy necessary to overcome the activation energy barrier. Thus, an enzyme-catalyzed reaction rate decreases at low temperatures.

18. An enzyme-catalyzed rate decreases at high temperatures because the enzyme loses its natural shape (denatures). As an enzyme denatures it ceases to be an effective catalyst.

19. If the V_{max} for an enzyme is decreased, the enzyme has lost some of its catalytic efficiency. Any process that decreases an enzyme's catalytic abilities is termed inhibition.

20. If the turnover number for an enzyme is increased, the enzyme can convert more reactants to products per unit time; the enzyme is a better catalyst. Any process that increases an enzyme's catalytic abilities is termed activation.

21. No. Enzyme A is more effective than enzyme B based on a comparison of turnover numbers (225/s for enzyme A vs. 120/s for enzyme B).

22. Yes. Enzyme B is more effective than enzyme A based on a comparison of turnover numbers (9.8×10^{-1}/s for enzyme B vs. 0.05/s for enzyme A).

23. Glutamine is the product of this liver metabolic pathway. Thus, glutamine control of this pathway must be "feed back". Since an increase in glutamine concentration causes the metabolic pathway to slow down, this control must be feedback inhibition. Feedback inhibition means that when a large amount of product has been formed, the beginning of a process will slow down.

24. Glutamine is a starting material (reactant) for the liver metabolic pathway which produces urea. Thus, glutamine control of this pathway must be "feedforward". Since an increase in glutamine concentration causes the metabolic pathway to speed up, this control must be feedforward activation.

25. The turnover number measures the number of substrate molecules converted to product by one enzyme molecule under optimum conditions. Chymotrypsin can convert one glycine-containing substrate to product every twenty seconds while the enzyme can convert two hundred L-tyrosine containing substrates to products each second. Chymotrypsin is a much more efficient catalyst for the L-tyrosine-containing substrate.

26. Feedback inhibition means that when a large amount of product has been formed, the beginning of a process will slow down. Thus, feedback inhibition protects against overproduction. "Feedback activation" means that when a large amount of product is formed, the beginning of the process will accelerate. A state of overproduction will worsen and might lead to cell death.

27.

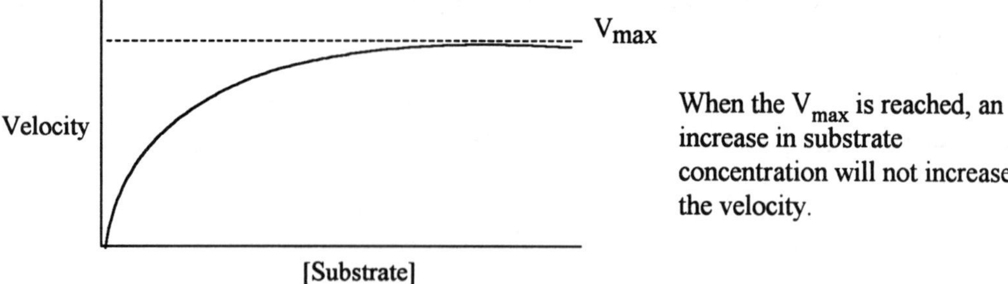

When the V_{max} is reached, an increase in substrate concentration will not increase the velocity.

28. The enzyme lactase catalyzes hydrolysis of the disaccharide lactose to produce two monosaccharides, D-glucose and D-galactose.

29. The enzymes used by paper manufacturers for wood fiber digestion are cellulase and hemicellulase. The enzyme used by clothing manufacturers for "biostoning" is cellulase. In both cases, enzyme treatment leads to a breakdown of the polysaccharide, cellulose.

CHAPTER 32

NUCLEIC ACIDS AND HEREDITY

1. The five nitrogen bases found in nucleotides:

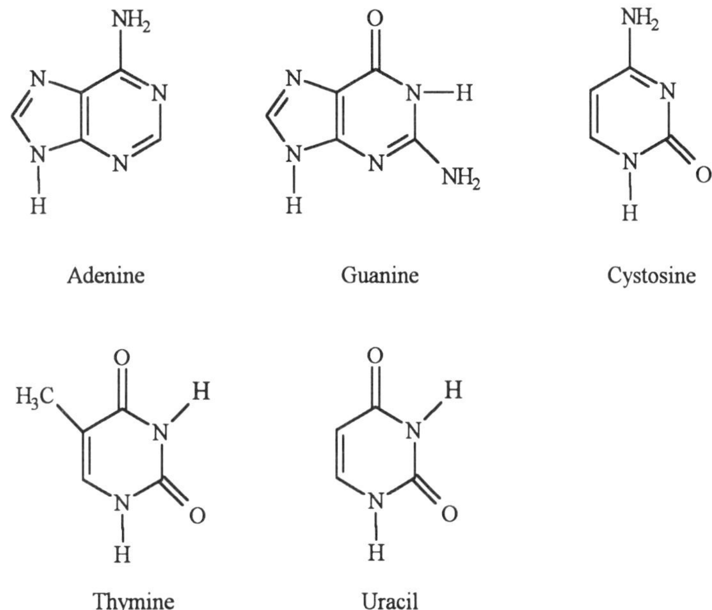

2. A nucleoside is a purine or pyrimidine base linked to a sugar molecule, usually ribose or deoxyribose. A nucleotide is a purine or pyrimidine base linked to a ribose or deoxyribose sugar which in turn is linked to a phosphate group.

3. There are three structural differences between DNA and RNA.

   (a) In RNA the sugar molecule is always ribose. In DNA, the sugar molecule is always deoxyribose, which has H instead of OH at carbon number two.

   (b) Both molecules use a mixture of four nitrogen bases. Both use cytosine, adenine, and guanine. In DNA, the fourth base is thymine. In RNA, the fourth base is uracil.

   (c) DNA exists as a double helix whereas RNA is a single strand of nucleotides.

4. The major function of ATP in the body is to store chemical energy, and to release it when called upon to carry out many of the complex reactions that are essential to most of our life processes. An equilibrium exists with ADP.

$$\text{ATP} + H_2O \underset{\text{energy storage}}{\overset{\text{energy utilization}}{\rightleftharpoons}} \text{ADP} + P_i + 35\text{ kJ}$$

5. The structure of DNA as proposed by Watson and Crick is a double-stranded helix. Each strand has a backbone of alternating phosphate and deoxyribose units. Each deoxyribose unit has one of the four nitrogen bases attached, but coming off the backbone, not part of the backbone. These nitrogen bases thus link to their complementary nitrogen base on the other strand of the double helix.

6. Complementary bases are the pairs that "fit" to hydrogen bond to each other between the two helixes of DNA. For DNA the complementary pairs are thymine with adenine, and cytosine with guanine, or T-A and C-G. For RNA it is U-A and C-G.

7. The genetic code is the sequence of nitrogen bases in a strand of DNA. The code determines the sequence of amino acids in protein molecules which will be assembled according to the code. It takes three nucleotides to make a codon, which determines one amino acid in a sequence.

8. Since there are only four different bases to make up the code, one nucleotide could only specify four possible amino acids; two nucleotides could specify 16 amino acids; 3 nucleotides could specify 64 amino acids. Since there are at least 20 amino acids needed, three nucleotides are required.

9. A brief outline of the biosynthesis of proteins:

   (a) A DNA strand produces a complementary mRNA strand which leaves the nucleus and travels to the cytoplasm where it becomes associated with a cluster of ribosomes, binding to five or more ribosomes.

   (b) With the aid of an enzyme, the proper amino acid attaches to a tRNA molecule by an ester linkage.

   (c) The amino acids are brought to the protein synthesis site by tRNA.

   (d) The initiation of a polypeptide chain always uses the mRNA codon AUG or GUG, which ties to the tRNA anticodon UAC. This code brings N-formylmethionine for procaryotic cells. The amino group is blocked by the formyl group, leaving the carboxyl group available to react with the amino group of the next amino acid.

   (e) The next tRNA, bringing an amino acid, comes in to the mRNA and links up, anticodon to codon. The peptide linkage is then made between amino acids.

   (f) The first tRNA is ejected, and a third enters the ribosome.

(g) The polypeptide chain terminates at a nonsense or termination-codon. The protein molecule breaks free.

10. A codon is a triplet of three nucleotides, and each codon specifies one amino acid. The cloverleaf model of transfer RNA has an anticodon loop consisting of seven unpaired nucleotides. Three of these make up the anticodon, which is complementary to, and hydrogen-bonds to the codon on mRNA.

11. The role of N-formylmethionine in procaryotic protein synthesis is to start the polypeptide chain so it goes in the right direction. It can only build from the carboxyl end. After the synthesis, the N-formylmethionine breaks loose from the protein.

12. From time to time a new trait appears in an individual that is not present in either parents or ancestors. These traits which are generally the result of genetic or chromosomal changes are called mutations.

13. A "DNA fingerprint" is a pattern of tagged DNA fragments on an electrophoretic gel which can be used to identify possible suspects.

14. The correct statements are: a, b, d, g, h, j, k, m, p, and r. The others are incorrect for these reasons:

(c) Only DNA is responsible for transmitting genetic information from parent to daughter cells.
(e) Codons are combinations of three nucleotides in an mRNA molecule.
(f) The double helical structure of DNA is held together by hydrogen bonds between nitrogen bases.
(i) The letters ATP stand for adenosine triphosphate.
(l) These are not the pairs of bases complementary to each other. See statement 14.b. for the correct pairings.
(n) Humans have 23 pairs, or 46 chromosomes.
(o) In meiosis the sperm and egg cells split in such a way that each cell has only half of the normal number of chromosomes, 23 in humans. After the two cells join, the fertilized cell again will have 46 chromosomes.
(q) Messenger RNA carries the code for the synthesis of proteins.
(s) DNA contains deoxyribose, not ribose.
(t) RNA's are single stranded but they do not always form a single helix conformation. For example, ribosomal RNA assumes a shape which is determined by the structure of the ribosome.

15. The letters are associated with compound names as follows:

(a) A, adenosine
(b) AMP, adenosine-5'-monophosphate
(c) dADP, deoxyadenosine-5'-diphosphate
(d) UTP, uridine-5'-triphosphate

16. The letters are associated with compound names as follows:

(a) G, guanosine
(b) GMP, guanosine-5'-monophosphate
(c) dGDP, deoxyguanosine-5'-diphosphate
(d) CTP, cytidine-5'-triphosphate

17. Structural formulas

(a) A

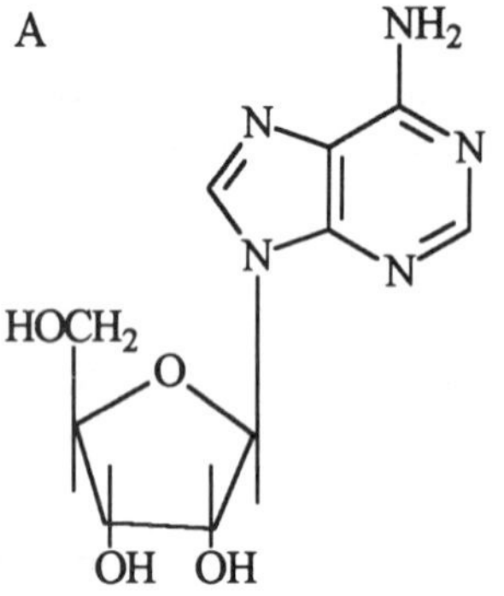

(b) AMP

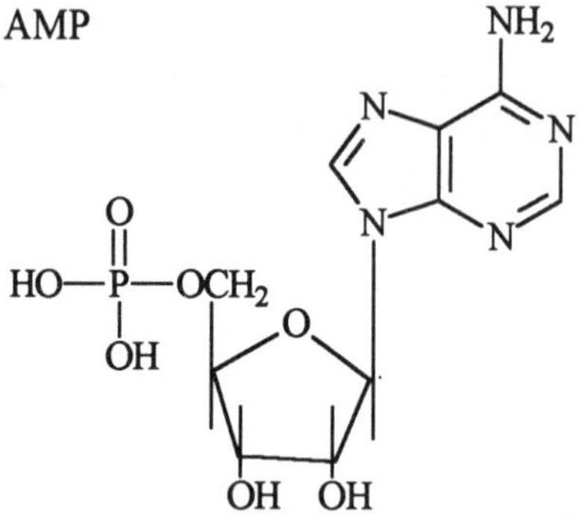

(c) CDP

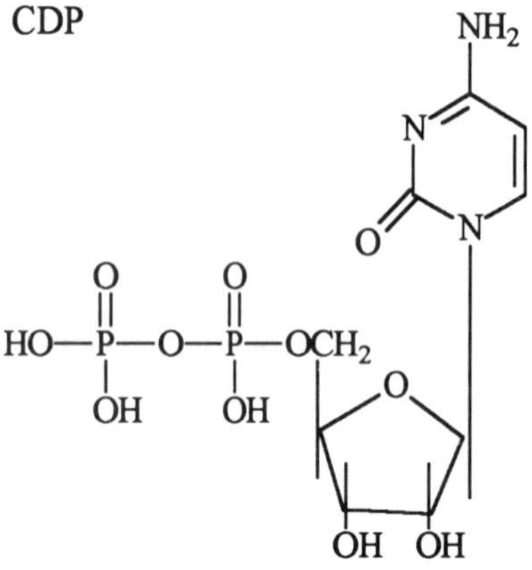

(d) dGMP

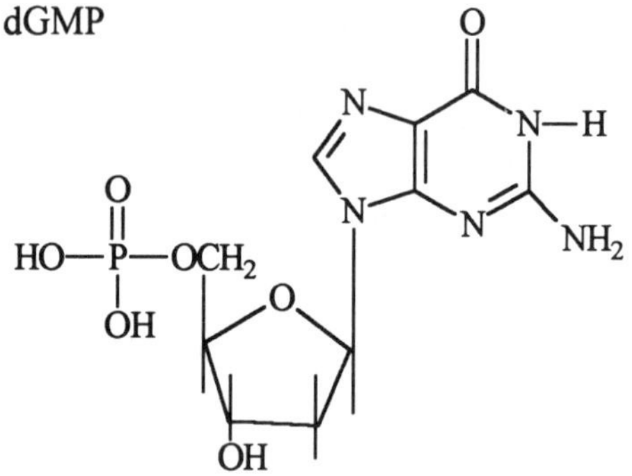

18. Structural formulas

(a) U

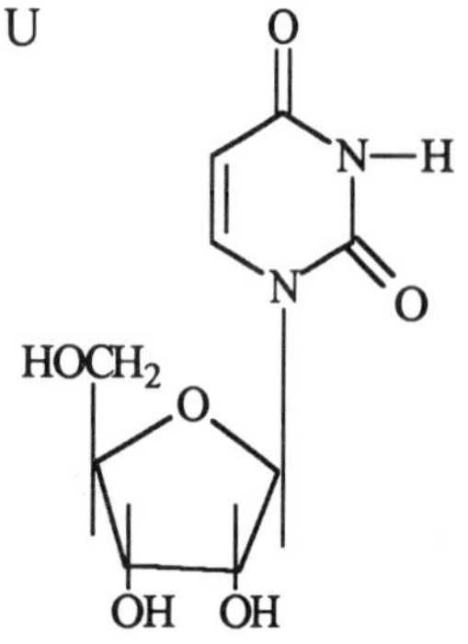

(b) UMP

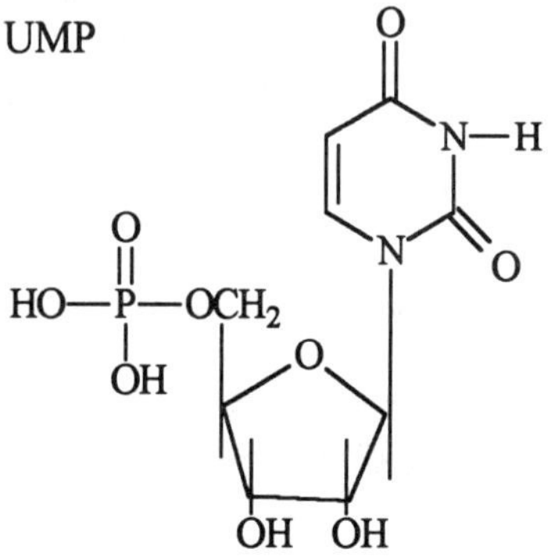

(c) CTP

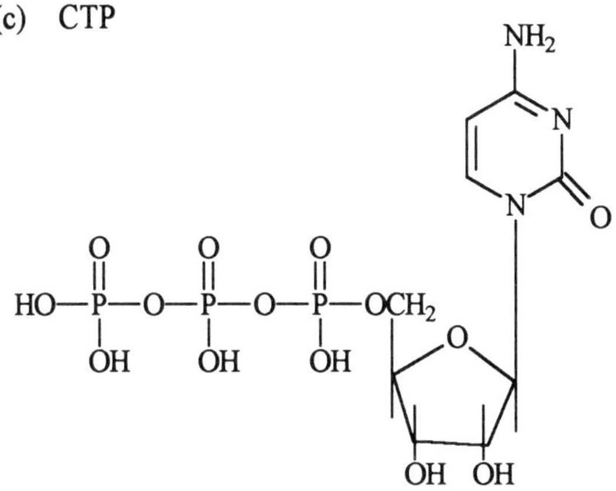

(d) dTMP

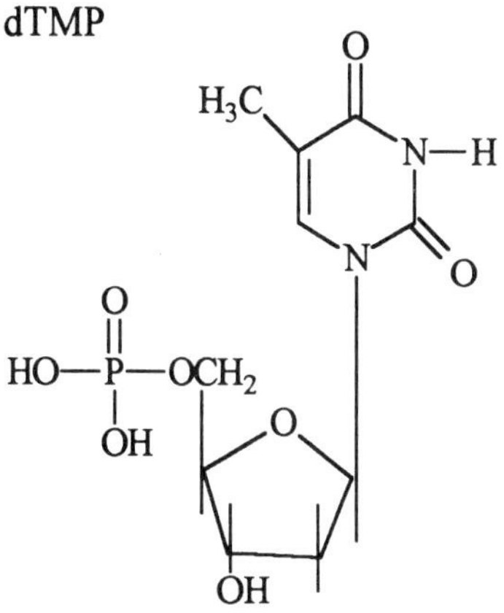

19. There are several sequences possible for the three-nucleotide, single-stranded DNA. One possible structure follows:

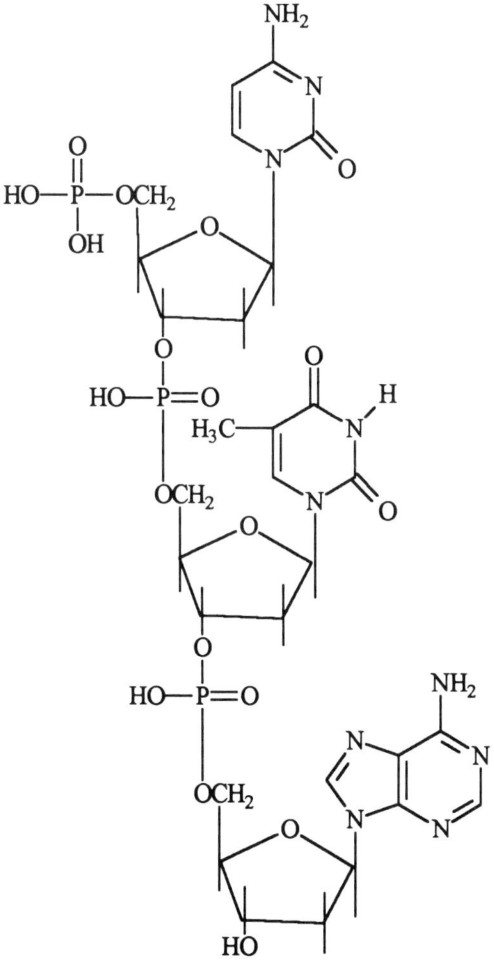

20. There are several possible sequences for the three-nucleotide, single-stranded RNA. One possible structure follows:

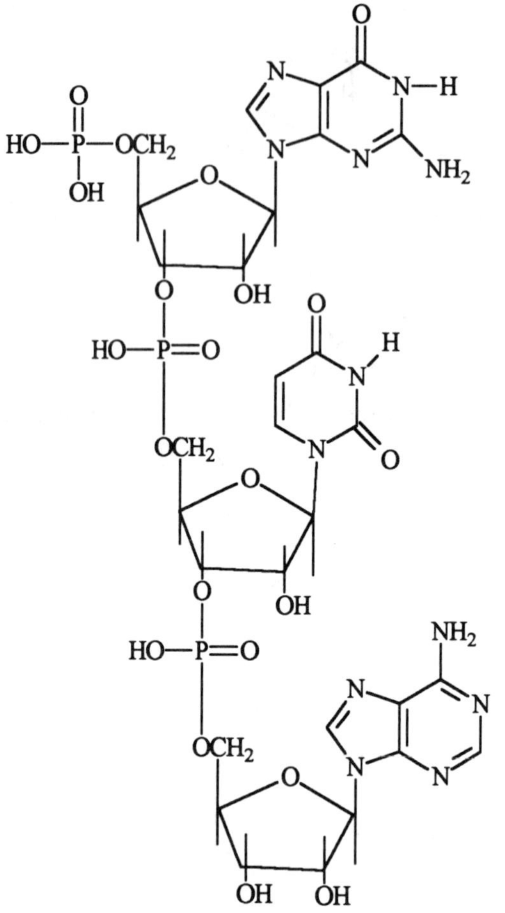

O
N
N—H
HO—P—OCH$_2$
OH
N
N
NH$_2$
O
O
OH
O
HO—P=O
H
N
OCH$_2$
N
O
O
O
OH
NH$_2$
HO—P=O
N
N
OCH$_2$
N
N
O
OH OH

21. The hydrogen bonding between adenine and uracil:

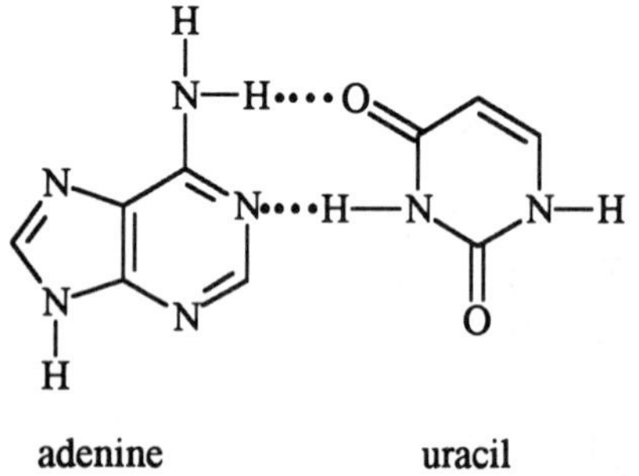

H
N—H••••O
N
N••••H—N
N—H
N
N
O
H

adenine uracil

(dotted lines - hydrogen bonds)

22. The hydrogen bonding between guanine and cytosine:

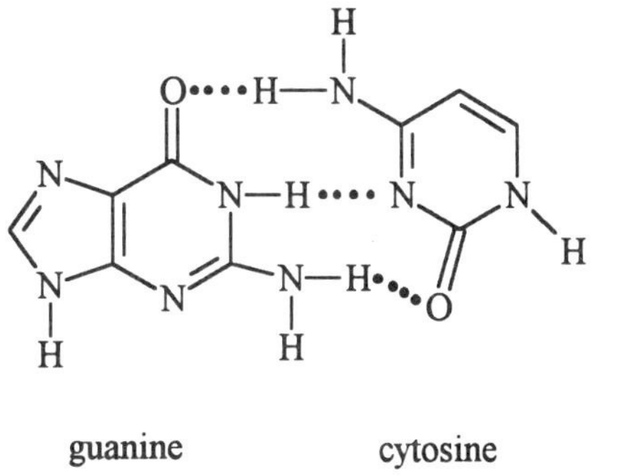

(dotted lines - hydrogen bonds)

guanine cytosine

23. RNA is responsible for moving genetic information to where it can be used for protein synthesis. There are three kinds of RNA:

    (a) Ribosomal RNA (rRNA) is found in the ribosomes where it is associated with protein roughly in the proportion 60-65% protein, 30-35% rRNA.

    (b) Messenger RNA (mRNA) carries genetic information from DNA to the ribosomes. It is a template made from DNA and carries the codons that direct the synthesis of proteins.

    (c) Transfer RNA (tRNA) is used to bring amino acids to the ribosomes for incorporation into protein molecules.

24. DNA is considered to be the genetic substance of life, because it contains the sequence of bases that carries the code for genetic characteristics.

25. Replication is the biological process of making DNA using a DNA template while the transcription process refers to the making of RNA using a DNA template.

26. Transcription is the process of making RNA using a DNA template while translation is a process for making protein using a mRNA template.

27. tRNA binds specific amino acids and brings them to the ribosome for protein synthesis. mRNA carries genetic information from DNA to the ribosome and serves as a template for protein synthesis.

28. Both mRNA and rRNA can be found in the ribosomes. rRNA serves as part of the ribosome structure while mRNA serves as a template for protein synthesis.

29. Three nucleotides are required to specify one amino acid. If there are 146 amino acid residues in the beta chain of hemoglobin, then the number of required nucleotides in mRNA is $3 \times 146 = 438$.

30. Three nucleotides are required to specify one amino acid. For the 573 amino acid residues in the phosphoglycerate kinase enzyme, the number of required nucleotides in mRNA is $3 \times 573 = 1719$.

31. For a DNA sequence, TCAATACCCGCG,

    (a) the complementary mRNA will be: AGUUAUGGGCGC.

    (b) the anticodon order in tRNA will be: UCAAUACCCGCG.

    (c) the sequence of amino acids coded by the DNA will be: Ser-Tyr-Gly-Arg

32. A segment of DNA strand consists of GCTTAGACCTGA.

    (a) The order in the complementary mRNA will be: CGAAUCUGGACU.

    (b) The anticodon order in tRNA will be: GCUUAGACCUGA.

    (c) The sequence of amino acids coded by the DNA is Arg-Ile-Trp-Thr.

33. Transcription makes a polymer of nucleotides by forming phosphate ester bonds to connect the nucleotides to each other. The phosphate ester combines a phosphoric acid with an alcohol.

34. Translation makes a polymer of amino acids by forming amide bonds to connect the amino acids to each other. The amide bond combines an amine with a carboxylic acid.

35. Translation termination occurs when the ribosome reaches a "nonsense" or termination codon along the mRNA. No tRNA (in normal cells) has the anticodon to match the termination codon and so no more amino acids are added to the newly synthesized protein chain. The peptidyl-tRNA connection is broken and the protein chain is released from the ribosome.

36. Translation initiation occurs when the ribosome reaches a special AUG or GUG codon along the mRNA. Since there is commonly more than one AUG or GUG codon, the ribosome must use other information to choose the special AUG or GUG. This codon is the starting point for protein synthesis and is bound by either a special tRNA carrying N-formylmethionine (in procaryotes) or a tRNA carrying methionine (in eucaryotes).

37. mRNA codons and corresponding tRNA anticodons follow:

    (a) GUC, CAG

    (b) AGG, UCC

    (c) UUU, AAA

    (d) CCA, GGU

38. mRNA codons and corresponding tRNA anticodons follow:

(a) CGC, GCG

(b) ACA, UGU

(c) GAU, CUA

(d) UUC, AAG

39. Thymine and adenine are complementary bases in DNA; therefore each time one appears, its complement appears. There is no fixed relationship between the amount of thymine and the amount of guanine, since they are not complementary. Table 32.2 shows that the one set of bases is often far different from the other set, varying by as much as a 3:2 ratio. The complementary bases "fit" with each other with respect to H-bonding between them; also with respect to the size of the molecules.

40. In RNA the guanine content does not have to be equal to the cytosine content, because RNA is a single strand. Its complements are on the DNA template, not on the RNA. They have to be equal in DNA, because it is a double helix, with each nucleotide hydrogen-bonded to its complement.

41. A substitution mutation will change one codon. An insertion mutation will shift the sequence position of all bases following the insertion by one. Each base will take the position of its neighbor. For example, if base C is inserted into the sequence ... UUC ACG GCC ..., every codon will change, ... UCU CAC GGC C... In general, every codon following an insertion is changed.

42. For the mRNA segment, UUUCAUAAG,

(a) the coded amino acids are: Phe-His-Lys.
(b) the sequence of DNA for this mRNA is: AAAGTATTC.

43. In DNA fingerprinting, DNA polymerase is used to copy very small samples of DNA so that enough will be available to separate and visualize on gel electrophoresis.

44. The Human Genome Project is a cooperative effort by leading scientific laboratories to sequence the entire human genome (approximately 3 billion base pairs).

CHAPTER 33

NUTRITION

1. Nutrients are components of food that provide for body growth, maintenance, and repair. Food is the material we eat.

2. The "energy allowance" represents the recommended dietary energy supply needed to maintain health.

3. Marasmus is a chronic calorie deficiency while kwashiorkor represents a protein deficiency condition.

4. Candy contains primarily sucrose, a source of calories, but little or no other nutrients. Thus, candy is said to provide "empty calories."

5. The essential fatty acids have 18 or more carbon atoms per molecule and are polyunsaturated.

6. Essential fatty acids are required for normal growth and development. On a fat-free diet many deleterious physiological changes occur, including skin lesions, kidney damage, poor growth, and impaired fertility.

7. There are eight essential amino acids. They are isoleucine, leucine, lysine, methionine, phenylalanine, threonine, tryptophan, and valine.

8. Animal proteins generally are a source of all twenty common amino acids while vegetable proteins may be deficient in one or several of the amino acids.

9. The fat-soluble vitamins include A, D, E and K. The water-soluble vitamins are C and the B vitamins.

10. Vitamin D: functions as a regulator of calcium metabolism; Vitamin K: enables blood clotting to occur normally; Vitamin A: functions to furnish the pigment that makes vision possible. Many vitamins act as coenzymes.

11. Major elements are needed in relatively large quantities by the body while only small amounts of the trace elements are required.

12. Calcium is a major constituent of bones and teeth and is also important in nerve transmission among other functions.

13. Solid foods provide about 30-40% of the average daily water consumption while liquids yield about another 50% of the total.
14. Some common categories of food additives are: (1) nutrients; (2) preservatives; (3) anticaking agents; (4) emulsifiers; (5) thickeners; (6) flavor enhancers; (7) colors; (8) nonsugar sweeteners.

15. The five principal digestive juices are: saliva, gastric juice, pancreatic juice, bile and intestinal juice.

16. The U. S. Food and Drug Administration is responsible for regulating the use of food additives.

17. Enzymes present in the digestive juices are: **saliva**: salivary amylase; **gastric juice**: pepsin, and gastric lipase; **pancreatic juice**, trypsinogen, chymotrypsinogen, procarboxypeptidase, amylopsin, steapsin; **bile**: no enzymes; **intestinal juice**: sucrase, maltase, lactase, aminopeptidase, dipeptidase, nucleases, phosphatase, and intestinal lipase.

18. Chyme is the material of liquid consistency found in the stomach consisting of food particles reduced to small size and mixed with gastric juices.

19. The digestive function of the liver is the production of bile, one of the important digestive juices. The bile is stored in the gall bladder.

20. Intestinal mucosal cells produce many enzymes needed to complete digestion, e.g., disaccharidases, aminopeptidases, and dipeptidases.

21. The liver takes up foreign compounds which are then oxidized by enzymes on the endoplasmic reticulum.

22. The correct statements are: a, d, e, g, i, j, l, m, and o. The others are not correct for these reasons:

    (b) Specific vegetable proteins may be deficient in one or several amino acids.
    (c) Oleic acid, which contains only one double bond, is not an essential fatty acid.
    (f) Vitamins are organic nutrients in contrast to trace elements which are inorganic.
    (h) Calcium is an important major element.
    (k) Most digestion occurs in the small intestine.
    (n) Bile acids are needed to emulsify fats not carbohydrates.

23. (a) (23 g protein) + (19 g fat) = 42 g

    (b) (52 mg sodium) + (297 mg potassium) + (2.3 mg magnesium) + (2.5 mg iron) + (4.7 mg zinc) + (9.0 mg calcium) = 388 mg

    (c) (0.015 mg Vitamin A) + (0.092 mg thiamin) + (0.218 mg riboflavin) + (3.2 mg niacin) + (0.330 mg Vitamin B_6) + (0.007 mg folic acid) + (0.002 mg vitamin B_{12}) = 3.9 mg

24. (a) (70 g protein) + (25 g carbohydrate) + (37 g fat) = 132 g

(b) (770 mg sodium) + (564 mg potassium) + (68 mg magnesium) + (3.5 mg iron) + (2.7 mg zinc) + (56 mg calcium) = 1464 mg

(c) (0.0565 mg Vitamin A) + (0.322 mg thiamin) + (0.408 mg riboflavin) + (29.5 mg niacin) + (1.2 mg Vitamin B_6) + (0.016 mg folic acid) + (0.00082 mg Vitamin B_{12}) = 31.5 mg

25. (a) (19 g fat) (9 kcal/g fat) = 2×10^2 kcal

(b) $$\left[\frac{(2 \times 10^2 \text{ kcal})}{(270 \text{ kcal})}\right](100) = 70\%$$

26. (a) (37 g fat) (9 kcal/g fat) = 3×10^2 kcal

(b) $$\left[\frac{(3 \times 10^2 \text{ kcal})}{(728 \text{ kcal})}\right](100) = 40\ \%$$

27. $$\left[\frac{(9.0 \text{ mg})\left(\frac{1 \text{ g}}{1000 \text{ mg}}\right)}{1.20 \text{ g}}\right](100) = 0.75\%$$

28. $$\left[\frac{(56 \text{ mg})\left(\frac{1 \text{ g}}{1000 \text{ mg}}\right)}{0.80 \text{ g}}\right](100) = 7.0\%$$

29. $$\left[\frac{(2\ \mu\text{g})}{(2.0\ \mu\text{g})}\right](100) = 10\ \%$$

30. $$\left[\frac{(0.82\ \mu\text{g})}{(2.0\ \mu\text{g})}\right](100) = 41\%$$

31. The three macronutrient classes are proteins, carbohydrates and lipids (fats).

32. The two micronutrient classes are minerals and vitamins.

33. The three classes of nutrients which do not commonly supply energy for the cells are (a) vitamins, (b) minerals and (c) water.

34. The three classes of nutrients which commonly supply energy for the cells are proteins, carbohydrates and lipids.

35. Starch, as a complex carbohydrate, provides dietary carbohydrate in a form which is slowly digested, enabling the body to control distribution of this energy nutrient.

36. Cellulose is important as dietary fiber. Although it is not digested, cellulose absorbs water and provides dietary bulk which helps maintain a healthy digestive tract.

37. Saliva is slightly acidic with a pH below 7.

38. Pancreatic juice is slightly basic with a pH above 7.

39. Proteins are digested in the stomach and small intestine.

40. Carbohydrates are digested in the mouth and small intestine.

41. Pancreatic amylase digests carbohydrates.

42. Pepsin digests proteins.

43. Milk contains (a) lipids, (b) minerals (calcium), (c) water and (d) carbohydrates.

44. 9 kg of body fat represents about 80,000 kcal of energy. Complete starvation for about 30 days would cause a net loss of about 9 kg of fat. Thus, this new diet is unreasonable.

45. Both vitamins and trace elements are needed in only small amounts by the body and cannot be synthesized within the body's cells. Vitamins are organic molecules and are often modified by the body. Trace elements are inorganic and, in general, are not modified before use.

46. Fat (9 kcal/g) contains more than twice as much energy as compared with carbohydrate (4 kcal/g). Thus, a tablespoon of butter will greatly increase the calorie content of a medium size baked potato.

47. (a) $\left[\dfrac{\text{50 fat Cal per serving}}{\text{100 Calories per serving}}\right](100) = 50\%$

(b) (3 g protein per serving) (2 servings) = 6 g

(c) $\left[\dfrac{\text{100 Calories per serving}}{\text{1000 Calories per day}}\right](100) = 10\%$

(d) $\left[\dfrac{\text{2 g fiber per serving}}{\text{10 g total carbohydrate per serving}}\right](100) = 20\%$

48. The foods of higher animals must be digested before they can be utilized because foods are primarily large molecules. Foods must be broken down to much smaller molecules in order to pass through the intestinal walls into the blood and lymph systems where they can be utilized in the metabolic process of animals.

49. Galactose is a component of the carbohydrate lactose which is found in milk. Thus, milk must commonly be changed to eliminate lactose from a baby's diet.

50. Nutritionists recommend less than 300 mg of dietary cholesterol per day. One large egg (270 mg cholesterol) represents almost the total daily recommended amount (about 90% of the RDA).

CHAPTER 34

BIOENERGETICS

1. As more carbon dioxide is formed, it will react with water to form carbonic acid.

$$H_2O + CO_2 \rightleftharpoons H_2CO_3$$

Carbonic acid will create a more acidic environment for the muscle tissue.

$$H_2CO_3 \rightleftharpoons H^+ + HCO_3^-$$

2. Hemoglobin contains amine groups which can react with carbon dioxide to form carbamino ions,

$$R{-}NH_2 + CO_2 \longrightarrow [R{-}NH_2{-}CO_2]^- + H^+$$

3. Fats and carbohydrates are good sources of cellular energy because they contain many reduced carbon atoms.

4. The most common high energy phosphate bond in the cell is the phosphate anhydride bond.

$$^{-}O{-}\overset{\overset{O}{\|}}{\underset{\underset{O^-}{|}}{P}}{\sim}O{-}\overset{\overset{O}{\|}}{\underset{\underset{O^-}{|}}{P}}{-}O{-}R$$

It is generally associated with adenosine triphosphate (ATP).

5. ATP is known as the "common energy currency" of the cell because energy from many different catabolic processes is stored in this molecule. In turn, most anabolic pathways draw energy from this common pool.

6. An oxidation-reduction coenzyme is a reusable organic compound which helps an enzyme carry out an oxidation-reduction reaction.

7. The nicotinamide ring of NAD^+ becomes reduced.

$$NAD^+ \text{ (pyridinium ring with } {-}CONH_2\text{, } N^+{-}R) + 2\,e^- + H^+ \longrightarrow NADH \text{ (1,4-dihydropyridine ring, } CH_2 \text{ at C-4, with } {-}CONH_2\text{, } N{-}R)$$

NAD+ NADH

8. Oxidative phosphorylation uses the mitochondrial electron transport system to directly produce ATP from oxidation-reduction reactions. Substrate-level phosphorylation does not use the electron transport system, but, instead involves transfer of a phosphate group from a substrate to ADP to form ATP.

9. Procaryotes (procaryotic cells) are simple cells without internal membrane-bound bodies. In contrast, eucaryotes (eucaryotic cells) contain internal membrane-bound bodies (organelles) which allow for spatial organization of the cells' metabolism.

10. Oxidative phosphorylation takes place in the mitochondria.

11. Both chloroplasts and mitochondria are organelles which are bounded by two membranes. In both cases they contain much folded internal membrane (which facilitates oxidation-reduction reactions).

12. Chloroplast pigments trap light to provide energy for photosynthesis.

13. The overall photosynthetic reaction in higher plants is as follows:

$$6\,CO_2 + 6\,H_2O + 2820\text{ kJ} \longrightarrow C_6H_{12}O_6 + 6\,CO_2$$

14. The correct statements are: a, c, e, f, h, i, j, k and l. The others are not correct for these reasons:

    (b) Anabolic processes form complex biological substances from simpler compounds.

    (d) As a typical carbohydrate carbon is less reduced than a typical fatty acid carbon, the carbohydrate carbon will supply less energy.

    (g) Carbon dioxide contains highly oxidized carbons and will supply no cellular energy.

    (m) $NADP^+$ is reduced during photosynthesis.

15. $CO_2(g) + H_2O(l) \rightleftharpoons H_2CO_3(aq) \rightleftharpoons H^+(aq) + HCO_3^-(aq)$

    As more carbon dioxide is produced by the muscle cells, this gas reacts with water to make carbonic acid. The carbonic acid decreases the pH of the surrounding solution which causes the hemoglobin to bind less tightly to oxygen. Hydrogen ions stimulate release of oxygen from hemoglobin:

    $H^+ + HbO_2 \rightleftharpoons HbH^+ + O_2$

16. $CO_2(g) + H_2O(l) \rightleftharpoons H_2CO_3(aq) \rightleftharpoons H^+(aq) + HCO_3^-(aq)$

    The reaction between carbon dioxide and water is reversible. As carbon dioxide is exhaled from the lungs, carbonic acid is converted to carbon dioxide and the surrounding solution's pH increases.

Since the hydrogen ion concentration decreases, hemoglobin binds to oxygen more tightly. The following reaction is reversed:

$$H^+ + HbO_2 \rightleftharpoons HbH^+ + O_2$$

17. The conversion of glucose to carbon dioxide is catabolic because:

    (a) a larger molecule (glucose) is converted to a smaller molecule (carbon dioxide);

    (b) the carbons from glucose become oxidized as they are converted to carbon dioxide.

18. The conversion of acetate to long-chain fatty acids is anabolic because:

    (a) a smaller acetate is converted to a larger long-chain fatty acid;

    (b) The carbons from acetate become reduced (on the average) as they are converted to long-chain fatty acids.

19. Procaryotic cells contain no chloroplasts.

20. Chloroplasts carry out photosynthesis, not oxidative phosphorylation.

21. The chemical changes in the mitochondria are said to be catabolic because:

    (a) the mitochondrial processes convert many different molecules to smaller carbon dioxide molecules;

    (b) during these chemical changes, the carbons become progressively more oxidized;

    (c) as a result, ATP is produced.

22. The chemical changes in the chloroplasts are said to be anabolic because:

    (a) smaller carbon dioxide molecules are converted to larger glucose molecules;

    (b) as this transformation takes place, the carbons become progressively more reduced;

    (c) this process requires an input of energy (from sunlight).

23. Three important characteristics of an anabolic process are: (1) simpler substances are built up into complex substances; (2) often carbons are reduced; (3) often cellular energy is consumed.

24. Three important characteristics of a catabolic process are: (1) complex substances are broken down into simpler substances; (2) often carbons are oxidized; (3) often cellular energy is produced.

25. $2\ FADH_2 + O_2 \longrightarrow 2\ FAD + 2\ H_2O$

$$(2.38 \text{ mol } FADH_2)\left(\frac{1 \text{ mol } O_2}{2 \text{ mol } FADH_2}\right) = 1.19 \text{ mol } O_2$$

26. $2\ NADH + 2H^+ + O_2 \longrightarrow 2\ NAD^+ + 2H_2O$

$$(0.67 \text{ mol NADH})\left(\frac{1 \text{ mol } O_2}{2 \text{ mol NADH}}\right) = 0.34 \text{ mol } O_2$$

27. $NAD^+ + 2H^+ + 2\ e^- \longrightarrow NADH$

$$(11.75 \text{ mol } NAD^+)\left(\frac{2 \text{ mol } e^-}{1 \text{ mol } NAD^+}\right) = 23.50 \text{ mol } e^-$$

28. $FAD + 2\ H^+ + 2e^- \longrightarrow FADH_2$

$$(0.092 \text{ mol FAD})\left(\frac{2 \text{ mol } e^-}{1 \text{ mol FAD}}\right) = 0.18 \text{ mol } e^-$$

29. The molecule shown contains two high energy phosphate anhydride bonds.

30. The molecule shown contains two high energy phosphate anhydride bonds.

31. $ADP + P_i + 35\ kJ \longrightarrow ATP + H_2O$

$$(0.55 \text{ mol ADP})\left(\frac{35 \text{ kJ}}{\text{mol ADP}}\right) = 19 \text{ kJ}$$

32. $ATP + H_2O \longrightarrow ADP + P_i + 35\ kJ$

$$(1.65 \text{ mol ATP})\left(\frac{35 \text{ kJ}}{\text{mol ATP}}\right) = 58 \text{ kJ}$$

33. Since the phosphorylated substrate phosphoenolpyruvate contains only one high energy phosphate bond, substrate-level phosphorylation can produce only one ATP.

34. Since the phosphorylated substrate 1,3-diphosphoglycerate contains only one high energy phosphate bond, substrate-level phosphorylation can produce only one ATP.

35. Mitochondrial electron transport and oxidative phosphorylation produces 3 ATP per NADH and 2 ATP per $FADH_2$. Thus,

$$(4\text{ NADH})\left(\frac{3\text{ ATP}}{\text{NADH}}\right) + (2\text{ FADH}_2)\left(\frac{2\text{ ATP}}{\text{FADH}_2}\right) = 16\text{ ATP}$$

36. Mitochondrial electron transport and oxidative phosphorylation produces 3 ATP per NADH and 2 ATP per $FADH_2$. Thus,

$$(2\text{ NADH})\left(\frac{3\text{ ATP}}{\text{NADH}}\right) + (3\text{ FADH}_2)\left(\frac{2\text{ ATP}}{\text{FADH}_2}\right) = 12\text{ ATP}$$

37. Rapid breathing expels more carbon dioxide causing more hydrogen carbonate ion to react with hydrogen ions

$$HCO_3^-(aq) + H^+(aq) \rightleftharpoons H_2CO_3(aq) \rightleftharpoons H_2O(l) + CO_2(g)$$

Thus, the blood will become less acidic and the pH will rise.

38. The mitochondria carry out most of the oxidation-reduction reactions for the cell. Because these reactions involve electron movement, they are more easily controlled in a non-aqueous environment. Thus, mitochondria contain 90% membrane by mass.

39. No. Higher plant cells need mitochondria as well as chloroplasts in order to oxidize energy storage molecules to provide for the cellular energy needs.

40. Both ATP and NAD^+ contain a ribose which (1) is linked to an adenine at carbon one and (2) is linked to a phosphate at carbon five.

41. No, photosynthesis is an anabolic process by which carbon atoms are reduced. (Carbon dioxide is converted to glucose.)

CHAPTER 35

CARBOHYDRATE METABOLISM

1. Carbohydrates are considered energy storage molecules because they contain biologically-usable, reduced carbon atoms.

2. Glucose catabolism breaks down glucose to the smaller carbon dioxide molecule in an oxidative process. This produces energy for the cell and is catabolic. Photosynthesis produces glucose from carbon dioxide via a reductive path. Light energy is required and this is an anabolic process.

3. Each mole of glucose will yield 2820 kJ of energy when oxidized. Thus, 3 moles of glucose will provide 3 moles × 2820 kJ/mole or 8460 kJ.

4. Enzymes are needed so that biochemical reactions will proceed fast enough to keep the cell alive. Metabolism is often controlled by controlling enzyme activity.

5. Initial muscle contraction requires a readily and quickly available form of energy, ATP. Once ATP stores have been depleted, muscle cells turn to muscle glycogen as a source of chemical energy.

6. The liver replenishes muscle glucose (via the blood) by (a) releasing glucose from liver glycogen stores and (b) by converting lactic acid back into glucose.

7. The final reactions in anaerobic glucose metabolism convert NADH back to NAD^+. This coenzyme can then be reused to oxidize more glucose.

8. One high energy phosphate bond (in GTP) is formed directly in the citric acid cycle. In addition, one $FADH_2$ and three NADHs are produced.

9. Acetyl-CoA is an acetyl group bonded to a coenzyme A group. Coenzyme A contains adenine, ribose, diphosphate, pantothenic acid, and thioethanolamine. In the citric acid cycle, the acetyl group enters the cycle, leaving the coenzyme. In the process of the cycle, the acetyl group is oxidized to two molecules of carbon dioxide. In the process, energy is transferred, creating 11 ATP molecules and 1 GTP molecule. Acetyl-CoA serves to tie metabolism of carbohydrates, fats and certain amino acids to the citric acid cycle.

10. Hormones are chemical substances that act as control or regulatory agents in the body. Hormones are secreted by the endocrine, or ductless glands directly into the bloodstream and are transported to various parts of the body to exert specific control functions.

11. (a) The glucose concentration in blood under normal fasting conditions is about 70 to 90 mg/100 mL of blood.

(b) The blood glucose concentrations considered as (1) hyperglycemic is 90 to 140 mg/100 mL of blood; and (2) hypoglycemic is 50 to 70 mg/100 mL of blood.

12. The renal threshold is the concentration of a substance in the blood above which the kidneys begin to excrete that substance.

13. The blood glucose concentration is maintained within certain definite limits by three hormones. If blood glucose concentration is high, insulin stimulates glycogenesis, the formation of glycogen from glucose. If blood glucose concentration is low, adrenalin (epinephrine) and glucagon stimulate glycogenolysis, the formation of glucose from glycogen.

14. The correct statements are: a, c, h, i, j, l, and m. The others are not correct for these reasons:

(b) Almost no biological reaction will proceed at a sufficient rate to maintain life in the absence of an enzyme (catalyst).

(d) Glycolysis refers to the anaerobic oxidation of glucose to produce lactic acid.

(e) The Embden-Meyerhof pathway uses substrate-level phosphorylation to form ATP.

(f) Lactic acid is used in gluconeogenesis and is formed by glycolysis.

(g) ATP is produced when the citric acid cycle, electron transport, and oxidative phosphorylation operate in concert.

(k) Hormones are secreted into the blood stream to control metabolism.

15. Glycogenolysis breaks down glycogen to form glucose.

```
   H—C=O
     |
   H—C—OH
     |
  HO—C—H          D-glucose
     |
   H—C—OH
     |
   H—C—OH
     |
     CH2OH
```

16. Glycogenesis is the process by which glucose is used to produce glycogen.

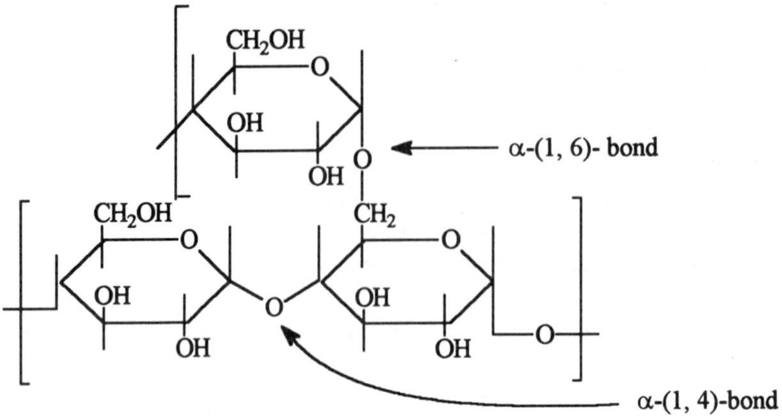

17. Glycogenesis is anabolic because this process starts with smaller precursors (glucose molecules) and builds-up large products (glycogen molecules).

18. Glycogenolysis is catabolic because it starts with larger precursors (glycogen molecules) and forms smaller products (glucose molecules).

19. The ATP production which is associated with the citric acid cycle is oxidative phosphorylation. ATP is formed as O_2 is reduced during electron transport.

20. The Embden-Meyerhof pathway uses substrate-level phosphorylation to produce ATP. That is, following carbon oxidation a phosphate group is transferred from a substrate to ADP, forming ATP.

21.

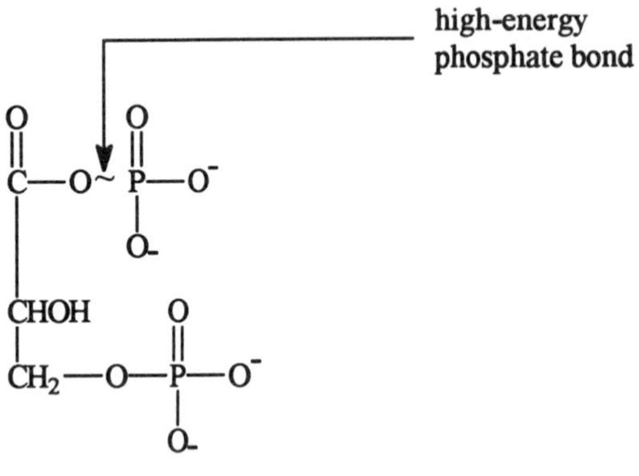

22.

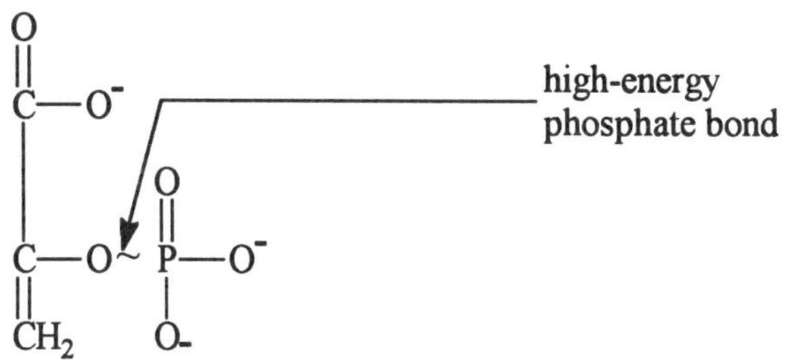

$$\begin{array}{l} O \\ \| \\ C-O^- \\ | \\ C-O\sim P(=O)(-O^-)-O^- \\ \| \\ CH_2 \end{array}$$

high-energy phosphate bond

23. Although glucose is oxidized, molecular oxygen (O_2) is not used in the Embden-Meyerhof pathway. Thus, this pathway is anaerobic.

24. Although no molecular oxygen (O_2) is used in the citric acid cycle, reduced coenzymes (NADH and $FADH_2$) are produced. These reduced coenzymes must be oxidized by electron transport so that the citric acid cycle can continue. And, it is electron transport which uses molecular oxygen. Thus, the citric acid cycle depends on the presence of molecular oxygen and is considered to be aerobic.

25. Glycolysis produces 2 moles of lactate per mole of glucose. Thus, 2.8 moles of glucose will yield 5.6 moles of lactate.

26. Glycolysis yields 2 moles of ATP per mole of glucose. Thus, 0.85 moles of glucose will yield 1.7 moles of ATP.

27. The end products of the anaerobic catabolism of glucose in yeast cells are ethanol and carbon dioxide.

28. The end product of the anaerobic catabolism of glucose in muscle tissue is lactate.

29. The Embden-Meyerhof pathway is defined as catabolic because (a) glucose is broken down to smaller compounds, (b) the carbons of glucose are oxidized and, (c) this pathway causes production of ATP.

30. The gluconeogenesis pathway is considered to be anabolic because (a) smaller, non-carbohydrate precursors such as lactate are converted to the larger product, glucose and (b) as this conversion takes place, the carbons become progressively more reduced.

31.

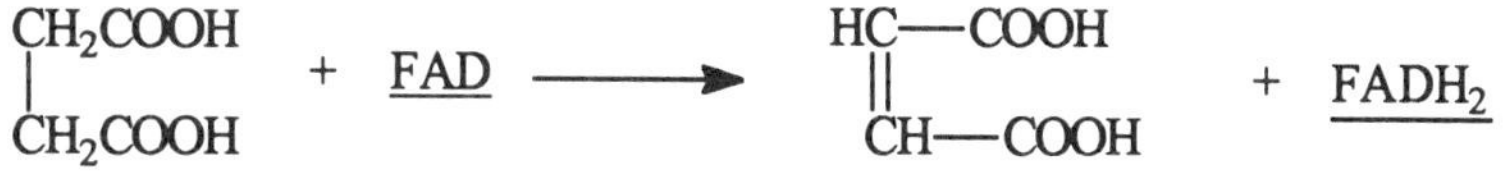

$$\begin{array}{l} CH_2COOH \\ | \\ CH_2COOH \end{array} + \underline{FAD} \longrightarrow \begin{array}{l} HC-COOH \\ \| \\ CH-COOH \end{array} + \underline{FADH_2}$$

32. $$\underset{\displaystyle CH_2COOH}{HO{-}\overset{|}{C}HCOOH} + \underline{NAD^+} \longrightarrow \underset{\displaystyle CH_2{-}COOH}{O{=}\overset{|}{C}{-}COOH} + \underline{NADH} + H^+$$

33. Electron transport and oxidative phosphorylation are needed to convert the reduced coenzyme products of the citric acid cycle into oxidized coenzymes with ATP formation.

34. The glycolysis pathway uses substrate-level phosphorylation to produce ATP and, thus, does not need electron transport or oxidative phosphorylation.

35. Like vitamins, hormones are generally needed in only minute amounts.

36. Unlike vitamins which must be supplied in the diet, hormones are produced in the body.

37. The normal blood glucose concentration is 70-90 mg/100 mL of blood under fasting conditions. Hypoglycemia occurs when the blood glucose concentration drops below this range. Thus, 65 mg/100 mL of blood is hypoglycemic.

38. The normal blood glucose concentration is 70-90 mg/100 mL of blood under fasting conditions. Hyperglycemia occurs when the blood glucose concentration rises above this range. Thus, 350 mg/100 mL of blood is hyperglycemic.

39. The citric acid cycle depends on the presence of molecular oxygen. Large amounts of free oxygen were probably not available on earth until photosynthetic organisms had evolved. (O_2 is a product of photosynthesis.)

40. Hormones function as chemical messengers. They are chemical substances that act as control or regulatory agents in the body. Enzymes are catalysts for specific reactions, allowing these reactions to occur faster and under milder conditions than would otherwise be possible.

41. Insulin is not effective when given orally because it is a protein, and would be hydrolyzed to amino acids in the gastrointestinal tract.

42. If a large overdose of insulin was taken by accident, the blood glucose concentration would drop to a low level, probably in the hypoglycemic range. This could result in fainting, convulsions, and unconsciousness.

43. Epinephrine (adrenalin) is sometimes called the emergency or crisis hormone because it stimulates glycogenolysis, which raises the concentration of glucose in the blood, giving the body additional energy for an emergency or crisis situation.

44. CH_3CHCOO^- (with OH on the central carbon)

lactate

All three carbons of lactate are more reduced than the carbon in CO_2. Thus, lactate can provide more cellular energy upon further oxidation of the carbons.

45. A multireaction system or pathway is more efficient because (a) intermediate compounds may be used in more than one pathway and (b) energy is either produced or used more effectively when released or consumed in small increments. Energy produced in one large increment may possibly destroy the cell.

46. In aerobic catabolism, pyruvate is converted to carbon dioxide yielding many ATPs while consuming molecular oxygen. In anaerobic catabolism, pyruvate is used to recycle NADH back to NAD^+.

CHAPTER 36

METABOLISM OF LIPIDS AND PROTEINS

1. A fatty acid contains many reduced carbons which can be oxidized to supply cellular energy.

2. Knoop prepared a homologous series of straight chain fatty acids with a phenyl group at one end and a carboxyl group at the other end. He fed each of these acids to test animals and analyzed the urine of the animals for metabolic products of the acid. Those animals which had been fed acids with an even number of carbons excreted phenylaceturic acid. Those animals which had been fed acids with an odd number of carbon atoms excreted hippuric acid. The experiment showed that the metabolism of the acid must occur by shortening the chain two carbon atoms at a time.

3. In the oxidation of fatty acids, the carbon atom beta to the carboxyl group is oxidized forming a β-keto acid which is then cleaved between the α and β carbon atoms leaving a new fatty acid that is two carbons shorter than the original fatty acid.

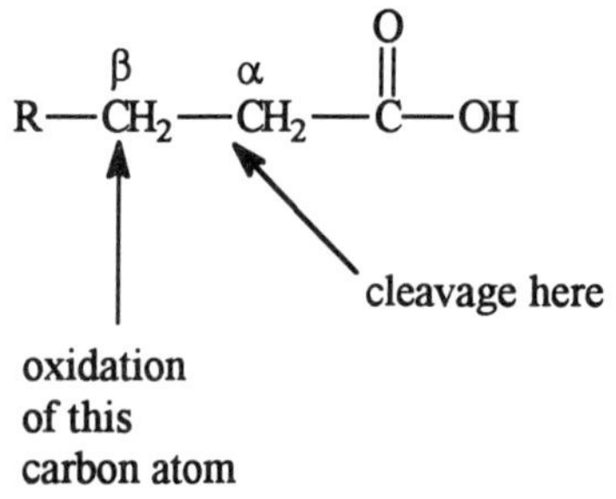

$$R{-}\overset{\beta}{C}H_2{-}\overset{\alpha}{C}H_2{-}\overset{O}{\overset{\|}{C}}{-}OH$$

cleavage here

oxidation of this carbon atom

4. Ketone bodies are relatively soluble in the blood stream and can be transported to energy-deficient cells in time of need. Ketone bodies supply energy via β-oxidation and the citric acid cycle.

5. Ketosis is the accelerated production of ketone bodies leading to high blood levels of these compounds with a corresponding increase in blood acidity.

6. In addition to being a food reserve, fat has two principal functions in the body. It acts as a cushion or shock absorber around internal organs, and it acts as an insulating blanket.

7. Fatty acid synthesis is not the reverse of β-oxidation. β-oxidation occurs in the mitochondria while fatty acids are synthesized in the cytoplasm. In fatty acid synthesis, the growing molecule is linked to an acyl carrier protein while in β-oxidation the fatty acid being broken down is linked to a coenzyme A. Finally, fatty acid synthesis uses some different reactions and a different coenzyme ($NADP^+$) which are not found in β-oxidation.

8. The obese protein signals the body to (a) increase catabolism of fats and (b) decrease the appetite for food.

9. In the enzymatic oxidation of fatty acids from fats, the fatty acids are oxidized, ultimately forming acetyl-CoA. Acetyl-CoA enters the citric acid cycle to produce energy.

10. Soybeans enrich the soil because their roots serve as host to the symbiotic, nitrogen-fixing bacteria.

11. L-glutamic acid is involved in the majority of transaminations and is, thus, a central compound in nitrogen transfer. L-glutamic acid is also the major nitrogen source for the urea cycle.

12. The possible metabolic fates of amino acids are:

    (a) incorporation into a protein;

    (b) utilization in the synthesis of other nitrogenous compounds such as nucleic acids;

    (c) deamination to a keto acid which can be (1) utilized to synthesize other compounds, or (2) oxidized to carbon dioxide and water.

13. The nitrogen in L-glutamine is less basic than ammonia and, thus, is less toxic to biological organisms.

14. Nitrogen is excreted as (a) ammonia by fish and (b) uric acid by birds.

15. The structure of urea is

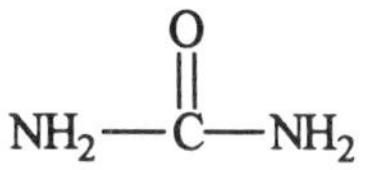

$$NH_2-\overset{\displaystyle O}{\overset{\|}{C}}-NH_2$$

and the structure of uric acid is

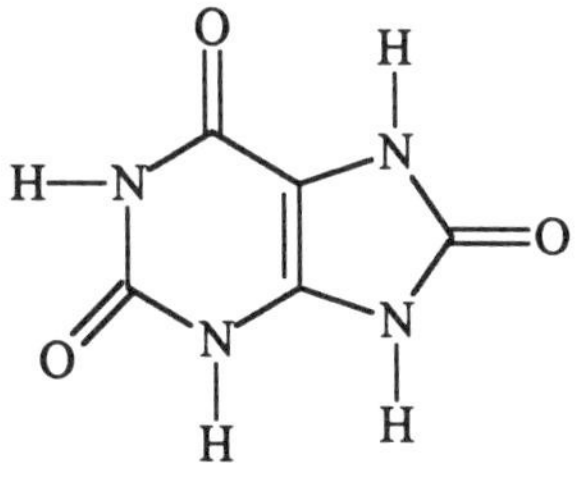

16. Three ATP's are used for every urea produced. The production of urea is important because this molecule provides a nontoxic means for excreting nitrogen.

17. Acetyl-CoA is considered to be an important central intermediate in metabolism for the following reasons:

    (a) Complete catabolism of most energy containing nutrients is achieved by conversion to acetyl-CoA and oxidation via the citric acid cycle.

    (b) Fats and amino acids can be synthesized from acetyl-CoA.

    (c) Carbohydrates, fats and amino acids can be converted to acetyl-CoA.

    (d) Acetyl-CoA is the central intermediate in converting (1) carbohydrates to fats or amino acids, (2) fats to amino acids or, (3) amino acids to fats.

18. The correct statements are: c, f, g, h, j, l, m, n, o, r, s, t, v. The others are not correct for these reasons:

    (a) Palmitic acid, as with all fatty acids, is a good source of stored metabolic energy because it contains many reduced carbons.

    (b) Fatty acids provide more energy than carbohydrates on a per-gram basis because fatty acids contain more carbons on a per-gram basis and these carbons are more reduced than those in carbohydrates.

    (d) Fatty acid oxidation can only continue if molecular oxygen is present to allow for mitochondrial electron transport and oxidative phosphorylation. Thus, fatty acid oxidation is aerobic.

    (e) β-oxidation directly produces no ATPs but depends on mitochondrial electron transport and oxidative phosphorylation to form ATP.

    (i) Lipogenesis is a process which forms lipids. Catabolism of fatty acids is the process which breaks down these compounds.

    (k) The acyl carrier protein is used only in fatty acid synthesis. Beta oxidation uses coenzyme A instead.

    (p) Nitrogen fixation occurs in lower organisms but is not carried out by higher plants.

    (q) A positive nitrogen balance means that more nitrogen is consumed than is excreted.

    (u) The acetyl group is linked to coenzyme A via a thioester bond in acetyl-CoA.

    (w) Urea is an especially nontoxic form of nitrogen because this compound is not a strong base.

19. The two carbons that will be found in acetyl-CoA after one pass through the beta oxidation pathway by butyric acid are circled.

$CH_3CH_2\boxed{CH_2COOH}$

20. The two carbons that will remain after two passes through the beta oxidation pathway by caproic acid are circled.

$\boxed{CH_3CH_2}CH_2CH_2CH_2COOH$

21. The β carbon of palmitic acid is marked with a star.

$CH_3(CH_2)_{12}\overset{*}{C}H_2CH_2COOH$

22. The β carbon of myristic acid is marked with a star.

$CH_3(CH_2)_{10}\overset{*}{C}H_2CH_2COOH$

23. Lauric acid ($CH_3(CH_2)_{10}COOH$) has 12 carbons and will yield six molecules of acetyl-CoA.

24. Palmitic acid ($CH_3(CH_2)_{14}COOH$) has 16 carbons and will yield eight molecules of acetyl-CoA.

25. Lauric acid ($CH_3(CH_2)_{10}COOH$) will be broken down to six acetyl-CoA molecules by five passes through the beta oxidation pathway. Thus, five $FADH_2$ molecules will be produced.

26. Palmitic acid ($CH_3(CH_2)_{14}COOH$) will be broken down to eight acetyl-CoA molecules by seven passes through the beta oxidation pathway. Thus, seven NADH molecules will be produced.

27. Complete conversion of myristic acid ($CH_3(CH_2)_{12}COOH$) to seven acetyl-CoA molecules involves six passes through the beta oxidation pathway. Six NADH molecules and six $FADH_2$ molecules will be produced. Therefore:

$$(6\ \text{NADH})\left(\frac{3\ \text{ATP}}{\text{NADH}}\right) + (6\ \text{FADH}_2)\left(\frac{2\ \text{ATP}}{\text{FADH}_2}\right) = 30\ \text{ATP}$$

28. Complete conversion of lauric acid ($CH_3(CH_2)_{10}COOH$) to six acetyl-CoA molecules involves five passes through the beta oxidation pathway. Five NADH molecules and five $FADH_2$ molecules will be produced. Therefore:

$$(5\ \text{NADH})\left(\frac{3\ \text{ATP}}{\text{NADH}}\right) + (5\ \text{FADH}_2)\left(\frac{2\ \text{ATP}}{\text{FADH}_2}\right) = 25\ \text{ATP}$$

29. The acyl carrier protein (ACP) is similar to coenzyme A (CoA) in that

    (a) both molecules have –SH groups which form thioesters with a carboxylic acids;

    (b) both molecules act as carriers for fatty acids during metabolism, ACP during anabolism and CoA during catabolism.

30. The acyl carrier protein (ACP) differs from coenzyme A (CoA) in that

    (a) ACP is a protein while CoA is a coenzyme;

    (b) ACP is used in fatty acid synthesis while CoA is used in the beta oxidation pathway.

31. A negative nitrogen balance means more nitrogen is excreted than ingested. A low protein diet will further decrease the amount of nitrogen ingested and should cause the nitrogen balance to become more negative.

32. A positive nitrogen balance means more nitrogen is ingested than excreted. A rapid growth spurt commonly means the body is using ingested amino acids to build new proteins. Thus, less nitrogen will be excreted and the nitrogen balance should become more positive.

33. L-leucine is a ketogenic amino acid because a diet high in L-leucine causes an increase in blood levels of the ketone body, acetoacetic acid.

34. L-aspartic acid is not a ketogenic amino acid because a diet high in L-aspartic acid caused no increase in blood levels of the ketone body β-hydroxy-butyric acid. (Other ketone bodies would not increase their blood levels also.) Because L-aspartic acid is not a ketogenic amino acid, it must be a glucogenic amino acid.

35. During transamination,

    (a) L-alanine yields $CH_3C(=O)COOH$

    (b) L-serine yields $HOCH_2C(=O)COOH$

36. During transamination,

    (a) L-aspartic acid yields $HOOCCH_2C(=O)COOH$

    (b) L-phenyl alanine yields $C_6H_5{-}CH_2C(=O)COOH$

37. The portion of carbamoyl phosphate which is incorporated into urea is circled.

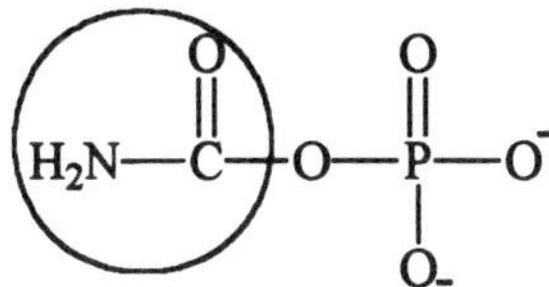

38. The portion of L-aspartic acid which is incorporated into urea is circled.

COOH
|
CH_2
|
(H_2N)—CH—COOH

39. It is possible to become obese despite a low fat diet because the body is capable of synthesizing fats from nonfat starting materials.

40. β-hydroxybutyric acid is a ketone body which does not contain a ketone group.

CH_3CHCH_2COOH
|
OH

41. β-oxidation is a pathway with relatively few reactions, requiring FAD, NAD^+, ATP, and coenzyme A. This process forms no ATP directly but depends upon mitochondrial electron transport and oxidative phosphorylation to make use of the reduced coenzymes, NADH and $FADH_2$ to form ATP.

42. In metabolism, most ATP is formed as acetyl-CoA is catabolized by the citric acid cycle, electron transport and oxidative phosphorylation.

    (1) One six-carbon hexose (glucose) will be converted to two molecules of pyruvate by the Embden-Meyerhof pathway which, in turn, are converted to two acetyl-CoA molecules.

    (2) A six-carbon fatty acid will yield three molecules of acetyl-CoA via the beta oxidation pathway.

    The six-carbon fatty acid yields more ATP than the six carbon hexose during catabolism primarily because more acetyl-CoA molecules are formed and can feed into the citric acid cycle, electron transport and oxidative phosphorylation.

43. $CH_3\overset{O}{\overset{\|}{C}}—SCoA \; + \; 6\,HOOCCH_2\overset{O}{\overset{\|}{C}}—SCoA \; + \; 12\;NADPH \; + \; 12\;H^+ \longrightarrow$

$$CH_3(CH_2)_{12}COOH \; + \; 6CO_2 \; + \; 5\,H_2O \; + \; 7\;CoA\text{-}SH \; + \; 12\;NADP^+$$

44. Transamination:

$$\underset{\displaystyle \text{}}{\overset{\displaystyle \quad\ CH_3\ \ O}{CH_3CH_2CH{-}\overset{\|}{C}{-}COOH}} + \underset{\text{L-glutamic acid}}{HOOCCH_2CH_2\underset{\displaystyle NH_2}{\underset{|}{C}}HCOOH} \xrightarrow{\text{enzyme}} \underset{\text{isoleucine}}{CH_3CH_2\overset{\displaystyle CH_3}{\overset{|}{C}}H\underset{\displaystyle NH_2}{\underset{|}{C}}HCOOH}$$

$$+ \underset{\alpha\text{-ketoglutaric acid}}{HOOCCH_2CH_2\overset{\displaystyle O}{\overset{\|}{C}}COOH}$$

45. Malonyl CoA is used in fatty acid synthesis (anabolism) but not in beta oxidation (catabolism). Thus, the lecture would probably focus on fatty acid anabolism.

46. Dehydration refers to the removal of hydrogen. In step 2 of beta oxidation,

$$RCH_2CH_2CH_2\overset{\displaystyle O}{\overset{\|}{C}}{-}SCoA + FAD \longrightarrow RCH_2CH{=}CH\overset{\displaystyle O}{\overset{\|}{C}}{-}SCoA + FADH_2$$

The fatty acid carbons at the alpha and beta positions change their oxidation numbers from -2 to -1; they are oxidized.

47. $(1.0 \text{ mol palmitic acid})\left(\dfrac{256.4 \text{ g}}{\text{mol palmitic acid}}\right) = 2.6 \times 10^2 \text{ g}$

$$\left(\frac{9 \text{ kcal}}{\text{g}}\right)(2.6 \times 10^2 \text{ g}) = 2 \times 10^3 \text{ kcal}$$